UNDERSTANDING
INTERMEDIATE ALGEBRA

UNDERSTANDING INTERMEDIATE ALGEBRA

Robert G. Moon
Gus Klentos
Fullerton College

Merrill Publishing Company
A Bell & Howell Information Company
Columbus   Toronto   London   Melbourne

Cover Art: Michael Linley Illustration
Published by Merrill Publishing Company
A Bell & Howell Information Company
Columbus, Ohio 43216
This book was set in Optima
Administrative Editor: John Yarley
Production Coordinator: Ben Shriver
Cover Designer: Cathy Watterson

Copyright © 1987 by Merrill Publishing Company. All rights reserved. No part of this book may be reproduced in any form, electronic or mechanical, including photocopy, recording or any information storage and retrieval system, without permission in writing from the publisher. "Merrill Publishing Company" and "Merrill" are registered trademarks of Merrill Publishing Company.

Library of Congress Catalog Card Number: 86-62991
International Standard Book Number: 0-675-20268-X
Printed in the United States of America
1 2 3 4 5 6 7 8 9—91 90 89 88 87

To Marilyn and Maria

PREFACE

This book is a sequel to *Understanding Elementary Algebra,* Second Edition, by Robert G. Moon and is designed for students taking a second course in algebra. It is assumed that students have successfully completed a course in elementary or basic algebra.

The main purpose of this book is to develop understanding and computational skills with the techniques usually found in a second-level algebra course. To reach this goal, the authors have written a text that students can read and understand. The treatment of each topic is straightforward, and mathematically sound. However, it is not so rigorous that students will be confused. Every concept is carefully developed and followed by examples. Important terms appear in color when first introduced. Color is also used in examples and illustrations to highlight important features.

Since many students do not necessarily take a second course in algebra directly after the first course, there is often a significant time lag in which students forget basic algebraic concepts. Thus, the first chapter of this text contains a comprehensive review of basic algebraic concepts.

Equations and applied problems are heavily emphasized throughout the text.

Each section contains many examples with detailed step-by-step solutions, illustrations, and other aids. Important concepts and rules are enclosed in boxes for easy identification.

Each section also contains a set of exercises suitable for homework. The exercises contain numerous problems that range from the simple to the more complex. Answers to the odd-numbered problems are included at the back of the book. Full solutions to all the problems are found in the Solutions Manual, available from the publisher. The text can easily be adapted to fit a variety of instructional programs offered at self-study centers and learning laboratories.

A summary and a set of self-checking exercises appear at the end of each chapter. Students should be encouraged to study these before a chapter test. They also constitute an excellent review for the final examination.

Calculator use is integrated throughout the book. The purpose of the book is not to each the use of a calculator—this is best done in the classroom. However, where appropriate, the authors have introduced the use of certain keys and have provided calculator key schematics. Also, many exercise sets contain a calculator section.

TEXT ORGANIZATION

Review Chapter 1 is primarily a review of basic algebra. It covers the basic definitions and properties of real numbers. The number line is developed with emphasis on absolute value, inequality, and intervals. Order of operation and the use of a calculator are addressed. Operations with signed numbers are reviewed together with algebraic expressions, natural number exponents, and roots.

Equations and Inequalities Chapter 2 discusses linear equations and inequalities together with absolute value equations and inequalities. Formulas are solved for a specified variable. Students also receive experience solving applied problems with linear equations and inequalities.

Polynomials Chapter 3 deals with operations with polynomial expressions. Special products are developed. Factoring, for greater emphasis and understanding, is divided into three parts: common factors, binomial factoring, and trinomial factoring. Factoring is then used to solve higher degree equations.

Rational Expressions Chapter 4 discusses operations on rational expressions. Equations containing rational expressions are solved. Students gain experience by solving applied problems involving work and motion.

Exponents, Roots, and Radicals Chapter 5 relates exponents and roots. Negative exponents are developed and applied to scientific notation. Operations are performed on radical expressions. And, students are taught to solve equations containing radicals.

Quadratic Equations and Inequalities In Chapter 6 quadratic equations are introduced and solved by three methods: factoring, completing the square, and the quadratic formula. Quadratic inequalities are solved by using the signs of their factors as they apply to intervals. The set of complex numbers is also developed.

Relations and Functions Chapter 7 discusses relations, functions, and their inverses. Particular emphasis is placed on the linear function, including the point-slope and slope-intercept forms. Linear inequalities in two variables are also discussed and graphed. Forms of variation are developed, including solution of applied problems.

Systems of Linear Equations Chapter 8 deals with systems of linear equations in two and three variables. Systems are solved by elimination and substitution. Determinants are developed from matrices in order to develop Cramer's Rule. Applied problems are then solved using systems of linear equations.

Conic Sections Chapter 9 develops the conic sections: parabolas, circles, ellipses, and hyperbolas. Students learn to solve non-linear systems of equations as well as to graph second-degree inequalities.

Exponential and Logarithmic Functions Chapter 10 discusses the interrelationships of exponential and logarithmic functions. Particular attention is paid to applied problems using the calculator. Computation with common logarithms is done primarily to illustrate the properties of logarithms. This chapter also introduces the base e.

Natural Number Functions Chapter 11 discusses sequences, series, progressions, and the binomial theorem. Applied problems are emphasized.

SUPPLEMENTARY MATERIAL

The following supplementary material is available from Merrill Publishing:

Instructor's Manual and Test Bank.
   Contains numerous test questions for each chapter. Also has two mid-term examinations and two final examinations.

Teacher's Solutions Manual.
   Contains detailed solutions to all sets of exercises.

Answer Key.
   Answers to all exercise questions.

Student's Study Manual.
   Contains a synopsis and examples covering each section of the text. Detailed solutions of selected exercises are also included.

(IBM-PC) *Computerized Test Bank.*
(Apple IIe) *Computerized Test Bank.*
   Contains the questions found in the *Instructor's Manual and Test Bank,* cross-referenced to the section numbers in the text. Available for both Apple and IBM-PC computers. Questions can be either multiple choice or free-response.

ACKNOWLEDGEMENTS

The authors wish to thank the following reviewers for their helpful comments: John C. Anderson; Dale E. Boye, Schoolcraft College; Dawson Carr, Sandhills Community College; Douglas Cook, Owens Technical College; Grace P. Foster; Beaufort County Community College; Marceline C. Gratiaa, Winona State University; Richard A. Langlie, North Hennepin Community College; Robert A. Nowlan, Southern Connecticut State University; and Wesley W. Tom.

Robert G. Moon
Gus Klentos

CONTENTS

1 BASIC PROPERTIES OF REAL NUMBERS   1

- **1.1** Basic Definitions and Symbols   2
- **1.2** Inequality, Absolute Value, and the Number Line   7
- **1.3** Order of Operations and the Calculator   15
- **1.4** Properties of Real Numbers   21
- **1.5** Operations with Signed Numbers   27
- **1.6** Natural Number Exponents and Roots   37
- **1.7** Algebraic Expressions   45
  Summary   51

2 SOLVING FIRST DEGREE EQUATIONS AND INEQUALITIES   57

- **2.1** Solving Linear Equations   58
- **2.2** Solving Linear Inequalities   65
- **2.3** Solving Absolute Value Equations   71
- **2.4** Solving Absolute Value Inequalities   76
- **2.5** Working with Formulas   80
- **2.6** Applied Problems Involving Linear Equations and Inequalities   85
  Summary   96

3 POLYNOMIAL EXPRESSIONS — 101

- **3.1** Polynomials   102
- **3.2** Multiplication and Division of Polynomials   106
- **3.3** Special Products of Binomials   111
- **3.4** Common Factors   116
- **3.5** Factoring Binomials   119
- **3.6** Factoring Trinomials   124
- **3.7** Solving Equations by Factoring   130
- Summary   134

4 RATIONAL EXPRESSIONS — 139

- **4.1** Reduction of Rational Expressions   140
- **4.2** Multiplication and Division of Rational Expressions   144
- **4.3** Addition and Subtraction of Rational Expressions   149
- **4.4** Simplifying Complex Fractions   154
- **4.5** Equations and Formulas Containing Rational Expressions   158
- **4.6** Applied Problems Involving Work and Motion   164
- Summary   169

5 EXPONENTS, ROOTS, AND RADICALS — 173

- **5.1** Roots and Radicals   174
- **5.2** Negative Exponents and Scientific Notation   178
- **5.3** Rational Exponents   186
- **5.4** Simplifying Radicals   190
- **5.5** Operating with Radical Expressions   194
- **5.6** Rationalizing Denominators   198
- **5.7** Solving Equations Containing Radicals   203
- Summary   206

6 SOLVING QUADRATIC EQUATIONS AND INEQUALITIES — 211

- **6.1** Solving Quadratic Equations by Factoring   212
- **6.2** Solving Quadratic Equations by Completing the Square   216

- **6.3** The Quadratic Formula 222
- **6.4** Solving Quadratic Inequalities 230
- **6.5** The Set of Complex Numbers 236
- **6.6** Operating with Complex Numbers 241
  Summary 247

7 RELATIONS, FUNCTIONS, AND THEIR GRAPHS 253

- **7.1** Ordered Pairs, Relations, and Graphs 254
- **7.2** Functions 262
- **7.3** Inverse Relations and Functions 270
- **7.4** The Linear Function 278
- **7.5** Equations of the Straight Line 289
- **7.6** Linear Inequalities 296
- **7.7** Variation 303
  Summary 310

8 SYSTEMS OF LINEAR EQUATIONS 321

- **8.1** Systems of Linear Equations in Two Variables 322
- **8.2** Systems of Linear Equations in Three Variables 330
- **8.3** Second and Third Order Determinants 335
- **8.4** Cramer's Rule for Solving Systems of Linear Equations 341
- **8.5** Using Systems of Linear Equations to Solve Applied Problems 346
  Summary 353

9 THE CONIC SECTIONS 359

- **9.1** Graphing Quadratic Equations in Two Variables— Parabolas 360
- **9.2** The Quadratic Function 365
- **9.3** The Circle 373
- **9.4** The Ellipse 380
- **9.5** The Hyperbola 384
- **9.6** Conic Sections and Curve Recognition 392
- **9.7** Non-Linear Systems of Equations 396
- **9.8** Second Degree Inequalities 401
  Summary 406

10 EXPONENTIAL AND LOGARITHMIC FUNCTIONS — 415

- **10.1** Exponential Functions 416
- **10.2** Applications of Exponential Functions and the Base e 423
- **10.3** Logarithmic Functions 428
- **10.4** Properties of Logarithms 433
- **10.5** Computing with Common Logarithms 439
- **10.6** Logarithms, Exponentials, and the Calculator 445
- **10.7** Exponential and Logarithmic Equations 450
- Summary 457

11 NATURAL NUMBER FUNCTIONS — 465

- **11.1** Sequences and Series 466
- **11.2** Arithmetic Progressions 470
- **11.3** Geometric Progressions 476
- **11.4** The Binomial Expansion 483
- Summary 491

CUMULATIVE REVIEW EXERCISE — 496

SOLUTIONS TO CUMULATIVE REVIEW EXERCISE — 498

APPENDIX/TABLES

Table I Powers and Roots
Table II Powers of e
Table III Natural Logarithms
Table IV Common Logarithms
Answers to Odd-Numbered Exercises

INDEX

BASIC PROPERTIES OF REAL NUMBERS

1

1.1 Basic Definitions and Symbols
1.2 Inequality, Absolute Value, and the Number Line
1.3 Order of Operations and the Calculator
1.4 Properties of Real Numbers
1.5 Operations with Signed Numbers
1.6 Natural Number Exponents and Roots
1.7 Algebraic Expressions

BASIC DEFINITIONS AND SYMBOLS

1.1

Algebra is a study of properties and patterns of numbers. Its applications are wide and varied, relating to most disciplines in our society. For convenience and ease of operation algebra depends upon the use of many symbols. In this chapter we will introduce many of the symbols to be used throughout this text.

We begin by discussing the term set. A set is a collection of objects. In algebra these objects are usually numbers and are called elements or members of the set. Sets are often represented by uppercase letters. A set can be defined by either describing its elements with a word statement or listing its elements within braces. Sets are said to be equal provided they contain the same elements.

EXAMPLE 1

Define this set by describing its elements:
$$S = \{a, b, c, d\}$$

Solution: S is the set consisting of the first four letters of the English alphabet.

EXAMPLE 2

Define this set by listing its elements:
C is the set consisting of the first five counting numbers.

Solution: $C = \{1, 2, 3, 4, 5\}$

It is possible for a set to have no elements whatsoever. Consider the set of counting numbers between 9 and 10. No such elements exist. The set is empty. The empty set is also called the null set and is symbolized by a pair of braces, { }.

The sets we have discussed thus far are finite sets since the number of elements is limited. Sets containing an unlimited number of elements are called infinite sets. The set of all counting numbers is an example of an infinite set. It is called the set of natural numbers and is denoted by the letter N.

$$N = \{1, 2, 3, 4, 5, 6, 7, 8, 9, 10, 11, \ldots\}$$

The three dots show that the established pattern continues indefinitely.

Membership in a given set is denoted by the Greek letter epsilon, ϵ. The symbolic statement, $2 \epsilon N$, is read, 2 is an element or a member of the set of natural numbers. To indicate that an object is not an element of a specified set, a slash is placed through the epsilon, $\not\epsilon$. Thus, $0 \not\epsilon N$, denotes that 0 is not an element of the natural numbers.

A variable is a symbol representing any element of a set containing more than one element. If $S = \{1, 2, 3, 4, 5\}$ and $x \epsilon S$, then x is a variable representing any of the first five natural numbers. The set S is called the replacement set or domain of x.

If a symbol has a replacement set consisting of only one element, then it is called a constant. If $A = \{5\}$ and $k \in A$, then k is a constant representing only the number 5.

Combining variables and constants with set notation forms a convenient method of specifying sets called set builder notation. This notation, as seen below, specifies a symbol and follows with a descriptive statement.

$$T = \{x \mid x \in N \text{ and } x \text{ is less than } 4\}$$

This set is read: T is the set of all x such that x is an element of the natural numbers and x is also less than 4. Listing the elements, we see that $T = \{1, 2, 3\}$. Note that the vertical line in set builder notation is read, "such that."

List the elements of A, where
$A = \{y \mid y \in N \text{ and } y \text{ is between 6 and 10}\}$

EXAMPLE 3

Solution: $A = \{7, 8, 9\}$

Sets of numbers can be conveniently illustrated by using a number line. To construct a number line we first draw a horizontal line. Then we select an arbitrary point and label it zero. This point is called the origin since it is the originating point of the number line. Next, as seen in Figure 1.1, a scale is established locating points equally spaced to the left and right of the origin. Positive numbers name points to the right of the origin while negative numbers name points to the left. Zero is neither positive nor negative.

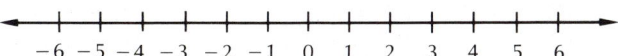

FIGURE 1.1 Number Line

Each number is said to be a coordinate of the associated point, while the point itself is called the graph of the number.

Pairs of coordinates that are equidistant from the origin, but on opposite sides of the origin, are called additive inverses (or opposites) of each other. For example, 3 is the additive inverse of -3, and -3 is the additive inverse of 3. The additive inverse of zero is zero itself.

To graph sets of numbers we construct a number line and locate the associated points.

Graph the set T, where $T = \left\{1, 3, -3, -\dfrac{1}{2}, 0\right\}$

EXAMPLE 4

Solution:

4   BASIC PROPERTIES OF REAL NUMBERS

EXAMPLE 5   Graph $\{x \mid x \in N \text{ and } x \text{ is less than } 2\}$

Solution: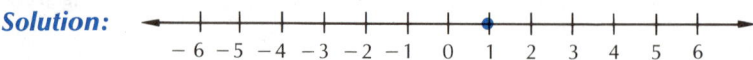

In the study of algebra we are particularly interested in the following six sets of numbers:

1. Natural Numbers: The set of natural numbers is denoted by N. It is simply the set of counting numbers:

$$N = \{1, 2, 3, 4, 5, 6, 7, 8, 9, 10, 11, 12, \ldots\}$$

Notice that zero is not a natural number.

2. Whole Numbers: The set of whole numbers is denoted by W. The whole numbers include all of the natural numbers together with zero.

$$W = \{0, 1, 2, 3, 4, 5, 6, 7, 8, 9, 10, 11, 12, \ldots\}$$

3. Integers: The set of integers is denoted by I. It consists of all the whole numbers together with their additive inverses.

$$I = \{\ldots, -5, -4, -3, -2, -1, 0, 1, 2, 3, 4, 5, \ldots\}$$

4. Rational Numbers: The set of rational numbers is denoted by Q. It consists of all the numbers that can be expressed as a ratio or fraction of two integers.

$$Q = \left\{ \frac{a}{b} \;\middle|\; a, b, \in I \text{ and } b \neq 0 \right\}$$

Each of these numbers is rational: $\frac{1}{2}$, $-\frac{2}{3}$, $\frac{11}{5}$, -3, 1, and 0. The set of rational numbers includes all of the natural numbers, whole numbers and integers. It also includes positive and negative common fractions. Since a fraction can be considered as an indicated division, each rational number can be divided out to obtain a decimal. It can be shown that a rational number will produce either a terminating or a repeating decimal. For example, the rational number $\frac{1}{2}$, when divided out, produces the terminating decimal, .5. But, the rational number $\frac{5}{11}$ will produce .454545 . . ., a repeating decimal. As a matter of convenience, repeating decimals are usually written with a bar over the repetitive block of numerals. Thus, .454545 . . . = $.\overline{45}$. Each terminating decimal represents a rational number and each repeating decimal is a rational number.

5. Irrational Numbers: The set of irrational numbers is denoted by H. It consists of all numbers with decimal representations that do not terminate and do not repeat. For example, .303003000300003000003 . . . is an irrational number; it does not terminate and the pattern is nonrepetitive because the frequency of zeros continually increases by one.

$$H = \{x \mid x \text{ is a nonterminating, nonrepeating decimal}\}$$

BASIC DEFINITIONS AND SYMBOLS 5

Each of these numbers is also irrational: π, $\frac{\pi}{2}$, .123456789101112 . . ., $\sqrt{2}$, and $-\sqrt{23}$.

By definition, it is impossible for any number to be both rational and irrational.

6. Real Numbers: The set of real numbers is denoted by R. It contains all of the rational numbers and irrational numbers. More simply, the set of real numbers contains all of the numbers having a decimal representation. Every point on a number line can be labeled with a real number.

$$R = \{x \mid x \text{ represents a decimal numeral}\}$$

or

$$R = \{x \mid x \text{ is a coordinate of a point on a number line}\}$$

An example of a number which is not real is $\sqrt{-1}$. This number, called the imaginary unit, will be discussed in a later chapter.

These six special sets of numbers are illustrated in Figure 1.2. Observe that the set of real numbers includes each of the other sets.

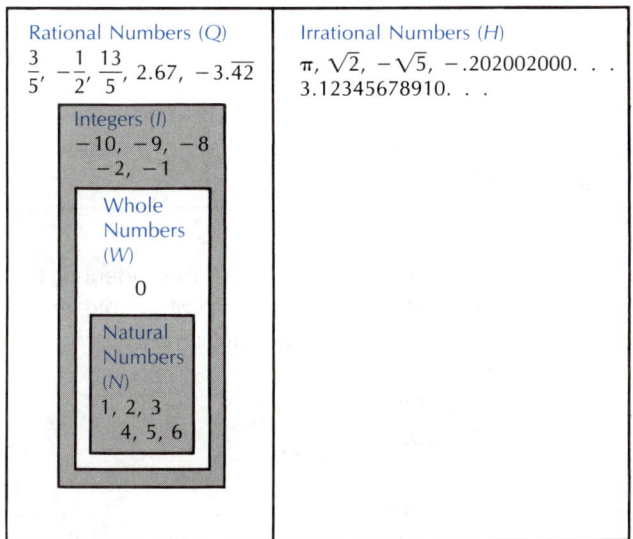

FIGURE 1.2

Exercise 1.1   True or false.

1. Zero is a natural number.

2. $-\frac{2}{3}$ is a rational number.

3. $0.\overline{45}$ is an irrational number.

4. -5 is the additive inverse of 5.
5. $\sqrt{-1}$ is a real number.
6. An irrational number is not a real number.

Define each set by describing its elements. State if the set is finite or infinite.
7. {a, b, c}
8. {1, 2, 3, 4, 5}
9. {2, 4, 6, 8, . . .}
10. {1, 3, 5, 7, . . .}
11. {5, 10, 15, 20, . . .}
12. {10, 20, 30}
13. {Monday, Tuesday, Wednesday, Thursday, Friday, Saturday, Sunday}
14. { }

Define these sets by listing the elements.
15. The set of all the days in the week whose names begin with "S".
16. The set of all months which have exactly 30 days.
17. {x | x ∈ N and x is exactly divisible by 3}
18. {x | x ∈ W and x is less than 5}
19. {x | x ∈ N and x is less than 1}
20. {x | x is a number that is both rational and irrational}
21. {x | x is an even integer between -2 and 2}
22. {natural number multiples of 6}
23. {x | x + 2 = 6 and x ∈ N}

In Exercises 24–27, each problem describes a certain number. Identify that number as one of the following: (a) positive integer, (b) negative integer, (c) rational number which is not an integer, (d) irrational number.

24. $-\dfrac{4}{5}$ **25.** $\dfrac{18}{2}$
26. $\sqrt{2}$ **27.** 4.2

Graph the following sets on the real number line.
28. {-3, -1, 3}
29. {x | x ∈ N and x is less than 2}
30. {integers between -3 and 3, inclusive}
31. {x | x ∈ I and x is less than 2}
32. {odd whole numbers between 1 and 5}
33. {x | x ∈ W and x + 3 = 5}
34. {x | x ∈ R and x · x = 4}

Put a check mark in each box that describes the given number.

	Natural	Integer	Rational	Irrational	Real
35. -2					
36. 0					
37. .026					
38. $1 + \sqrt{2}$					
39. $\dfrac{2}{3}$					
40. 0.7070070007 . . .					

Problems 41–45. True or false.

41. (a) $\dfrac{6}{35} \in Q$ (b) $3.14 \in H$ (c) $\{1, 2\} = \{2, 1\}$
42. (a) $\{1, 2, 4\} = \{2, 4, 1\}$ (b) $0 = \{\ \}$ (c) $.\overline{48} \in H$
43. (a) Zero is an integer.
    (b) Zero is a rational number.
    (c) Zero is a real number.
44. (a) Every rational number can be written as a terminating decimal.
    (b) Every rational number is also an integer.
    (c) A real number can be thought of as any number which can be represented by a decimal.
45. (a) All whole numbers are included in the set of integers.
    (b) All irrational numbers are real numbers.
    (c) All rational numbers are included in the set of integers.

INEQUALITY, ABSOLUTE VALUE AND THE NUMBER LINE

1.2

You should recall from section 1.1 that real numbers are coordinates of points on the number line. In this section we will use the number line to show graphs of inequality and absolute value.

Inequality

A statement that two numbers are not equal is called an inequality. For example, $8 \neq 5$, read 8 does not equal 5, is an inequality. If two numbers are not equal, then one of the numbers must be greater than the other. The symbol $>$ means "is greater than." Thus, the inequality $8 > 5$ is read "8 is greater than 5." The symbol $<$ means "is less than." Hence, the inequality $0 < 6$ is read, 0 is less than 6. These two symbols are called symbols of order and are defined using the number line.

8 BASIC PROPERTIES OF REAL NUMBERS

> **Definition of Symbols of Order**
> For $a, b, \in R$,
> (1) $a > b$ means the graph of a lies to the right of b on the number line.
> (2) $b < a$ means the graph of b lies to the left of a on the number line.

EXAMPLE 1 Use the correct symbol of order to describe the coordinates m and n on the following number line.

Solution: The graph of n lies to the left of m. Thus, $n < m$. It is also true that the graph of m lies to the right of n. Therefore, it is also correct to write $m > n$.

To be used correctly, the symbols of order must always point toward the smaller number.

EXAMPLE 2 The symbols of order have been used correctly in the following statements:

(a) $5 > -1$
(b) $3 > -10$
(c) $0 < 2$
(d) $-15 < -12$
(e) $\dfrac{1}{2} > \dfrac{1}{3}$

As illustrated in Table 1.1, several other symbols are also used to denote inequality.

TABLE 1.1 Inequality Symbols

Symbol	Meaning	Examples
$\geq$	is greater than or equal to	$8 \geq -1$, since 8 is greater than -1.
		$-3 \geq -3$, since -3 is equal to -3.
$\leq$	is less than or equal to	$5 \leq 9$, since 5 is less than 9.
		$5 \leq 5$, since 5 is equal to 5.
$\not>$	is not greater than (means $\leq$)	$-3 \not> 4$ since -3 is less than 4.
$\not<$	is not less than (means $\geq$)	$0 \not< -2$ since 0 is greater than -2

Inequality can be used with set builder notation to form intervals. An interval is a set of coordinates used to form a graph on the number line.

Graph A = {x | x > 2} **EXAMPLE 3**

Solution: Show the graph of all points which lie to the right of 2. Place an open circle on the point named by 2, since 2 ∉ A.

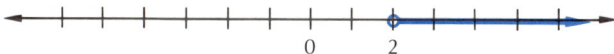

Graph B = {x | x ≤ −3} **EXAMPLE 4**

Solution: This is the graph of all real numbers less than or equal to −3. The graph consists of all points to the left of −3 together with −3, itself. A solid dot is placed on the point named by −3 because −3 ϵ B.

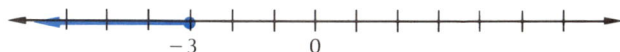

Compound Statements

Intervals can also be formed by compound statements. A compound statement imposes more than one condition on a variable. If two or more conditions are imposed simultaneously we use the word "and." The solution is composed of those values held in common and is called the intersection of the conditions.

Graph C = {x | x ϵ W and x < 5} **EXAMPLE 5**

Solution: The members of C must satisfy two conditions simultaneously. They must be whole numbers and they must be less than 5. The solution (or intersection) is simply {0, 1, 2, 3, 4}. These coordinates are then graphed to form the following five points:

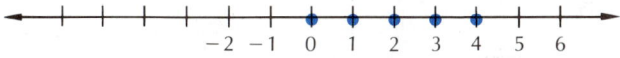

Graph D = {x | x > −3 and x < 2} **EXAMPLE 6**

Solution: The members of D must be greater than −3 and, at the same time, less than 2. The solution (or intersection) consists of all real numbers between −3 and 2. Open circles are used to show that −3 ∉ D and 2 ∉ D.

Compound statements also connect conditions with the word "or." The solution to this type of statement is called the union and is composed of those values satisfying at least one of the conditions.

10   BASIC PROPERTIES OF REAL NUMBERS

EXAMPLE 7   Graph $E = \{x \mid x > 3 \text{ or } x \leq -1\}$

Solution: The solution (or union) of these two conditions consists of all real numbers greater than 3 together with all real numbers less than or equal to -1. The interval is described by two arrows. An open circle shows that $3 \notin E$, while a solid dot indicates that $-1 \in E$.

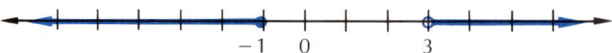

Often, compound statements contain neither the connective "and" nor the connective "or." In this case, the word "and" is assumed, producing a solution which is the intersection of the conditions.

EXAMPLE 8   Graph $\{x \mid x \geq 0, x < 3\}$

Solution: We assume the connective "and." The graph below consists of all points whose coordinates are greater than or equal to 0 and, at the same time, less than 3.

It is common in algebra to use this notation for an interval occuring between two points: $\{x \mid a < x < b\}$. We read the condition from inside out, beginning with the variable: "x is greater than a and x is less than b." The solution contains all real numbers between a and b. The interval is described by the following graph:

EXAMPLE 9   Graph $\{x \mid -5 \leq x \leq 2\}$

Solution: The graph contains all points whose coordinates are between -5 and 2. Also, solid dots are used to show that the interval includes the end points -5 and 2.

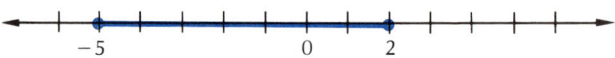

EXAMPLE 10   Graph $\{x \mid -1 < x \leq 4\}$

Solution: The graph contains all points between -1 and 4, excluding -1 and including 4.

Absolute Value

The absolute value of a real number a, written $|a|$, is the distance on the number line from the origin to the point named by a. Thus, as seen in the following figure, the absolute value of −4 is 4 and the absolute value of 4 is 4.

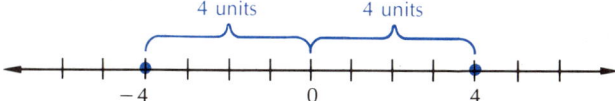

Since absolute value refers to a distance, it is never negative.

The following expressions containing absolute values have been evaluated. **EXAMPLE 11**

(a) $|6| = 6$
(b) $|-3| = 3$
(c) $|0| = 0$
(d) $|-3| + |10| = 13$
(e) $-|5| = -5$
(f) $-|-5| = -5$

The formal definition of absolute value is given in two parts as follows:

Definition of Absolute Value

For $a \in R$,
(1) If $a \geq 0$, $|a| = a$.
(2) If $a < 0$, $|a| = -a$.

The first part of the defintion is simple. It states that when the number inside the absolute value symbol is positive or zero, then the absolute value is equal to the original number. Hence, $|8| = 8$ and $|0| = 0$. The second part of this definition is somewhat more difficult. It says that when the number inside the absolute value symbol is negative, then the result is equal to the additive inverse (or opposite) of the original negative number. The answer is therefore positive. For example, $|-9| = -(-9) = 9$.

Intervals involving absolute values can be graphed by locating the corresponding points on the number line.

Graph $\{x \mid |x| = 3\}$ **EXAMPLE 12**

Solution: Determine all numbers that have an absolute value of 3. Both 3 and −3 have an absolute value of 3. The associated points are then located on the number line.

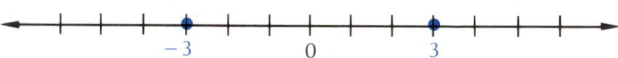

Absolute Value Inequalities

In the last example we found that the graph of the absolute value equation $|x| = 3$ contained only two points, -3 and 3. Let us now examine an absolute value inequality defined by a "less than" symbol, such as $|x| < 3$. As we learned earlier, absolute value represents the distance between a given number and the origin. Thus, the condition $|x| < 3$ must represent all numbers whose distance from the origin is less than 3 units. As seen in the following graph, this includes all numbers between -3 and 3.

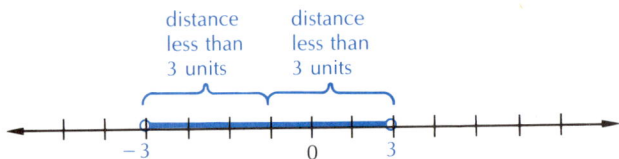

EXAMPLE 13   Graph $\{x \mid |x| < 5\}$

Solution: The graph consists of all points within 5 units of the origin. It can also be described as $\{x \mid -5 < x < 5\}$.

EXAMPLE 14   Graph $\{x \mid |x| \leq 2\}$

Solution: The graph consists of all points within 2 units of the origin together with the end points -2 and 2. It can also be described as $\{x \mid -2 \leq x \leq 2\}$.

We now consider an absolute value inequality defined by a "greater than" symbol, such as $|x| > 3$. This condition represents all numbers whose distance from the origin is more than 3 units. As seen in the following graph, this includes all numbers less than -3 together with those numbers greater than 3.

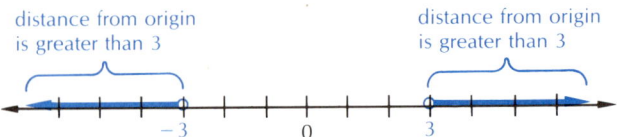

EXAMPLE 15   Graph $\{x \mid |x| > 5\}$

Solution: The graph consists of all points located more than 5 units from the origin. It can also be described as $\{x \mid x < -5 \text{ or } x > 5\}$.

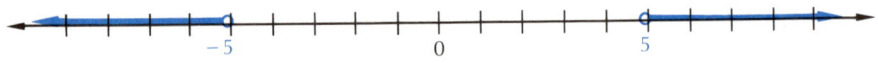

INEQUALITY, ABSOLUTE VALUE AND THE NUMBER LINE   13

Graph $\{x \mid |x| \geq 2\}$ **EXAMPLE 16**

Solution: The graph consists of all points located more than 2 units from the origin, together with the end points 2 and -2. The interval can also be described as $\{x \mid x \leq -2 \text{ or } x \geq 2\}$.

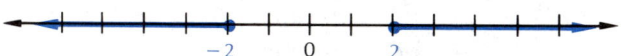

To help you understand the various intervals discussed in this section, we have summarized the types in Table 1.2.

TABLE 1.2 Summary of Intervals

Intervals $a, b, c, \in R$ with $a < b$ and $c > 0$	Graph
(1) $\{x \mid a < x < b\}$	$a \quad b$
(2) $\{x \mid x < a \text{ or } x > b\}$	$a \quad b$
(3) $\{x \mid \|x\| = c\}$ or $\{c, -c\}$	$-c \quad 0 \quad c$
(4) $\{x \mid \|x\| < c\}$ or $\{x \mid -c < x < c\}$	$-c \quad 0 \quad c$
(5) $\{x \mid \|x\| > c\}$ or $\{x \mid x < -c \text{ or } x > c\}$	$-c \quad 0 \quad c$

Express each of the following by using the order symbols.   **Exercise 1.2**
1. -6 is less than -3
2. 4 is not greater than 7
3. x is positive
4. -4 is greater than -5
5. 0 is less than or equal to 3
6. 5 is between 3 and 6

Fill in each blank with an appropriate order symbol to form a true statement.
7. -7 _____ -2 8. 0 _____ -3
9. -7 _____ -12 10. $\dfrac{1}{2}$ _____ 0

BASIC PROPERTIES OF REAL NUMBERS

11. $-\dfrac{5}{2}$ _____ $-\dfrac{5}{4}$
12. 2 _____ 5 _____ 7
13. $|3|$ _____ $|-2|$
14. $-|-6|$ _____ $|6|$

Evaluate each of the following.
15. $-|-3|$
16. $|-4| + |-2|$
17. $|0|$
18. $|-5| - |5|$
19. $-(|-3| \cdot |-2|)$
20. $12 - |-4|$
21. If $a < 0$, simplify $|a|$.
22. If $b > 0$, simplify $|-b|$.
23. If $m > 0$, simplify $|-m|$.
24. If $m < 0$, simplify $|-m|$.

In problems 25–30, determine which integers can be used in place of x to make the statement true.
25. $-6 < x < -3$
26. $0 < x < 4$
27. $-4 < x \leq 2$
28. $-3 \leq x \leq 3$
29. $-1 \leq x < 4$
30. $-2 < |x| < 3$

Graph each of the following sets on the real number line.
31. $\{x \mid x \leq 1\}$
32. $\{x \mid x > -3\}$
33. $\{x \mid x \leq 4 \text{ and } x \in W\}$
34. $\{x \mid x > -2 \text{ and } x < 3\}$
35. $\{x \mid -1 \leq x \leq 6\}$
36. $\{x \mid x \geq 4 \text{ or } x < -2\}$
37. $\{x \mid -3 < x \leq 3, x \text{ is an integer}\}$
38. $\{x \mid x > -5 \text{ and } x \leq 1\}$
39. $\{x \mid 2 < x \leq 6, x \in N\}$
40. $\{x \mid x \geq 4 \text{ or } x < -2\}$
41. $\{x \mid -3 < x \leq 0\}$
42. $\{x \mid |x| \leq 3\}$
43. $\{x \mid |x| > 3\}$
44. $\{x \mid |x| > 1 \text{ or } |x| = 1\}$
45. $\{x \mid x > -1 \text{ and } x \leq 4\}$
46. $\{x \mid |x| \geq 1\}$
47. $\{x \mid x > 2 \text{ or } x < 2\}$
48. $\{x \mid x < 5, x > 2, x \in W\}$
49. $\{x \mid -2 < x \leq 0 \text{ or } x > 2\}$
50. $\{x \mid x > -1 \text{ and } x \leq 4, x \in I\}$
51. $\{x \mid x \text{ is not negative}\}$
52. $\{x \mid x > 2 \text{ or } x < -2\}$

An alternate symbol for $\{x \mid x > a \text{ or } x < b\}$ is $\{x \mid x > a\} \cup \{x \mid x < b\}$. An alternate symbol for $\{x \mid x > a \text{ and } x < b\}$ is $\{x \mid x > a\} \cap \{x \mid x < b\}$. Graph each of the following sets on the real number line.
53. $\{x \mid x < -2\} \cup \{x \mid x \geq 0\}$
54. $\{x \mid x \leq -1\} \cup \{x \mid x > 2\}$
55. $\{x \mid x > -1\} \cap \{x \mid x \geq 4\}$
56. $\{x \mid x > -2\} \cap \{x \mid x < 3\}$
57. $\{x \mid 4 < x\} \cup \{x \mid 0 > x\}$
58. $\{x \mid 5 \geq x \geq 2\}$
59. $\{x \mid -1 \geq x \geq -3\}$
60. $\{x \mid 2 \geq |x|\}$

ORDER OF OPERATIONS AND THE CALCULATOR

1.3

The four fundamental operations of real numbers are addition, subtraction, multiplication and division.

Addition is symbolized by the plus sign, $+$. Thus, the sum of the addends x and 3 is written, $x + 3$.

Subtraction is denoted by the minus sign, $-$. If y is subtracted from x, the difference is written, $x - y$.

Multiplication is indicated by either a dot, parenthesis, or no symbol at all. The latter is used only if there is no confusion. The numbers being multiplied are called factors. The result is called the product. Hence, the product of 3 and x can be written, $3 \cdot x$, $(3)(x)$, or simply $3x$.

Division is symbolized by either the fraction bar or this familiar symbol, $\div$. Thus, x divided by y is written, $\dfrac{x}{y}$ or $x \div y$. x is called the dividend and y is called the divisor. The result is called the quotient. You should recall that division is checked by multiplying the quotient times the divisor to obtain the dividend. Consequently,

$$\frac{15}{3} = 5 \text{ since } 5 \cdot 3 = 15$$

As illustrated below, the check for division shows that dividing by zero is impossible:

(a) $\dfrac{5}{0}$ produces no answer since no number times zero equals 5.

(b) $\dfrac{0}{0}$ does not produce a single (or unique) answer because any number times zero is zero.

Thus, we say division by zero is undefined.

Order of Operations

When evaluating expressions containing more than one operation, it is necessary to have an agreement governing the order in which the operations are performed. Otherwise, as shown in the following example, a single expression may give rise to more than one answer.

Evaluate the expression $6 \cdot 3 + 5$. **EXAMPLE 1**

Solution a: Add first; multiply second.

$6 \cdot \underline{3 + 5}$
$6 \cdot 8$
48

Solution b: Multiply first; add second.

$$6 \cdot 3 + 5$$
$$\underbrace{}_{18} + 5$$
$$23$$

Obviously it is not desirable for this to happen. Thus, mathematicians have agreed to perform operations in a certain order. This is accomplished using grouping symbols consisting of parentheses ( ), brackets [ ], and braces { }, together with the following order of operations agreement:

The Order of Operations Agreement

Step 1. Evaluate all grouping symbols.*
Step 2. Perform all multiplications and divisions in the order they appear, working from left to right.
Step 3. Perform all additions and subtractions in the order they appear, working from left to right.

*In the event a group contains more than one operation, evaluate it using Step 2 and then Step 3.

EXAMPLE 2

Evaluate $36 \div (10 - 8) - 4 \cdot (1 + 2) + (2 \cdot 5)$.

Solution:

Step 1. Evaluate all grouping symbols.

$$36 \div \underbrace{(10 - 8)}_{2} - 4 \cdot \underbrace{(1 + 2)}_{3} + \underbrace{(2 \cdot 5)}_{10}$$

Step 2. Do all multiplications and divisions in the order they appear, working from left to right.

$$\underbrace{36 \div 2}_{18} - \underbrace{4 \cdot 3}_{12} + 10$$

Step 3. Do all additions and subtractions in the order they appear, working from left to right.

$$\underbrace{18 - 12}_{6} + 10$$
$$\underbrace{6 + 10}_{16}$$

Classifying Expressions

Expressions are classified according to the final operation performed. For example, $6 + 3 \cdot 5$ is classified as a basic sum since addition is the final operation to be performed. This means the expression, $6 + 3 \cdot 5$, is basically the sum of 6 and 15.

EXAMPLE 3

Evaluate and classify $0 \div 3 + 6 \cdot 2 \div 3 - 4 \cdot 0$.

Solution: No grouping symbols are present, so begin with Step 2 of the order of operations agreement.

Step 1. Multiply and divide in order of their appearance.

$$\underbrace{0 \div 3}_{0} + \underbrace{6 \cdot 2 \div 3}_{4} - \underbrace{4 \cdot 0}_{0}$$

Step 2. Add and subtract in order of their appearance.

$$\underbrace{0 + 4}_{4} - 0$$
$$\underbrace{4 - 0}_{0}$$

Thus, $0 \div 3 + 6 \cdot 2 \div 3 - 4 \cdot 0 = 4$. The order of operations is indicated by the circled numerals:

$$\overset{①}{0} \div \overset{⑤}{3} + \overset{②}{6} \cdot \overset{③}{2} \div \overset{⑥}{3} - \overset{④}{4} \cdot 0 = 4$$

The final operation performed was subtraction. This expression is a basic difference.

Evaluate and classify $(10 - 2) \cdot (8 + 4)$. **EXAMPLE 4**

Solution: These are independent grouping symbols. That is, one set does not appear inside the other. Therefore, it does not matter which is evaluated first.

$$\underbrace{(10 - 2)}_{8} \cdot \underbrace{(8 + 4)}_{12}$$
$$96$$

Thus, $(10 - 2)(8 + 4) = 96$. The two choices for performing the order of operations are:

$$\overset{①}{(10} - \overset{③}{2)} \cdot \overset{②}{(8 + 4)} = 96 \text{ or } \overset{②}{(10} - \overset{③}{2)} \cdot \overset{①}{(8 + 4)} = 96$$

This expression is a basic product.

Evaluate and classify $5 \cdot [38 - (10 + 5) \cdot 2] + 6 \cdot 0$. **EXAMPLE 5**

Solution: When grouping symbols occur within other grouping symbols, it is easier to begin on the inside. Also, use the order of operations agreement to evaluate ungrouped pairs of numbers.

$5 \cdot [38 - \underline{(10 + 5)} \cdot 2] + 6 \cdot 0$
$5 \cdot [38 - \underline{15 \cdot 2}] + 6 \cdot 0$
$5 \cdot [38 - \underline{30}] + 6 \cdot 0$
$\underline{5 \cdot 8} + \underline{6 \cdot 0}$
$40 + 0$
40

$$\overset{④}{5} \cdot [\overset{③}{38} - \overset{①}{(10 + 5)} \cdot \overset{②}{2}] + \overset{⑥}{6} \cdot \overset{⑤}{0} = 40$$

Thus, $5 \cdot [38 - (10 + 5) \cdot 2] + 6 \cdot 0 = 40$.
The expression is a basic sum.

A fraction bar, as shown in the next example, is considered to be a grouping symbol denoting division. It indicates that the numerator and denominator are to be evaluated first, prior to performing the division.

EXAMPLE 6

Evaluate and classify $\dfrac{6 + 12}{2 \cdot 3}$.

Solution: The fraction bar is both a grouping symbol and an indicated division.

$$\dfrac{6 + 12}{2 \cdot 3} = (6 + 12) \div (2 \cdot 3)$$
$$= 18 \div 6$$
$$= 3$$

The expression is a basic quotient.

Expressions containing variables can be evaluated by first substituting values from the variables' domain. The resulting expression is then simplified using your knowledge of grouping symbols and order of operations. These expressions can also be classified according to the last operation performed.

EXAMPLE 7

Evaluate and classify $\dfrac{x + 3[y - (z + 1)]}{(x + 1)(x + 2)}$, where $x = 0$, $y = 5$, and $z = 2$.

Solution: Substitute and evaluate.

$$\dfrac{x + 3[y - (z + 1)]}{(x + 1)(x + 2)} = \dfrac{0 + 3[5 - (2 + 1)]}{(0 + 1)(0 + 2)}$$
$$= \dfrac{0 + 3[5 - 3]}{1 \cdot 2}$$
$$= \dfrac{0 + 3 \cdot 2}{1 \cdot 2}$$
$$= \dfrac{6}{2}$$
$$= 3$$

The expression is a basic quotient.

Calculators

With your instructor's approval, you may wish to use a calculator to evaluate certain expressions. To successfully operate a calculator, you must know how it "thinks." You must know what logic it uses. Most calculators utilize one of three types of logic systems: RPN (reverse Polish notation), algebraic logic, or arithmetic logic.

RPN calculators do not have an "equal" key. If your calculator is an RPN type, read the manual carefully and if necessary obtain extra help from your instructor.

ORDER OF OPERATIONS AND THE CALCULATOR 19

To determine if your calculator is an algebraic or arithmetic type, evaluate the expression $3 + 4 \cdot 5$ by pressing the keys in this sequence:

$$\boxed{3} \boxed{+} \boxed{4} \boxed{\times} \boxed{5} \boxed{=}$$

If your calculator gives the correct answer 23, then it is algebraic and follows the order of operations agreement. However, if your calculator gives an answer of 35 it uses arithmetic logic. For an arithmetic calculator you will have to remember the order of operations agreement and press the keys accordingly. For this example, $3 + 4 \cdot 5$, the keys should be pressed in this sequence:

$$\boxed{4} \boxed{\times} \boxed{5} \boxed{+} \boxed{3} \boxed{=}$$

Throughout the rest of this text, calculator key sequences will be shown for algebraic logic.

Evaluate using your calculator. Round the answer to three significant digits. **EXAMPLE 8**

$$\frac{12.56}{3.82} + (6.92)(5.013)$$

Solution: $\boxed{12.56} \boxed{\div} \boxed{3.82} \boxed{+} \boxed{6.92} \boxed{\times} \boxed{5.013} \boxed{=}$ 37.977918
Rounded answer is 38.0.

Evaluate using your calculator. Round the answer to four significant digits. **EXAMPLE 9**

$$\frac{8.903 + 2.0271}{(2.957)(0.8512)}$$

Solution: Remember, the fraction bar is also a grouping symbol. Thus:

$$\frac{8.903 + 2.0271}{(2.957)(0.8512)} = \frac{[8.903 + 2.0271]}{[(2.957)(0.8512)]}$$

$\boxed{(} \boxed{8.903} \boxed{+} \boxed{2.0271} \boxed{)} \boxed{\div} \boxed{(} \boxed{2.957} \boxed{\times} \boxed{.8512} \boxed{)} \boxed{=}$ 4.3425136
Rounded answer is 4.343.

Simplify each expression by using the order of operations agreement. Classify each as a basic sum, basic difference, basic product or basic quotient. **Exercise 1.3**

1. $5 + 3 \cdot 4$
2. $3 \cdot 4 - 2$
3. $6 + 8 \div 2$
4. $8 \div 4 - 4$
5. $12 + 3 \cdot 2 - 12 \div 3$
6. $3 + 4 \cdot 3 + 8 - 2 \cdot 3$
7. $4 \cdot 5 + 10 \div 2 - 2 \cdot 5$
8. $\frac{18}{8 - 2} - 4$
9. $\frac{3(8 - 6)}{2} + \frac{8}{4}$
10. $3(4 - 1) + \frac{3(2 + 4)}{6}$
11. $5(3 + 5) \div [4 + 2(3)]$
12. $(40 \div 4) \div 2$
13. $0 \div 4 + 6 \cdot 2 \div 3 - 2$
14. $3 \cdot [4(6 - 2) + 2(6 + 2)]$

BASIC PROPERTIES OF REAL NUMBERS

15. 4{22 − [(13 − 3) − 5]}
16. 3{12 + [(13 − 2) − 4]}
17. 17 − {6 − [13 − 2(7 − 2)]}
18. 0 ÷ 3 + 4 · 3 ÷ 2 ÷ 2 · 0
19. 48 ÷ 4 · 2(6)
20. $\dfrac{18 - 3(4)}{9 - 6}$
21. [18 ÷ 3(6)][14 · 2 ÷ 4]

Insert grouping symbols so that each of the following statements is true.
22. 3 · 2 + 2 · 5 = 60
23. 3 · 2 + 2 · 5 = 36
24. 3 · 2 + 2 · 5 = 16
25. 3 + 9 ÷ 3 · 2 = 12
26. 60 − 3 + 4 · 8 ÷ 2 = 2
27. 8 ÷ 2 + 4 − 6 ÷ 2 = 1
28. 5 − 4 − 3 − 2 − 1 = 3
29. $\dfrac{5 - 3 - 1}{2(4) - 1} = \dfrac{1}{2}$

Evaluate the following. Show the order of operations by using circled numerals.
30. [(6 ÷ 2 + 6) − 7] ÷ 2
31. 15 ÷ 3 · (5 − 2) · 3
32. 3(7 + 8) − [2 · (5 + 1) + 4]
33. 4{16 − [(13 − 2) − 4]}

Evaluate and classify. Let x = 1, y = 4, z = 0 and w = 2.
34. 3x + 2(y − 2z + 3)
35. 5yw − 3z + 4x
36. (x + 2y)(y + 3w)
37. $\dfrac{3[y - (z + w)]}{y - x}$
38. $\dfrac{y}{w - 1}$ − (w + 2x)
39. 3{2x + x[y − z(3w + x)]}

Use your calculator to evaluate the following. Round answers to three significant digits.
40. (.069)(8.015)(4.91)
41. (81.6)(93.94) ÷ 72.4
42. (2.61)(3.04) + (1.06)(3.79)
43. 18.6 − [9.2 + 4.6(.85 − .73)]
44. .73 ÷ (2.61)(.58)
45. 2.63 + 4.36 ÷ (2)(.624)
46. $\dfrac{11.1 - 3.28}{2.98(5.23)}$
47. $\dfrac{14.97 - 9.22}{15.27(3.49)}$
48. [87 + 43(79 − 27)] · [91 ÷ (56 − 49)]
49. [(197)(13.29 + 11.79) + .14] · 19
50. $\dfrac{21.7 - 3.14(2.63)}{48.06(.4301)}$
51. $\dfrac{(32.8)(16.9) + (8.07)(0.671)}{15.8 + 12.9}$

PROPERTIES OF REAL NUMBERS

1.4

The rules governing the set of real numbers and its operations are called the axioms or properties of the real numbers. They are assumed to be true and constitute the foundation of algebra.

Properties of Equality

For $a, b, c \in R$, we assume the following properties of the equality relationship symbolized by the familiar equal sign, $=$.

---
1. The Reflexive Property

$$a = a$$

The reflexive property states that each real number is equal to itself.

---
2. The Symmetric Property

If $a = b$, then $b = a$.

The symmetric property states that the left and right members of the equality can be interchanged. If $6 = x$, then it is also correct to write $x = 6$.

---
3. The Substitution Property

$a = b$, then b may be replaced for a in any statement without altering its truth or falsity.

For example, if $x = y$ and $3x = 10$, then the equation could also be written as $3y = 10$.

---
4. The Addition Property

If $a = b$, then $a + c = b + c$.

The addition property states that any real number can be added to both sides of an equation. For example, if $x - 3 = 10$, then $x - 3 + 3 = 10 + 3$ or $x = 13$.

---
5. The Multiplication Property

If $a = b$ and $c \neq 0$, then $ac = bc$.

The multiplication property allows us to multiply both sides of an equation by the same non-zero real number. For example, if $3x = 12$, then $\frac{1}{3} \cdot 3x = \frac{1}{3} \cdot 12$ or $x = 4$.

Now let us examine some properties involving the operations of addition and multiplication as they apply to the set of real numbers.

Properties of the Set of Real Numbers

Given the operations of addition and multiplication we assume the following properties where $a, b, c \in R$.

---

1. Closure Properties for Addition and Multiplication

$$a + b \in R \text{ and } ab \in R.$$

A set of numbers is said to be closed for a given operation provided that applying the operation to members of the set produces a result which is also a member of that set. For example, the sum of two natural numbers is always a natural number. Thus, the set of natural numbers is closed for addition. It is also closed for multiplication. However, the natural numbers are not closed for subtraction since $3 - 3 = 0$, and zero is not a natural number. Nor are the natural numbers closed for division since $1 \div 2 = .5$ and $.5$ is not a natural number.

The closure properties for addition and multiplication of real numbers state that the sum of two real numbers is a real number, and that the product of two real numbers is also a real number. It is also true that the set of real numbers is closed for subtraction and non-zero division.

---

2. Commutative Properties of Addition and Multiplication

$$a + b = b + a \text{ and } ab = ba$$

The commutative properties state that the order of the addends may be interchanged and the order of factors may be interchanged. Thus, $2 + 3 = 3 + 2$ and $3 \cdot 5 = 5 \cdot 3$.

---

3. Associative Properties of Addition and Multiplication

$$(a + b) + c = a + (b + c) \text{ and } (ab)c = a(bc).$$

PROPERTIES OF REAL NUMBERS

The associative properties state that addends may be regrouped and factors may be regrouped, allowing the operations to be performed in a different order. For example, $(2 + 3) + 4 = 2 + (3 + 4) = 9$ and $(4 \cdot 2) \cdot 5 = 4 \cdot (2 \cdot 5) = 40$.

---

4. Properties of Identity

There is a unique real number 0, called the additive identity, such that
$$0 + a = a \text{ and } a + 0 = a$$
There is a unique real number 1, called the multiplicative identity, such that
$$1 \cdot a = a \text{ and } a \cdot 1 = a$$

The term unique means exactly one. There is exactly one real number that does not produce a change when added. That number is the additive identity, 0. For example, $0 + 8 = 8$.

Similarly, in multiplication there is exactly one real number that produces no change. It is the multiplicative identity, 1. For example, $1 \cdot 6 = 6$. It will also be helpful to remember that this property allows us to think of a variable such as x as being $1x$.

---

5. Inverse Properties

For each real number a, there is a unique real number $-a$, called the additive inverse (or opposite) of a, such that
$$a + (-a) = 0$$
For each non-zero real number a, there is a unique real number $\frac{1}{a}$, called the multiplicative inverse (or reciprocal) of a, such that
$$a \cdot \frac{1}{a} = 1$$

Every real number has an additive inverse, and their respective sum is zero. For example, the additive inverse of 8 is -8, and $8 + (-8) = 0$. Zero is its own additive inverse because $0 + 0 = 0$.

Every real number except zero has a multiplicative inverse. For example, the multiplicative inverse of 5 is $\frac{1}{5}$ and $5 \cdot \frac{1}{5} = 1$. Zero has no multiplicative inverse because $\frac{1}{0}$ is not defined.

> **6. Distributive Property of Multiplication with Respect to Addition**
> $a(b + c) = ab + ac$ and $(b + c)a = ba + ca$

The distributive property involves both multiplication and addition. It states that a number, when multiplied times a sum, must be shared with each addend. For example, $5 \cdot (x + 3) = 5x + 5 \cdot 3$ or $5x + 15$.

It is also important to know that the distributive property will change a basic product to a basic sum.

$$\underset{\text{basic product}}{5 \cdot (x + 3)} = \underset{\text{basic sum}}{5x + 15}$$

Using the symmetric property of equality, we can reverse the distributive property to change a basic sum to a basic product.

$$ab + ac = a(b + c)$$

This procedure is called factoring.

EXAMPLE 1 The following expressions have been factored using the distributive property. Each problem can be checked by multiplying out the right side to obtain the left.

(a) $5x + 5y = 5(x + y)$
(b) $6m + 12n = 6(m + 2n)$
(c) $14a + 7 = 7(2a + 1)$

Theorems of Real Numbers

The properties of equality together with the properties of real numbers imply other statements called theorems. A theorem is a statement of fact that follows logically from previous definitions, properties, or theorems.

> **1. Extending the Distributive Property**
> For $a, b, c, d \in R$, $a(b + c + d) = ab + ac + ad$
> Proof: $a(b + c + d) = a[b + (c + d)]$ [associative property]
> $= ab + a(c + d)$ [distributive property]
> $= ab + ac + ad$ [distributive property]

This process can be carried on indefinitely. Thus, the distributive property of multiplication with respect to addition holds true for any number of addends.

EXAMPLE 2

The following have been multiplied according to the distributive property or its extension.
(a) $5(3x + 4) = 15x + 20$
(b) $3(2x + 4y + 3z) = 6x + 12y + 9z$
(c) $7ab(c + 2d + 3e) = 7abc + 14abd + 21abe$

EXAMPLE 3

The distributive property or its extension has been used to factor the following expressions.
(a) $12xy + 6y = 6y(2x + 1)$
(b) $3a + 9b + 12c = 3(a + 3b + 4c)$
(c) $8mn + 12m + 2n + 6 = 2(4mn + 6m + n + 3)$

---

2. The Multiplication by Zero Property

For $a \in R$, $0 \cdot a = a \cdot 0 = 0$

$$\begin{aligned} \text{Proof: } a \cdot 0 &= a(0 + 0) &&\text{[additive identity]} \\ &= a \cdot 0 + a \cdot 0 &&\text{[distributive property]} \end{aligned}$$

Since $a \cdot 0 + a \cdot 0 = a \cdot 0$, $a \cdot 0$ is the additive identity. Since the additive identity is unique, $a \cdot 0 = 0$.

This theorem states that the product of zero and any real number is zero. For example, $0(-12) = 0$.

---

3. The Double Opposite Law

For $a \in R$, $-(-a) = a$

$$\begin{aligned} \text{Proof: } a + (-a) &= 0 &&\text{[additive inverse]} \\ \text{but, } (-a) + [-(-a)] &= 0 &&\text{[additive inverse]} \end{aligned}$$

Thus, a and $-(-a)$ are both additive inverses of $-a$. Since additive inverses are unique, $-(-a) = a$.

The double opposite law states that the opposite of an opposite is the original number. Thus, $-(-2) = 2$ or $-[-(-5)] = -5$. This leads to the conclusion that an even number of opposites is equivalent to a single plus sign, while an odd number is equivalent to a single minus sign.

EXAMPLE 4

The double opposite law has been used to simplify these expressions to a single sign.
(a) $-\{-[-(-6)]\} = +6$ or 6
(b) $-[-(-4)] = -4$

Exercise 1.4

Name the property or theorem which justifies each of the following. Assume all letters represent real numbers.

1. $7 \cdot 8 = 8 \cdot 7$
2. $4 + 0 = 4$
3. $4 + 0 = 0 + 4$
4. $3 + (4 + 5) = 3 + (5 + 4)$
5. $(6 + 4) + 8 = 6 + (4 + 8)$
6. $3 \cdot 1 = 3$
7. $8 \cdot 0 = 0$
8. $-(-6) = 6$
9. $y + (-y) = 0$
10. $(x + 6) + 2 = 2 + (x + 6)$
11. $\frac{1}{6}(6) = 1$
12. $(3 + x) \cdot 0 = 0$
13. $4 + [a + (-a)] = 4 + 0$
14. $5(c + 2) = 5c + 5 \cdot 2$
15. If $x = 5$, then $x + 3 = 5 + 3$
16. $1 \cdot (x + y) = 1 \cdot x + 1 \cdot y$
17. If $b = x$, then $x = b$
18. $0 + 0 = 0$
19. If $a = b$ and $b = 2$, then $a = 2$
20. $3x = 3x$
21. $(x + y) + [-(x + y)] = 0$
22. $5(x + 3) = 5x + 5 \cdot 3$
23. $-4 + [k + (-k)] = -4 + 0$
24. $-(-x) = x$

In problems 25–29 use the distributive property to fill in the blanks.

25. $5(c + d) = $ _____
26. $(a + b) \cdot 3 = $ _____
27. $4(2m + n) = $ _____
28. $3(3a + 4b + 2c) = $ _____
29. $5ab(2c + d + 3e) = $ _____

In problems 30–34 use the distributive property to factor each expression.

30. $6ab + 6b = $ _____
31. $\sqrt{2} \cdot a + \sqrt{2} \cdot b = $ _____
32. $7x + 21 = $ _____
33. $4ab + 8ac + 8a = $ _____
34. $abx + axy + xz = $ _____

Give the reciprocal of each number. (If none exists, so state.)

35. 4
36. 1
37. $\frac{1}{3}$
38. 0
39. $\frac{7}{2}$

Are the following sets closed under (a) addition? (b) multiplication?

40. $\{0, 2, 4, 6, 8, \ldots\}$
41. $\{1\}$
42. $\{0, 1\}$
43. $\{1, 2, 3\}$

Problems 44–49. True or false.

44. (a) $\frac{10}{0} = 0$
    (b) $-\{-[-6]\} = -6$

45. (a) The integers are closed under addition.
  (b) The operation of division is not commutative.
46. (a) The distributive property changes a basic product into a basic sum.
  (b) Every real number as a multiplicative inverse.
47. (a) Zero is the additive identity.
  (b) If the dividend and divisor are both zero, then the quotient is zero.
48. (a) One is the multiplicative identity.
  (b) The whole numbers are closed under subtraction.
49. (a) The operation of subtraction is commutative.
  (b) The operation of subtraction is associative.

Give a reason for each step in the following proofs.
50. Prove: For all real numbers x and y,
$$(x + y) + (-y) = x$$
  Step 1. $(x + y) + (-y) = x + [y + (-y)]$ _____
  Step 2. $\qquad\qquad\qquad = x + 0$ _____
  Step 3. $\qquad\qquad\qquad = x$ _____
51. Prove: For all real numbers a and b,
$$a + [b + (-a)] = b$$
  Step 1. $a + [b + (-a)] = a + [(-a) + b]$ _____
  Step 2. $\qquad\qquad\qquad = [a + (-a)] + b$ _____
  Step 3. $\qquad\qquad\qquad = 0 + b$ _____
  Step 4. $\qquad\qquad\qquad = b$ _____

OPERATIONS WITH SIGNED NUMBERS

1.5

In this section we review the four fundamental operations as they apply to signed real numbers: addition, subtraction, multiplication and division.

Addition

We use the number line to add two numbers having like signs, -2 and -3. As shown in Figure 1.3, we start at the origin and draw an arrow representing -2 by extending it 2 units to the left. Next, we draw a second arrow representing -3. This arrow begins at the tip of the first arrow and extends 3 units to the left. It terminates at the point having a coordinate of -5. Thus, $(-2) + (-3) = -5$.

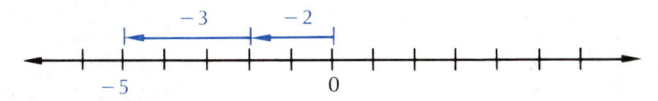

FIGURE 1.3

BASIC PROPERTIES OF REAL NUMBERS

This procedure suggests the following rule:

> To add two real numbers having like signs, add their absolute values and use the common sign of the original numbers.

EXAMPLE 1

(a) $(-6) + (-2) = -8$
(b) $(+8) + (+3) = +11$ or 11
(c) $-12 + (-5) = -17$

We now use the number line to add two numbers having unlike signs, -5 and $+2$. As shown in Figure 1.4, we again start at the origin and draw an arrow extending 5 units to the left. Then, from the tip of this arrow we draw a second arrow extending 2 units to the right. This arrow terminates at the point having a coordinate of -3. Thus, $(-5) + (+2) = -3$.

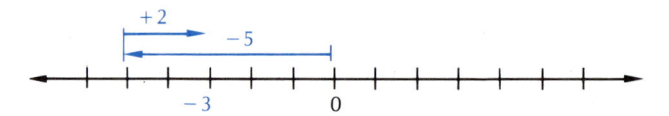

FIGURE 1.4

> To add two different real numbers having unlike signs, subtract the smaller absolute value from the larger and use the sign of the number having the greater absolute value.

EXAMPLE 2

Add $(-12) + 7$

Solution:

Step 1. The absolute values are 12 and 7.

Step 2. Subtract 7 from 12, obtaining 5.

Step 3. The sum is negative since -12 has the greater absolute value.

Thus, $(-12) + 7 = -5$

To be successful in algebra, you must be able to work addition problems likes these rapidly and mentally.

EXAMPLE 3

(a) $-16 + 5 = -11$
(b) $10 + (-3) = 7$
(c) $-1 + 12 = 11$

It is also important to be able to add a series containing three or more signed numbers. When evaluating a series, the order of operations agreement instructs us to add from left to right. However, since addition is commutative and associative, we are free to add in any order that is convenient.

Add $-10 + 8 + (-2)$ **EXAMPLE 4**

Solution 1: Use the order of operations agreement.
$$-10 + 8 + (-2)$$
$$-2 + (-2)$$
$$-4$$

Solution 2: Since addition is commutative and associative we can add the negative numbers first.
$$-10 + 8 + (-2)$$
$$-12 + 8$$
$$-4$$

Add $16 + 12 + (-16) + (-12) + 3$ **EXAMPLE 5**

Solution: It is convenient to add opposites first, obtaining the zeros.
$$16 + 12 + (-16) + (-12) + 3$$
$$0 + 0 + 3$$
$$3$$

When a series contains grouping symbols, do the work inside first. Then, addition may be performed in any order you wish.

Add $-15 + (-10 + 8) + 4$ **EXAMPLE 6**

Solution:
$$-15 + (-10 + 8) + 4$$
$$-15 + (-2) + 4$$
$$-17 + 4$$
$$-13$$

Subtraction

Let us now examine the operation of subtraction. If we begin with 9 and subtract 5, the answer is 4. Thus, $9 - 5 = 4$. The same answer is obtained by adding 9 and -5 as follows:
$$9 - 5 = 9 + (-5)$$
$$= 4$$

In a similar manner we can subtract 3 from 10:
$$10 - 3 = 10 + (-3)$$
$$= 7$$

This procedure of adding opposites suggests the following definition:

Definition of Subtraction

For $a, b \in R$,
$$a - b = a + (-b)$$
(Subtracting a number is the same as adding its opposite.)

The definition of subtraction tells us to make two sign changes.

subtraction is changed to addition

$$a - b = a + (-b)$$

subtrahend is changed to its opposite

Since subtraction is defined as a sum, we make two sign changes and then use the rules for addition.

EXAMPLE 7 Subtract $-7 - (-10)$

Solution: Change subtraction to addition and change -10 to $+10$.

$$-7 - (-10) = -7 + (+10)$$
$$= 3$$

EXAMPLE 8
(a) $12 - (-2) = 12 + 2 = 14$
(b) $-6 - (+3) = -6 + (-3) = -9$
(c) $-13 - 4 = -13 + (-4) = -17$
(d) $10 - 18 = 10 + (-18) = -8$
(e) $0 - 10 = 0 + (-10) = -10$

It is often necessary to evaluate expressions that contain more than one subtraction. Also, many expressions contain both subtraction and addition. Such expressions are evaluated by converting each subtraction to addition. The result is a series where the additions may be performed in any convenient sequence. Remember, however, that only numbers following subtraction symbols have their signs changed.

EXAMPLE 9 Evaluate $10 - (-2) - 4$

Solution: Convert each subtraction to addition, then add in any convenient sequence.
$$10 - (-2) - 4 = 10 + 2 + (-4)$$
$$= 8$$

OPERATIONS WITH SIGNED NUMBERS   31

EXAMPLE 10

Evaluate $-6 + (-10) - (-3) - 1$

Solution: Convert each subtraction to addition. Do not change the sign of the addend, -10.

$$-6 + (-10) - (-3) - 1 = -6 + (-10) + 3 + (-1)$$
$$= -14$$

EXAMPLE 11

Evaluate $10 - (15 - 18) - (-2)$

Solution: Evaluate grouping symbols first.

$$10 - (15 - 18) - (-2) = 10 - [15 + (-18)] - (-2)$$
$$= 10 - (-3) - (-2)$$
$$= 10 + 3 + 2$$
$$= 15$$

Multiplication

You should recall that a product is the answer to a multiplication problem. We now review the rules for finding products of signed real numbers.

---
1. The product of zero and any real number is zero.
2. If two real numbers have like signs their product is positive.
3. If two real numbers have unlike signs their product is negative.

EXAMPLE 12

The preceding rules have been used to multiply these expressions:
(a) $0(-6) = 0$
(b) $2 \cdot 10 = 20$
(c) $(-9)(-3) = 27$
(d) $-8 \cdot 3 = -24$

Multiplication is commutative and associative. Therefore, expressions involving only products can be evaluated either by the order of operations agreement or by any convenient sequence.

EXAMPLE 13

Evaluate $3 \cdot (-2)(-1)(-6)$

Solution 1: Use the order of operations agreement.

$$3 \cdot (-2)(-1)(-6)$$
$$-6(-1)(-6)$$
$$6(-6)$$
$$-36$$

Solution 2: We find it convenient to multiply the negative numbers first.
$$3\underbrace{(-2)(-1)(-6)}$$
$$3 \cdot (-12)$$
$$-36$$

Division

As discussed in section 1.3, division is closely related to multiplication. The product of the quotient and divisor is equal to the dividend. For example, $\frac{15}{3} = 5$ since $5 \cdot 3 = 15$. Because of this relationship, the rules for dividing two real numbers are similar to the rules for multiplication.

> 1. Division by zero is undefined.
> 2. Zero divided by a non-zero real number is zero.
> 3. If two real numbers have like signs their quotient is positive.
> 4. If two real numbers have unlike signs their quotient is negative.

EXAMPLE 14   The preceding rules have been applied to the following:

(a) $\frac{-6}{0}$ is undefined

(b) $\frac{0}{0}$ is undefined

(c) $\frac{0}{-5} = 0$

(d) $\frac{-10}{-2} = 5$

(e) $\frac{-12}{4} = -3$

Expressions containing combinations of the fundamental operations can be evaluated by using your knowledge of grouping symbols and the order of operations agreement.

EXAMPLE 15   Evaluate $\frac{-3(10 - 18)}{(-2 - 4)(12 - 10)}$

Solution: Evaluate grouping symbols first.
$$\frac{-3(10 - 18)}{(-2 - 4)(12 - 10)} = \frac{-3(-8)}{(-6)(2)}$$
$$= \frac{24}{-12}$$
$$= -2$$

Variables

As illustrated by the following theorems, multiplication and division of variables obey the same rules as signed numbers.

Opposite Law

$-a = -1a$, where $a \in R$.

Proof: $a + (-1) \cdot a = 1 \cdot a + (-1) \cdot a$ [multiplicative identity]
$ = a[1 + (-1)]$ [distributive property]
$ = a \cdot 0$ [additive inverse property]
$ = 0$ [multiplication by zero]

Since $a + (-1) \cdot a = 0$, $-1 \cdot a$ must be the additive inverse of a. But, the additive inverse of a is $-a$ and is unique. Thus, $-a$ must equal $-1 \cdot a$.

The opposite law states that the opposite of a real number is equivalent to multiplying it by negative one.

EXAMPLE 16

(a) $-12 = -1 \cdot 12$
(b) $-x = -1 \cdot x$
(c) $-1 \cdot m = -m$

Multiplying Variables

1. $a(-b) = (-a)b = -(ab)$, where $a, b, \in R$.

Proof: $a(-b) = a \cdot (-1 \cdot b)$ [opposite law]
$ = (-1 \cdot a) \cdot b$ [associative and commutative]
$ = (-a) \cdot b$ [opposite law]

And, $a(-b) = a \cdot (-1 \cdot b)$ [opposite law]
$ = -1 \cdot (a \cdot b)$ [associative and commutative]
$ = -(a \cdot b)$ [opposite law]

2. $(-a)(-b) = ab$, where $a, b \in R$.

Proof: $(-a)(-b) = (-1 \cdot a)(-1 \cdot b)$ [opposite law]
$ = (-1)(-1)(ab)$ [associative and commutative]
$ = 1(ab)$ [$(-1)(-1) = 1$]
$ = ab$ [multiplicative identity]

This theorem shows that multiplication of variables follows the same rules as multiplication of signed numbers.

34 BASIC PROPERTIES OF REAL NUMBERS

EXAMPLE 17
(a) $2(-x) = -2x$
(b) $-(mn) = -mn$
(c) $(-3)(-y) = 3y$

> **Distributive Property of Multiplication with Respect to Subtraction**
> $a(b - c) = ab - ac$, where $a, b, c \in R$.
> Proof: $a(b - c) = a[b + (-c)]$ [definition of subtraction]
> $ = ab + a(-c)]$ [distributive property]
> $ = ab + [-(ac)]$ [multiplying variables]
> $ = ab - ac$ [definition of subtraction]

This theorem shows us that multiplication is distributive over subtraction as well as addition. For example:

$$5(2x - 3) = 10x - 15$$

The distributive properties of multiplication with respect to addition and subtraction can be combined and extended to any number of terms, as illustrated in these examples.

EXAMPLE 18   The following have been multiplied according to the distributive properties:
(a) $3(2x - 3y + 4z) = 6x - 9y + 12z$
(b) $-2a(3b - 4c + d - e) = -6ab + 8ac - 2ad + 2ae$

EXAMPLE 19   The following have been factored according to the distributive properties:
(a) $5x - 10y + 15 = 5(x - 2y + 3)$
(b) $12mn - 8m + 4n - 4 = 4(3mn - 2m + n - 1)$

> **Division**
> $$\frac{a}{b} = a \cdot \frac{1}{b}, \text{ where } a, b \in R \text{ and } b \neq 0$$
> Proof: The division check states that the product of the quotient and divisor must equal the dividend. Thus, the quotient $\left(a \cdot \frac{1}{b}\right)$, when multiplied by the divisor b, must produce the dividend, a.
> $\left(a \cdot \frac{1}{b}\right) \cdot b = a \cdot \left(\frac{1}{b} \cdot b\right)$ [associative]
> $\phantom{\left(a \cdot \frac{1}{b}\right) \cdot b} = a \cdot 1$ [multiplicative inverse]
> $\phantom{\left(a \cdot \frac{1}{b}\right) \cdot b} = a$ [multiplicative identity]

OPERATIONS WITH SIGNED NUMBERS

We see from this theorem that dividing by a real number is equivalent to multiplying by its reciprocal.

EXAMPLE 20

(a) $\dfrac{12}{3} = 12 \cdot \dfrac{1}{3}$

(b) $\dfrac{2}{5} = 2 \cdot \dfrac{1}{5}$

(c) $\dfrac{a}{7} = a \cdot \dfrac{1}{7}$

Dividing Variables

1. $\dfrac{-a}{b} = \dfrac{a}{-b} = -\dfrac{a}{b}$, where $a, b, \in R$ and $b \neq 0$.

2. $\dfrac{-a}{-b} = \dfrac{a}{b}$, where $a, b \in R$ and $b \neq 0$.

The proof is similar to the proof for multiplying variables, and is left as problems 51 and 52 in Exercise 1.5.

Division with variables follows the same procedure as division of signed numbers.

EXAMPLE 21

(a) $\dfrac{-x}{2} = \dfrac{x}{-2}$

(b) $\dfrac{-x}{2} = -\dfrac{x}{2}$

(c) $\dfrac{-m}{-3} = \dfrac{m}{3}$

Exercise 1.5

Add or subtract as indicated.

1. $-8 + 6$
2. $6 + (-9)$
3. $-4 + (-3)$
4. $3 - 8$
5. $-2 - 9$
6. $-6 - 16$
7. $-5 - (-3)$
8. $4 - (-9)$
9. $11 - (-3) - (-4)$
10. $10 - (-6) - (-8)$
11. $-8 - (-6) - (4 - 6)$
12. $-4 - (-8) + (-4 + 9)$
13. $27 + (-39 + 12) - 11$
14. $-13.6 + 10.2 - 5.8$

Multiply or divide as indicated, if possible.

15. $(-16)(-4)$
16. $\dfrac{-10}{0}$

17. $\dfrac{(-8 + 5) \cdot (-4)}{-1 - 5}$

18. $-3[-2 + (6 - 1)] \div [-18 - 2]$

19. $-4[6 + (-4)] \div [-5 - (-3)]$

20. $[-2(6 - 2) + 3] \cdot [12 \div (2 - 8)]$

21. $(-|-8|)(|-7| + |-5|)$

22. $(-3)(-3)(5) - (-6)(-6) \div (-2)(-2)$

23. $\dfrac{(-5) \cdot 4 - (-8)}{(-4) + 1}$

Evaluate each of the following, if possible. Let $a = 3$, $b = 4$, and $c = -5$.

24. $3a - 2c$
25. $3b(a + 5c)$
26. $-2c(3a - 2c)$
27. $-4(-b - 2c)$
28. $(a - 2c)(b - 2c)$
29. $\dfrac{2c - a}{4b + 3c}$
30. $\dfrac{a + 2b + 3c}{2b + c - a}$
31. $|ab - 3a - abc|$
32. $|ab| \cdot |c - a|$
33. $(|a| + b) - (-a + |c|)$
34. $(a + b + c) \div (a + c + 2b)$

Determine which of the following properties, A through F, is being used to write the given statement.

A. $-a = -1 \cdot a$

B. $\dfrac{a}{b} = a \cdot \dfrac{1}{b}$

C. $a(-b) = (-a)b = -(ab)$

D. $(-a)(-b) = ab$

E. $\dfrac{-a}{b} = \dfrac{a}{-b} = -\dfrac{a}{b}$

F. $\dfrac{-a}{-b} = \dfrac{a}{b}$

35. $(-2)x = -(2x)$
36. $(-x)(-y) = xy$
37. $-\dfrac{1}{4} = \dfrac{1}{-4}$
38. $-1 \cdot x = -x$
39. $\dfrac{-2}{5} = -2 \cdot \dfrac{1}{5}$
40. $\dfrac{(x - 1)}{-y} = \dfrac{-(x - 1)}{y}$

Multiply the following according to the distributive properties.

41. $5(2x + 3y - 4z)$
42. $-3a(2b - 5c - d + e)$
43. $-2ab(x - 2y + 3z)$
44. $(-3a + b - c)4$
45. $(x - y - 2z)(-2a)$
46. $-xy(ab - c + 2e)$

Use the distributive properties to factor the following.
47. $4x - 8y + 12$
48. $8ab - 4a + 16b - 4$
49. $27xy - 6x - 12y$
50. $-mn + 2n - xn$

Prove the following using the theorems and definitions previously presented in the text.

51. $\dfrac{-a}{b} = -\dfrac{a}{b}$

> Hint: Follow the form of the proof under multiplying variables given in the text.

52. $\dfrac{-a}{-b} = \dfrac{a}{b}$

Use your calculator to evaluate the following. Round your answers to three significant digits.
53. $-6.8 - [2.9 - 4.6(.86 - .92) - 6.8]$
54. $(-46.2)(-.034) - (8.1)(-.65) - (27.1)(3.6)$
55. $(14.34) - [(6.823 - 83.2) \div 2.15]$
56. $(-4.835)(2.62)(73.46 - 5[13.24])$
57. $(17.32 - 43.86)(4.76 \div 3.48 \cdot 5.2)$

NATURAL NUMBER EXPONENTS AND ROOTS

1.6

Numbers multiplied together are called *factors*. For example, 3 and 5 are factors of 15 because $3 \cdot 5 = 15$. It is also common to have expressions where a factor is repeated, such as $2 \cdot 2 \cdot 2 \cdot 2$ or $x \cdot x$, or $(a + b)(a + b)(a + b)$. This occurs so frequently in algebra that a shortened form, called *exponential notation*, has been developed. Thus, $2 \cdot 2 \cdot 2 \cdot 2 = 2^4$, where 2 is the *base* and 4 is the *exponent* or *power*. The symbol 2^4 is called an *exponential* and is read, "2 to the fourth power." It should be apparent that a natural number exponent indicates the number of times the base is used as a factor.

Definition

If $a \in R$ and if n is any natural number, then

$$a^n = \underbrace{a \cdot a \cdot a \cdots a}_{n \text{ factors of } a}$$

n is the exponent, a is the base, and a^n is an exponential.

BASIC PROPERTIES OF REAL NUMBERS

EXAMPLE 1 Here are more illustrations of exponential notation:

(a) $(-4)(-4) = (-4)^2$; -4 is the base and 2 is the exponent or power. The exponential is $(-4)^2$ and is read "negative four to the second power" or "negative four squared."

(b) $(a + b)(a + b)(a + b) = (a + b)^3$; $(a + b)$ is the base and 3 is the exponent. The exponential is $(a + b)^3$ and is read "the group (a plus b) to the third power" or "the group (a plus b) cubed."

If an expression is written without an exponent, we assume that the exponent is 1. Thus, $x = x^1$.

When evaluating expressions containing exponentials, it is important to know that an exponent applies only to the nearest symbol on its left unless grouping symbols indicate otherwise. Thus, -3^2 means $-(3)(3)$ or -9. It does not mean $(-3)(-3)$.

EXAMPLE 2 The following exponential expressions have been evaluated.

(a) $-2^4 = -(2)(2)(2)(2) = -16$

(b) $(-2)^4 = (-2)(-2)(-2)(-2) = 16$

(c) $2 \cdot 5^2 = 2 \cdot 5 \cdot 5 = 50$

(d) $(2 \cdot 5)^2 = (2 \cdot 5)(2 \cdot 5) = 100$

We now develop four properties of exponents that will enable us to operate with exponentials. First we examine multiplication of exponentials having identical bases. Consider $x^4 \cdot x^2$.

$$x^4 \cdot x^2 = (x \cdot x \cdot x \cdot x)(x \cdot x)$$
$$= \underbrace{x \cdot x \cdot x \cdot x}_{4} \cdot \underbrace{x \cdot x}_{2} \quad \text{[six factors of x]}$$
$$= x^6$$

The sum of the exponents, $4 + 2$, gives the new exponent, 6. This example suggests the following property.

Addition Property of Exponent

If $a \in R$ and m, n are natural numbers, then
$$a^m \cdot a^n = a^{m+n}$$
(To multiply exponentials having identical bases, add their exponents.)

EXAMPLE 3 These equations illustrate the addition property of exponents.

(a) $5^2 \cdot 5^8 = 5^{2+8} = 5^{10}$

(b) $x^2 \cdot x^3 \cdot x^6 = x^{11}$

(c) $y \cdot y^8 = y^1 \cdot y^8 = y^9$

NATURAL NUMBER EXPONENTS AND ROOTS

EXAMPLE 4

Multiply $(2x^2y)(-3x^3y^2)$

Solution: Multiply the numbers and add the exponents which appear on identical bases.

$$(2x^2y)(-3x^3y^2) = 2(-3)(x^2 \cdot x^3)(y^1 \cdot y^2)$$
$$= -6x^5y^3$$

Next, we examine division of exponentials having the same bases. Consider $\dfrac{x^5}{x^2}$, where $x \neq 0$.

$$\frac{x^5}{x^2} = \frac{x \cdot x \cdot x \cdot x \cdot x}{x \cdot x}$$

$$= \frac{x \cdot x \cdot x \cdot \boxed{x \cdot x}}{\boxed{x \cdot x}}$$

$$= x \cdot x \cdot x \cdot 1$$

$$= x^3$$

The difference of the exponents, $5 - 2$, gives the exponent, 3. This example enables us to state the following property.

Subtraction Property of Exponents

If $a \in R$ and m, n are natural numbers where $m \geq n$, then

$$\frac{a^m}{a^n} = a^{m-n}, \quad \text{where } a \neq 0$$

(To divide exponentials having identical bases, subtract their exponents.)

The statement $m \geq n$ says the exponent m must be greater than or equal to the exponent n. (In section 5.2, we will discuss what happens if m is allowed to be less than n.)

An interesting and useful consequence occurs when m and n are equal:

$$\text{If } m = n, \text{ then } \frac{a^m}{a^n} = \frac{a^m}{a^m}$$
$$= a^{m-m}$$
$$= a^0$$

But for $a \neq 0$, $\dfrac{a^m}{a^m} = 1$ because any non-zero number divided by itself is 1. Thus, $a^0 = 1$ when $a \neq 0$.

EXAMPLE 5　The following equations illustrate the subtraction property of exponents.

(a) $\dfrac{10^8}{10^5} = 10^{8-5} = 10^3$

(b) $\dfrac{x^7}{x^3} = x^4$, where $x \neq 0$

(c) $\dfrac{y^2}{y} = \dfrac{y^2}{y^1} = y^1$ or y, where $y \neq 0$

(d) $\dfrac{x^6}{x^6} = x^{6-6} = x^0 = 1$, where $x \neq 0$

EXAMPLE 6　Divide $\dfrac{-15x^6y^3}{5x^2y}$, where $x, y \neq 0$.

Solution:　Divide the numbers and subtract the exponents appearing on identical bases.

$$\dfrac{-15x^6y^3}{5x^2y} = -3x^4y^2$$

When working with exponents, we will often need to raise an exponential to a power. For example, consider $(x^5)^2$. The exponential x^5 is raised to the second power. This is called a power to a power. To simplify we use the definition of exponents together with the addition property of exponents.

$$\begin{aligned}(x^5)^2 &= x^5 \cdot x^5 &&\text{(definition)}\\ &= x^{5+5} &&\text{(addition property of exponents)}\\ &= x^{10}\end{aligned}$$

This process can be shortened by simply multiplying the exponents 5 and 2 to obtain the new exponent 10. This suggests the following property.

Power to a Power Property of Exponents

If $a \in R$ and m, n are natural numbers, then

$$(a^m)^n = a^{mn}$$

(To simplify an exponential raised to a power, multiply the exponents.)

EXAMPLE 7　These equations illustrate the power to a power property.

(a) $(10^2)^6 = 10^{12}$

(b) $(y^3)^4 = y^{12}$

(c) $[(a+b)^2]^3 = (a+b)^6$

As stated in the next property, the power to a power property can be extended so that it applies to any base that is either a product or a quotient.

NATURAL NUMBER EXPONENTS AND ROOTS

Distributive Property of Exponents

If $a, b \in R$ and n is a natural number, then

$$(ab)^n = a^n b^n$$

$$\left(\frac{a}{b}\right)^n = \frac{a^n}{b^n}, \text{ where } b \neq 0$$

(An exponent may be shared with each factor of a basic product or basic quotient.)

This property is true only for bases that are products or quotients. It does **not** hold true for sums and differences. As shown by the following, an exponent must never be distributed over a basic sum or a basic difference.

$$(3 + 4)^2 \neq 3^2 + 4^2$$
$$7^2 \neq 9 + 16$$
$$49 \neq 25$$

Each of these equations illustrates a correct use of the distributive property of exponents.

EXAMPLE 8

(a) $(mn)^3 = m^3 n^3$

(b) $(2x)^3 = 2^3 x^3 = 8x^3$

(c) $(-3a^2 b^4)^2 = (-3)^2 \cdot (a^2)^2 \cdot (b^4)^2 = 9a^4 b^8$

(d) $\left(\frac{5x^3}{3y^2}\right)^2 = \frac{5^2 (x^3)^2}{3^2 (y^2)^2} = \frac{25x^6}{9y^4}$, where $y \neq 0$

Summary of the Properties of Exponents

For $a, b \in R$ and $m, n \in N$

(1) Addition Property of Exponents

$$a^m \cdot a^n = a^{m+n}$$

(2) Subtraction Property of Exponents

$$\frac{a^m}{a^n} = a^{m-n}, \text{ where } a \neq 0 \text{ and } m > n$$

(3) Power to a Power Property of Exponents

$$(a^m)^n = a^{mn}$$

(4) Distributive Property of Exponents

$$(ab)^n = a^n b^n$$

$$\left(\frac{a}{b}\right)^n = \frac{a^n}{b^n}, \text{ where } b \neq 0$$

Roots

From our discussion of exponents we know that 7 squared is 49. The opposite of squaring a number is called taking a square root. For example, a square root of 49 is 7. Another square root of 49 is -7 because $(-7)^2 = 49$. As implied by this example, every positive real number has two square roots. Zero has only one square root, namely zero. Negative real numbers have no real square roots since any non-zero real number multiplied by itself is positive.

The nonnegative square root is called the principal square root and is indicated by the radical sign, $\sqrt{}$. Thus, the positive (or principal) square root of 49 is written, $\sqrt{49} = 7$. The negative square root of 49 is written, $-\sqrt{49} = -7$.

EXAMPLE 9 The following square roots have been evaluated.
(a) $\sqrt{100} = 10$
(b) $-\sqrt{100} = -10$
(c) $\sqrt{0} = 0$
(d) $\sqrt{\dfrac{25}{36}} = \dfrac{5}{6}$

Even roots are square roots, fourth roots, sixth roots, and so on. Even roots behave similarly to square roots. For example, the positive fourth rooth of 16 is written $\sqrt[4]{16} = 2$, while the negative fourth root of 16 is written $-\sqrt[4]{16} = -2$.

EXAMPLE 10 The following even roots have been evaluated.
(a) $\sqrt[4]{81} = 3$
(b) $-\sqrt[4]{81} = -3$
(c) $\sqrt[6]{64} = 2$

Odd roots are cube roots, fifth roots, seventh roots, and so on. Odd roots behave quite differently from even roots. For each odd root, every real number has exactly one root. Consequently, there is no need to define principal odd roots.

EXAMPLE 11 The following odd roots have been evaluated.
(a) $\sqrt[3]{8} = 2$
(b) $\sqrt[3]{-8} = -2$
(c) $\sqrt[5]{32} = 2$
(d) $\sqrt[5]{-32} = -2$
(e) $\sqrt[7]{0} = 0$

Many expressions contain exponentials and roots in combination with the four basic operations. To evaluate such expressions, it is necessary to follow

NATURAL NUMBER EXPONENTS AND ROOTS

43

the rules for grouping symbols and the order of operations agreement as it applies to exponentials and roots.

Order of Operations

If grouping symbols are present:
Evaluate each set of grouping symbols by using the rules below. Begin on the inside and work outward. Remember, fraction bars are grouping symbols indicating division.

If no grouping symbols are present:
(1) Evaluate all exponentials and roots as they occur, working left to right.
(2) Perform all multiplications and divisions as they occur, working left to right.
(3) Perform all additions and subtractions as they occur, working left to right.

Evaluate $\dfrac{2^3(\sqrt[3]{-27} + \sqrt{16})}{-3^2 + 2^3}$

EXAMPLE 12

Solution:
$$\frac{2^3(\sqrt[3]{-27} + \sqrt{16})}{-3^2 + 2^3} = \frac{8(-3 + 4)}{-9 + 8}$$
$$= \frac{8 \cdot 1}{-1}$$
$$= \frac{8}{-1}$$
$$= -8$$

Calculator

With your instructor's approval, you may wish to evaluate certain expressions with your calculator. To evaluate exponentials and roots, we generally use these keys (variations may exist among brands):

$\boxed{x^2}$ square

$\boxed{\sqrt{x}}$ square root

$\boxed{y^x}$ exponential

$\boxed{\text{INV}}$ or $\boxed{\text{2nd}}$ inverse operation or second function

Evaluate $(2.3)(3.1)^4$. Round the answer to two significant digits.

EXAMPLE 13

Calculator Sequence: $\boxed{2.3}$ $\boxed{\times}$ $\boxed{3.1}$ $\boxed{y^x}$ $\boxed{4}$ $\boxed{=}$ 212.40983
The rounded answer is 210

EXAMPLE 14   Evaluate $\sqrt[3]{98.6}$. Round the answer to three significant digits.

Calculator Sequence: (98.6) (INV) (y^x) (3) (=) 4.6198262
The rounded answer is 4.62

EXAMPLE 15   Evaluate $\dfrac{\pi\sqrt{6.51} - \sqrt[5]{16.4}}{(8.43)^2 - (2.07)^3}$. Round the answer to three significant digits.

Calculator Sequence: (( ) (π) ($\times$) (6.51) ($\sqrt{x}$) (−) (16.4) (INV) (y^x) (5) ( ))
($\div$) (( ) (8.43) (x^2) (−) (2.07) (y^x) (3) ( )) (=)
.10074674
The rounded answer is 0.101

Exercise 1.6   Evaluate.

1. $(-1)^{19}$
2. $-3^2 + 4$
3. $-2^3 + 6$
4. $-2^2 - (-3)^2$
5. $(-2)^2 + (-1)^{18}$
6. $(-2 \cdot 3)^3$
7. $[(-3)(-2)]^3$
8. $-(-4)^3$
9. $-(-2)^5$
10. $\left(-\dfrac{2}{3}\right)^4$
11. $-\sqrt{100}$
12. $(-\sqrt{49})(-\sqrt{36})$
13. $(\sqrt[4]{16})(\sqrt{16})$
14. $\sqrt[3]{-27}$
15. $(-\sqrt[4]{16})(\sqrt[3]{8})$
16. $-\sqrt[5]{-32}$
17. $(-\sqrt[5]{-1})^2$
18. $(\sqrt[4]{81})(\sqrt[3]{-8})$
19. $-\sqrt[3]{-125}$
20. $\sqrt[5]{243}$

Apply the properties of exponents and simplify. (Assume denominators are not zero.) There should be no grouping symbols and each base should be written only once.

21. $2^3 \cdot 2^4$
22. $3^3 \cdot 3^2$
23. $\dfrac{5^7}{5^2}$
24. $\dfrac{3^6}{3^6}$
25. $x^2 \cdot x^5$
26. $x^6 \cdot y^4$
27. $\dfrac{x^8}{x^8}$
28. $\dfrac{x^6}{x^4}$
29. $x^4 \cdot x \cdot x^3$
30. $(3xy^2)(-2x^2y^3)$
31. $(-xy)^3(-2x)^2$
32. $(x^2y)(x^3y^2)(xy)$
33. $(a^2b)^2 \cdot (a^3b^2)$
34. $(x^3y)(y^2z)^2(xz^2)$
35. $(x^3)^2 \cdot x^4$
36. $(6x^3y^2)(3x^2y)$
37. $(x^2y)^2(x^2y^3)(x)$
38. $-(a^3b^2)^2 \cdot (ab^3) \cdot (ab)$

ALGEBRAIC EXPRESSIONS

39. $\dfrac{15x^6y^2}{-5x^2y}$

40. $\dfrac{a^3 \cdot b^4}{a^2 \cdot b^2}$

41. $\dfrac{6x^2yz^3}{-6xyz}$

42. $\dfrac{18x^3y^8}{6xy^4}$

43. $\left(\dfrac{x^4y^2}{x}\right)^2$

44. $\dfrac{x^4 \cdot z^3}{(xy)^2}$

45. $\dfrac{(4x^3)^2}{8x^6}$

46. $(xy^2z^3)^3 \cdot (x^3z)^2$

47. $-(3x)(x^2)(-3x)^2$

Evaluate if possible.

48. $\dfrac{3^2(\sqrt[3]{-64} + \sqrt[3]{8})}{-2 + 2^3}$

49. $\dfrac{-3^2[\sqrt[4]{16} - (-2)^2]}{\sqrt[3]{-1}}$

50. $\dfrac{(-\sqrt{25})^2 - (\sqrt{36})}{\sqrt[3]{1} + \sqrt[3]{-1}}$

51. $\dfrac{-\sqrt{1} + \sqrt[3]{8} - \sqrt[4]{16}}{-(3)^2}$

Use your calculator to evaluate the following. Round your answers to three significant digits.

52. $\dfrac{(15.851)^2(3.941)^2}{(2.68)^2}$

53. $[(2.03)(4.91)]^2 - [(3.06)(9.031)]^2$

54. $(2.658)^5$

55. $(-6.912)^4$

56. $\dfrac{(.468)^3}{(2.66)^2}$

57. $\pi\sqrt[3]{26.4}$

58. $\dfrac{\sqrt{98.6} - \sqrt[3]{61.4}}{(3.01)^3}$

59. $\dfrac{\sqrt[5]{\pi} - \sqrt[4]{\pi}}{\sqrt[3]{\pi}}$

60. $\dfrac{\sqrt[3]{6.51} - \sqrt{.424}}{(6.23)^3 - (2.07)^2}$

ALGEBRAIC EXPRESSIONS

1.7

Any meaningful combination of constants, variables and basic operations is called an *algebraic expression*. To evaluate an algebraic expression means to replace the variables with their corresponding values and to simplify the resulting expression.

EXAMPLE 1 Evaluate $3x^2y + 6xy^2 - 4$, where $x = 3$ and $y = -2$.

Solution: $\begin{aligned}3x^2y + 6xy^2 - 4 &= 3 \cdot 3^2 \cdot (-2) + 6 \cdot 3 \cdot (-2)^2 - 4 \\ &= 3 \cdot 9 \cdot (-2) + 6 \cdot 3(4) - 4 \\ &= -54 + 72 - 4 \\ &= 14\end{aligned}$

The terms of an algebraic expression are the quantities which have been either added or subtracted. For example, the algebraic expression, $3x^2y + 6xy^2 - 4$, has three terms: $3x^2y$, $6xy^2$, and 4.

Any factor or group of factors in a term is said to be the coefficient of the remaining factors. The numerical factor in a term is called the numerical coefficient. If a term does not contain a numerical coefficient, it is understood to be 1.

EXAMPLE 2 Consider the algebraic expression $6x^2 + x$.

(a) x^2 is the coefficient of 6.
(b) 6 is the coefficient of x^2.
(c) 6 is the numerical coefficient of $6x^2$.
(d) 1 is the numerical coefficient of x.

Terms that differ only in their numerical coefficients are called like terms or similar terms. Like terms can be combined by applying the distributive property.

EXAMPLE 3 Combine $5x + 3x$

Solution: Use the distributive property to factor out the x.
$$\begin{aligned}5x + 3x &= (5 + 3)x \\ &= 8x\end{aligned}$$

EXAMPLE 4 Combine $4a^3 - a^3 + 2a^3$

Solution: $\begin{aligned}4a^3 - a^3 + 2a^3 &= (4 - 1 + 2)a^3 \\ &= 5a^3\end{aligned}$

By studying Examples 3 and 4, it is obvious that like terms can be added or subtracted by combining their respective numerical coefficients.

EXAMPLE 5 Combine $7x^2y + 4xy^2 - 3x^2y - 5xy^2$

Solution: Combine numerical coefficients of like terms.

$$7x^2y + 4xy^2 - 3x^2y - 5xy^2 = 4x^2y - xy^2$$

Algebraic expressions can be added vertically if like terms are arranged in the same columns.

Add $(3x^2 + 5x - 3) + (2x^2 - 3x + 1) + (-x + 4)$
EXAMPLE 6

Solution: Arrange like terms in the same columns and combine their numerical coefficients.

$$\begin{array}{r} 3x^2 + 5x - 3 \\ 2x^2 - 3x + 1 \\ \underline{- x + 4} \\ 5x^2 + x + 2 \end{array}$$

You should recall that subtraction is performed by converting to addition according to this definition: $a - b = a + (-b)$. Thus, algebraic expressions can be subtracted vertically by changing each sign in the subtrahend and then combining like terms.

Subtract $(3x^2 - 8x + 5) - (x^2 + 3x - 9)$
EXAMPLE 7

Solution: Arrange vertically, change signs in the subtrahend, and add coefficients of like terms.

$$\begin{array}{r} 3x^2 - 8x + 5 \\ \underline{x^2 + 3x - 9} \end{array} \quad \xrightarrow{\text{change signs}} \quad \begin{array}{r} 3x^2 - 8x + 5 \\ \underline{-x^2 - 3x + 9} \\ 2x^2 - 11x + 14 \end{array}$$

Often when simplifying algebraic expressions it will be necessary to remove grouping symbols preceded by either a plus sign or a minus sign. Therefore, we must examine such expressions to see how a plus or minus sign affects each term within a group. First, consider a group preceded by a plus sign and follow this argument:

$$\begin{aligned} +(a + b) &= +1 \cdot (a + b) \quad &&\text{[coefficient of 1]} \\ &= 1a + 1b \quad &&\text{[distributive property]} \\ &= a + b \quad &&\text{[multiplicative identity]} \end{aligned}$$

Since $+(a + b) = a + b$, we conclude that a plus sign in front of a grouping symbol does not affect any term within the group.

Next, we consider a group preceded by a minus sign:

$$\begin{aligned} -(a + b) &= -1(a + b) \quad &&\text{[opposite law]} \\ &= -1a + (-1)b \quad &&\text{[distributive property]} \\ &= -a + (-b) \quad &&\text{[opposite law]} \\ &= -a - b \quad &&\text{[definition of subtraction]} \end{aligned}$$

Since $-(a + b) = -a - b$, we conclude that a minus sign in front of a group changes the sign of each term within the group.

Removing Group Symbols
1. Removing a grouping symbol preceded by a plus sign does not affect any term within the group.
$$+(a + b) = a + b$$
2. Removing a grouping symbol preceded by a minus sign changes the sign of each term within the group.
$$-(a + b) = -a - b$$

EXAMPLE 8   Simplify $5x + (3y - 7)$.

Solution: Make no changes within the group.
$$5x + (3y - 7) = 5x + 3y - 7$$

EXAMPLE 9   Simplify $3m^3 - (m^2 - m + 3)$.

Solution: Change the sign of each term within the group.
$$3m^3 - (m^2 - m + 3) = 3m^3 - m^2 + m - 3$$

EXAMPLE 10   Simplify $2x^3 - [x - (y^2 - 3y) + 4]$.

Solution: When groups occur within groups, you will find it easier to remove the innermost first and work your way to the outside.
$$2x^3 - [x - (y^2 - 3y) + 4] = 2x^3 - [x - y^2 + 3y + 4]$$
$$= 2x^3 - x + y^2 - 3y - 4$$

After removing grouping symbols, many algebraic expressions can be further simplified by combining like terms. We say an algebraic expression is in **simplest form** when all grouping symbols have been removed and all like terms have been combined.

EXAMPLE 11   Write in simplest form $(x - 2) - [3x - (x + 1) + 2]$.

Solution: Remove grouping symbols and combine like terms. The absence of a sign in front of a group is treated as a plus sign.
$$(x - 2) - [3x - (x + 1) + 2] = +(x - 2) - [3x - (x + 1) + 2]$$
$$= +(x - 2) - [3x - x - 1 + 2]$$
$$= x - 2 - 3x + x + 1 - 2$$
$$= -x - 3$$

Multiplication and division by single terms can be performed using the properties of exponents and (when necessary) the distributive property.

EXAMPLE 12   The following have been multiplied.
(a) $(5x^3y)(-2xy^2)(4xy) = -40x^5y^4$
(b) $3x^2y(4x^3y^2 - 2x^2y + xy) = 12x^5y^3 - 6x^4y^2 + 3x^3y^2$

ALGEBRAIC EXPRESSIONS

EXAMPLE 13

The following have been divided.

(a) $\dfrac{-36a^4b^3c^2}{3a^2bc^2} = -12a^2b^2$

(b) $\dfrac{3x^3y^2 - 12x^2y + 6x}{3xy} = \dfrac{3x^3y^2}{3xy} - \dfrac{12x^2y}{3xy} + \dfrac{6x}{3xy}$

$\qquad = x^2y - 4x + \dfrac{2}{y}$

Algebraic expressions containing more than one term can be multiplied using the distributive properties repetitively. One expression is considered to be a single number and is distributed with each term of the second expression. The resulting expression is simplified by distributing again.

EXAMPLE 14

Multiply using the distributive property:
$(x + 1)(2x + 3)$

Solution: Consider $(x + 1)$ to represent one number and distribute it with each term of $2x + 3$.

$(x + 1)(2x + 3) = (x + 1) \cdot 2x + (x + 1) \cdot 3$

$\qquad = (x + 1)2x + (x + 1) \cdot 3$ [distribute again]

$\qquad = 2x^2 + 2x + 3x + 3$

$\qquad = 2x^2 + 5x + 3$

EXAMPLE 15

Multiply using the distributive property:
$(2a - 1)(3a^2 - 2a + 5)$

Solution: Consider $(2a - 1)$ to represent one number, then use the distributive property.

$(2a - 1)(3a^2 - 2a + 5) = (2a - 1) \cdot 3a^2 - (2a - 1) \cdot 2a + (2a - 1) \cdot 5$

$\qquad = [6a^3 - 3a^2] - [4a^2 - 2a] + [10a - 5]$

$\qquad = 6a^3 - 3a^2 - 4a^2 + 2a + 10a - 5$

$\qquad = 6a^3 - 7a^2 + 12a - 5$

Exercise 1.7

Evaluate the following expressions if $m = -1$ and $n = -3$.
1. $6n^2 + 3n - 5$
2. $n^3 + 2n^2 - 3m - 1$
3. $m^3 - 3m^2 + 1$
4. $-2m^2 - n + 2$

Evaluate each expression for the given values of the variable.
5. $\dfrac{5(F - 32)}{9}$; $F = 104$

6. $\dfrac{x^2 + y}{3x + y}$; $x = 3$ and $y = -2$

7. $(x + y)^2 + (x - y)^2$; $x = 2$ and $y = -3$

8. $\dfrac{Rr}{R + r}$; $R = 50$ and $r = 25$

Add or subtract as indicated.

9. $(5x^2 + 3x - 6) + (2x^2 - x + 7)$
10. $(3a^2 - 2a - 1) + (a^2 + a - 6)$
11. $(5x^2 - 3x + 1) - (7x^2 + 4x + 3)$
12. $(a^2 - a + 1) - (a^2 + 2a + 2)$
13. $(6x^2 - 8x) - (4x^2 - 2x - 3)$
14. $(x^3 - x) - (x^2 - x) + (x^3 - 1)$
15. $(x^2 - 8xy + y^2) + (2x^2 - 6xy - y^2)$
16. $(6x^3 - 3x + 1) + (2x^3 - 2x - 5) - (3x^3 - x + 4)$
17. $(6x^4 - 2x^3 - 6) - (-8x^4 - 4x^3 + 6)$
18. $(3x^2 - 2) + (5x - 7) - (4x^3 + x) - (2x + 1)$
19. $4x + 5x^2 - 3x^2 + x - 3x^2 + 4x - 6x^2$
20. $3a^2 - 6a^3 - 5a^2 + 11a^3 - 4a + 8$

Remove grouping symbols and combine like terms.

21. $2 - [3x - (x + 1) + 2]$
22. $4 - [3x + (x - 4) - 1]$
23. $x - \{2y - [x - (x - y)] + x\}$
24. $-(x - 1) + [2x - (3 + x) - 2]$
25. $x - 6[3x - (2 - x)]$
26. $x - \{x - [y - (y + x)] - y\}$
27. $3 - \{2x - [1 - (x + y)] + [x - 2y]\}$
28. $2x - \{x - [x - (x - y)] + 2x\}$
29. $x^2 - \{x - [3x - y] - y^2\}$
30. Find the sum of $3x - 2x^2$, $6x + 2x^2$, and $-3x + 2x^2 + 2$.
31. Subtract $6ab - 4b^2$ from $5ab + 8b^2$.
32. Subtract $x^2 + 3x - 1$ from the sum of $2x^2 - 3x + 1$ and $x^2 + x + 2$.
33. Subtract $x^2 - x$ from the sum of $x^2 + x$ and $2x^2 - 2$.

Multiply using the distributive property and collect any like terms.

34. $(x + 2)(2x + 3)$
35. $(2x - 1)(3x + 4)$
36. $(3x + 1)(4x - 3)$
37. $(x^2 + 1)(x^2 - 1)$
38. $(x^2 + 1)(x + 1)$
39. $(x + 3)(x^2 + x + 5)$
40. $(x - 2)(x^2 + 2x + 4)$

41. $(3x - 2)(x^2 - x + 2)$
42. $(x + 2)(x^2 - 2x + 4)$
43. $(x + 1)(x^2 - x + 1)$
44. $(x^2 - 2)(x^2 + x + 1)$
45. $(x^2 + x - 2)(x^2 - x + 2)$
46. $(3x^2 - x + 2)(2x^2 + 2x + 3)$

Simplify.
47. $(6x^4y^2)(-3x^2y)(2xy)$
48. $-2x^3y(4x^2y^3 - 3x^2y + xy^2)$
49. $-xy^2(-3x^2 + 4xy - 6x - 2y)$
50. $\dfrac{8x^4y^2 - 4x^2y + x^2y}{x^2y}$
51. $\dfrac{(2x^7)(3x^2) + x^4(2x^2)}{x^3}$

Summary

SYMBOLS

$\{a, b\}$	The set containing a and b
$\{\ \}$	Empty set
$\in$	Is a member of or belongs to
N	The set of natural numbers
W	The set of whole numbers
I	The set of integers
Q	The set of rational numbers
H	The set of irrational numbers
R	The set of real numbers
$=$	Is equal to
$\neq$	Is not equal to
$<$	Is less than
$\leq$	Is less than or equal to
$>$	Is greater than
$\geq$	Is greater than or equal to
$\cup$	Union
$\cap$	Intersection
$\|x\|$	Absolute value of x

TERMS

Variable—a symbol representing any element of a set containing more than one element.

Reciprocal—the multiplicative inverse

Exponent—a number that gives the number of times the base is used as a factor.

Factors—numbers that are multiplied together.

BASIC PROPERTIES OF REAL NUMBERS

PROPERTIES OF EQUALITY

Reflexive $\quad a = a$
Symmetric $\quad$ If $a = b$, then $b = a$.
Substitution $\quad$ If $a = b$, then b may be replaced for a in any statement without altering its truth or falsity.
Addition $\quad$ If $a = b$, then $a + c = b + c$.
Multiplication $\quad$ If $a = b$ and $c \neq 0$, then $ac = bc$.

PROPERTIES OF REAL NUMBERS

Closure for Addition and Multiplication
$$a + b \in R, ab \in R$$
Commutative Properties of Addition and Multiplication
$$a + b = b + a$$
$$ab = ba$$
Associative Properties of Addition and Multiplication
$$(a + b) + c = a + (b + c)$$
$$(ab)c = a(bc)$$
Properties of Identity
$$\text{Additive: } 0 + a = a \text{ and } a + 0 = a$$
$$\text{Multiplicative: } 1 \cdot a = a \text{ and } a \cdot 1 = a$$
Inverse
$$\text{Additive: } a + (-a) = 0$$
$$\text{Multiplicative: } a \cdot \frac{1}{a} = 0, \text{ where } a \neq 0$$
Distributive Property of Multiplication with Respect to Addition
$$a(b + c) = ab + ac \text{ and } (b + c)a = ba + ca$$
Multiplication by Zero
$$0 \cdot a = a \cdot 0 = 0$$
Opposite Law
$$-a = -1 \cdot a$$
Double Opposite Law
$$-(-a) = a$$
Properties of Exponents
$$a^0 = 1$$
$$\frac{a^m}{a^n} = a^{m-n}, \text{ where } a \neq 0 \text{ and } m \geq n$$
$$(a^m)^n = a^{mn}$$
$$a^m \cdot a^n = a^{m+n}$$
$$(ab)^n = a^n b^n$$
Definition of Subtraction
$$a - b = a + (-b)$$

41. $(3x - 2)(x^2 - x + 2)$
42. $(x + 2)(x^2 - 2x + 4)$
43. $(x + 1)(x^2 - x + 1)$
44. $(x^2 - 2)(x^2 + x + 1)$
45. $(x^2 + x - 2)(x^2 - x + 2)$
46. $(3x^2 - x + 2)(2x^2 + 2x + 3)$

Simplify.

47. $(6x^4y^2)(-3x^2y)(2xy)$
48. $-2x^3y(4x^2y^3 - 3x^2y + xy^2)$
49. $-xy^2(-3x^2 + 4xy - 6x - 2y)$
50. $\dfrac{8x^4y^2 - 4x^2y + x^2y}{x^2y}$
51. $\dfrac{(2x^7)(3x^2) + x^4(2x^2)}{x^3}$

Summary

SYMBOLS

$\{a, b\}$	The set containing a and b
$\{\ \}$	Empty set
$\in$	Is a member of or belongs to
N	The set of natural numbers
W	The set of whole numbers
I	The set of integers
Q	The set of rational numbers
H	The set of irrational numbers
R	The set of real numbers
$=$	Is equal to
$\neq$	Is not equal to
$<$	Is less than
$\leq$	Is less than or equal to
$>$	Is greater than
$\geq$	Is greater than or equal to
$\cup$	Union
$\cap$	Intersection
$\|x\|$	Absolute value of x

TERMS

Variable—a symbol representing any element of a set containing more than one element.

Reciprocal—the multiplicative inverse

Exponent—a number that gives the number of times the base is used as a factor.

Factors—numbers that are multiplied together.

BASIC PROPERTIES OF REAL NUMBERS

PROPERTIES OF EQUALITY

Reflexive $a = a$
Symmetric   If $a = b$, then $b = a$.
Substitution   If $a = b$, then b may be replaced for a in any statement without altering its truth or falsity.
Addition   If $a = b$, then $a + c = b + c$.
Multiplication   If $a = b$ and $c \neq 0$, then $ac = bc$.

PROPERTIES OF REAL NUMBERS

Closure for Addition and Multiplication
$$a + b \in R,\ ab \in R$$

Commutative Properties of Addition and Multiplication
$$a + b = b + a$$
$$ab = ba$$

Associative Properties of Addition and Multiplication
$$(a + b) + c = a + (b + c)$$
$$(ab)c = a(bc)$$

Properties of Identity
$$\text{Additive: } 0 + a = a \text{ and } a + 0 = a$$
$$\text{Multiplicative: } 1 \cdot a = a \text{ and } a \cdot 1 = a$$

Inverse
$$\text{Additive: } a + (-a) = 0$$
$$\text{Multiplicative: } a \cdot \frac{1}{a} = 0, \text{ where } a \neq 0$$

Distributive Property of Multiplication with Respect to Addition
$$a(b + c) = ab + ac \text{ and } (b + c)a = ba + ca$$

Multiplication by Zero
$$0 \cdot a = a \cdot 0 = 0$$

Opposite Law
$$-a = -1 \cdot a$$

Double Opposite Law
$$-(-a) = a$$

Properties of Exponents
$$a^0 = 1$$
$$\frac{a^m}{a^n} = a^{m-n}, \text{ where } a \neq 0 \text{ and } m \geq n$$
$$(a^m)^n = a^{mn}$$
$$a^m \cdot a^n = a^{m+n}$$
$$(ab)^n = a^n b^n$$

Definition of Subtraction
$$a - b = a + (-b)$$

Definition of Division

$$\frac{a}{b} = a \cdot \frac{1}{b}, b \neq 0$$

Definition of Absolute Value

$$|x| = x \text{ if } x \geq 0, \text{ or}$$
$$-x \text{ if } x < 0$$

If grouping symbols are present, begin on the inside and work outward.

1. Evaluate all exponentials and roots as they occur, working left to right.
2. Perform all multiplications and divisions as they occur, working left to right.
3. Perform all additions and subtractions as they occur, working left to right.

ORDER OF OPERATIONS AGREEMENT

Self-Checking Exercise

True or false.
1. $x^0 = 0$, where $x \neq 0$.
2. $5x$ and $5x^2$ are like terms.
3. The expression $-x + y - z$ can be written as $-(x - y + z)$.
4. $2 \in \{1, 2, 4\}$
5. All integers are included in the set of real numbers.
6. The inequalities $|x| < 4$ and $-4 < x < 4$ are equivalent.
7. If a is any real number, $\dfrac{0}{a} = 0$.
8. The reciprocal of -5 is $-\dfrac{1}{5}$.
9. The expression $5x + 3$ is a basic sum.
10. $4^4 \cdot 4^8 = 4^{12}$

State which property of equality or real numbers is being used to write the statement.

11. $5x - 5y = 5(x - y)$
12. $4x + 0 = 4x$
13. If $y = a^2$ and if $a = x + 1$, then $y = (x + 1)^2$
14. $4 + (x + 2) = 4 + (2 + x)$
15. $5 \cdot 3$ is a real number
16. If $6 = x$, then $x = 6$
17. $0(x + 3) = 0$
18. $-[-(-3)] = -3$
19. Pick out the irrational numbers from this list:

$$\sqrt{5}, \; -\frac{22}{7}, \; \sqrt{100}, \; .\overline{41}, \; 0.135, \; \pi$$

20. Pick out the rational numbers from this list:

$$\sqrt{5}, \; -\frac{22}{7}, \; \sqrt{100}, \; .\overline{41}, \; 0.135, \; \pi$$

Classify as a basic sum, difference, product or quotient.
21. $14 \div 2 \cdot 7 + 3 \cdot 4$
22. $(3x^2 + x - 1)(x + 3)$

Evaluate if $x = -2$, $y = 1$, $z = 0$.
23. $-2\{x + x[y - z(3x + y)]\}$
24. $\dfrac{x^2 - y}{xy + 2z} + \dfrac{y}{x}$

Factor:
25. $8xy - 8y$
26. $4ab - 16ac + 8a$

SOLUTIONS TO SELF-CHECKING EXERCISE

Evaluate:
27. $-3 - (-6) + (-4 + 9)$
28. $-2[-1 + (4 - 2)] \div [-4 + 2]$
29. $(-|-6|)(|2 - 4| - |-2|)$
30. $(-2)(2)(-4) - (-4)(-4) \div (-2)(-2)$

Multiply using the distributive property and simplify.
31. $-3xy(a - 2b + 3c)$
32. $(x - 1)(2x + 3)$
33. $(x + 2)(x^2 - 2x + 4)$
34. $(x^2 + 3)(x^2 - 3)$

Evaluate or simplify. Assume that denominators do not represent zero.
35. $(x^3y)(x^4y^3)(xy)$
36. $\dfrac{-18x^4y^6}{6xy^3}$
37. $(x^3)^4 \cdot x^2$
38. $\sqrt[4]{16} + \sqrt[3]{-8}$
39. $-2^2[\sqrt[3]{-8} - (-2)^2]$
40. $(-\sqrt{25} - \sqrt{36})^2$
41. $0 \div 2 + 4 \cdot 3 \div 3 - 2 \cdot 0$
42. $(2x^2y^4)^3$
43. $(x^6)^3$
44. $4x^6 \cdot x^3$
45. $\dfrac{-4x^{10}}{(2x)^2}$
46. $(-2x^2y^3)^3$

Remove grouping symbols and combine like terms.
47. $3 - [2x - (x + 3) + 1]$
48. $1 - \{x - [1 - (2x + y)] + [x - y]\}$

Add or subtract as indicated.
49. $(3x^2 - 2x + 3) + (4x^2 - 3x - 5)$
50. $(2y^2 - y + 2) - (4y^2 + y + 2)$
51. $3a + 4a^2 - a^2 + a - 3a^2 + 5a - 6a^2$

Do as indicated.
52. Graph $\{x | -2 \leq x < 3\}$
53. Graph $\{x | \, |x| > 2\}$
54. Graph $\{x | \, x < 4 \text{ and } x \in W\}$

---

1. false ($x^0 = 1$)
2. false (exponents differ)
3. true
4. true
5. true
6. true
7. false $\left(\dfrac{0}{0} \neq 0\right)$
8. true
9. true

Solutions to Self-Checking Exercise

10. true
11. distributive
12. additive identity
13. substitution
14. commutative property for addition
15. closure property for multiplication
16. symmetric
17. zero multiplication
18. double opposite
19. $\sqrt{5}, \pi$
20. $-\frac{22}{7}, \sqrt{100}, .\overline{41}, 0.135$
21. basic sum
22. basic product
23. 8
24. -2
25. $8y(x-1)$
26. $4a(b-4c+2)$
27. 8
28. 1
29. 0
30. 0
31. $-3axy + 6bxy - 9cxy$
32. $2x^2 + x - 3$
33. $x^3 + 8$
34. $x^4 - 9$
35. x^8y^5
36. $-3x^3y^3$
37. x^{14}
38. 0
39. 24
40. 121
41. 4
42. $8x^6y^{12}$
43. x^{18}
44. $4x^9$
45. $-x^8$
46. $-8x^6y^9$
47. $5 - x$
48. $2 - 4x$
49. $7x^2 - 5x - 2$
50. $-2y^2 - 2y$
51. $-6a^2 + 9a$

52. [number line from −2 to 3, open circles at 0 and 3]

53. [number line with open circle at 1, shaded to left and right extending outward]

54. [number line with points at 0, 2, 3]

SOLVING FIRST DEGREE EQUATIONS AND INEQUALITIES

2

2.1 Solving Linear Equations
2.2 Solving Linear Inequalities
2.3 Solving Absolute Value Equations
2.4 Solving Absolute Value Inequalities
2.5 Working with Formulas
2.6 Applied Problems Involving Linear Equations and Inequalities

SOLVING LINEAR EQUATIONS

2.1

An *equation* is a symbolic statement of equality, such as $3x + 4 = 10$. A *solution* or *root* of an equation is any value from the variable's domain that makes the equation true.

EXAMPLE 1 2 is a solution of the equation $3x + 4 = 10$, since $3 \cdot 2 + 4 = 10$ is true.

There are two important types of equations, identities and conditional equations. An *identity* is an equation that is true for every value of the variable's domain.

EXAMPLE 2 Each of these equations is an identity for the given domain of the variable:
(a) $x + 3 = 3 + x$, where $x \in R$
(b) $y \cdot 0 = 0$, where $y \in R$
(c) $\dfrac{n}{n} = 1$, where $n \in R$ and $n \neq 0$

A *conditional equation* is any equation which is false for at least one value of the variable's domain. (Unless stated otherwise, we will assume that the variable's domain is the set of real numbers.) The truth of a conditional equation depends upon the value chosen for the variable. For example, the equation, $3x = -12$, is a conditional equation. It is true when x is -4, but false for all other real numbers.

EXAMPLE 3 Each of these equations is conditional:
(a) $5x - 10 = -20$; true when x is -2, but false for all other values.
(b) $3(y - 4) = -12$; true when y is 0, but false for all other values.
(c) $n^2 = 9$; true when n is either 3 or -3, but false for all other values.

Conditional equations are said to be *equivalent* when they have exactly the same solutions.

EXAMPLE 4 These three equations are equivalent to each other because they all have exactly the same solution, -1.
(a) $2x + 3 = 1$
(b) $2x = -2$
(c) $x = -1$

EXAMPLE 5 None of these equations is equivalent to any other because each has a different solution.
(a) $3y = -12$ [solution is -4]
(b) $y^2 = 16$ [solutions are -4 and 4]
(c) $y - 4 = 0$ [solution is 4]

The following properties of equivalent equations will enable us to change a given equation into a simpler equivalent form.

SOLVING LINEAR EQUATIONS 59

> **Properties of Equivalent Equations**
> 1. **Addition–Subtraction Property**
>   If the same number or expression is added to (or subtracted from) both sides of an equation, then the resulting equation is equivalent to the original.
> 2. **Multiplication–Division Property**
>   If the same number or expression, except zero, is multiplied times (or divided into) both sides of an equation, then the resulting equation is equivalent to the original.

Change the equation $x - 6 = 1$ into the equivalent form of $x = 7$. **EXAMPLE 6**

Solution: $x - 6 = 1$
$x - 6 + 6 = 1 + 6$ [add 6 to both sides]
$x = 7$

Change the equation $-6x = 18$ into the equivalent form of $x = -3$. **EXAMPLE 7**

Solution: $-6x = 18$
$\dfrac{-6x}{-6} = \dfrac{18}{-6}$ [divide both sides by -6]
$x = -3$

In this section we will use the properties of equivalent equations to solve linear (or first degree) equations.

> **Definition of Linear Equations**
> A linear (or first degree) equation in one variable is any equation which has an equivalent form fitting the pattern,
> $$ax + b = c, \text{ where } a, b, c \in R \text{ and } a \neq 0$$
> (Linear equations are also said to be first degree because the variable has an exponent of 1.)

$6x + 2 = 8$ is a linear equation, where $a = 6$, $b = 2$, and $c = 8$. **EXAMPLE 8**

To solve an equation means to find all of its solutions (or roots). This will be accomplished by using the properties of equivalent equations to isolate the variable. That is, the equation will be transformed into the form $x = k$, where k is a real number.

Solve $5x - 3 = 12$ **EXAMPLE 9**

Solution: Isolate the variable by using the addition–subtraction property. Then solve for x using the multiplication–division property.

$$5x - 3 = 12$$
$$5x - 3 + 3 = 12 + 3 \quad \text{[add 3 to both sides]}$$
$$5x = 15$$
$$\frac{5x}{5} = \frac{15}{5} \quad \text{[divide both sides by 5]}$$
$$x = 3$$

The solution is 3. Check by substituting 3 for x in the original equation.

$$5x - 3 = 12$$
$$5(3) - 3 = 12$$
$$15 - 3 = 12$$
$$12 = 12$$

EXAMPLE 10

Solve $\frac{2x}{3} + 5 = 1$

Solution: First, eliminate the fraction by multiplying both sides by 3. Then, isolate the variable.

$$\frac{2x}{3} + 5 = 1$$
$$3 \cdot \left(\frac{2x}{3} + 5\right) = 3 \cdot 1 \quad \text{[multiply both sides by 3]}$$
$$2x + 15 = 3$$
$$2x + 15 - 15 = 3 - 15 \quad \text{[subtract 15 from both sides]}$$
$$2x = -12$$
$$\frac{2x}{2} = \frac{-12}{2} \quad \text{[divide both sides by 2]}$$
$$x = -6$$

The solution is -6. Check by substituting -6 for x in the original equation.

$$\frac{2x}{3} + 5 = 1$$
$$\frac{2(-6)}{3} + 5 = 1$$
$$\frac{-12}{3} + 5 = 1$$
$$-4 + 5 = 1$$
$$1 = 1$$

SOLVING LINEAR EQUATIONS

EXAMPLE 11

Solve $2(m - 5) - 7m = 3(m + 2)$

Solution: Multiply and combine like terms. Then isolate the variable.

$2(m - 5) - 7m = 3(m + 2)$
$2m - 10 - 7m = 3m + 6$
$-10 - 5m = 3m + 6$
$-10 - 5m + 10 = 3m + 6 + 10$ [add 10 to both sides]
$-5m = 3m + 16$
$-5m - 3m = 3m + 16 - 3m$ [subtract $3m$ from both sides]
$-8m = 16$
$\dfrac{-8m}{-8} = \dfrac{16}{-8}$ [divide both sides by -8]
$m = -2$

The solution is -2. Be sure to check it in the original equation.

Using the preceding examples as a guide, we list the steps to follow when solving linear equations.

To Solve a Linear Equation
1. Eliminate any fractions by multiplying both sides by the same number.
2. Simplify each side by eliminating grouping symbols and combining like terms.
3. Use the addition–subtraction property to obtain an equivalent form having the term containing a variable on one side and the number on the other.
4. Use the multiplication–division property to obtain an equivalent equation of the form $x = k$, where k is a real number.
5. Check by substituting back into the original equation.

EXAMPLE 12

Solve $\dfrac{x}{2} + \dfrac{x}{3} = x - 2$

Step 1. Eliminate the fractions.

$\dfrac{x}{2} + \dfrac{x}{3} = x - 2$

$6\left(\dfrac{x}{2} + \dfrac{x}{3}\right) = 6(x - 2)$ [multiply both sides by 6]

$6 \cdot \dfrac{x}{2} + 6 \cdot \dfrac{x}{3} = 6x - 12$

$3x + 2x = 6x - 12$

Step 2. Combine like terms.
$$3x + 2x = 6x - 12$$
$$5x = 6x - 12$$

Step 3. Obtain the variable term on one side and the number on the other.
$$5x = 6x - 12$$
$$5x - 6x = 6x - 12 - 6x \quad \text{[subtract } 6x \text{ from both sides]}$$
$$-x = -12$$

Step 4. Obtain a form of $x = a$.
$$-x = -12$$
$$(-1)(-x) = (-1)(-12) \quad \text{[multiply both sides by } -1\text{]}$$
$$x = 12$$

Step 5. Check the solution.
$$\frac{x}{2} + \frac{x}{3} = x - 2$$
$$\frac{12}{2} + \frac{12}{3} = 12 - 2$$
$$6 + 4 = 10$$
$$10 = 10$$

EXAMPLE 13 Solve $y - (2y + 1) = 7 - 3(y + 1)$

Step 1. The equation contains no fractions.

Step 2. Eliminate grouping symbols and combine like terms.
$$y - (2y + 1) = 7 - 3(y + 1)$$
$$y - 2y - 1 = 7 - 3y - 3$$
$$-y - 1 = 4 - 3y$$

Step 3. Obtain the variable term on one side and the number on the other.
$$-y - 1 = 4 - 3y$$
$$-y - 1 + 1 = 4 - 3y + 1 \quad \text{[add 1 to both sides]}$$
$$-y = 5 - 3y$$
$$-y + 3y = 5 - 3y + 3y \quad \text{[add } 3y \text{ to both sides]}$$
$$2y = 5$$

Step 4. Divide by the numerical coefficient of y to obtain a form of $y = a$.
$$2y = 5$$
$$\frac{2y}{2} = \frac{5}{2} \quad \text{[divide both sides by 2]}$$
$$y = \frac{5}{2}$$

Step 5. Check the solution.
$$y - (2y + 1) = 7 - 3(y + 1)$$
$$\frac{5}{2} - \left(2 \cdot \frac{5}{2} + 1\right) = 7 - 3\left(\frac{5}{2} + 1\right)$$
$$\frac{5}{2} - (5 + 1) = 7 - \frac{15}{2} - 3$$
$$\frac{5}{2} - 6 = 4 - \frac{15}{2}$$
$$\frac{5}{2} - \frac{12}{2} = \frac{8}{2} - \frac{15}{2}$$
$$-\frac{7}{2} = -\frac{7}{2}$$

With practice you will learn to apply these steps mentally and your work will resemble the following two examples.

Solve $5(n - 2) + 1 = 8n + 2 - n$ **EXAMPLE 14**

Solution:
$$5(n - 2) + 1 = 8n + 2 - n$$
$$5n - 10 + 1 = 8n + 2 - n$$
$$5n - 9 = 7n + 2$$
$$-2n = 11$$
$$n = -\frac{11}{2}$$

Solve $\frac{y - 2}{3} - \frac{1}{2} = -\frac{y}{4}$ **EXAMPLE 15**

Solution:
$$\frac{y - 2}{3} - \frac{1}{2} = -\frac{y}{4}$$
$$12\left[\frac{y - 2}{3} - \frac{1}{2}\right] = 12\left[-\frac{y}{4}\right]$$
$$12\left(\frac{y - 2}{3}\right) - 12\left(\frac{1}{2}\right) = 12\left(-\frac{y}{4}\right)$$
$$4(y - 2) - 6 = -3y$$
$$4y - 8 - 6 = -3y$$
$$4y - 14 = -3y$$
$$-14 = -7y$$
$$2 = y$$

When solving linear equations, you may encounter two special cases. These cases are easily recognized during the solution process because the variable disappears, producing either a true statement or a false statement.

EXAMPLE 16

Solve $3x + 1 = 3(x + 2)$

Solution: Multiply, then attempt to isolate the variable.

$$3x + 1 = 3(x + 2)$$
$$3x + 1 = 3x + 6$$
$$3x + 1 - 3x = 3x + 6 - 3x \quad \text{[subtract } 3x \text{ from both sides]}$$
$$1 = 6 \quad \text{[false]}$$

The variable disappeared, leaving a false statement. When this happens, the original equation has no solution.

EXAMPLE 17

Solve $3x + 2 = 3(x + 2) - 4$

Solution: Multiply and combine like terms. Then attempt to isolate the variable.

$$3x + 2 = 3(x + 2) - 4$$
$$3x + 2 = 3x + 6 - 4$$
$$3x + 2 = 3x + 2$$
$$3x + 2 - 3x = 3x + 2 - 3x \quad \text{[subtract } 3x \text{ from both sides]}$$
$$2 = 2 \quad \text{[true]}$$

The variable disappeared, leaving a true statement. This means that the original equation is an identity and every number in the variable's domain is a solution. Thus, every real number is a solution of the original equation.

Exercise 2.1

Determine whether the given equation is an identity or conditional equation for the given domain.

1. $2x + 3 = 3 + 2x$, where $x \in R$
2. $2(x - 1) = 2x - 2$, where $x \in R$
3. $4y = 0$, where $y \in R$
4. $y^2 + y = y(y + 1)$, where $y \in R$

Solve the following equations and check by substituting.

5. $x - 5 = 8$
6. $2 - 2x = 10$
7. $5y = 0$
8. $6 = -y - 2$
9. $3y - 6 = -12$
10. $9 - 3x = 0$
11. $19 - 2m = 13$
12. $3m - 5 = 2m + 5$
13. $11 - 8x = 15 - 6x$
14. $3(2x + 1) = -15$
15. $5(2n + 6) = 10$
16. $4(2n - 3) = 3(2n + 4)$
17. $4n + 3 = 2n - 1$
18. $3x + 4 = 13 + x$
19. $5 + 3x = 7x + 5$
20. $2 - x = 3 - 5x$
21. $3 - 2y = 10$
22. $3y + 1 = 7y - 2$
23. $5y = 4y + 3 + y$
24. $10(a + 1) = -12$
25. $-3a - 2(3a + 1) = -7a - 22$

26. $3a + 2 = 3a$
27. $2(a - 4) = 5a + 2$
28. $-\dfrac{7x}{9} = 2$
29. $\dfrac{5x}{11} = -3$
30. $\dfrac{6x}{5} = -1$
31. $3 + \dfrac{x}{3} = \dfrac{7}{12}$
32. $2 + \dfrac{x}{2} = \dfrac{7}{8}$
33. $2 + \dfrac{x}{4} = \dfrac{x}{3}$
34. $2 - \dfrac{x}{4} = \dfrac{x}{6}$
35. $3x - 9 + 5x = 4 - x - 3$
36. $-4(x - 8) + 3(2x + 1) = 7$
37. $x - (2x + 1) = 8 - 3(x + 1)$
38. $3[x - 2(5x - 1)] = 2$
39. $2 - [3 + 2(x - 1)] = 4x$
40. $7 = 5x - 3(x - 3)$
41. $(2x - 3)(x + 2) - 4 = x(2x - 1)$
42. $(x + 4)(2x - 1) - 6 = x(2x + 1)$
43. $\dfrac{x}{10} - 1 = 2\left(\dfrac{1}{20} - x\right)$
44. $6x - 3(3x + 2) = 4 - 5x$
45. $(x + 2)(x + 3) - 2x = 3 + x^2$
46. $\dfrac{x}{2} - \dfrac{x}{3} = x - 2$
47. $\dfrac{y + 2}{3} - \dfrac{1}{2} = -\dfrac{y}{2}$
48. $\dfrac{2y}{3} + \dfrac{5y}{4} = \dfrac{y}{2}$

SOLVING LINEAR INEQUALITIES

2.2

An inequality is a mathematical statement of order where expressions are related by any of these symbols:

    $<$ is less than
    $\leq$ is less than or equal to
    $>$ is greater than
    $\geq$ is greater than or equal to

Each of the following is an inequality.   **EXAMPLE 1**

(a) $2x < 6$
(b) $3y - 5 \geq 7$
(c) $-3 < x < 5$

SOLVING FIRST DEGREE EQUATIONS AND INEQUALITIES

To solve an inequality means to find every real number which makes the inequality true. Solutions of inequalities can be illustrated graphically on the number line.

EXAMPLE 2 Graph $\{x \mid x \geq -2\}$

Solution:

EXAMPLE 3 Graph $\{x \mid -3 \leq x < 2\}$

Solution:

Inequalities are solved by using two important properties that enable us to change a given inequality into a simpler equivalent form.

The Addition–Subtraction Property of Inequalities

Given an inequality, the same real number may be added to (or subtracted from) both sides without changing the solutions.
For $a, b, c \in R$,

$$\text{If } a < b, \text{ then } a + c < b + c \text{ and } a - c < b - c$$

(This property holds true for all inequality symbols.)

EXAMPLE 4 Simplify the inequality $x - 5 \geq 7$ by adding 5 to both sides.

Solution:
$$x - 5 \geq 7$$
$$x - 5 + 5 \geq 7 + 5$$
$$x \geq 12$$

The Multiplication–Division Property of Inequalities

(1) Given an inequality, the same positive real number may be multiplied times both sides (or divided into both sides) without changing the solutions.
For $a, b, c \in R$ and $c > 0$,

$$\text{If } a < b, \text{ then } ac < bc \text{ and } \frac{a}{c} < \frac{b}{c}$$

(2) Given an inequality, the same negative real number may be multiplied times both sides (or divided into both sides) without changing the solutions, provided the inequality symbol is reversed.
For $a, b, c \in R$ and $c < 0$,

$$\text{If } a < b, \text{ then } ac > bc \text{ and } \frac{a}{c} > \frac{b}{c}$$

(This property holds true for all inequality symbols.)

SOLVING LINEAR INEQUALITIES

EXAMPLE 5

Simplify the inequality $3x \leq -15$ by dividing both sides by 3.

Solution: $3x \leq -15$

$$\frac{3x}{3} \leq \frac{-15}{3}$$

$x \leq -5$

EXAMPLE 6

Simplify the inequality $-4y > 20$ by dividing both sides by -4.

Solution: $-4y > 20$

$$\frac{-4y}{-4} < \frac{20}{-4} \quad \text{[reverse inequality symbol]}$$

$y < -5$

We now use the properties of inequality to develop techniques for solving linear (or first degree) inequalities.

Definition of a Linear Inequality

A linear (or first degree) inequality in one variable has an equivalent form fitting one of these patterns:

$ax + b < c$ or $ax + b \leq c$, where $a, b, c \in R$ and $a \neq 0$

(Linear inequalities are also said to be first degree because the variable has an exponent of 1.)

EXAMPLE 7

$5x + 12 < 7$ is a linear inequality where $a = 5$, $b = 12$ and $c = 7$.

Linear inequalities are solved in the same manner as linear equations except the inequality symbol must be reversed when each side is multiplied or divided by a negative number.

To Solve a Linear Inequality

(1) Eliminate any fractions by multiplying both sides by the same number.
(2) Simplify each side by eliminating grouping symbols and combining like terms.
(3) Use the addition–subtraction property to obtain an equivalent form having the term containing a variable on one side and the number on the other.
(4) Use the multiplication–division property to obtain an inequality of the form $x < k$ or $x > k$, where k is a real number.

EXAMPLE 8

Solve $x - \dfrac{2}{3} \leq \dfrac{2x}{3} + \dfrac{x}{5}$ and graph the solutions.

Step 1. Eliminate the fractions:
$$15\left(x - \frac{2}{3}\right) \leq 15\left(\frac{2x}{3} + \frac{x}{5}\right)$$
$$15x - 15 \cdot \frac{2}{3} \leq 15 \cdot \frac{2x}{3} + 15 \cdot \frac{x}{5}$$
$$15x - 10 \leq 10x + 3x$$

Step 2. Combine like terms:
$$15x - 10 \leq 10x + 3x$$
$$15x - 10 \leq 13x$$

Step 3. Obtain the variable term on one side and the number on the other:
$$15x - 10 \leq 13x$$
$$15x - 10 + 10 \leq 13x + 10 \quad \text{[add 10 to both sides]}$$
$$15x \leq 13x + 10$$
$$15x - 13x \leq 13x + 10 - 13x \quad \text{[subtract 13x from both sides]}$$
$$2x \leq 10$$

Step 4. Obtain a form of $x \leq k$.
$$2x \leq 10$$
$$\frac{2x}{2} \leq \frac{10}{2} \quad \text{[divide both sides by 2]}$$
$$x \leq 5$$

Graph the interval $\{x \mid x \leq 5\}$

As with linear equations, these steps can be applied mentally, shortening the solution process.

EXAMPLE 9

Solve $2(x + 3) \geq x + 3x - 1$. Graph the solutions on the number line.

Solution:
$$2(x + 3) \geq x + 3x - 1$$
$$2x + 6 \geq x + 3x - 1 \quad \text{[multiply]}$$
$$2x + 6 \geq 4x - 1 \quad \text{[combine like terms]}$$
$$2x + 6 - 6 \geq 4x - 1 - 6 \quad \text{[subtract 6 from both sides]}$$
$$2x \geq 4x - 7$$
$$2x - 4x \geq -7 \quad \text{[subtract 4x from both sides]}$$
$$-2x \geq -7$$
$$\frac{-2x}{-2} \leq \frac{-7}{-2} \quad \left[\begin{array}{l}\text{divide both sides by } -2 \text{ and} \\ \text{reverse inequality symbol}\end{array}\right]$$
$$x \leq \frac{7}{2}$$

SOLVING LINEAR INEQUALITIES

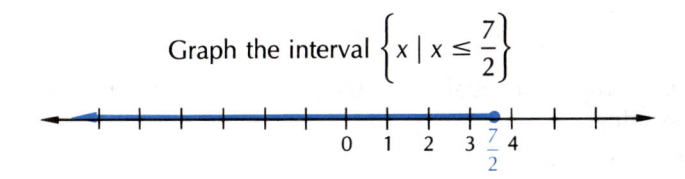

Graph the interval $\left\{x \mid x \leq \dfrac{7}{2}\right\}$

[number line with closed circle at 7/2 between 3 and 4, shaded to the left]

As happened with equations, you may encounter two special cases of inequalities. Similarly, during the solution process the variable disappears, resulting in either a true inequality or a false inequality.

Solve $3(m + 2) > 4m - (m + 1)$ **EXAMPLE 10**

Solution:

$3(m + 2) > 4m - (m + 1)$

$3m + 6 > 4m - m - 1$

$3m + 6 > 3m - 1$

$3m + 6 - 6 > 3m - 1 - 6$ [subtract 6 from both sides]

$3m > 3m - 7$

$3m - 3m > 3m - 7 - 3m$ [subtract 3m from both sides]

$0 > -7$ [true]

The variable m disappeared, leaving a true inequality. This tells us that every number in the variable's domain is a solution. Thus, every real number is a solution of the original inequality.

Solve $\dfrac{x}{2} + 3 \leq \dfrac{2x}{3} - \dfrac{x}{6}$ **EXAMPLE 11**

Solution:

$6\left(\dfrac{x}{2} + 3\right) \leq 6\left(\dfrac{2x}{3} - \dfrac{x}{6}\right)$

$6 \cdot \dfrac{x}{2} + 6 \cdot 3 \leq 6 \cdot \dfrac{2x}{3} - 6 \cdot \dfrac{x}{6}$

$3x + 18 \leq 4x - x$

$3x + 18 \leq 3x$

$3x + 18 - 3x \leq 3x - 3x$ [subtract 3x from both sides]

$18 \leq 0$ [false]

The variable x disappeared, leaving a false inequality. This tells us that the original inequality has no solutions.

Compound Inequalities

In mathematics it is often necessary to work with compound inequalities. A compound inequality such as, $2 < x < 5$, is a combination (or string) of inequalities imposing two or more restrictions on the variable. As you recall, this inequality restricts x to be between 2 and 5. Compound inequalities can be solved in a manner similar to other inequalities. You should note that whatever is done to one member is done to all three members of the string.

EXAMPLE 12   Solve $-4 < 2x + 6 < 10$ and graph the solutions.

Solution: Use the properties of inequality to isolate the variable in the middle of the string.

$$-4 < 2x + 6 < 10$$
$$-4 - 6 < 2x + 6 - 6 < 10 - 6 \quad \text{[subtract 6 from all members]}$$
$$-10 < 2x < 4$$
$$-\frac{10}{2} < \frac{2x}{2} < \frac{4}{2} \quad \text{[divide all members by 2]}$$
$$-5 < x < 2$$

Graph the interval $\{x \mid -5 < x < 2\}$

EXAMPLE 13   Solve $-3 \leq 5 - 4m \leq 15$

Solution:
$$-3 \leq 5 - 4m \leq 15$$
$$-3 - 5 \leq 5 - 4m - 5 \leq 15 - 5 \quad \text{[subtract 5 from all members]}$$
$$-8 \leq -4m \leq 10$$
$$\frac{-8}{-4} \geq \frac{-4m}{-4} \geq \frac{10}{-4} \quad \begin{bmatrix}\text{divide all members} \\ \text{by } -4 \text{ and reverse} \\ \text{inequality symbols}\end{bmatrix}$$
$$2 \geq m \geq -\frac{5}{2}$$
or
$$-\frac{5}{2} \leq m \leq 2$$

Exercise 2.2

Solve and also show the graph of the solutions.

1. $4x < 12$
2. $6a > 24$
3. $-2y > 12$
4. $3 - x < -4$
5. $2m \leq -4$
6. $-x < -10$
7. $2 - 2x > 12$
8. $5y \geq -10$
9. $-4a - 3 < 13$
10. $3n + 1 \leq 16$
11. $-6x - 2 \leq 16$
12. $3y - 4 \geq 17$
13. $-3y - 4 \geq 2$
14. $-x + 13 \geq 3$
15. $4m + 9 < 6 + 4m$
16. $3y - (1 - y) \geq 0$
17. $4x - 8 < x + 1$
18. $15 - 6a > 9a - 15$

Solve the following inequalities
19. $4 - 5y < 2y + 7$
20. $2(x + 1) - 3(2x - 4) < 0$
21. $3(x + 2) - 4(2x - 4) > 0$
22. $2x - (4x + 12) < x + 2$
23. $3y - (5y - 3) < y + 4$
24. $2(4x - 1) \leq 12 - 2(x + 8)$
25. $7x - 2(3 - x) < 4(1 + x)$
26. $5x \leq 6x$
27. $\dfrac{2x - 6}{3} \geq 0$
28. $\dfrac{3a - 1}{2} < 4$
29. $\dfrac{5x - 6}{8} > 3$
30. $1 < \dfrac{2x + 3}{10}$
31. $\dfrac{x + 2}{3} < \dfrac{x - 2}{4}$
32. $\dfrac{7 + 5x}{3} - \dfrac{2 - x}{2} \geq 0$
33. $\dfrac{x}{5} - 1 > x + \dfrac{2}{5}$
34. $\dfrac{1}{4}(2x - 1) - \dfrac{1}{2}(4x + 3) \leq 0$
35. $x - 3 \leq \dfrac{-2}{3}(x + 4)$
36. $5 < x + 4 < 6$
37. $-3 \leq x - 3 \leq 6$
38. $0 < 3x < 5$
39. $-8 < 2y + 4 < 12$
40. $4 < 2 - x < 6$
41. $2 < 4 - x < 8$
42. $-3 < 3 - x < 9$
43. $-1 \leq 1 - 2x \leq 3$
44. $-5 \leq 3 - 4y \leq 10$
45. $3 < 5x + 3 < 23$

SOLVING ABSOLUTE VALUE EQUATIONS

2.3

An **absolute value equation** is any equation where an expression containing the variable is enclosed by an absolute value symbol.

Each of the following is an absolute value equation.

EXAMPLE 1

(a) $|x| = 5$
(b) $|3x + 2| = 7$
(c) $|2x - 5| = |3x + 4|$

In section 1.2 we found that the absolute value of a real number a is written $|a|$, and represents the distance on the number line from the origin to the point named by a. The definition was stated in two parts as follows:

Definition of Absolute Value

For $a \in R$,
(1) If $a \geq 0$, $|a| = a$.
(2) If $a < 0$, $|a| = -a$.

We also found that the interval $\{x \mid |x| = c,$ where $c \geq 0\}$ is equivalent to $\{c, -c\}$.

EXAMPLE 2 Graph $\{x \mid |x| = 5\}$

Solution: Since both 5 and -5 have an absolute value of 5, the given interval is equivalent to $\{5, -5\}$. The associated points are then located on the number line

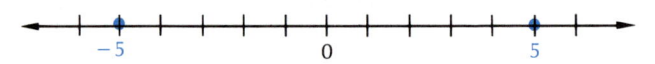

```
    •                       •
---+---+---+---+---+---+---+---+---+---+---
   -5              0               5
```

This suggests the following property that we will use to solve absolute value equations.

Equality Property of Absolute Value

For $a \in R$ and $a \geq 0$,

If $|x| = a$, then $x = a$ or $x = -a$.

EXAMPLE 3 Solve $|x| = 7$

Solution: Use the equality property of absolute value. If $|x| = 7$, then $x = 7$ or $x = -7$. Thus, the solutions are 7 and -7.

Check. Substitute each solution for x in the original equation.

$|x| = 7$ $|x| = 7$
$|7| = 7$ $|-7| = 7$
$7 = 7$ $7 = 7$

EXAMPLE 4 Solve $|y| = -3$

Solution: Since the absolute value of a number is never negative, this equation has no solution.

To Solve an Absolute Value Equation of the Form $|ax + b| = c$
(1) Be sure the absolute value equation has solutions. Remember, the absolute value of a number is never negative.
(2) Use the equality property of absolute value to remove the absolute value symbols and form two equations. One is equal to the constant while the other is equal to the opposite of the constant.
(3) Solve the resulting equations.
(4) Check the solutions by substituting back into the original equation.

SOLVING ABSOLUTE VALUE EQUATIONS

Solve $|3x| = 12$ **EXAMPLE 5**

Step 1. The equation has solutions since 12 is positive:

Step 2. Use the equality property of absolute value to form two equations:
If $|3x| = 12$, then $3x = 12$ or $3x = -12$

Step 3. Solve the resulting equations:
$$3x = 12 \qquad\qquad 3x = -12$$
$$x = 4 \qquad\qquad x = -4$$
The solutions are 4 and -4.

Step 4. Check the solutions:

$$|3x| = 12 \qquad\qquad |3x| = 12$$
$$|3 \cdot 4| = 12 \qquad\qquad |3 \cdot (-4)| = 12$$
$$|12| = 12 \qquad\qquad |-12| = 12$$
$$12 = 12 \qquad\qquad 12 = 12$$

This solution technique can be shortened by applying the steps mentally as illustrated in the following examples:

Solve $|5x - 2| = 8$ **EXAMPLE 6**

Solution:
$$5x - 2 = 8 \qquad\text{or}\qquad 5x - 2 = -8$$
$$5x - 2 + 2 = 8 + 2 \qquad\qquad 5x - 2 + 2 = -8 + 2$$
$$5x = 10 \qquad\qquad 5x = -6$$
$$\frac{5x}{5} = \frac{10}{5} \qquad\qquad \frac{5x}{5} = -\frac{6}{5}$$
$$x = 2 \qquad\qquad x = -\frac{6}{5}$$

The solutions are 2 and $-\frac{6}{5}$.

Check.
$$|5x - 2| = 8 \qquad\qquad |5x - 2| = 8$$
$$|5 \cdot 2 - 2| = 8 \qquad\qquad \left|5 \cdot \left(-\frac{6}{5}\right) - 2\right| = 8$$
$$|10 - 2| = 8 \qquad\qquad |-6 - 2| = 8$$
$$|8| = 8 \qquad\qquad |-8| = 8$$
$$8 = 8 \qquad\qquad 8 = 8$$

SOLVING FIRST DEGREE EQUATIONS AND INEQUALITIES

EXAMPLE 7   Solve $2|3 - 8m| - 7 = 5$

Solution: Isolate the absolute value on one side in order to apply the equality property.

$$2|3 - 8m| - 7 = 5$$
$$2|3 - 8m| = 12$$
$$|3 - 8m| = 6$$

$3 - 8m = 6$ or $3 - 8m = -6$
$-8m = 3$ $-8m = -9$
$m = -\dfrac{3}{8}$ $m = \dfrac{9}{8}$

The solutions are $-\dfrac{3}{8}$ and $\dfrac{9}{8}$.

Check.

$2|3 - 8m| - 7 = 5$ $2|3 - 8m| - 7 = 5$
$2\left|3 - 8\left(-\dfrac{3}{8}\right)\right| - 7 = 5$ $2\left|3 - 8 \cdot \dfrac{9}{8}\right| - 7 = 5$
$2|3 + 3| - 7 = 5$ $2|3 - 9| - 7 = 5$
$2|6| - 7 = 5$ $2|-6| - 7 = 5$
$2 \cdot 6 - 7 = 5$ $2 \cdot 6 - 7 = 5$
$12 - 7 = 5$ $12 - 7 = 5$
$5 = 5$ $5 = 5$

This final example illustrates a technique for solving certain equations containing two absolute values.

EXAMPLE 8 $|y + 7| = |2y - 5|$

Solution: The equation is satisfied if $y + 7$ and $2y - 5$ are equal to each other or if $y + 7$ and $2y - 5$ are opposites of each other. This produces the following two equations which are solved below:

$y + 7 = 2y - 5$ or $y + 7 = -(2y - 5)$
$-y = -12$ $y + 7 = -2y + 5$
$y = 12$ $3y = -2$
                            $y = -\dfrac{2}{3}$

The solutions are 12 and $-\dfrac{2}{3}$.

Check. $|y + 7| = |2y - 5|$
$|12 + 7| = |2 \cdot 12 - 5|$
$|19| = |24 - 5|$
$19 = 19$

$|y + 7| = |2y - 5|$
$\left|-\dfrac{2}{3} + 7\right| = \left|2\left(-\dfrac{2}{3}\right) - 5\right|$
$\left|\dfrac{19}{3}\right| = \left|-\dfrac{4}{3} - 5\right|$
$\left|\dfrac{19}{3}\right| = \left|\dfrac{-19}{3}\right|$
$\dfrac{19}{3} = \dfrac{19}{3}$

Solve and check: **Exercise 2.3**

1. If $x \in R$, give the definition of $|x|$.
2. True or false.
    (a) If $x \in R$, $x \geq 0$ and $|x| = 6$, then $x = 6$ or $x = -6$.
    (b) If $y \in R$ and $|y| = -4$, then $y = 4$ or $y = -4$.
3. $|x| = 1$
4. $|x| = 6$
5. $|a| = 0$
6. $|y| = -2$
7. $|x - 3| = 8$
8. $|z + 2| = 6$
9. $|3a + 1| = 12$
10. $|2y - 1| = 4$
11. $|4x - 3| = 0$
12. $|3x + 5| = 0$
13. $\left|x - \dfrac{1}{2}\right| = \dfrac{1}{4}$
14. $\left|m + \dfrac{2}{3}\right| = \dfrac{1}{3}$
15. $\left|\dfrac{1}{2}x + 2\right| = 4$
16. $\left|\dfrac{1}{5}x - 3\right| = 5$
17. $|y| - 7 = 8$
18. $|a| + 3 = 6$
19. $|b + 4| = -2$
20. $|x| + 5 = 5$
21. $|4 - 2x| = 1$
22. $|6 - 3x| = 4$
23. $|1 - 2x| = 9$
24. $\left|1 - \dfrac{1}{4}y\right| = 4$
25. $|2x + 3| + 4 = 7$
26. $5|y - 6| - 3 = 2$
27. $|4 - 3x| + 8 = -2$
28. $2|2 - x| + 5 = 11$
29. $|x + 3| = |3x + 3|$
30. $|m + 2| = |2m - 1|$
31. $|2y - 1| = |y + 4|$
32. $|4z - 2| = |z - 1|$
33. $\Big||x + 2| + 3\Big| = 4$
34. $\Big||x - 1| - 3\Big| = 5$
35. $\Big||x + 1| - 5\Big| = 2$
36. $3|x + 1| + 5 = 2|x + 1| + 8$

SOLVING ABSOLUTE VALUE INEQUALITIES

2.4

An *absolute value inequality* is any inequality where an expression containing the variable is enclosed by an absolute value symbol.

EXAMPLE 1   Each of the following is an absolute value inequality.
(a) $|x| < 3$
(b) $|2x + 5| \leq 7$
(c) $|5 - 3x| > 10$

In section 1.2 we found that the graph of the interval $\{x \mid |x| > c, c \geq 0\}$ is equivalent to that of $\{x \mid x > c \text{ or } x < -c\}$.

EXAMPLE 2   Graph $\{x \mid |x| > 4\}$

Solution: This interval describes all points that lie more than 4 units from the origin and is equivalent to $\{x \mid x > 4 \text{ or } x < -4\}$. This interval is graphed on the number line as follows:

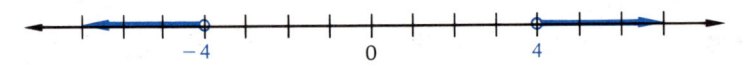

This suggests the following property for solving absolute value inequalities defined by the "greater than" symbol.

Inequality Property for Absolute Value ($>$)

For $a \in R$ and $a \geq 0$,

If $|x| > a$, then $x > a$ or $x < -a$

EXAMPLE 3   Solve and graph $|2x| > 6$

Solution: Use the above property to form two inequalities. If $|2x| > 6$, then $2x > 6$ or $2x < -6$. Solve both inequalities and graph.

$$2x > 6 \quad \text{or} \quad 2x < -6$$
$$x > 3 \quad \text{or} \quad x < -3$$

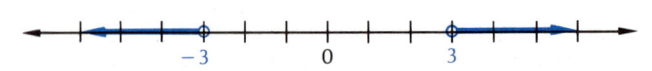

EXAMPLE 4   Solve and graph $|3x - 5| \geq 10$

SOLVING ABSOLUTE VALUE INEQUALITIES

Solution: Form two inequalities. If $|3x - 5| \geq 10$, then $3x - 5 \geq 10$ or $3x - 5 \leq -10$. Solve both inequalities and graph.

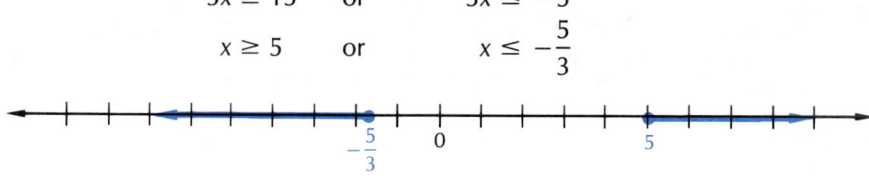

$$3x - 5 \geq 10 \quad \text{or} \quad 3x - 5 \leq -10$$
$$3x \geq 15 \quad \text{or} \quad 3x \leq -5$$
$$x \geq 5 \quad \text{or} \quad x \leq -\frac{5}{3}$$

Solve and graph $|7 - 2x| > 5$ **EXAMPLE 5**

Solution: If $|7 - 2x| > 5$, then $7 - 2x > 5$ or $7 - 2x < -5$. Remember to reverse the inequality symbol when dividing both sides by a negative number.

$$7 - 2x > 5 \quad \text{or} \quad 7 - 2x < -5$$
$$-2x > -2 \quad \text{or} \quad -2x < -12$$
$$\frac{-2x}{-2} < \frac{-2}{-2} \quad \text{or} \quad \frac{-2x}{-2} > \frac{-12}{-2} \quad \text{[reverse inequality]}$$
$$x < 1 \quad\quad\quad\quad x > 6$$

In section 1.2 we also found that the graph of the interval $\{x \mid |x| < c, c > 0\}$ is equivalent to that of $\{x \mid -c < x < c\}$.

Graph $\{x \mid |x| < 5\}$ **EXAMPLE 6**

Solution: This interval describes all points that lie within 5 units of the origin and is equivalent to $\{x \mid -5 < x < 5\}$. The graph consists of all points between -5 and 5.

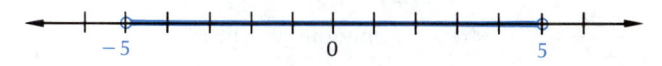

This example enables us to state the following property for solving absolute value inequalities defined by the "less than" symbol.

Inequality Property for Absolute Value ($<$)

For $a \in R$ and $a > 0$,

$$\text{If } |x| < a, \text{ then } -a < x < a$$

---

78   SOLVING FIRST DEGREE EQUATIONS AND INEQUALITIES

EXAMPLE 7 Solve and graph $|6x| < 18$

Solution: Use the above property to form a compound inequality. If $|6x| < 18$, then $-18 < 6x < 18$. Solve and graph:

$$-18 < 6x < 18$$

$$\frac{-18}{6} < \frac{6x}{6} < \frac{18}{6} \quad \text{[divide all members by 6]}$$

$$-3 < x < 3$$

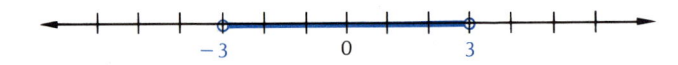

EXAMPLE 8 Solve and graph $|3x - 7| \leq 11$

Solution: If $|3x - 7| \leq 11$, then $-11 \leq 3x - 7 \leq 11$. Solve and graph the compound inequality.

$$-11 \leq 3x - 7 \leq 11$$

$$-4 \leq 3x \leq 18 \quad \text{[add 7 to all members]}$$

$$-\frac{4}{3} \leq x \leq 6 \quad \text{[divide all members by 3]}$$

EXAMPLE 9 Solve and graph $|5 - 3x| < 8$

Solution: If $|5 - 3x| < 8$, then $-8 < 5 - 3x < 8$. Solve and graph the compound inequality, remembering to reverse the inequality symbols when dividing by a negative number.

$$-8 < 5 - 3x < 8$$

$$-13 < -3x < 3$$

$$\frac{-13}{-3} > \frac{-3x}{-3} > \frac{3}{-3} \quad \text{[reverse inequality symbols]}$$

$$\frac{13}{3} > x > -1 \text{ or } -1 < x < \frac{13}{3}$$

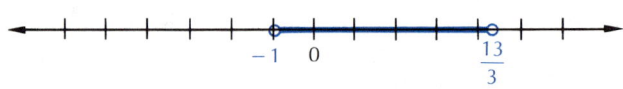

In summary, the following chart should help you to solve absolute value inequalities where the expression contained within the absolute value symbol is of the form $ax + b$, where $a \neq 0$.

Type of Inequality $c > 0$	Find Solution by Solving:	Type of Interval
$\lvert ax + b \rvert > c$	$ax + b > c$ or $ax + b < -c$	split interval
$\lvert ax + b \rvert < c$	$-c < ax + b < c$	one interval

Exercise 2.4

Graph:
1. $\{x \mid \lvert x \rvert \geq 3\}$
2. $\{x \mid \lvert x \rvert > 2\}$
3. $\{x \mid \lvert x \rvert \leq 4\}$
4. $\{x \mid \lvert x \rvert < 3\}$

Solve and also show the graph of the solution.
5. $\lvert x \rvert < 5$
6. $\lvert y \rvert < 4$
7. $\lvert z \rvert \leq 3$
8. $\lvert 5x \rvert < 15$
9. $\lvert x + 1 \rvert < 3$
10. $\lvert y + 2 \rvert \leq 4$
11. $\lvert z - 2 \rvert < 7$
12. $\lvert y - 6 \rvert \leq 6$
13. $\lvert x + 1 \rvert < -2$
14. $\lvert x \rvert > 0$
15. $\lvert z \rvert \geq 3$
16. $\lvert y \rvert > 1$
17. $\lvert 3x \rvert > 3$
18. $\lvert y + 2 \rvert > 4$
19. $\lvert z - 2 \rvert > 6$
20. $\lvert 5x \rvert \leq 10$

Solve:
21. $\lvert 3x - 1 \rvert < 5$
22. $\lvert 2x - 3 \rvert \leq 3$
23. $\left\lvert \dfrac{x}{2} - 1 \right\rvert \leq 1$
24. $\lvert x + 3 \rvert < 5$
25. $\lvert 1 - 2x \rvert < 5$
26. $\lvert 2x + 1 \rvert > 5$
27. $\lvert 3 - x \rvert \geq 6$
28. $\lvert 4x + 2 \rvert \geq 2$

80   SOLVING FIRST DEGREE EQUATIONS AND INEQUALITIES

29. $|4 - 2x| \geq 10$
30. $|1 - x| > 2$
31. $|4 - x| > 6$
32. $|2 - x| < 4$
33. $\left|2x + \dfrac{1}{3}\right| < 2$
34. $|3x - 1| \geq 5$
35. $\left|\dfrac{1}{2}x - 1\right| < 3$
36. $\left|2x + \dfrac{1}{2}\right| \leq \dfrac{1}{2}$

WORKING WITH FORMULAS

2.5

Equations containing more than one variable expressing relationships or dependencies among physical quantities are called formulas. Some formulas you may be familiar with are listed below.

1. $F = \dfrac{9}{5}C + 32$ converts Celsius (C°) temperature to Fahrenheit (F°).
2. $d = rt$ gives the distance (d) traveled by a moving object when the average rate (r) and time (t) are known.
3. $s = 16t^2$ gives the distance (s) in feet that an object will free fall on Earth in t seconds.
4. $A = lw$ gives the area of a rectangle when the length (l) and width (w) are known.
5. $A = \dfrac{1}{2}bh$ gives the area of a triangle when the base (b) and height (h) are known.
6. $A = \dfrac{h(b + c)}{2}$ gives the area of a trapezoid when the height (h), upper base (b), and lower base (c) are known.

To apply a formula, substitute the known quantities for the corresponding variables and evaluate.

EXAMPLE 1   Use the formula $F = \dfrac{9}{5}C + 32$ to convert a temperature of 20° Celsius to Fahrenheit.

Solution:
$$F = \dfrac{9}{5}C + 32$$
$$= \dfrac{9}{5} \cdot 20 + 32$$
$$= 36 + 32$$
$$= 68°F$$

EXAMPLE 2

Use the formula $A = \dfrac{h(b + c)}{2}$ to find the area of this trapezoid.

$b = 10$ cm
$h = 5$ cm
$c = 16$ cm

Solution:
$$A = \frac{h(b + c)}{2}$$
$$= \frac{5 \cdot (10 + 16)}{2}$$
$$= \frac{5 \cdot 26}{2}$$
$$= 65 \text{ cm}^2$$

EXAMPLE 3

Use the formula $d = rt$ to find the average rate of an airliner that traveled 3000 miles in 6 hours.

Solution:
$$d = rt$$
$$3000 = r \cdot (6)$$
$$\frac{3000}{6} = r \quad \text{[divide both sides by 6]}$$
$$500 = r$$
$$r = 500 \text{ miles per hour}$$

In many disciplines, such as in the sciences or economics, it is necessary to use a single formula to solve several problems. For instance, as in Example 3, you might be given the formula $d = rt$ and be asked to solve several problems for the rate, r. You could save time by rearranging the formula and solving for r. This means to use the properties of equivalent equations to obtain r, alone, on one side of the formula.

EXAMPLE 4

Solve the formula $d = rt$ for r. Then, find the average rate of an automobile that traveled 600 miles in 12 hours.

Solution: Divide both sides of the formula by t.
$$d = rt$$
$$\frac{d}{t} = \frac{rt}{t}$$
$$\frac{d}{t} = r \text{ or } r = \frac{d}{t}, \text{ where } t \neq 0$$

To find r, substitute 600 for d and 12 for t.

$$r = \frac{d}{t}$$

$$= \frac{600}{12}$$

$$= 50 \text{ miles per hour}$$

This procedure of rearranging a formula to isolate a symbol on one side is called **solving a formula for a specified symbol**.

EXAMPLE 5 The following triangle has an area of 20 square inches and a base of 5 inches. Solve the formula $A = \frac{1}{2}bh$ for h, then substitute and find the height of the triangle.

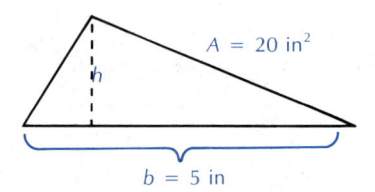

$A = 20 \text{ in}^2$

$b = 5 \text{ in}$

Solution: Eliminate the fraction by multiplying both sides by 2. Then, solve for h.

$$A = \frac{1}{2}bh$$

$$2A = 2 \cdot \frac{1}{2}bh$$

$$2A = bh$$

$$\frac{2A}{b} = \frac{bh}{b} \quad \text{[divide both sides by } b\text{]}$$

$$\frac{2A}{b} = h \text{ or } h = \frac{2A}{b}, \text{ where } b = \neq 0.$$

Substitute 20 for A and 5 for b.

$$h = \frac{2A}{b}$$

$$= \frac{2 \cdot 20}{5}$$

$$= 8 \text{ inches}$$

EXAMPLE 6 Solve the formula $A = \frac{h(b + c)}{2}$ for b.

WORKING WITH FORMULAS   83

Solution: Eliminate the fraction by multiplying both sides by 2.

$$A = \frac{h(b + c)}{2}$$

$$2A = 2 \cdot \frac{h(b + c)}{2}$$

$$2A = h(b + c)$$

$$2A = hb + hc \qquad \text{[multiply]}$$

$$2A - hc = hb \qquad \begin{bmatrix}\text{isolate the term}\\ \text{containing } b.\end{bmatrix}$$

$$\frac{2A - hc}{h} = \frac{hb}{h} \qquad \text{[divide both sides by } h\text{]}$$

$$\frac{2A - hc}{h} = b \text{ or } b = \frac{2A - hc}{h}, \text{ where } h \neq 0$$

Solve $k^2 a = 2a + 2c$ for a. **EXAMPLE 7**

Solution: All terms containing a must be brought together on one side. Then, factor out the a.

$$k^2 a = 2a + 2c$$

$$k^2 a - 2a = 2c$$

$$a(k^2 - 2) = 2c \qquad \text{[factor out the } a\text{]}$$

$$\frac{a(k^2 - 2)}{(k^2 - 2)} = \frac{2c}{(k^2 - 2)} \qquad \text{[divide both sides by } k^2 - 2\text{]}$$

$$a = \frac{2c}{k^2 - 2}, \text{ where } k^2 \neq 2$$

Solve $\dfrac{1}{f} = \dfrac{1}{p} + \dfrac{1}{q}$ for q. **EXAMPLE 8**

Solution: Eliminate fractions by multiplying both sides by fpq.

$$\frac{1}{f} = \frac{1}{p} + \frac{1}{q}$$

$$fpq \cdot \frac{1}{f} = fpq \cdot \frac{1}{p} + fpq \cdot \frac{1}{q}$$

$$pq = fq + fp$$

$$pq - fq = fp \qquad \text{[obtain terms containing } q \text{ on one side]}$$

$$q(p - f) = fp \qquad \text{[factor out the } q\text{]}$$

$$\frac{q(p - f)}{(p - f)} = \frac{fp}{(p - f)} \qquad \text{[divide both sides by } p - f\text{]}$$

$$q = \frac{fp}{p - f}, \text{ where } p - f \neq 0$$

Exercise 2.5

Use the following formulas to find the indicated value.

1. $d = r \cdot t$ — Find the distance traveled if the rate r is 35 miles per hour and the time t is $2\frac{1}{2}$ hours.

2. $A = \pi \cdot r^2$ — Find the area of a circle if the radius r is 4.3 inches and $\pi \approx 3.14$.

3. $C = \dfrac{5F - 160}{9}$ — Find the Celsius temperature for a Fahrenheit temperature of 95°.

4. $t = \pi \sqrt{\dfrac{l}{g}}$ — Find the time of a pendulum swing in seconds if the length $l = 128.64$ centimeters, $g = 32.16$ and $\pi \approx 3.14$.

5. $I = prt$ — Find the interest on an account if the principal p is \$830, the rate r is 8% and the time t is 2 years.

6. $S = R\left[\dfrac{(1 + i)^n - 1}{i}\right]$ — Find S if R is 100, n is 2, and i is .01.

Solve the following formulas for the indicated letter.

7. $D = r \cdot t$ for t
8. $I = prt$ for p
9. $I = prt$ for r
10. $E = mc^2$ for m
11. $A = \dfrac{1}{2}bh$ for b
12. $A = \dfrac{1}{2}bh$ for h
13. $C = 2\pi r$ for r
14. $P = 2l + 2w$ for l
15. $P = 2l + 2w$ for w
16. $s = \dfrac{1}{2}gt^2$ for g
17. $P = a + b + c$ for c
18. $P = a + b + c$ for b
19. $I = \dfrac{E}{R}$ for E
20. $I = \dfrac{E}{R}$ for R
21. $A = \dfrac{1}{2}h(b + c)$ for h
22. $A = \dfrac{1}{2}h(b + c)$ for b
23. $C = \dfrac{5}{9}(F - 32)$ for F
24. $l = a + (n - 1)d$ for d
25. $l = a + (n - 1)d$ for n
26. $S = \dfrac{a}{1 - r}$ for a
27. $S = \dfrac{a}{1 - r}$ for r
28. $V = \dfrac{1}{3}\pi r^2 h$ for h
29. $y = mx + b$ for x
30. $A = 2\pi rh + 2\pi r^2$ for h
31. $\dfrac{1}{p} + \dfrac{1}{q} = \dfrac{1}{f}$ for f
32. $CF = \dfrac{wv^2}{gr}$ for r

33. $hp = \dfrac{Fs}{550t}$ for s, where hp represents horsepower

34. $V = Kt + g$ for K

35. $K^2 a = 2a + 2c$ for a

36. $c = s - rs$ for s

37. $S = \dfrac{a(1 - r^n)}{1 - r}$ for a

38. $hp = \dfrac{Fs}{550t}$ for t, where hp represents horsepower

39. $m + n = a - xa + ya$ for a

APPLIED PROBLEMS INVOLVING LINEAR EQUATIONS AND INEQUALITIES

2.6

Equations and inequalities are used to solve applied problems. To do this, it is necessary to change verbal statements into mathematical sentences. This is accomplished by translating key word phrases into algebraic expressions involving the operations of addition, subtraction, multiplication, or division. These algebraic expressions are called mathematical models. For example, the phrase "5 more than twice a number" can be represented by the mathematical model $2x + 5$. "More than" is translated to addition, x represents the unknown number, and "twice" is translated to multiplication by 2. Any letter may be used to represent an unknown number. If we had selected n, then the mathematical model would be $2n + 5$.

The following tables give common word phrases and their corresponding mathematical models.

Addition	
Word Phrase	Mathematical Model
the sum of a number and 7	$x + 7$
a number plus 8	$x + 8$
a number added to 15	$x + 15$
12 more than a number	$x + 12$
a number increased by 4	$x + 4$
the total or sum of two numbers	$x + y$

Subtraction

Word Phrase	Mathematical Model
3 subtracted from a number	$x - 3$
5 less than a number	$x - 5$
5 less a number	$5 - x$
10 fewer than a number	$x - 10$
a number decreased by 2	$x - 2$
8 minus a number	$8 - x$
a number minus 9	$x - 9$
the difference between two numbers	$x - y$

Multiplication

Word Phrase	Mathematical Model
10 times a number	$10x$
a number multiplied by 3	$3x$
the product of a number and 6	$6x$
twice a number	$2x$
a number is tripled	$3x$
a number is quadrupled	$4x$
$\frac{3}{5}$ of some number	$\frac{3}{5} \cdot x$ or $\frac{3x}{5}$
the product of two numbers	xy

Division

Word Phrase	Mathematical Model
a number divided by 6	$\frac{x}{6}$
12 divided by some number	$\frac{12}{x}$
the quotient of some number and 3	$\frac{x}{3}$
the ratio of some number to 9	$\frac{x}{9}$
half of a number	$\frac{x}{2}$ or $\frac{1}{2}x$
the quotient of two numbers	$\frac{x}{y}$

APPLIED PROBLEMS INVOLVING LINEAR EQUATIONS AND INEQUALITIES

Combinations

Word Phrase	Mathematical Model
7 more than twice a number	$2x + 7$
5 less than three times a number	$3x - 5$
the ratio of twice a number to 7	$\dfrac{2x}{7}$
a number added to 4 times itself	$4x + x$
the number of cents in x nickels	$5x$
the number of cents in x quarters	$25x$
the sum of three consecutive integers	$x + (x + 1) + (x + 2)$
the sum of two consecutive even integers	$x + (x + 2)$
the sum of three consecutive odd integers	$x + (x + 2) + (x + 4)$

To solve applied problems, each word phrase is translated to its corresponding mathematical model. These models are then related by either an equation or an inequality, as illustrated in the following table.

Verbal Problem	Equation or Inequality
Twice a number increased by 5 is 17.	$2x + 5 = 17$
The sum of three consecutive integers is 33.	$x + (x + 1) + (x + 2) = 33$
If 4 times a certain number is decreased by 32, the result is twice the number.	$4x - 32 = 2x$
George needs at least 360 points to obtain a grade of A in his algebra class. His scores on three tests have been 82, 94, and 88. What can his score be on the fourth test to assure him of an A?	$82 + 94 + 88 + x \geq 360$

You should note that the equal sign, $=$, is used as a translation for the word "is" or for any phrase denoting the idea of "sameness." The inequality symbol for "greater than or equal," $\geq$, is used to describe the phrase, "at least." Similarly, the symbol for "less than or equal," $\leq$, is used to describe the phrase, "at most."

When solving applied problems it is helpful to follow these steps.

To Solve a Word Problem
1. Read the problem carefully. Determine what is given and what is to be found. Select a variable and write down what it represents. Then translate the word phrases into mathematical models. If possible, sketch the diagram and label the parts.
2. Relate the mathematical models by forming an equation or inequality.
3. Solve the equation or inequality.
4. Interpret the solution. Often it is not the answer to the written problem. Something else may have to be done.
5. Check your answer back in the original written problem.

EXAMPLE 1   The length of a rectangle is 16 inches more than the width. The perimeter (distance around) is 88 inches. Find the length and width of the rectangle.

Step 1.   Since the length is defined in terms of the width, we let x represent the width. The length is then $x + 16$ because the length is 16 inches more than the width. We draw a diagram and label the parts.

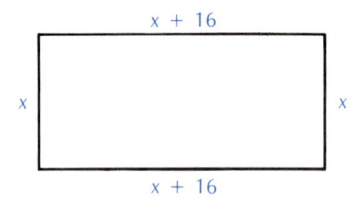

$x + 16$

x [rectangle] x

$x + 16$

Step 2.   The equation is formed by noting that the perimeter is 88 inches.
$x + (x + 16) + x + (x + 16) = 88$

Step 3.   Solve the equation.
$$x + (x + 16) + x + (x + 16) = 88$$
$$x + x + 16 + x + x + 16 = 88$$
$$4x + 32 = 88$$
$$4x = 56$$
$$x = 14$$

Step 4.   Interpret the solution. The width is 14 inches. The length is $x + 16$, or 30 inches.

Step 5.   Check the answers back into the original written problem. The length of 30 inches is 16 inches more than the width of 14 inches. The perimeter is 88 inches.
$14 + 30 + 14 + 30 = 88$

APPLIED PROBLEMS INVOLVING LINEAR EQUATIONS AND INEQUALITIES

EXAMPLE 2

A man invests $6,000 in two accounts. One account earns 8% interest and the other earns 10% interest. If the total annual yield is $550, how much money is invested at each rate?

Step 1. Let x represent the amount invested at 8%; then 6,000 − x represents the amount invested at 10%. The annual interest generated by each account is determined by multiplying the principal times the rate. The total interest is the sum from both accounts as shown in this diagram.

| interest at 8% $(p \cdot r)$ | + | interest at 10% $(p \cdot r)$ | = | total interest |

$8\% \cdot x$ $\quad\quad 10\% \cdot (6,000 - x) \quad\quad 550$

Step 2. The equation is formed by adding the interest from both accounts to form the total of $550.

$8\% \cdot x + 10\% \cdot (6,000 - x) = 550$

Step 3. Solve the equation. Both sides of the equation will be multiplied by 100 to convert percents to whole numbers.

$$8\% \cdot x + 10\% \cdot (6,000 - x) = 550$$
$$.08x + .10(6,000 - x) = 550$$
$$8x + 10(6,000 - x) = 55,000 \quad \text{[multiply by 100]}$$
$$8x + 60,000 - 10x = 55,000$$
$$-2x + 60,000 = 55,000$$
$$-2x = -5,000$$
$$x = 2,500$$

Step 4. Interpret the solution. $2,500 is invested at 8%. 6,000 − x, or $3,500, is invested at 10%.

Step 5. Check the answers back into the original written problem.

$$8\%x \quad + 10\%(6,000 - x) = 550$$
$$.08(2,500) + .10(3,500) \quad\quad = 550$$
$$200 + 350 \quad\quad\quad\quad\quad\quad = 550$$

EXAMPLE 3

A person made a trip at an average speed of 45 miles per hour. A second person made the same trip in one hour less time at an average speed of 60 miles per hour. Find the distance of the trip.

Solution: Problems of this type are called rate problems. They utilize the formula $d = rt$ (distance equals rate multiplied by time).

Step 1. Let t represent the time in hours necessary for the first person to make the trip. t − 1 would then represent the time necessary for the second person to make the same trip. Since distance is the product of rate and time, 45t represents the distance traveled by the first person and 60(t − 1) represents the distance traveled by the second person. Study the accompanying diagram illustrating the trip.

1st person: time is t hours
rate is 45 miles per hour
distance is 45t miles

Trip •————————————→————•

2nd person: time is t − 1 hours
rate is 60 miles per hour
distance is 60(t − 1) miles

Step 2. The equation is formed by noting that the distance was the same for both persons.

45t = 60(t − 1)

Step 3. Solve the equation.

45t = 60(t − 1)
45t = 60t − 60
−15t = −60
t = 4 hours

Step 4. Interpret the solution. We wish to find the distance which is described by either 45t or 60(t − 1). Substitute 4 for t and evaluate either expression.

45t = 45 · 4 = 180 miles
60(t − 1) = 60(4 − 1) = 180 miles

Step 5. Check the answer. Since d = rt,
180 = 45 · 4 and 180 = 60 · 3

EXAMPLE 4

How much pure alcohol must be added to 12 ounces of a 45% alcohol solution to obtain a 60% alcohol solution?

Solution: Problems of this type are called mixture problems. They are solved by forming an equation involving only one portion of the mixture—in this example, the alcohol. The amount of alcohol in the pure solution together with the amount of alcohol in the 45% solution must be equal to the amount of alcohol in the resulting 60% solution. This relationship is diagrammed as follows:

| amount of alcohol in pure solution | + | amount of alcohol in 45% solution | = | amount of alcohol in 60% solution |

APPLIED PROBLEMS INVOLVING LINEAR EQUATIONS AND INEQUALITIES 91

Step 1. Let x represent the number of ounces of the pure solution. Since there are 12 ounces of the 45% solution, there are x + 12 ounces of the 60% solution. The amount of alcohol in each solution is described by multiplying the percentage of alcohol times the number of ounces of solution. This information is described in the following diagram.

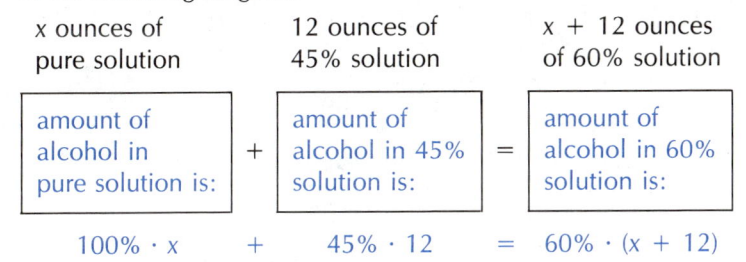

x ounces of pure solution	12 ounces of 45% solution	x + 12 ounces of 60% solution
amount of alcohol in pure solution is:	amount of alcohol in 45% solution is:	amount of alcohol in 60% solution is:
100% · x	+ 45% · 12	= 60% · (x + 12)

Step 2. Form the equation by referring to the diagram.

100% · x + 45% · 12 = 60% · (x + 12)

Step 3. Solve the equation. Both sides of the equation will be multiplied by 100 to convert percents to whole numbers.

$$100\% \cdot x + 45\% \cdot 12 = 60\% \cdot (x + 12)$$
$$1.00x + .45(12) = .60(x + 12)$$
$$100x + 45(12) = 60(x + 12) \quad \text{[multiply by 100]}$$
$$100x + 540 = 60x + 720$$
$$40x = 180$$
$$x = \frac{180}{40}$$
$$x = \frac{9}{2} \text{ or } 4.5$$

Step 4. Interpret the solution. 4.5 ounces of pure alcohol must be added to the 45% solution.

Step 5. Check the solution.

amount of alcohol in pure solution is 100% · (4.5)	amount of alcohol in 45% solution is 45% · 12	amount of alcohol in 60% solution is 60% · (4.5 + 12)
4.5	+ 5.4	= 9.9

EXAMPLE 5

A student in an algebra class received grades of 88, 76, and 82 on three tests. What score is necessary on the fourth test if the average is to be at least 84?

Step 1. Let x represent the score on the fourth test.

Step 2. Form the inequality by noting that the average is to be at least 84. Remember, the phrase "at least" is represented by the "greater than or equal" symbol, $\geq$.

$$\frac{88 + 76 + 82 + x}{4} \geq 84$$

Step 3. Solve the inequality. Multiply each side by 4 to clear the inequality of fractions.

$$\frac{88 + 76 + 82 + x}{4} \geq 84$$

$$88 + 76 + 82 + x \geq 84 \cdot 4 \quad \text{[multiply by 4]}$$

$$246 + x \geq 336$$

$$x \geq 90$$

Step 4. Interpret the solution. A score of 90 or higher is necessary for the average to be at least 84.

Step 5. Check the solution by noting that the average is 84 when the fourth test score is 90.

$$\frac{88 + 76 + 82 + 90}{4} = 84$$

As you become more familiar with applied problems, the five-step solution procedure can be condensed as illustrated in the following examples.

EXAMPLE 6

One number is ten more than another. When the larger is subtracted from twelve times the smaller, the difference is 45. Find both numbers.

Solution: Since the larger number is defined in terms of the smaller, we let x represent the smaller number. The larger number is then represented by $x + 10$. The following equation is formed by subtracting the larger number from 12 times the smaller and setting the difference equal to 45.

$$12x - (x + 10) = 45$$
$$12x - x - 10 = 45$$
$$11x - 10 = 45$$
$$11x = 55$$
$$x = 5$$

The smaller number is 5. The larger number is $x + 10$ or 15.

Check. 15 is ten more than 5, and $12 \cdot 5 - 15 = 45$.

EXAMPLE 7

Find three consecutive odd integers whose sum is 45.

APPLIED PROBLEMS INVOLVING LINEAR EQUATIONS AND INEQUALITIES

Solution: Let n represent the first odd integer. $n + 2$ and $n + 4$ would then represent the next two consecutive odd integers. The following equation is formed by setting the sum equal to 45.

$$n + (n + 2) + (n + 4) = 45$$
$$n + n + 2 + n + 4 = 45$$
$$3n + 6 = 45$$
$$3n = 39$$
$$n = 13$$

The three consecutive odd integers are 13, 15, and 17.

Check. Each of the numbers is an odd integer. And, $13 + 15 + 17 = 45$.

A car rental agency has two weekly rental plans. Plan A charges $130 plus 25¢ per mile. Plan B charges $280 with unlimited mileage. After how many miles will the charge for plan A exceed that for plan B?

EXAMPLE 8

Solution: Let m represent the number of miles to be driven. The charge for plan A is $130 + (.25)m$. The charge for plan B is 280. To find when the charge for plan A exceeds that for plan B, we solve the following inequality:

$$130 + (.25)m > 280$$
$$13{,}000 + 25m > 28{,}000 \quad \text{[multiply by 100]}$$
$$25m > 15{,}000$$
$$m > \frac{15{,}000}{25}$$
$$m > 600$$

After 600 miles the charge for plan A is greater.

Check. Charge for 601 miles:

plan A   plan B
$130 + (.25)601 > \quad 280$
$\$280.25 \quad\quad > \$280$

Translate the following statements into mathematical expressions.

Exercise 2.6

1. The product of a number and six.
2. A number decreased by three.
3. Five less than a number.
4. Four less than twice a number.

5. Ten decreased by twice a number.
6. Four more than three times a number.
7. The sum of two consecutive odd integers.
8. The sum of three consecutive even integers.
9. A number increased by 10%.
10. The total value of x dimes and y quarters.

Solve the following applied problems by translating to the corresponding mathematical model.

11. The sum of a number and 8 is 26. Find the number.
12. If $\frac{5}{3}$ of a number is added to 4, the result is 24. Find the number.
13. When a number is increased by 8%, the result is 27. Find the number.
14. If a number is multiplied by 4, the difference of this product and 6 is twice the number. Find the number.
15. Four times the sum of a number and 2 is equal to 4 less than twice the number. Find the number.
16. Joe wants to buy a tablet that costs 11 cents more than twice as much money as he now has. How much does he have if the price of the tablet is 95 cents?
17. The width of a garden is two-thirds of its length. If its perimeter is 80 feet, find its dimensions.
18. Find two consecutive even integers whose sum is 54.
19. Find three consecutive even integers whose sum is 60.
20. Find two consecutive odd intergers whose sum is 76.
21. Find three consecutive integers whose sum is 75.
22. Find three consecutive odd integers such that three times the sum of the first two exceeds the third by 47.
23. If the larger of two consecutive odd integers is subtracted from twice the smaller, the result is 21. Find the larger number.
24. The sum of two consecutive even integers is 16 more than their difference. Find the numbers.
25. The length of a rectangle exceeds twice its width by 3 inches. If the perimeter is 90 inches, find the dimensions.
26. What number added to 46 gives a sum which is two more than 5 times the original number?
27. If the length of a rectangular plot is 24 feet more than its width and the perimeter is 176 feet, what are the dimensions of the plot?
28. A man has $1,200 invested, part at 10% and part at 8%. The total interest is $103. How much is invested at 8%?
29. A sum of money is invested at 6% simple interest. A second sum, which is $350 greater than the first, is invested at 8%. If the total annual income from both investments is $280, find the amount invested at each rate.

APPLIED PROBLEMS INVOLVING LINEAR EQUATIONS AND INEQUALITIES

30. A total of $5,075 is invested, part at 6% and part at 8%. If the interest from the two investments is the same, how much is invested at each rate?
31. A car travels 30 miles an hour slower than a train. If the train covers 120 miles in the same time the car covers 80 miles, how fast does the train travel?
32. A man made a trip at an average rate of 50 miles per hour. A second man made the same trip in one hour less time at an average rate of 60 miles per hour. What was the distance of the trip?
33. An airplane traveled to a certain city at 200 miles per hour and returned at 300 miles per hour. If the total traveling time is 2 hours, how far away is the city?
34. Two cars start at the same spot and travel in opposite directions. At the end of 4 hours they are 440 miles apart. If one car travels 15 miles per hour faster than the other, what are their speeds?
35. A boat leaves a harbor traveling due north at 36 knots. One hour later, another boat traveling at 45 knots leaves the harbor to overtake the first boat. How far out to sea will the second boat overtake the first boat?
36. How many liters of 30% acid solution must be added to 40 liters of a 12% acid solution to obtain a 20% solution?
37. How much 25% antifreeze and 60% antifreeze should be combined to give 30 gallons of 40% antifreeze?
38. How many ounces of a 5% acid solution must be added to 10 ounces of a 20% solution to make a 10% acid solution?
39. How many gallons of 30% alcohol and 80% alcohol should be mixed together to obtain 5 gallons which is 50% alcohol?
40. A collection of nickels and quarters has a value of $10. How many nickels and quarters are in the collection if there are 16 fewer nickels than quarters? Hint: Let x represent the number of quarters and $x - 16$ the number of nickels. The quarters are worth $25x$ cents and the nickels are worth $5(x - 16)$ cents. The entire collection is worth 1,000 cents.
41. A collection of 15 coins is made up of quarters and dimes. The collection is worth $2.70. How many quarters and dimes are there? (See hint in problem 40.)
42. If 800 tickets were sold for a game for a total of $1,000, and if adults' tickets cost $2 each and children's tickets $1 each, how many of each kind of ticket were sold?
43. A savings bank contains $4.50 in nickels, dimes, and quarters. It contains 4 more dimes than quarters and 3 times as many nickels as dimes. How many coins of each kind are there in the bank?
44. Bob bought twice as many pounds of apples at 50¢ a pound as peaches at 60¢ a pound. He spent a total of $3.20. How many pounds of apples did he buy?
45. A salesman is paid $600 a month plus a 3% commission on sales. On the average, how much must he sell each month to earn at least $25,200 for the year?

46. In a certain class a student earns test grades of 75, 80, and 79. What score must he earn on the last test in order to come out with a B average (a B average is at least 80 and less than 90)?
47. A bowler scores 138, 144, 130, 155, and 148 in five games. What range of scores must be bowled in the next game to attain an average between 140 and 150?

Summary

KEY WORDS AND PHRASES

Equation—a symbolic statement of equality.
Solution or *Root of an Equation*—any value from the variable's domain that makes the equation true.
Identity Equation—an equation that is true for every value of the variable's domain.
Conditional Equation—any equation which is false for at least one value of the variable's domain.
Equivalent Equations—equations that have exactly the same solutions.
Linear Equation—an equation which has an equivalent form of $ax + b = c$, where $a, b, c \in R$ and $a \neq 0$.
Absolute Value Equation—an equation containing the variable enclosed by an absolute value symbol.
Inequality—a mathematical statement described by the symbols $<, \leq, >$, or $\geq$.
Linear or *First Degree Inequality*—an inequality which has an equivalent form of $ax + b < c$ or $ax + b \leq c$, where $a, b, c, \in R$ and $a \neq 0$.
Compound Inequality—a combination of inequalities imposing two or more restrictions on the variable.
Absolute Value Inequality—an inequality containing the variable enclosed by an absolute value symbol.
Formula—an equation containing more than one letter expressing relationships or dependencies among physical quantities.

PROPERTIES

Addition–Subtraction Property of Equivalent Equations
   If the same number or expression is added to (or subtracted from) both sides of an equation, then the resulting equation is equivalent to the original.

Multiplication–Division Property of Equivalent Equations
   If the same number or expression, except zero, is multiplied times (or divided into) both sides of an equation, then the resulting equation is equivalent to the original.

Addition–Subtraction Property of Inequalities
   The same real number may be added to (or subtracted from) both sides of an inequality without changing the solutions.

Multiplication–Division Property of Inequalities
    The same positive real number may be multiplied times both sides (or divided into both sides) of an inequality without changing the solutions; the same negative real number may be multiplied times both sides (or divided into both sides) of an inequality without changing the solutions, provided the inequality symbol is reversed.

Equality Property of Absolute Value
    For $a \in R$ and $a \geq 0$, if $|x| = a$, then $x = a$ or $x = -a$.

Inequality Property for Absolute Value ($>$)
    For $a \in R$ and $a \geq 0$, if $|x| > a$, then $x > a$ or $x < -a$.

Inequality Property for Absolute Value ($<$)
    For $a \in R$ and $a > 0$, if $|x| < a$, then $-a < x < a$.

Self-Checking Exercise

Solve the following equations.
1. $3x - 3 = 15$
2. $2(x + 1) = 2$
3. $2x - 4 = 2(x - 2)$
4. $\dfrac{2x}{3} - 5 = 1$
5. $y - (3y + 1) = 6 - 2(y + 1)$
6. $\dfrac{y}{3} + \dfrac{y}{2} = y + 1$

Solve the following inequalities and show the graphs of the solutions.
7. $2x < 3x$
8. $5x - 4 > 7x + 8$
9. $2(x - 2) \geq x + 2(x - 1)$
10. $x - \dfrac{1}{3} \leq \dfrac{2x}{3} + \dfrac{x}{4}$
11. $-2 < 2x + 6 < 8$
12. $-2 \leq 4 - 3x \leq 13$

Solve the following absolute value equations.
13. $|x| = 8$
14. $|x - 1| = 5$
15. $|4x - 3| = 7$
16. $|x + 3| = -4$
17. $|1 - 4x| = 3$
18. $|x + 4| = |2x - 3|$

Solve the following absolute value inequalities.
19. $|x| < 2$
20. $|x + 2| \geq 2$
21. $|x - 3| \leq 4$
22. $|3x - 1| > 5$
23. $|3x - 2| \leq 7$
24. $|2 - x| > 2$

Solve for the indicated letter.
25. $I = prt$ for t
26. $V = Kt + g$ for g
27. $y = mx + b$ for x
28. $F = \dfrac{9}{5}C + 32$ for C
29. $A = \dfrac{h(b + c)}{2}$ for c
30. $l = a + (n - 1)d$ for d

Write a mathematical model for each statement.
31. The sum of two consecutive integers.
32. Twice a number increased by six.
33. Five less than twice a number.
34. The sum of three consecutive odd integers.
35. The number of cents in x nickels and y dimes.
36. A number subtracted from twelve.

Solve each of the following problems. Give the mathematical model and the solution.
37. One number is two less than three times another. Their sum is 30. Find the numbers.

38. The length of a rectangle exceeds twice its width by 4 inches. If the perimeter is 62 inches, find the length and width.
39. The sum of three consecutive integers is 60. Find the three numbers.
40. $6000 more is invested at 11% than at 9%. The two investments produce an annual income of $2660. How much is invested at each rate?
41. A man made a certain trip at an average rate of 45 miles per hour. A second man made the same trip in one hour less time at an average rate of 60 miles per hour. What was the distance of the trip?
42. How many gallons of a 10% salt solution should be added to 20 gallons of a 60% salt solution to obtain a 50% salt solution?

Solutions to Self-Checking Exercise

1. $x = 6$
2. $x = 0$
3. all real numbers
4. $x = 9$
5. no solution
6. $y = -6$
7. $x > 0$
8. $x < -6$
9. $x \leq -2$
10. $x \leq 4$
11. $-4 < x < 1$
12. $-3 \leq x \leq 2$
13. $x = 8$ or $x = -8$
14. $x = 6$ or $x = -4$
15. $x = \dfrac{5}{2}$ or $x = -1$
16. no solution
17. $x = -\dfrac{1}{2}$ or $x = 1$
18. $x = 7$ or $x = -\dfrac{1}{3}$
19. $-2 < x < 2$
20. $x \geq 0$ or $x \leq -4$
21. $-1 \leq x \leq 7$
22. $x > 2$ or $x < -\dfrac{4}{3}$
23. $-\dfrac{5}{3} \leq x \leq 3$
24. $x < 0$ or $x > 4$
25. $t = \dfrac{I}{pr}$
26. $g = V - Kt$
27. $x = \dfrac{y - b}{m}$
28. $C = \dfrac{5F - 160}{9}$
29. $c = \dfrac{2A - bh}{h}$
30. $d = \dfrac{l - a}{n - 1}$
31. $x + (x + 1)$
32. $2x + 6$
33. $2x - 5$
34. $x + (x + 2) + (x + 4)$
35. $5x + 10y$
36. $12 - x$
37. $x + (3x - 2) = 30$
   the numbers are 8 and 22

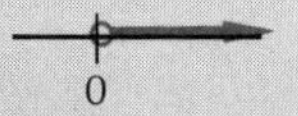

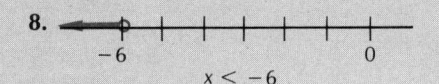

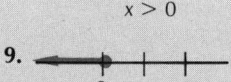

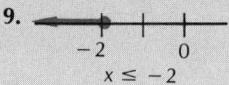

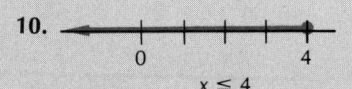

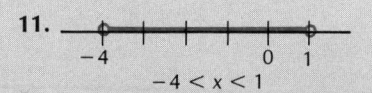

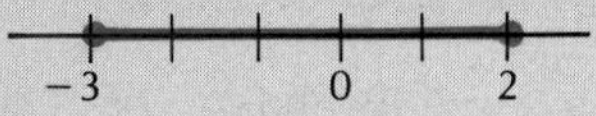

38. $x + (2x + 4) + x + (2x + 4) = 62$
length is 22 inches, width is 9 inches

39. $x + (x + 1) + (x + 2) = 60$
the numbers are 19, 20, and 21

40. $.09x + .11(x + 6,000) = 2,660$
$10,000 at 9%
$16,000 at 11%

41. $60(x - 1) = 45x$
180 miles

42. $.10x + .60(20) = .50(x + 20)$
5 gallons

POLYNOMIAL EXPRESSIONS

3

3.1 Polynomials
3.2 Multiplication and Division of Polynomials
3.3 Special Products of Binomials
3.4 Common Factors
3.5 Factoring Binomials
3.6 Factoring Trinomials
3.7 Solving Equations by Factoring

POLYNOMIALS

3.1

A *polynomial* is an algebraic expression of one or more terms, each of which is a product of a constant and a variable raised to a whole number power. That is, each term of a polynomial must fit the pattern ax^n, where $a \in R$ and $n \in W$.

EXAMPLE 1

Each of these expressions is a polynomial.
(a) $5m^3$
(b) $-3y$ $[-3y = -3y^1]$
(c) 8 $[8 = 8x^0$, where $x \neq 0]$
(d) $3x^4 + 2x^3 - 7x^2 + x - 4$

EXAMPLE 2

None of these expressions is a polynomial.
(a) $\dfrac{5}{x^2}$
(b) $3y^2 - \dfrac{1}{y^3}$
(c) $3\sqrt{x}$
(d) $\dfrac{x^2 - 5}{x + 1}$

Polynomials may also contain more than one variable. However, each term must still be a product, where each variable is raised to a whole number power.

EXAMPLE 3

$x^4 - 2x^3y + 3x^2y^2 - 4xy^3 - y^4$ is said to be a polynomial in x and y.

Most polynomials in this course will contain only one variable, and can be written in descending powers of the variable by arranging the terms so that the exponents decrease from left to right.

EXAMPLE 4

Write each polynomial in descending powers of the variable.
(a) $3 + 2x^2 - x$ would be written as $2x^2 - x + 3$.
(b) $m^2 - 3m^4 + 2m - 1$ would be written as $-3m^4 + m^2 + 2m - 1$.

The *degree of a polynomial* in one variable is the largest exponent used on the variable. A *constant polynomial* other than zero has a degree of zero. The *zero polynomial* has no degree whatsoever.

EXAMPLE 5

Determine the degree of each polynomial.
(a) $4x^3 - x^2 + 2x - 1$ has a degree of 3.
(b) $m^5 - 1$ has a degree of 5.

(c) $x + 3$ has a degree of 1.
(d) 4 has a degree of 0, since $4 = 4x^0$, where $x \neq 0$.
(e) 0 has no degree whatsoever, because 0 could be written as $0x^n$, where n is any whole number.

In mathematics, it is common for several polynomials to appear in one problem. Thus, it is convenient to identify these polynomials by using upper case letters together with the variable. For example, the polynomial $2x^4 - x^3 + x^2 - 4$ might be symbolized by $P(x)$ (read "P of x") and written $P(x) = 2x^4 - x^3 + x^2 - 4$. This notation enables us to indicate the value of the polynomial for specific values of the variable. Hence, as illustrated in Example 6, the symbol $P(-2)$ would represent the value of the polynomial when $x = -2$.

If $P(x) = 2x^4 - x^3 + x^2 - 4$, find $P(-2)$. **EXAMPLE 6**

Solution: Let $x = -2$ and evaluate the polynomial.
$$P(x) = 2x^4 - x^3 + x^2 - 4$$
$$P(-2) = 2(-2)^4 - (-2)^3 + (-2)^2 - 4$$
$$= 2(16) - (-8) + 4 - 4$$
$$= 32 + 8 + 4 - 4$$
$$= 40$$

If $Q(m) = 3m^2 - m + 2$, find $Q\left(\dfrac{1}{2}\right)$. **EXAMPLE 7**

Solution: Let $m = \dfrac{1}{2}$ and evaluate the polynomial.
$$Q(m) = 3m^2 - m + 2$$
$$Q\left(\dfrac{1}{2}\right) = 3\left(\dfrac{1}{2}\right)^2 - \left(\dfrac{1}{2}\right) + 2$$
$$= 3 \cdot \dfrac{1}{4} - \dfrac{1}{2} + 2$$
$$= \dfrac{3}{4} - \dfrac{1}{2} + 2$$
$$= \dfrac{9}{4}$$

Certain types of polynomials appear so frequently that they are given special names describing the number of terms. A polynomial containing one term is called a monomial. A polynomial containing two terms is a binomial, and a polynomial of three terms is a trinomial.

EXAMPLE 8   Classify these polynomials according to the number of terms.

(a) $5y^2 - y + \dfrac{1}{3}$ [trinomial]

(b) $x^2 + 3$ [binomial]

(c) $3m^4$ [monomial]

(d) -7 [monomial]

Polynomials can be added or subtracted by combining the numerical coefficients of like terms. As you learned in section 1.8, this can be accomplished either vertically or horizontally by removing grouping symbols.

EXAMPLE 9   Add $(5x^2 - 7x + 2) + (3x^2 + 8x - 10)$
Vertical Solution:

$$\begin{array}{r} 5x^2 - 7x + 2 \\ 3x^2 + 8x - 10 \\ \hline 8x^2 + x - 8 \end{array}$$

Horizontal Solution:

$$\begin{aligned}(5x^2 - 7x + 2) + (3x^2 + 8x - 10) &= 5x^2 - 7x + 2 + 3x^2 + 8x - 10 \\ &= 5x^2 + 3x^2 - 7x + 8x + 2 - 10 \\ &= 8x^2 + x - 8\end{aligned}$$

EXAMPLE 10   Subtract $(7y^2 - 3y - 6) - (3y^2 - 5y + 4)$
Vertical Solution:

$$\begin{array}{l}7y^2 - 3y - 6 \\ 3y^2 - 5y + 4\end{array} \quad \xrightarrow{\text{change signs}} \quad \begin{array}{r}7y^2 - 3y - 6 \\ -3y^2 + 5y - 4 \\ \hline 4y^2 + 2y - 10\end{array}$$

Horizontal Solution:

$$\begin{aligned}(7y^2 - 3y - 6) - (3y^2 - 5y + 4) &= 7y^2 - 3y - 6 - 3y^2 + 5y - 4 \\ &= 7y^2 - 3y^2 - 3y + 5y - 6 - 4 \\ &= 4y^2 + 2y - 10\end{aligned}$$

Calculator

With your instructor's approval, it is convenient to evaluate polynomial expressions using a calculator. This is easily done by using the memory (or storage) key (STO) and the recall key (RCL).

EXAMPLE 11   If $P(x) = 3x^5 - 4x^3 - x^2 + 3$, find $P(2.56)$.
Round the answer to three significant digits.

Solution: Place 2.56 into the memory and recall it for each substitution.

(2.56) (STO) (y^x) (5) (×) (3) (−) (4) (×) (RCL) (y^x) (3) (−) (RCL) (x^2) (+) (3) (=) 259.19103

Rounded answer is 259.

POLYNOMIALS

Exercise 3.1

Classify the following as a monomial, binomial, or trinomial and give the degree.

1. $x^2 + x - 1$
2. $3x^2 + \frac{1}{2}x$
3. $1 - x$
4. $1 + x + x^3$
5. 12
6. $\frac{1}{2}x - \frac{3}{2}$

Write the following in descending powers of x.

7. $3x^3 - 2x - 4 + 5x^5$
8. $10 - 2x^5 + 4x^2 - 2x^3$
9. $x^2 - 3 + x^3 - x^4$
10. $-y^2 + y - 5^3 + 2y^4$

Find the indicated values.

11. If $P(x) = 2x^2 - 3x + 7$, find $P(2)$ and $P(-2)$.
12. If $P(x) = x^2 - x + 4$, find $P(-3)$ and $P(1)$.
13. If $P(x) = x^2 + 3x - 4$, find $P(-3)$, $P(2)$, and $P(3)$.
14. If $P(x) = (x - 3)^2 - x^2$, find $P(-2)$ and $P(1)$.
15. If $P(x) = x^3 - 5x^2 - x + 2$, find $P(1)$, $P(-2)$, and $P(0)$.
16. If $P(x) = 2x^3 - 3x^2 - 2x + 4$, find $P(2)$, $P(-2)$ and $P(0)$.
17. If $P(x) = x^3 - x^2 + 3x + 1$, find $P(1)$ and $P(-3)$.
18. If $P(x) = x^2 - 3x + 1$ and $Q(x) = x^3 + 2$, find $P(3)$ and $Q(-2)$.

Combine the following polynomials and write the result in descending powers of x.

19. $(3x + 2) + (4x - 7)$
20. $(x^2 + x + 5) + (x^2 + 2)$
21. $(x^2 - 7x + 8) + (x^2 + 2x - 3)$
22. $(3x + 5) - (2x - 2)$
23. $(3x^2 + 2x + 7) - (2x^2 - 6x + 1)$
24. $(2 - 6x + 5x^2) - (3x^2 - 2x^2 - 2x - 6)$
25. $4x + 2x^2 + 3x^2 - 3x - x + 2x - 6x^2$
26. $(3x^5 - 2x^4 + x^3 - 2) + (x^4 - 3x^2 + 3)$
27. $(5 - 6y^3 + 5y^2 + 4y) + (9y^3 - 6y^2 + 4y + 6)$
28. $(-3x^4 - x^3 + 5) - (4x^4 + 2x^2 - 5)$
29. Given the polynomials $x^2 - 5x - 3$, $-2x^2 + x - 3$, and $4x^2 - 2x + 1$, subtract the sum of the first two from the sum of the last two.
30. Subtract $3x^2 - x + 2$ from the sum of $x + 3$ and $x^2 - 4x + 3$.

Use the calculator to evaluate the following. Round answers to three significant digits.

31. If $P(x) = 3x^3 + x^2 + 4$, find $P(1.65)$.
32. If $P(x) = 2x^4 + 3x^2 - x - 5$, find $P(2.83)$.
33. If $P(x) = x^5 - 5x^4 + 2x^3 - 1$, find $P(-2.67)$.
34. If $P(x) = 3 - x - x^3$, find $P\left(-\frac{3}{4}\right)$.
35. If $P(x) = 1.87x^3 - 2.83x^2 + 1.66x + 0.89$, find $P(3.24)$.
36. If $P(x) = -0.83x^4 + 0.66x^2 - x + 5$, find $P(2.83)$.

MULTIPLICATION AND DIVISION OF POLYNOMIALS

3.2

Polynomials are multiplied and divided using the procedures studied for algebraic expressions in section 1.7.

Multiplication and division of monomials is performed using the addition and subtraction properties of exponents.

EXAMPLE 1 Multiply $(-3a^2b)(-4bc^3)(2a^3b^2)$

Solution: Multiply numerical coefficients and add exponents appearing on identical bases.
$$(-3a^2b)(-4bc^3)(2a^3b^2) = 24a^5b^4c^3$$

EXAMPLE 2 Divide $\dfrac{-36x^4y^5z^2}{3xy^3z^2}$, where $x, y, z \neq 0$

Solution: Divide numerical coefficients and subtract exponents appearing on identical bases.
$$\frac{-36x^4y^5z^2}{3xy^3z^2} = -12x^3y^2z^0$$
$$= -12x^3y^2$$

By applying the distributive properties, this procedure can be extended to polynomials containing more than one term.

EXAMPLE 3 Multiply $-3xy^2(2x^3 - xy + y^3)$

Solution: Use the distributive property to multiply $-3xy^2$ times each term within the group.
$$-3xy^2(2x^3 - xy + y^3) = -6x^4y^2 + 3x^2y^3 - 3xy^5$$

EXAMPLE 4 Divide $\dfrac{6a^4b^3 - 10a^3b^2 + 2a^2b - 4a^2}{2a^2b}$, where $a, b \neq 0$

Solution: Divide $2a^2b$ into each term of the dividend.
$$\frac{6a^4b^3 - 10a^3b^2 + 2a^2b - 4a^2}{2a^2b} = \frac{6a^4b^3}{2a^2b} - \frac{10a^3b^2}{2a^2b} + \frac{2a^2b}{2a^2b} - \frac{4a^2}{2a^2b}$$
$$= 3a^2b^2 - 5ab + 1 - \frac{2}{b}$$

EXAMPLE 5 Use the distributive property to multiply $(x + 2)(2x + 3)$.

Solution: Consider $(x + 2)$ to represent one number, then multiply it times each term of the polynomial $(2x + 3)$.

$(x + 2)(2x + 3) = (x + 2) \cdot 2x + (x + 2) \cdot 3$
$ (x + 2) \cdot 2x + (x + 2) \cdot 3 \quad \text{[distribute again]}$
$ 2x^2 + 4x + 3x + 6$
$ 2x^2 + 7x + 6 \quad \text{[combine like terms]}$

Use the distributive property to multiply $(3a - 2)(4a^2 + 5a - 1)$. **EXAMPLE 6**

Solution: Multiply the binomial $(3a - 2)$ times each term of the trinomial.
$(3a - 2)(4a^2 + 5a - 1) = (3a - 2) \cdot 4a^2 + (3a - 2) \cdot 5a - (3a - 2) \cdot 1$
$ = 12a^3 - 8a^2 + 15a^2 - 10a - 3a + 2$
$ = 12a^3 + 7a^2 - 13a + 2$

It is often easier to multiply polynomials using a vertical form similar to the process of multiplying two natural numbers such as 123 × 56. When multiplying vertically, each term of one polynomial is multiplied times each term of the other polynomial. Like terms are then placed in the same columns and added to produce the final product. We illustrate the vertical method in Example 7 using the same problem as Example 6. Be sure to compare the two methods.

Multiply vertically $(3a - 2)(4a^2 + 5a - 1)$. **EXAMPLE 7**

Solution: Arrange vertically and multiply each term of the trinomial by each term of the binomial.

$$
\begin{array}{r}
4a^2 + 5a - 1 \\
3a - 2 \\
\hline
12a^3 + 15a^2 - 3a \quad \longleftarrow \quad 3a(4a^2 + 5a - 1) \\
- 8a^2 - 10a + 2 \quad \longleftarrow \quad -2(4a^2 + 5a - 1) \\
\hline
12a^3 + 7a^2 - 13a + 2 \quad \longleftarrow \quad \text{add the columns}
\end{array}
$$

Multiply vertically $(m^2 + m - 1)(m^2 - m + 2)$. **EXAMPLE 8**

Solution:
$$
\begin{array}{r}
m^2 + m - 1 \\
m^2 - m + 2 \\
\hline
m^4 + m^3 - m^2 \\
- m^3 - m^2 + m \\
2m^2 + 2m - 2 \\
\hline
m^4 + 0m^3 + 0m^2 + 3m - 2 \quad \text{or} \quad m^4 + 3m - 2
\end{array}
$$

Polynomials containing more than one term can be divided using a procedure similar to the long division process of arithmetic. We review the arithmetic process by dividing 73 by 6.

$$
\begin{array}{r}
12 \quad \longleftarrow \text{[quotient]} \\
\text{[divisor]} \longrightarrow 6\overline{)73} \quad \longleftarrow \text{[dividend]} \\
\underline{6} \\
13 \\
\underline{12} \\
1 \quad \longleftarrow \text{[remainder]}
\end{array}
$$

POLYNOMIAL EXPRESSIONS

The answer is 12 R 1 or $12 + \frac{1}{6}$, written more simply as $12\frac{1}{6}$. This answer may be checked by adding the remainder to the product of the quotient and divisor to produce the dividend. Thus, $12 \cdot 6 + 1 = 73$.

To illustrate long division of polynomials, we will divide $2x^2 + 11x + 16$ by $x + 2$ and compare it to the division of 742 by 23.

Step (1): Indicate the division by this symbol:

$$23\overline{)742}$$

Step (1): Write each polynomial in descending powers of the variable and indicate the division by this symbol:

$$x + 2\overline{)2x^2 + 11x + 16}$$

Step (2): Divide 2 into 7.

$$\begin{array}{r} 3 \\ 23\overline{)742} \end{array}$$

Step (2): Divide x into $2x^2$.

$$\begin{array}{r} 2x \phantom{{}+11x+16} \\ x + 2\overline{)2x^2 + 11x + 16} \end{array}$$

Step (3): Subtract the product of (3)(23) from 74.

$$\begin{array}{r} 3 \\ 23\overline{)742} \\ \underline{69} \\ 5 \end{array}$$

Step (3): Subtract the product of $(2x)(x + 2)$ from $2x^2 + 11x$.

$$\begin{array}{r} 2x \phantom{{}+11x+16} \\ x + 2\overline{)2x^2 + 11x + 16} \\ \underline{2x^2 + 4x} \\ 7x \end{array}$$

Step (4): Bring down the 2.

$$\begin{array}{r} 3 \\ 23\overline{)742} \\ \underline{69\downarrow} \\ 52 \end{array}$$

Step (4): Bring down the 16.

$$\begin{array}{r} 2x \phantom{{}+11x+16} \\ x + 2\overline{)2x^2 + 11x + 16} \\ \underline{2x^2 + 4x}\downarrow \\ 7x + 16 \end{array}$$

Step (5): Divide 2 into 5 and place the result, 2, in the quotient.

$$\begin{array}{r} 32 \\ 23\overline{)742} \\ \underline{69} \\ 52 \end{array}$$

Step (5): Divide x into 7x and place the result, 7, in the quotient.

$$\begin{array}{r} 2x + 7 \\ x + 2\overline{)2x^2 + 11x + 16} \\ \underline{2x^2 + 4x} \\ 7x + 16 \end{array}$$

Step (6): Subtract the product of (2)(23) from 52.

$$\begin{array}{r} 32 \\ 23\overline{)742} \\ \underline{69} \\ 52 \\ \underline{46} \\ 6 \end{array}$$

Step (6): Subtract the product of $(x + 2)(7)$ from $7x + 16$.

$$\begin{array}{r} 2x + 7 \\ x + 2\overline{)2x^2 + 11x + 16} \\ \underline{2x^2 + 4x} \\ 7x + 16 \\ \underline{7x + 14} \\ 2 \end{array}$$

Step (7): Write the answer.

$$32 + \frac{6}{23}$$

Step (7): Write the answer.

$$2x + 7 + \frac{2}{x + 2}$$

MULTIPLICATION AND DIVISION OF POLYNOMIALS

Step (8) Check the answer.

$$\begin{array}{r} 32 \leftarrow \text{[quotient]} \\ \times\ 23 \leftarrow \text{[divisor]} \\ \hline 96 \\ 64 \\ \hline 736 \\ +\ \ 6 \leftarrow \text{[add remainder]} \\ \hline 742 \leftarrow \text{[dividend]} \end{array}$$

Step (8): Check the answer.

$$\begin{array}{r} 2x + 7 \leftarrow \text{[quotient]} \\ x + 2 \leftarrow \text{[divisor]} \\ \hline 2x^2 + 7x \\ 4x + 14 \\ \hline 2x^2 + 11x + 14 \\ +\ \ 2 \leftarrow \left[\begin{array}{l}\text{add}\\ \text{remainder}\end{array}\right] \\ \hline 2x + 11x + 16 \leftarrow \text{[dividend]} \end{array}$$

Divide $16a - 17a^2 + 6a^3 - 6$ by $2a - 3$. **EXAMPLE 9**

Solution: Write in descending powers of the variable and use long division.

Step (1): $2a - 3\overline{)6a^3 - 17a^2 + 16a - 6}$

Step (2):
$$\begin{array}{r} 3a^2 \\ 2a - 3\overline{)6a^3 - 17a^2 + 16a - 6} \\ 6a^3 - 9a^2 \end{array}$$

Step (3):
$$\begin{array}{r} 3a^2 \\ 2a - 3\overline{)6a^3 - 17a^2 + 16a - 6} \\ \underline{6a^3 - 9a^2} \\ -8a^2 + 16a \end{array}$$ $\left[\begin{array}{l}\text{to subtract, change}\\ \text{signs and add}\end{array}\right]$

Step (4):
$$\begin{array}{r} 3a^2 - 4a \\ 2a - 3\overline{)6a^3 - 17a^2 + 16a - 6} \\ 6a^3 - 9a^2 \\ -8a^2 + 16a \\ \underline{-8a^2 + 12a} \\ 4a - 6 \end{array}$$ $\left[\begin{array}{l}\text{to subtract, change}\\ \text{signs and add}\end{array}\right]$

Step (5):
$$\begin{array}{r} 3a^2 - 4a + 2 \\ 2a - 3\overline{)6a^3 - 17a^2 + 16a - 6} \\ 6a^3 - 9a^2 \\ -8a^2 + 16a \\ \underline{-8a^2 + 12a} \\ 4a - 6 \\ \underline{4a - 6} \\ 0 \end{array}$$ $\left[\begin{array}{l}\text{to subtract, change}\\ \text{signs and add}\end{array}\right]$

Step (6): The answer is $3a^2 - 4a + 2$. Since the remainder is 0, the problem can be checked by simply multiplying the quotient times the divisor to obtain the dividend.

$$\begin{array}{r} 3a^2 - 4a + 2 \leftarrow \text{[quotient]} \\ 2a - 3 \leftarrow \text{[divisor]} \\ \hline 6a^3 - 8a^2 + 4a \\ -9a^2 + 12a - 6 \\ \hline 6a^3 - 17a^2 + 16a - 6 \leftarrow \text{[dividend]} \end{array}$$

Divide $2m^3 + 3m - 1$ by $m - 2$. **EXAMPLE 10**

Solution: The polynomial is missing the m^2 term. When this occurs, each missing term should be included with a coefficient of zero.

$$\begin{array}{r}
2m^2 + 4m + 11 \\
m - 2 \overline{\smash{)}\, 2m^3 + 0m^2 + 3m - 1} \\
\underline{2m^3 - 4m^2} \\
4m^2 + 3m \\
\underline{4m^2 - 8m} \\
11m - 1 \\
\underline{11m - 22} \\
21
\end{array}$$

[to subtract, change signs and add.]

The answer is $2m^2 + 4m + 11 + \dfrac{21}{m - 2}$.

Remember, each division problem can be checked by multiplying the quotient times the divisor and adding the remainder. The result should be the dividend.

Exercise 3.2

Multiply

1. $-2a^2b(-4bc^3)(2a^3b^2)$
2. $6r^3s^2t(-5r^2st^2)$
3. $(5ab)(2a^2b)(-ab^2)$
4. $(-x^3)(-x^2y^4)(-2xy^2)$
5. $-t^2(ty^2)(2t)(-3y^2)$
6. $x^3y^2z(-3xyz^2)(-x^4y^3z^3)$

Divide (assume that denominators do not represent zero).

7. $\dfrac{36x^4y^5}{3xy^3}$
8. $\dfrac{12m^4n^2p^3}{6mnp^2}$
9. $\dfrac{-x^4y^8z^{10}}{xy^7z^5}$
10. $\dfrac{4a^4b^6c^8}{-a^3b^4c^2}$

Use the distributive property and multiply.

11. $-4x^2y(x^2 - 3x + 2)$
12. $-6a^4(2a^2 + 4a - 5)$
13. $3x^2y(2xy^3 + 3x^2y - 4xy)$
14. $3xy^2(2x^3 - 4y + y^3)$
15. $2x^3(3x^2 - 4x^3 + 8x - 5)$
16. $(x + 2)(x + 4)$
17. $(x + 1)(2x + 3)$
18. $(x + 5)^2$
19. $(2x + 1)^2$
20. $(x - 3)(2x + 1)$
21. $(3a - 2)(4a^2 + 5a + 1)$
22. $(x + 1)(x^2 - 2x + 3)$
23. $(y + 2)(y^2 - 2y + 4)$
24. $(2x^2 + 3x - 1)(x^2 - 5x)$
25. $(x + 1)^3$
26. $(2x - 1)^3$

Multiply vertically.

27. $(x - 3y)(3x^2 + 5x - 2)$
28. $(x + y - 3)(2x - y + 2)$

29. $(m^2 - m + 2)(m^2 - m - 1)$
30. $(m^3 - m^2 + n - 1)(m^2 - m + 1)$

Divide by the long division method.
31. $(3x^2 + 2x - 8) \div (x + 2)$
32. $(3x^3 - x + 1) \div (x - 1)$
33. $(3x^3 - 5x^2 + 2x - 1) \div (x - 2)$
34. $(x^4 - 5x^2 + 12x - 1) \div (x + 3)$
35. $(x^4 + 1) \div (x + 1)$
36. $(2x^4 - x + 6) \div (x - 2)$
37. $(x^5 - 1) \div (x - 1)$
38. $(x^3 + x^2 - x + 1) \div (x^2 - x + 1)$
39. $(6x^3 + x^2 - 6x + 1) \div (3x^2 + 2x - 1)$

SPECIAL PRODUCTS OF BINOMIALS

3.3

Multiplication of polynomials can always be performed using the distributive property. Many polynomials in algebra are binomials, and it is convenient to develop shortcuts enabling us to find products quickly without listing all of the steps. The shortcuts are called special products of binomials and fall into these three categories:

(1) Product of Two Binomials (FOIL Shortcut).
   Example: $(3x + 4)(2x + 5)$

(2) Product of a Sum and a Difference.
   Example: $(2m + 3)(2m - 3)$

(3) Squaring a Binomial.
   Example: $(3a - 5)^2$

First we develop the FOIL Shortcut by using the distributive property to multiply two general binomials.

$$(a + b)(c + d) = (a + b) \cdot c + (a + b) \cdot d$$
$$= ac + bc + ad + bd$$
$$= ac + ad + bc + bd$$

Comparing the indicated product of $(a + b)(c + d)$ with the answer, we find the following pattern called the FOIL Shortcut.

$$(a + b)(c + d) = \underset{\text{first terms}}{ac} + \underset{\text{outer terms}}{ad} + \underset{\text{inner terms}}{bc} + \underset{\text{last terms}}{bd}$$

$$\ \ \text{F}\ \ \ \ \ \ \ \text{O}\ \ \ \ \ \ \ \text{I}\ \ \ \ \ \ \ \text{L}$$

First — Last
Inner — Outer

POLYNOMIAL EXPRESSIONS

> **FOIL Shortcut**
>
> To find the product of two binomials, multiply their first terms, multiply their outer terms, multiply their inner terms, and multiply their last terms. Then add these products together.
>
> $$ \overset{F}{} \ \ \overset{O}{} \ \ \overset{I}{} \ \ \overset{L}{}$$
> $$(a + b)(c + d) = ac + ad + bc + bd$$

EXAMPLE 1 Multiply $(x + 2)(x + 3)$

Solution:
$$(x + 2)(x + 3) = \overset{F}{x \cdot x} + \overset{O}{x \cdot 3} + \overset{I}{2 \cdot x} + \overset{L}{2 \cdot 3}$$
$$= x^2 + 3x + 2x + 6$$
$$= x^2 + \underbrace{5x} + 6$$

EXAMPLE 2 Multiply $(3m - 2n)(m + 4n)$

Solution:
$$(3m - 2n)(m + 4n) = \overset{F}{3m \cdot m} + \overset{O}{3m \cdot 4n} - \overset{I}{2n \cdot m} - \overset{L}{2n \cdot 4n}$$
$$= 3m^2 + \underbrace{12mn - 2mn} - 8n^2$$
$$= 3m^2 + 10mn - 8n^2$$

Often, as illustrated by the previous examples, the outer and inner products can be combined to produce the middle term of the answer. Thus, it is sometimes convenient to think of the FOIL Shortcut as $F + (O + I) + L$. You should learn to apply this shortcut mentally, obtaining the answer in one step.

EXAMPLE 3 Multiply $(2a + 1)(3a + 2)$

Solution:
$$(2a + 1)(3a + 2) = \overset{F}{6a^2} + \overset{(O+I)}{7a} + \overset{L}{2}$$

EXAMPLE 4 Multiply $(3x^2 + 2y)(x^2 - 3y)$

Solution:
$$(3x^2 + 2y)(x^2 - 3y) = \overset{F}{3x^4} - \overset{(O+I)}{7x^2y} - \overset{L}{6y^2}$$

The second special product is called a sum times a difference and is symbolized by $(a + b)(a - b)$. To be a sum times a difference, the two binomials must be identical except for the middle signs. Thus, one binomial is the sum of two terms while the other is the difference of the same two terms. The shortcut for finding the product of a sum times a difference is developed by using the FOIL method.

$$(a + b)(a - b) = a^2 - \underbrace{ab + ab} - b^2$$
$$= a^2 + 0 - b^2$$
$$= a^2 - b^2$$

SPECIAL PRODUCTS OF BINOMIALS

The result, $a^2 - b^2$, is called a difference of two squares.

> **Sum Times a Difference Shortcut**
>
> To multiply a sum times a difference, write the product of the first terms and subtract the product of the last terms.
>
> $$(a + b)(a - b) = \overset{F}{a^2} - \overset{L}{b^2}$$

Multiply $(3m + 5)(3m - 5)$ **EXAMPLE 5**

Solution: Multiply first terms and last terms to obtain a difference of two squares.

$$(3m + 5)(3m - 5) = (3m)(3m) - (5)(5)$$
$$= 9m^2 - 25$$

Multiply $\left(\frac{2}{3}x + \frac{1}{2}\right)\left(\frac{2}{3}x - \frac{1}{2}\right)$ **EXAMPLE 6**

Solution:
$$\left(\frac{2}{3}x + \frac{1}{2}\right)\left(\frac{2}{3}x - \frac{1}{2}\right) = \left(\frac{2}{3}x\right)\left(\frac{2}{3}x\right) - \left(\frac{1}{2}\right)\left(\frac{1}{2}\right)$$
$$= \frac{4}{9}x^2 - \frac{1}{4}$$

The third special product is a shortcut for squaring binomials and is obtained by using the FOIL method to multiply a binomial times itself.

$$(a + b)^2 = (a + b)(a + b)$$
$$= a^2 + \underbrace{ab + ba}_{2ab} + b^2$$
$$= a^2 + 2ab + b^2$$

Comparing the original binomial to the answer produces the following pattern:

$$(\overset{\downarrow\text{first term}}{a} + \overset{\text{last term}}{b})^2 = a^2 + 2ab + b^2$$

- square of the first term
- twice the product of the first and last terms
- square of the last term

Binomial Squared Shortcut

To square a binomial, form a trinomial by following this pattern: Square the first term, add twice the product of the first and last terms, then add the square of the last term.

$$(a + b)^2 = a^2 + \underbrace{2ab}_{ab \text{ times } 2} + b^2$$

EXAMPLE 7 Find $(m + 5)^2$

Solution:
$$(m + 5)^2 = m^2 + \underbrace{2(5m)}_{5m \text{ times } 2} + 25$$
$$= m^2 + 10m + 25$$

EXAMPLE 8 Find $(3x - 2y)^2$

Solution:
$$(3x - 2y)^2 = 9x^2 \underbrace{- 12xy}_{(3x)(-2y) \text{ times } 2} + 4y^2$$

As illustrated by the following examples, some polynomials of three or more terms can be regrouped so that a special product shortcut applies. Sometimes this is faster than using the vertical method, which would always apply.

EXAMPLE 9 Use the FOIL method to multiply $(x + y + 3)(x + y + 2)$.

Solution: Group $x + y$ as one term. Use FOIL, then square the resulting binomial.

$$(x + y + 3)(x + y + 2) = [(x + y) + 3][(x + y) + 2]$$
$$ \quad\ \ \text{F} \qquad (\text{O} + \text{I}) \qquad \text{L}$$
$$= (x + y)^2 + 5(x + y) + 6$$
$$= x^2 + 2xy + y^2 + 5x + 5y + 6$$

EXAMPLE 10 Use the sum times a difference shortcut to multiply $(2m + n + 3)(2m + n - 3)$.

Solution: Group $2m + n$ as one term, multiply to obtain a difference of two squares, then square the resulting binomial.

$$(2m + n + 3)(2m + n - 3) = [(2m + n) + 3][(2m + n) - 3]$$
$$ \qquad\qquad \text{F} \qquad\qquad \text{L}$$
$$= (2m + n)^2 - 9$$
$$= 4m^2 + 4mn + n^2 - 9$$

SPECIAL PRODUCTS OF BINOMIALS

In summary, we list the three special products for multiplying binomials and other selected polynomials.

Special Products

1. The FOIL Method
$$\;\;\;\text{F}\text{O}\text{I}\text{L}$$
$$(a + b)(c + d) = ac + ad + bc + bd$$

2. Sum Times a Difference
$$(a + b)(a - b) = a^2 - b^2$$

3. Squaring a Binomial
$$(a + b)^2 = a^2 + 2ab + b^2$$
$$(a - b)^2 = a^2 - 2ab + b^2$$

Exercise 3.3

Find each of the following products.

1. $(x + 9)(x + 3)$
2. $(y + 3)(2y + 5)$
3. $(5x + 2)(x + 3)$
4. $(x - 4)(x + 1)$
5. $(y - 2)(y + 5)$
6. $(z - 7)(z + 7)$
7. $(t - 2)(t - 6)$
8. $(r + 6)(r - 2)$
9. $(5x - 2y)(x - y)$
10. $(x + 3)^2$
11. $(y - 5)^2$
12. $(3x - 1)^2$
13. $(2n + 3)^2$
14. $(5x - 2y)^2$
15. $(3a + 8b)^2$
16. $(ax + by)^2$
17. $(2x - 1)(3x + 4)$
18. $(2x - 5y)(2x + 5y)$
19. $(4y + 3)(2y - 1)$
20. $(9x + 1)(4x - 5)$
21. $(3x - 2y)^2$
22. $(4x + 3)(4x - 3)$
23. $(2x + 5)(4x + 3)$
24. $(2x - 3)(3x - 5)$
25. $(6x - y)(6x + y)$
26. $(3x - 5d)(3x + 5d)$
27. $(-3t + 4)(-t + 5)$
28. $(8x - 9y)(8x - 9y)$
29. $(x^2 + 4)(x^2 + 3)$
30. $(x^2 + 5)(x^2 - 3)$
31. $(x^3 + 4)(x^3 - 6)$
32. $(x^3 - 1)(x^3 + 1)$
33. $(2x^3 + 5)(4x^3 - 1)$
34. $(x^3 - 1)(x^3 - 1)$
35. $(xy^2 + 10)(xy^2 - 2)$
36. $(6x - 1)(5 - x^2)$
37. $[(a + b) + 3][(a + b) + 4]$
38. $[(m + n) - 3][(m + n) + 3]$
39. $[(x + y) + 5][(x + y) - 5]$
40. $[(2h + k) + 8][(2h + k) + 2]$
41. $[(a + 2) + b]^2$
42. $[(2a + b) - 3]^2$
43. $[(x^2 - 3) - y]^2$
44. $(x^2 + 1)(x + 1)$
45. $(3x^2 - 2)(x + 3)$
46. $(x + 2)(x - 2)(3x - 1)$
47. $(x + y)(x - y)(5x + 2)$
48. $(2m + 3)(3m - 1)(2m - 3)$

COMMON FACTORS

3.4

In previous sections you learned how to multiply two polynomials. Now we reverse the process. We will take a single polynomial and learn to write it as a product of two or more simpler polynomials. This procedure is called factoring.

To factor a polynomial we begin by looking for the largest common factor. This is defined to be the largest expression which exactly divides into each term of the polynomial. For example, the polynomial $6x^2 + 12x$ has common factors of 2, 3, 6 and x. However, the largest common factor is $6x$.

EXAMPLE 1

Determine the largest common factor for each of these polynomials.

Solution:

Expression	Largest Common Factor
$3a - 3b$	3
$9x^2y + 36xy^2$	$9xy$
$12m^4 - 3m^3 + 9m^2$	$3m^2$

In section 1.4 you learned that the distributive property can be used to form a basic product. The expression common to each term is factored out. Thus,

$$ab + ac = a(b + c)$$

This property enables us to factor a polynomial by removing the largest common factor.

To remove the largest common factor follow these steps:
(1) Find the largest common factor.
(2) Rewrite the polynomial so that the largest common factor appears in each term.
(3) Remove the largest common factor using the distributive property.

EXAMPLE 2

Factor $12x^3 + 36x^2$

Solution:

Step 1. The largest common factor is $12x^2$. The numerical part is 12 since it is the largest number that divides into 12 and 36. Note that the variable portion is x^2 and it contains the smaller of the two exponents.

Step 2. Rewrite the polynomial so that $12x^2$ appears in each term.
$$12x^3 + 36x^2 = (12x^2) \cdot x + (12x^2) \cdot 3$$

Step 3. Remove $12x^2$ from each term.
$$(12x^2) \cdot x + (12x^2) \cdot 3 = 12x^2(x + 3)$$

COMMON FACTORS 117

Thus, $12x^3 + 36x^2 = 12x^2(x + 3)$. To check this answer, multiply it back out to obtain the original polynomial.

Check. $12x^2(x + 3) = 12x^3 + 36x^2$

Factor $8a^4b^3 - 16a^3b^2 + 4a^2b$ **EXAMPLE 3**

Solution:

Step 1. The largest common factor is $4a^2b$. 4 is the largest number that divides into 8, 16, and 4. The smallest exponents for the variables a and b are 2 and 1, respectively.

Step 2. Rewrite the polynomial so that $4a^2b$ appears in each term.
$8a^4b^3 - 16a^3b^2 + 4a^2b = (4a^2b)2a^2b^2 - (4a^2b)4ab + (4a^2b)1$

Step 3. Remove $4a^2b$ from each term.
$(4a^2b)2a^2b^2 - (4a^2b)4ab + (4a^2b)1 = 4a^2b(2a^2b^2 - 4ab + 1)$

Check. $4a^2b(2a^2b^2 - 4ab + 1) = 8a^4b^3 - 16a^3b^2 + 4a^2b$

With practice these steps can be done mentally and your work will follow the pattern of the next two examples.

Factor $12r^2s - 6rs + 18rs^2$ **EXAMPLE 4**

Solution: Remove the largest common factor of $6rs$ from each term.
$$12r^2s - 6rs + 18rs^2 = 6rs(2r) - 6rs(1) + 6rs(3s)$$
$$= 6rs(2r - 1 + 3s)$$

Factor $-2p^3 + 6p^2 - 4p$ **EXAMPLE 5**

Solution: It is correct to choose either $2p$ or $-2p$ for the largest common factor. Factoring out $2p$ gives:

$$-2p^3 + 6p^2 - 4p = 2p(-p^2) + 2p(3p) - 2p(2)$$
$$= 2p(-p^2 + 3p - 2)$$

Or, factoring out $-2p$ produces:
$$-2p^3 + 6p^2 - 4p = -2p(p^2) + (-2p)(-3p) + (-2p)2$$
$$= -2p(p^2 - 3p + 2)$$

Sometimes a particular situation may demand one form over the other. However, both are correct.

Sometimes the largest common factor may take the form of a group as in this example:

POLYNOMIAL EXPRESSIONS

EXAMPLE 6 Factor $x(x + 5) + 3(x + 5)$

Solution: The greatest common factor is $(x + 5)$. Think of it as one number and remove it from each term.

$$x(x + 5) + 3(x + 5) = (x + 5)(x + 3)$$

Often the polynomial does not have the greatest common factor within grouping symbols. When this occurs, the polynomial will have to be regrouped and factored as illustrated by the following examples. This process is called factoring by grouping.

EXAMPLE 7 Factor by grouping $2t - 2s + rt - rs$.

Solution: Group the first two terms and the last two terms:

$$2t - 2s + rt - rs = (2t - 2s) + (rt - rs)$$

Factor a 2 from the first group and an r from the second group:

$$(2t - 2s) + (rt - rs) = 2(t - s) + r(t - s)$$

The binomial $(t - s)$ is common to both terms, so it is factored out:

$$2(t - s) + r(t - s) = (t - s)(2 + r)$$

Thus, $2t - 2s + rt - rs = (t - s)(2 + r)$

Factoring by grouping works only if the same expression can be obtained within grouping symbols so it is common to both terms.

EXAMPLE 8 Factor by grouping $ab + ac + 3b + 3c$.

Solution: Group the first two terms and the last two terms:

$$ab + ac + 3b + 3c = (ab + ac) + (3b + 3c)$$
$$= a(b + c) + 3(b + c)$$
$$= (b + c)(a + 3)$$

EXAMPLE 9 Factor by grouping $m^3n^2 + m^3 - 4n^2 - 4$.

Solution: Group the first two terms and the last two terms. When grouping the last two terms it is important to remember that the minus sign changes the sign of each term within the parentheses.

$$m^3n^2 + m^3 - 4n^2 - 4 = (m^3n^2 + m^3) - (4n^2 + 4)$$
$$= m^3(n^2 + 1) - 4(n^2 + 1)$$
$$= (n^2 + 1)(m^3 - 4)$$

Exercise 3.4

Factor.
1. $10x + 20$
2. $2x^2 + 2$
3. $6y^2 + 3y$
4. $5y^2 + y^3$
5. $7a + 21$
6. $12a^3 - 9a^2$
7. $8xy^2 - 4x^2y^2$
8. $6t^2 - 12t^3$
9. $4ab - 5ab^2$
10. $2ab - 3a^2b^2$
11. $5ax + 15a^2$
12. $5t^3 + 15t^2$
13. $14x^6y^3 - 6x^4y^4$
14. $8a^4 - 6a^3 + 4a^2$
15. $x^2y - xy^2 + x^2y^2$
16. $6x^3 - 18x^2 + 9x^4$
17. $2b^4 + 4b^6 - 8b^3$
18. $15a^2b^2c^2 - 25ab^2c^2$
19. $24x^2y^2 + 36x^2y - 48xy$
20. $8t^3 - 12t^6 - 4t$
21. $3mn^2 - 6m^2n^2$
22. $12x^4 + 15x^3 - 9x^2$
23. $ay^2 + aby + ab$
24. $18xy^3 + 12x^2y^2 - 24x^3y^4$
25. $5x(a - b) - 3y(a - b)$
26. $3a(x - y) - 5b(x - y)$
27. $7(x - y)^3 + 11(x - y)^2$
28. $3(a + 2b)^2 + 9(a + 2b)$
29. $4a(x + y) - (x + y)$
30. $x^2(x^2 + 2) + (x^2 + 2)$
31. $4(a + b) - (a + b)^2$
32. $4(a - b)^2 - (a - b)$

Factor by grouping.
33. $x + y + 7ax + 7ay$
34. $x^3 + 2x^2 + 7x + 14$
35. $y^3 - 2y^2 + 5y - 10$
36. $2xy + 2y^2 + y + x$
37. $x^3 + 3x - 7x^2 - 21$
38. $ax^2 + 6x - ax - 6$
39. $am + bm + an + bn$
40. $x^2 - 2x - 3ax + 6a$
41. $x^2 + 3xy + 2x + 6y$
42. $mn^2 - n + mn - 1$
43. $(a + b)x^2 + (a + b)y^2$
44. $a(x + y)^2 + b(x + y)^2$
45. $a^2b^2 + b^2 + 4a^2 + 4$

FACTORING BINOMIALS

3.5

We now study factoring methods for these three special types of binomials:

(1) A difference of two squares: $a^2 - b^2$
(2) A sum of two cubes: $a^3 + b^3$
(3) A difference of two cubes: $a^3 - b^3$

Difference of Two Squares:

You should recall that a sum times a difference produces a difference of two squares. Thus,

$$\underbrace{(a + b)}_{\text{sum}}\underbrace{(a - b)}_{\text{difference}} = \underbrace{a^2 - b^2}_{\text{difference of two squares}}$$

Reversing this pattern shows that a difference of two squares must factor into a sum times a difference.

> **Factoring a Difference of Two Squares**
> $$a^2 - b^2 = (a + b)(a - b)$$

EXAMPLE 1 Factor $x^2 - 25$

Solution: $x^2 - 25 = x^2 - 5^2$
$= (x + 5)(x - 5)$

EXAMPLE 2 Factor $16m^2 - 49$

Solution: $16m^2 - 49 = (4m)^2 - 7^2$
$= (4m + 7)(4m - 7)$

When factoring it is important to remember two rules. First, always remove the largest common factor. And second, be sure to factor completely. That is, factor as far as possible.

EXAMPLE 3 Factor $5x^4y - 5y$

Solution: Remove the greatest common factor, $5y$. Then, factor completely.
$5x^4y - 5y = 5y(x^4 - 1)$
$= 5y(x^2 + 1)\ \underbrace{(x^2 - 1)}$
$= 5y(x^2 + 1)(x + 1)(x - 1)$

It is not uncommon for a difference of two squares to contain a group, as illustrated by this example:

EXAMPLE 4 Factor $(2m + n)^2 - 9$

Solution: Think of the binomial $(2m + n)$ as representing one number. Then, factor as a difference of two squares.
$(2m + n)^2 - 9 = [(2m + n) + 3][(2m + n) - 3]$
$= (2m + n + 3)(2m + n - 3)$

Sum and Difference of Two Cubes

Next we will develop a factoring technique for a sum or difference of two cubes. These binomials are characterized by being either a basic sum or difference where the numerical coefficients are perfect cubes and the exponents are divisible by 3.

FACTORING BINOMIALS

EXAMPLE 5

Each of these polynomials is either a sum or a difference of two cubes.
(a) $x^3 + 8$
(b) $27m^3 - 64$
(c) $(a + b)^3 + 1$
(d) $r^6 - 8t^3$

To develop a pattern for factoring a sum or difference of two cubes, we use the distributive properties to multiply out the following two products:

(1) $(a + b)(a^2 - ab + b^2) = (a + b)a^2 - (a + b)ab + (a + b)b^2$
$= a^3 + \underline{a^2b - a^2b} - \underline{ab^2 + ab^2} + b^3$
$= a^3 + 0 + 0 + b^3$
$= a^3 + b^3$

(2) $(a - b)(a^2 + ab + b^2) = (a - b)a^2 + (a - b)ab + (a - b)b^2$
$= a^3 - \underline{a^2b + a^2b} - \underline{ab^2 + ab^2} - b^3$
$= a^3 + 0 + 0 - b^3$
$= a^3 - b^3$

Reversing these two equalities provides us with a pattern for factoring a sum or difference of two cubes.

Factoring a Sum or Difference of Two Cubes

1. $a^3 + b^3 = (a + b)(a^2 - ab + b^2)$
2. $a^3 - b^3 = (a - b)(a^2 + ab + b^2)$

To use these patterns we note the following:

1. A sum or difference of two cubes produces two factors, one a binomial and the other a trinomial.

$a^3 + b^3 = (\underbrace{}_{\text{binomial}})(\underbrace{}_{\text{trinomial}})$

2. The binomial factor is formed by taking the cube roots of the corresponding terms in the original problem and by using the same sign.

$\overset{\text{cube roots}}{\overbrace{a^3 + b^3}} = (a + b)()$
$\underset{\text{same sign}}{\uparrow}$

3. The trinomial factor is formed by referring to the binomial factor, not the original problem. The first and last terms of the trinomial are formed by

POLYNOMIAL EXPRESSIONS

squaring the corresponding terms of the binomial factor. The middle term is the product of the two terms of the binomial with its sign changed.

$$a^3 + b^3 = (a + b)(a^2 - ab + b^2)$$

(square the terms; the middle term ab has its sign changed)

EXAMPLE 6 Factor $n^3 + 8$

Solution: $n^3 + 8$ is a sum of two cubes. Follow these three steps:

Step 1. $n^3 + 8 = (\underbrace{\quad\quad}_{\text{binomial}})(\underbrace{\quad\quad\quad}_{\text{trinomial}})$

Step 2. (cube roots, same sign)
$$n^3 + 8 = (n + 2)(\quad\quad)$$

Step 3. (square the terms; middle term $2n$ changes sign)
$$n^3 + 8 = (n + 2)(n^2 - 2n + 4)$$

Thus, $n^3 + 8 = (n + 2)(n^2 - 2n + 4)$

EXAMPLE 7 Factor $8x^3 - 27$

Solution: $8x^3 - 27$ is a difference of two cubes.

Step 1. $8x^3 - 27 = (\underbrace{\quad\quad}_{\text{binomial}})(\underbrace{\quad\quad\quad}_{\text{trinomial}})$

Step 2. (cube roots, same sign)
$$8x^3 - 27 = (2x - 3)(\quad\quad)$$

Step 3. (square the terms; middle term $-6x$ changes sign)
$$8x^3 - 27 = (2x - 3)(4x^2 + 6x + 9)$$

Thus, $8x^3 - 27 = (2x - 3)(4x^2 + 6x + 9)$

With practice you can abbreviate this factoring procedure by doing the three steps mentally.

EXAMPLE 8

Factor $125r^3 - 64s^3$

Solution: $125r^3 - 64s^3 = (5r - 4s)(25r^2 + 20rs + 16s^2)$

EXAMPLE 9

Factor $m^6 + 1$

Solution: This is a sum of two cubes because $m^6 + 1 = (m^2)^3 + 1^3$.
$m^6 + 1 = (m^2 + 1)(m^4 - m^2 + 1)$

As seen in the following example, a sum or difference of two cubes may contain common factors or groups. Remember, when factoring, the first step is to remove the greatest common factor. Groups can be thought of as one number and factored accordingly.

EXAMPLE 10

Factor $2(x - 1)^3 + 16y^3$

Solution: Remove the greatest common factor, 2. Think of $(x - 1)$ as representing one number.

$$2(x - 1)^3 + 16y^3 = 2[(x - 1)^3 + 8y^3]$$
$$= 2[(x - 1) + 2y][(x - 1)^2 - 2y(x - 1) + 4y^2]$$
$$= 2(x - 1 + 2y)(x^2 - 2x + 1 - 2xy + 2y + 4y^2)$$

In summary, when factoring binomials you should follow these steps:

Factoring Binomials
1. Remove the greatest common factor.
2. Determine whether the binomial is a difference of two squares. If so, factor completely.
$$a^2 - b^2 = (a + b)(a - b)$$
3. Determine whether the binomial is a sum or difference of two cubes. If so, factor completely.
$$a^3 + b^3 = (a + b)(a^2 - ab + b^2)$$
$$a^3 - b^3 = (a - b)(a^2 + ab + b^2)$$

EXAMPLE 11

Factor completely $2m^6p - 128n^6p$

Solution: Follow, in order, the three steps for factoring binomials:

Step 1. Remove the greatest common factor, $2p$.
$2m^6p - 128n^6p = 2p(m^6 - 64n^6)$

Step 2. Factor $m^6 - 64n^6$ as a difference of two squares:
$$2m^6p - 128n^6p = 2p(m^6 - 64n^6)$$
$$= 2p(m^3 + 8n^3)(m^3 - 8n^3)$$

Step 3. Factor the sum and difference of two cubes:
$$2m^6p - 128n^6p = 2p(m^3 + 8n^3)(m^3 - 8n^3)$$
$$= 2p(m + 2n)(m^2 - 2mn + 4n^2)(m - 2n)(m^2 + 2mn + 4n^2)$$

Exercise 3.5

Factor completely.

1. $a^2 - 49$
2. $x^2 - 9y^2$
3. $4 - 9y^2$
4. $x^4 - 1$
5. $9x^2 - 36$
6. $a^4b^2 - 4$
7. $\dfrac{1}{4} - 9x^2$
8. $1 - 16x^4$
9. $x^3 - x$
10. $3x^2 - 12$
11. $x^2 - 3$
12. $2x^4 - 18$
13. $x^3 - 36x$
14. $16x^2 - 4y^2$
15. $x^3 - 1$
16. $x^3 - 64$
17. $8x^3 + y^3$
18. $x^3 + 27$
19. $x^3 - 27y^3$
20. $a^3 + b^3c^3$
21. $(x - 2)^2 - y^2$
22. $(x - 1)^2 - 9y^2$
23. $5x^2 - 125$
24. $3x^4 - 3$
25. $8x^3 - 27y^3$
26. $8x^3 + 27$
27. $(x - 1)^3 + y^3$
28. $3x^2 - 12y^2$
29. $(s + t)^2 - 9a^2$
30. $x^2y^4 - z^2$
31. $8x^5 - 2x^3$
32. $16 - 9x^4y^4$
33. $(x + 1)^2 - 36y^2$
34. $2x^2 - 8y^4$
35. $16x^8 - 81y^{12}$
36. $t^3 - 16t$
37. $4a^4 - b^4$
38. $x^3 + 64y^3$
39. $8a^3 - 125b^3$

FACTORING TRINOMIALS

3.6

In section 3.3 you learned to multiply two binomials using the FOIL shortcut. You also found that the product of two binomials is often a trinomial, as illustrated next.

FACTORING TRINOMIALS

$$\overset{FOIL}{(x + 3)(x + 5) = x^2 + 5x + 3x + 15}$$
$$= x^2 + 8x + 15$$

To factor a trinomial such as $x^2 + 8x + 15$, we reverse the FOIL process by following this procedure:

Step 1. $x^2 + 8x + 15$ came from a product of two binomials. Thus, the factors are represented with two sets of grouping symbols.

$$() \cdot ()$$

Step 2. The x^2 term of the trinomial is the product of the first terms of the two binomials, $x \cdot x$. Thus, x is placed as the first term of each binomial factor.

$$(x) \cdot (x)$$

Step 3. 15 is the third term of the trinomial and represents the product of the last terms of the two binomials. Hence, factors of 15 must be placed for last terms. 15 has two sets of factors, $1 \cdot 15$ and $3 \cdot 5$. This gives the following two possibilities:

$(x + 1) \cdot (x + 15)$
or   —[factors of 15]
$(x + 3) \cdot (x + 5)$

Step 4. The correct combination is determined by taking the sum of the outer and inner products. This sum must produce the middle term of the original trinomial, $8x$.

$(x + 1) \cdot (x + 15)$ is incorrect since $15x + 1x \neq 8x$

$(x + 3) \cdot (x + 5)$ is correct since $5x + 3x = 8x$

Therefore, $x^2 + 8x + 15 = (x + 3) \cdot (x + 5)$

EXAMPLE 1

Factor $3m^2 + 10m - 8$

Solution:

Step 1. Write two sets of grouping symbols.

$$()()$$

Step 2. $3m^2$ came from the product of $3m$ and m.

$$(3m)(m)$$

Step 3. The third term of the trinomial is -8. Form last terms by considering all factors of -8.

$$(3m + 1)(m - 8)$$
$$(3m - 1)(m + 8)$$
$$(3m + 2)(m - 4)$$
$$(3m - 2)(m + 4)$$

[factors of -8]

Step 4. The sum of the outer and inner terms must produce the middle term of the trinomial, $10m$.

$(3m - 2)(m + 4)$ is correct since $12m - 2m = 10m$.

Therefore, $3m^2 + 10m - 8 = (3m - 2)(m + 4)$.

With practice, you will be able to shorten this procedure by finding the correct combination mentally. Your work will then parallel the following example.

EXAMPLE 2   Factor $2x^2 - 11xy + 5y^2$

Solution: $2x^2 - 11xy + 5y^2 = (\quad)(\quad)$
$= (2x\quad)(x\quad)$
$= (2x - y)(x - 5y)$

Check. $(2x - y)(x - 5y)$; $-10xy - xy = -11xy$

Factoring trinomials is much easier and faster if you memorize the various sign combinations described by the following three cases. Each case assumes that the leading coefficient of the trinomial is positive.

Case 1. If the trinomial contains only plus signs, then the binomial factors contain only plus signs.

$2a^2 + 7a + 3 = (2a + 1)(a + 3)$
  all plus   all plus

Case 2. If the middle term of the trinomial is minus and the last term is plus, then the middle sign of each binomial is minus.

$6r^2 - 11rs + 3s^2 = (2r - 3s)(3r - s)$
  minus plus   both minus

Case 3. If the last term of the trinomial is minus, then the binomials have opposite middle signs.

$$3m^2 - 7m - 5 = (3m - 5)(2m + 1)$$

minus   opposite signs

Use the pattern of signs to help factor $3n^2 - 11n + 10$.

EXAMPLE 3

Solution: The middle term is minus and the last term is plus. Thus, the middle sign of each binomial is minus.

$$3n^2 - 11n + 10 = (-)(-)$$
$$= (3n -)(n -)$$
$$= (3n - 5)(n - 2)$$

When factoring trinomials it is important to remember these points:
(1) Watch for common factors. Always remove the greatest common factor before doing anything else.
(2) Use the pattern of signs to set up the binomial factors. Check your result by taking the sum of the outer and inner products. This sum must equal the middle term of the trinomial.
(3) Always factor completely. Check each factor to see if it can be factored further.

Factor $2x^4 - 20x^2 + 18$

EXAMPLE 4

Solution: $2x^4 - 20x^2 + 18 = 2(x^4 - 10x^2 + 9)$
$$= 2(-)(-)$$
$$= 2(x^2 - 1)(x^2 - 9)$$
$$= 2(x + 1)(x - 1)(x + 3)(x - 3)$$

A special trinomial occurs frequently in algebra. It is called a *perfect square trinomial* and comes from squaring a binomial as follows:

$$(a + b)^2 = a^2 + 2ab + b^2$$
or
$$(a - b)^2 = a^2 - 2ab + b^2$$

A perfect square trinomial can be recognized by noting that the first and last terms are perfect squares and the middle term is twice the product of the first and last terms of the binomial. A perfect square trinomial will factor into the square of a binomial.

EXAMPLE 5 Factor $x^2 + 12x + 36$

Solution: The trinomial is a perfect square.

$$x^2 + 12x + 36$$
$$\downarrow \qquad \downarrow$$
$$(x)^2 \qquad 6^2$$

twice the product of x and 6

Thus, $x^2 + 12x + 36 = (x + 6)^2$

EXAMPLE 6 Each of the following is a perfect square trinomial and has been factored as the square of a binomial.

(a) $9m^2 + 6m + 1 = (3m + 1)^2$
   same sign

(b) $25p^2 - 20p + 4 = (5p - 2)^2$
   same sign

(c) $4x^2 + 12xy + 9y^2 = (2x + 3y)^2$

The following two tables summarize the types of factoring discussed in this chapter and provide a review of techniques to use according to the number of terms in the polynomial.

Factoring Types

1. Removing the greatest common factor.
$$ab + ac + ad = a(b + c + d)$$

2. Factoring by grouping
$$(a + b)c + (a + b)d = (a + b)(c + d)$$

3. Difference of two squares
$$a^2 - b^2 = (a + b)(a - b)$$

4. Sum or difference of two cubes
$$a^3 + b^3 = (a + b)(a^2 - ab + b^2)$$
$$a^3 - b^3 = (a - b)(a^2 + ab + b^2)$$

5. Reversing FOIL
$$a^2 + 3a + 2 = (a + 2)(a + 1)$$

6. Perfect square trinomial
$$a^2 + 2ab + b^2 = (a + b)^2$$
$$a^2 - 2ab + b^2 = (a - b)^2$$

Factoring Techniques

A. To factor a binomial:
  (1) Remove the greatest common factor, if any.
  (2) Check to see if the remaining binomial is a difference of two squares, sum of two cubes, or a difference of two cubes.
B. To factor a trinomial:
  (1) Remove the greatest common factor, if any.
  (2) See if the remaining trinomial is a perfect square or if it can be factored by reversing FOIL.
C. To factor four or more terms:
  (1) Remove the greatest common factor, if any.
  (2) Try to factor by grouping.

Remember, factor completely. Your final step should be to check each factor to see if any can be factored further.

Exercise 3.6

Factor the following trinomials.

1. $x^2 + 5x + 6$
2. $y^2 + 7y + 6$
3. $a^2 + 13a + 12$
4. $x^2 + 7x - 30$
5. $b^2 - 3a + 18$
6. $x^2 + 6xy + 9y^2$
7. $y^2 - 9y + 20$
8. $a^2 - 10a + 16$
9. $5x^2 + 9x + 4$
10. $3b^2 - 2b - 5$
11. $7y^2 + 23y + 6$
12. $3y^2 + 13y - 10$
13. $10b^2 - 3b - 18$
14. $16x^2 + 8xy + y^2$
15. $3a^2 - 25a - 18$
16. $15x^2 + 7x - 2$
17. $1 + 16y + 64y^2$
18. $9x^2 - 12xy + 4y^2$
19. $2x^2 - 9x - 5$
20. $2x^2 + 6x - 20$
21. $x^4 - 5x^2 + 4$
22. $2y^3 + 15y^2 + 7y$
23. $x^4 - 6x^3 - 7x^2$
24. $x^4 + 3x^2 - 4$
25. $x^4 - 6x^2 - 27$
26. $4 - 12x + 9x^2$
27. $10m^2 + mn - 3n^2$
28. $2x^3 - 4x^2 + 6x$
29. $16x^8 - 7x^6 + 6x^7$
30. $36x^6 - 13x^4 + x^2$
31. $6x^4 + 5x^2 - 25$
32. $81x^2 + 18xy + y^2$
33. $18x^2 - 15x - 18$
34. $12x^4 + 28x^2 - 5$
35. $25x^2y^2 - 20xy + 4$
36. $y^4 + 10y^2 - 24$
37. $49x^2 - b^2$
38. $8a^3 - 125$
39. $x^3 - 3x^2 - 10x$
40. $(x + a)^2 - 36$

41. $x^3 + 8y^3$

42. $ab - 2b + 5a - 10$

43. $2ax - 2ay + bx - by$

44. $16b^2 - 121$

45. $x^4 - 16y^4$

46. $6t^4 - 11t^2 - 10$

47. $3y^4 - 3y^3 - 90y^2$

48. $xy + wy + xz + wz$

49. $1 - x^3$

50. $6x^2 - 15x + 6$

51. $m^3n^2 - m^3 + 8n^2 - 8$

SOLVING EQUATIONS BY FACTORING

3.7

Some non-linear equations containing polynomials can be solved by factoring and by using a special property of zero called the zero factor property.

> **Zero Factor Property**
> If $a, b \in R$ and if $a \cdot b = 0$, then $a = 0$ or $b = 0$

This property indicates that whenever the product of two or more factors is zero, then at least one of the factors is zero.

EXAMPLE 1   Solve $(x + 2)(3x - 5) = 0$

Solution: This equation states that the product of $x + 2$ and $3x - 5$ is 0. According to the zero factor property, this can be true only if $x + 2 = 0$ or if $3x - 5 = 0$. Thus, we solve the two equations.

$x + 2 = 0 \qquad\qquad 3x - 5 = 0$

$\qquad x = -2 \qquad\qquad\qquad 3x = 5$

$\qquad\qquad\qquad\qquad\qquad\qquad x = \dfrac{5}{3}$

The solutions are -2 and $\dfrac{5}{3}$. They may be checked by substituting them, one at a time, into the original equation.

Check for $x = -2$

$(x + 2)(3x - 5) = 0$

$(-2 + 2)[3(-2) - 5] = 0$

$0 \cdot (-11) = 0$

Check for $x = \dfrac{5}{3}$

$(x + 2)(3x - 5) = 0$

$\left(\dfrac{5}{3} + 2\right)\left(3 \cdot \dfrac{5}{3} - 5\right) = 0$

$\dfrac{11}{3} \cdot 0 = 0$

EXAMPLE 2   Solve $(y + 3)(5y - 4)(y - 1) = 0$

Solution: Extending the zero factor property, we set each factor equal to zero and solve the resulting linear equations.

$$y + 3 = 0 \quad \text{or} \quad 5y - 4 = 0 \quad \text{or} \quad y - 1 = 0$$
$$y = -3 \qquad\qquad 5y = 4 \qquad\qquad y = 1$$
$$y = \frac{4}{5}$$

The solutions are -3, $\frac{4}{5}$, and 1.

To solve an equation such as $2m^2 - 5m = 12$ by using the zero factor property, we need to obtain a product which is equal to zero. Therefore, we add -12 to each side, producing $2m^2 - 5m - 12 = 0$. Then, the trinomial is factored and each factor is set equal to zero. This gives the two solutions $-\frac{3}{2}$ and 4, as shown in the following:

$$2m^2 - 5m = 12$$
$$2m^2 - 5m - 12 = 0$$
$$(2m + 3)(m - 4) = 0$$
$$2m + 3 = 0 \quad \text{or} \quad m - 4 = 0$$
$$2m = -3 \qquad\qquad m = 4$$
$$m = -\frac{3}{2}$$

Summarizing this procedure provides us with the following solution technique.

Solving Equations Using the Zero Factor Property

Step (1): Obtain all terms on one side of the equation. (The polynomial must be set equal to zero.)
Step (2): Factor the polynomial completely.
Step (3): Use the zero factor property to set each factor containing a variable equal to zero.
Step (4): Solve the resulting equations.

Solve $6n^2 - 26n = 20$

EXAMPLE 3

Solution:

Step 1. Obtain all terms on one side of the equation.
$$6n^2 - 26n = 20$$
$$6n^2 - 26n - 20 = 0 \qquad \text{[add } -20 \text{ to both sides]}$$

POLYNOMIAL EXPRESSIONS

Step 2. Factor the polynomial completely.
$$6n^2 - 26n - 20 = 0$$
$$2(3n^2 - 13n - 10) = 0$$
$$2(3n + 2)(n - 5) = 0$$

Step 3. Set each factor containing a variable equal to zero. (Note that $2 \neq 0$)
$$2(3n + 2)(n - 5) = 0$$
$$3n + 2 = 0 \quad \text{or} \quad n - 5 = 0$$

Step 4. Solve the resulting equations.
$$3n + 2 = 0 \qquad n - 5 = 0$$
$$3n = -2 \qquad n = 5$$
$$n = -\frac{2}{3}$$

The solutions are $-\frac{2}{3}$ and 5. Remember, the solutions can be checked by substituting, one at a time, into the original equation.

As you become more familiar with this solution technique, feel free to use a more abbreviated form as illustrated in the following examples.

EXAMPLE 4

Solve $6y^2 - 3 = 2(y - y^2)$

Solution: Multiply, then obtain all terms on one side.
$$6y^2 - 3 = 2(y - y^2)$$
$$6y^2 - 3 = 2y - 2y^2$$
$$8y^2 - 2y - 3 = 0$$
$$(4y - 3)(2y + 1) = 0$$
$$4y - 3 = 0 \quad \text{or} \quad 2y + 1 = 0$$
$$4y = 3 \qquad\qquad 2y = -1$$
$$y = \frac{3}{4} \qquad\qquad y = -\frac{1}{2}$$

The solutions are $\frac{3}{4}$ and $-\frac{1}{2}$.

EXAMPLE 5

Solve $2r^4 + 18 = 20r^2$

Solution: Obtain all terms on one side, then factor completely.
$$2r^4 + 18 = 20r^2$$
$$2r^4 - 20r^2 + 18 = 0$$
$$2(r^4 - 10r^2 + 9) = 0$$
$$2(r^2 - 9)(r^2 - 1) = 0$$
$$2(r + 3)(r - 3)(r + 1)(r - 1) = 0$$

$r + 3 = 0$	or	$r - 3 = 0$	or	$r + 1 = 0$	or	$r - 1 = 0$
$r = -3$		$r = 3$		$r = -1$		$r = 1$

The solutions are -3, 3, -1, and 1.

Factoring, together with zero factor property, can also be used to solve applied problems as seen in the next example.

EXAMPLE 6 The length of a rectangle is 12 feet more than its width. The area is 45 square feet. Find the dimensions of the rectangle.

Solution: Draw a diagram and label the parts. Remember, the area of a rectangle is the product of its length and width, $A = l \cdot w$.

```
   w  | Area = 45 ft.²
      |
        w + 12
```

$$A = l \cdot w$$
$$45 = (w + 12)w$$
$$45 = w^2 + 12w$$
$$0 = w^2 + 12w - 45$$
$$0 = (w + 15)(w - 3)$$
$$w + 15 = 0 \quad \text{or} \quad w - 3 = 0$$
$$w = -15 \qquad\qquad w = 3$$

A rectangle cannot have a negative width. The only solution is 3. Thus, the width is 3 feet and the length is $3 + 12$ or 15 feet.

Solve.

Exercise 3.7

1. $x^2 - 6x + 8 = 0$
2. $y^2 - y - 12 = 0$
3. $2x^2 - x = 0$
4. $3x^2 + 2x = 1$
5. $2y^2 = 2y$
6. $a^2 - 25 = 0$
7. $b^2 - 5b + 4 = 0$
8. $y^2 + 3y = 0$
9. $2x^2 - 3x - 2 = 0$
10. $x^2 - 4x + 4 = 0$
11. $2x^2 - 12x = -18$
12. $x^2 = 4$

13. $4 + 9y^2 = 12y$
14. $2a^2 = 9a + 5$
15. $2x^2 = 12x - 16$
16. $3x^2 + x - 4 = 0$
17. $2y^2 - 7y - 15 = 0$
18. $y^2 + 4y = 5$
19. $x^2 - 5x + 4 = 0$
20. $-3t^2 + 4t - 1 = 0$
21. $4a^2 = 9$
22. $x^2 - 6x = 16$
23. $x(2x - 3) = -1$
24. $x(x - 4) = 5$
25. $(x - 3)(x + 2) = 6$
26. $(2y - 3)(y + 1) = 3$
27. $2x^3 = 5x^2$
28. $x^4 - 5x^2 + 4 = 0$
29. $5x^3 + 5x^2 - 30x = 0$
30. $x^4 - 10x^2 + 9 = 0$
31. $2x - \dfrac{5}{3} = \dfrac{x^2}{3}$
32. $y^2 + \dfrac{1}{4}y - \dfrac{3}{8} = 0$
33. $x^4 - 13x^2 + 36 = 0$
34. $x^4 - 21x^2 = 100$

Solve the following applied problems.

35. The sum of a number and its square is 20. Find the number.
36. The product of two consecutive integers and their sum add up to be 55. Find the integers.
37. One number is 3 greater than another. Their product is 10. Find the numbers.
38. The product of two consecutive even integers is 80. Find the numbers.
39. One number is 2 less than twice another. The sum of their squares is 52. Find the numbers.
40. The length of a rectangle is 2 inches more than twice its width. The area is 24 square inches. Find the dimensions.
41. The length of a rectangle is 1 foot more than twice its width. If the area is 21 square feet, find the dimensions.
42. One square has a side 4 inches longer than the side of another square. The area of the larger square is 9 times as great as the area of the smaller square. Find the length of the side of each square.

Summary

KEY WORDS AND PHRASES

Polynomial—an algebraic expression of one or more terms, each of which is a product of a constant and a variable raised to a whole number power.
Degree of a Polynomial in One Variable—the largest exponent used on the variable.
Monomial—a polynomial containing one term.
Binomial—a polynomial containing two terms.
Trinomial—a polynomial containing three terms.
FOIL Shortcut—to find the product of two binomials, multiply their first

terms, multiply their outer terms, multiply their inner terms, multiply their last terms, and add the products.

$$\overset{\text{F}\text{O}\text{I}\text{L}}{(a + b)(c + d) = ab + ad + bc + bd}$$

Sum Times a Difference Shortcut—to find the product of a sum times a difference, multiply the first terms and multiply the last terms to obtain a difference of two squares.

$$(a + b)(a - b) = a^2 - b^2$$

Binomial Squared Shortcut—To square a binomial, square the first term, add twice the product of the first and last term, and add the square of the last term.

$$(a + b)^2 = a^2 + 2ab + b^2$$

Zero Factor Property—If $a, b \in R$ and if $a \cdot b = 0$, then $a = 0$ or $b = 0$.

FACTORING TYPES

1. Remove the greatest common factor:
$$ab + ac + ad = a(b + c + d)$$

2. Factoring by grouping:
$$(a + b)c + (a + b)d = (a + b)(c + d)$$

3. Difference of two squares:
$$a^2 - b^2 = (a + b)(a - b)$$

4. Sum or difference of two cubes:
$$a^3 + b^3 = (a + b)(a^2 - ab + b^2)$$
$$a^3 - b^3 = (a - b)(a^2 + ab + b^2)$$

5. Reversing FOIL:
$$a^2 + 3a + 2 = (a + 2)(a + 1)$$

6. Perfect square trinomials:
$$a^2 + 2ab + b^2 = (a + b)^2$$
$$a^2 - 2ab + b^2 = (a - b)^2$$

Self-Checking Exercise

Write in descending powers of the variable and state the degree.
1. $4x + 11x^3 - 3x^2$
2. $12x^6 - 2x^2 - x^3$
3. $12x^6 - 9x^8$
4. $2y^3 - 5y^4 + 7y^2 - 3y + 1$

Evaluate the polynomials for the indicated value.
5. If $P(x) = 3x^4 - x^3 + 2x - 1$, find $P(-2)$
6. If $Q(x) = (x - 2)^3 - x^2$, find $Q(-2)$
7. If $P(x) = 5x^3 - 2x^2 + 3x - 7$, find $P(0)$
8. If $Q(x) = x^4 - 2x^3 + 3x^2 + 2x - 1$, find $P(2)$

Classify these polynomials as a monomial, binomial, or trinomial.
9. $x^2 - \frac{1}{2}x$
10. $1 - 3x$
11. 10
12. $1 + x - x^2$

Add or subtract as indicated. Write the result in descending powers of the variable.
13. $(6x^2 - 4x^3 - 4x) + (10x^3 - 4x^2 + 3x)$
14. $(m^4 + 7 + 8m^3) - (2m^3 - 6m^2 + 2)$
15. $(3x + 4x^2 - 5) - (5x^2 - 4x^3 + 3x)$
16. $(x^3 - 2x^4 - 2) + (3 - 3x^2 + x^4) - (3x^2 - 2x - 6)$

Multiply or divide as indicated.
17. $(3x^2y)(2xy^4)(-x^2y^3)$
18. $\dfrac{4a^3b^4c^2}{-2a^2b^2}$
19. $-x^4(-x^3y^3)(-2xy^2)$
20. $\dfrac{x^6y^8z^{12}}{x^3y^4z^4}$

Multiply using the distributive property.
21. $-2x^2(x + 11)$
22. $5y^4(3y^3 - y^2 + 2y + 1)$
23. $(2x + 5)(x^2 + 1)$
24. $(x - 3)(x^2 - x + 4)$
25. $(3x + 1)(9x^2 - 3x + 1)$
26. $(x^2 + 1)(x - 1)$

Divide by the long division method.
27. $(2x^3 + x^2 + x - 1) \div (2x - 1)$
28. $(4x^3 - 7x - 3) \div (x + 1)$
29. $(3x^3 - 5x^2 - 2) \div (x + 2)$
30. $(x^4 + 2x^2 - x + 1) \div (x^2 + x - 2)$

Multiply. Use FOIL, sum times a difference, binomial squared, or any other applicable shortcut.
31. $(3x - 7)(2x + 5)$
32. $(5a + 2)(5a - 2)$
33. $(4x - 3)(2x + 1)$
34. $(4r^2 + 3)(4r^2 - 3)$
35. $(8x - 3)^2$
36. $(x^2 + 2)^2$
37. $(3y + 7)(y - 4)$
38. $(3a^2 + 2)(3a^2 - 2)$

SOLUTIONS TO SELF-CHECKING EXERCISE 137

Factor these binomials.
39. $6x - 24y$
40. $11k^2 + 12k^3$
41. $y^3 + 27$
42. $8 - x^3$
43. $100x^2 - 9y^4$
44. $2x^3 - 16y^3$
45. $x^4 - 1$
46. $(x - y)^2 - 25$

Factor these trinomials.
47. $18t^2 + 3ty - 3y^2$
48. $12x^2 - 23x + 10$
49. $4x^2 - 4xy + y^2$
50. $12a^3 + a^2 - 20a$

Factor by grouping.
51. $x^2 + xy + 3x + 3y$
52. $2b + 6 - ab - 3a$
53. $x^3 + y^3 - x^2y - xy^2$
54. $m^2 - n^2 + m - n$

Solve by factoring.
55. $8x^2 + 8x + 2 = 0$
56. $4a^2 = 3a$
57. $2a^2 - 3a - 5 = 0$
58. $2y^2 - y = 6$
59. $(2x + 1)(x - 2) = 18$
60. $x^4 - 10x^2 + 9 = 0$

Solve the following applied problems.
61. One number is two greater than another. The difference of their squares is 16. Find the two numbers.
62. The length of a rectangle is 2 inches more than three times the width. The area is 33 square inches. Find its dimensions.
63. The perimeter of a rectangle is 18 centimeters and the area is 18 square centimeters. Find the dimensions of the rectangle.

Solutions to Self-Checking Exercise

1. $11x^3 - 3x^2 + 4x$; degree 3
2. $12x^6 - x^3 - 2x^2$; degree 6
3. $-9x^8 + 12x^6$; degree 8
4. $-5y^4 + 2y^3 + 7y^2 - 3y + 1$; degree 4
5. 51
6. -68
7. -7
8. 15
9. binomial
10. binomial
11. monomial
12. trinomial
13. $6x^3 + 2x^2 - x$
14. $m^4 + 6m^3 + 6m^2 + 5$
15. $4x^3 - x^2 - 5$
16. $-x^4 + x^3 - 6x^2 + 2x + 7$
17. $-6x^5y^8$
18. $-2ab^2c^2$
19. $-2x^8y^5$
20. $x^3y^4z^8$
21. $-2x^3 - 22x^2$
22. $15y^7 - 5y^6 + 10y^5 + 5y^4$
23. $2x^3 + 5x^2 + 2x + 5$
24. $x^3 - 4x^2 + 7x - 12$
25. $27x^3 + 1$
26. $x^3 - x^2 + x - 1$
27. $x^2 + x + 1$
28. $4x^2 - 4x - 3$
29. $3x^2 - 11x + 22 + \dfrac{-46}{x + 2}$
30. $x^2 - x + 5 + \dfrac{-8x + 11}{x^2 + x - 2}$

31. $6x^2 + x - 35$
32. $25a^2 - 4$
33. $8x^2 - 2x - 3$
34. $16r^4 - 9$
35. $64x^2 - 48x + 9$
36. $x^4 + 4x^2 + 4$
37. $3y^2 - 5y - 28$
38. $9a^4 - 4$
39. $6(x - 4y)$
40. $k^2(11 + 12k)$
41. $(y + 3)(y^2 - 3y + 9)$
42. $(2 - x)(4 + 2x + x^2)$
43. $(10x + 3y^2)(10x - 3y^2)$
44. $2(x - 2y)(x^2 + 2xy + 4y^2)$
45. $(x^2 + 1)(x + 1)(x - 1)$
46. $(x - y + 5)(x - y - 5)$
47. $3(3t - y)(2t + y)$
48. $(4x - 5)(3x - 2)$
49. $(2x - y)^2$
50. $a(4a - 5)(3a + 4)$
51. $(x + y)(x + 3)$
52. $(b + 3)(2 - a)$
53. $(x + y)(x - y)^2$
54. $(m - n)(m + n + 1)$
55. $x = -\dfrac{1}{2}$
56. $a = 0$ or $a = \dfrac{3}{4}$
57. $a = -1$ or $a = \dfrac{5}{2}$
58. $y = 2$ or $y = -\dfrac{3}{2}$
59. $x = 4$ or $x = -\dfrac{5}{2}$
60. $x = 1, -1, 3,$ or -3
61. 3 and 5
62. length 11 inches, width 3 inches
63. length 6 centimeters, width 3 centimeters, or length 3 centimeters, width 6 centimeters

RATIONAL EXPRESSIONS

4

4.1   Reduction of Rational Expressions
4.2   Multiplication and Division of Rational Expressions
4.3   Addition and Subtraction of Rational Expressions
4.4   Simplifying Complex Fractions
4.5   Equations and Formulas Containing Rational Expressions
4.6   Applied Problems Involving Work and Motion

REDUCTION OF RATIONAL EXPRESSIONS

4.1

You learned in section 1.1 that a rational number is a fraction (or quotient) of two integers having a non-zero denominator. Similarly, a **rational expression** is defined to be a quotient of two polynomials. Since division by zero is undefined, the denominator of a rational expression can never be zero. Thus, the domain of a rational expression consists of all real numbers which do not make the denominator zero.

EXAMPLE 1

$\dfrac{x + 5}{x^2 - x - 6}$ is a rational expression. Find its domain.

Solution: The only real numbers not in the domain are those for which the denominator is zero.

$x^2 - x - 6 = 0$

$(x - 3)(x + 2) = 0$

$x - 3 = 0$ or $x + 2 = 0$

$x = 3 \qquad\qquad x = -2$

The domain of this rational expression consists of all real numbers except 3 and -2.

There are three signs associated with every fraction: a sign for the numerator, a sign for the denominator, and a sign for the fraction itself. You learned in section 1.5 that $\dfrac{-a}{b} = \dfrac{a}{-b} = -\dfrac{a}{b}$. Thus, any two signs of a fraction may be changed without affecting its value.

EXAMPLE 2

Given $\dfrac{2x - y}{-3}$, write an equivalent fraction with a numerator of $y - 2x$.

Solution: We change two signs: the numerator and denominator.

$$\dfrac{2x - y}{-3} = \dfrac{-(2x - y)}{+3}$$

$$= \dfrac{-2x + y}{3} \text{ or } \dfrac{y - 2x}{3}$$

In arithmetic you learned to reduce fractions to lowest terms. Thus, $\dfrac{10}{12}$ was reduced to $\dfrac{5}{6}$ by dividing the numerator and denominator by 2. In algebra, rational expressions are reduced to lowest terms in the same manner. The property used is called the fundamental principle of fractions.

REDUCTION OF RATIONAL EXPRESSIONS   141

Fundamental Principle of Fractions

If $a, b, c \in R$ and $b, c \neq 0$, then

$$\frac{a}{b} = \frac{ac}{bc} \text{ or } \frac{ac}{bc} = \frac{a}{b}$$

This property states that the numerator and denominator of a fraction may be multiplied or divided by the same non-zero number without changing the value of the fraction.

A rational expression is said to be reduced to lowest terms when the numerator and denominator do not share a common factor other than 1. To reduce a rational expression to lowest terms, we follow this procedure:

Reducing to Lowest Terms

(1) Factor the numerator and denominator completely.
(2) Divide the numerator and denominator by their greatest common factor.

Reduce to lowest terms, $\frac{5x^2y}{10xy}$.

EXAMPLE 3

Solution: The greatest common factor is $5xy$.

$$\frac{5x^2y}{10xy} = \frac{x \cdot 5xy}{2 \cdot 5xy} \quad \begin{bmatrix}\text{divide the numerator and}\\ \text{denominator by } 5xy.\end{bmatrix}$$

$$= \frac{x}{2}$$

Reduce to lowest terms, $\frac{2m^2 + m - 15}{4m - 10}$.

EXAMPLE 4

Solution: Factor the numerator and denominator completely, then divide the numerator and denominator by $2m - 5$.

$$\frac{2m^2 + m - 15}{4m - 10} = \frac{(m + 3)(2m - 5)}{2(2m - 5)}$$

$$= \frac{m + 3}{2}$$

The fundamental principle of fractions states that $\frac{a \cdot c}{b \cdot c} = \frac{a}{b}$. We say that the common factor of c is divided out. Thus, $\frac{a \cdot \cancel{c}}{a \cdot \cancel{c}} = \frac{a}{b}$. The slashes indicate a shortcut for dividing the numerator and denominator by the same number and can be used only under these conditions:

(1) Both the numerator and the denominator must be basic products.
(2) The expressions divided out must be factors of the numerator and denominator.

EXAMPLE 5

Reduce to lowest terms, $\dfrac{a^3 + b^3}{3a + 3b}$.

Solution: Factor the numerator and denominator, then divide out the common factor.

$$\dfrac{a^3 + b^3}{3a + 3b} = \dfrac{(a+b)(a^2 - ab + b^2)}{3(a+b)}$$

$$= \dfrac{a^2 - ab + b^2}{3}$$

Rational expressions often contain factors which are opposites (or additive inverses). These are expressions which differ only in sign, such as 8 and -8, or $x - y$ and $y - x$. A number (other than zero), when divided by its opposite, always produces -1. Hence, opposite factors in the numerator and denominator may be divided out by introducing a factor of -1. It may be placed in either the numerator or the denominator, but not in both.

EXAMPLE 6

Reduce to lowest terms, $\dfrac{2x^2 - 7x + 6}{6 - 4x}$.

Solution: Factor the numerator and denominator, then divide out the opposites by introducing a factor of -1.

$$\dfrac{2x^2 - 7x + 6}{6 - 4x} = \dfrac{(x - 2)(2x - 3)}{2(3 - 2x)} \quad \text{opposites}$$

$$= \dfrac{(x - 2)(-1)}{2}$$

$$= \dfrac{-x + 2}{2} \text{ or } \dfrac{2 - x}{2}$$

Exercise 4.1

Find the domain of each of these rational expressions.

1. $\dfrac{x}{x - 4}$
2. $\dfrac{4x - 1}{x^2}$
3. $\dfrac{4x - 3}{4}$
4. $\dfrac{5}{-y}$
5. $\dfrac{-3}{9a + 18}$
6. $\dfrac{x^2 + 1}{2x - 3}$
7. $\dfrac{5}{x^2 - 4}$
8. $\dfrac{x}{2x^2 + x - 3}$
9. $\dfrac{2x^2 + 11}{x^2 + 3x - 28}$

REDUCTION OF RATIONAL EXPRESSIONS   143

Write an equivalent fraction with the sign of the numerator changed.

10. $\dfrac{x}{3}$

11. $\dfrac{x-1}{4}$

12. $\dfrac{2x-1}{x+2}$

13. $\dfrac{x-3}{x-4}$

14. $\dfrac{x^2-3}{-2}$

15. $\dfrac{a+b}{a-b}$

Reduce to lowest terms. Assume denominators are not equal to zero.

16. $\dfrac{16x^3y}{4xy}$

17. $\dfrac{-6ab^4}{12ab}$

18. $\dfrac{24x^2y^2z}{-12x^4yz^3}$

19. $\dfrac{15a^2b^5}{12a^2b^3}$

20. $\dfrac{2x+2y}{x+y}$

21. $\dfrac{x^2+x-6}{x-2}$

22. $\dfrac{ax-a}{a}$

23. $\dfrac{y^2-y-2}{y+1}$

24. $\dfrac{x^2-7x+12}{x^2-9x+20}$

25. $\dfrac{a^2-4}{4}$

26. $\dfrac{2x^2-2x}{2x}$

27. $\dfrac{2x+4}{4}$

28. $\dfrac{x^2-1}{1+x}$

29. $\dfrac{a-b}{b-a}$

30. $\dfrac{x^3-xy^2}{4y-4x}$

31. $\dfrac{a^2-9}{3-a}$

32. $\dfrac{5a+5}{a^2-1}$

33. $\dfrac{(a-b)^2}{b-a}$

34. $\dfrac{x-2}{4-x^2}$

35. $\dfrac{1+x}{x^2-1}$

36. $\dfrac{(x-1)(x-3)}{(1-x)(3+x)}$

37. $\dfrac{2x-x^2}{4x-x^2}$

38. $\dfrac{x^3-1}{1-x^2}$

39. $\dfrac{x^2+xy-2y^2}{x^2-y^2}$

40. $\dfrac{x^2-3x-4}{x^2-x-12}$

41. $\dfrac{y^2-5y-14}{2-y}$

42. $\dfrac{4x^2-20x}{x^2-4x-5}$

43. $\dfrac{x^3+y^3}{x^2-y^2}$

44. $\dfrac{y-x}{x^2-y^2}$

45. $\dfrac{8x^2-4x^3}{2x-x^2}$

46. $\dfrac{4x^2-y^2}{y^3-8x^3}$

47. $\dfrac{x^2-4xy+3y^2}{y^2-x^2}$

48. $\dfrac{x^2+5x-14}{x-2}$

49. $\dfrac{3x-6x^2}{8x^2y-2y}$

144 RATIONAL EXPRESSIONS

50. $\dfrac{6x^2 - 3xy}{-4x^2y + 2xy^2}$

51. $\dfrac{2x^2 - 3x - 9}{12 - 7x + x^2}$

52. $\dfrac{3x^4 + 6x^3 - 9x}{3x}$

53. $\dfrac{ac + ad + bc + bd}{ab + ac + b^2 + bc}$

54. $\dfrac{xy - yw + xz - zw}{xy + yw + xz + zw}$

55. $\dfrac{2x^2 + 6x - xy - 3y}{2x - y}$

56. $\dfrac{x^3 - 3x^2 + 9x - 27}{x^3 - 27}$

57. $\dfrac{x^3 + 8y^3}{4y^2 - x^2}$

MULTIPLICATION AND DIVISION OF RATIONAL EXPRESSIONS

4.2

As you learned in arithmetic, the product of two fractions is formed by multiplying their numerators and multiplying their denominators. Rational expressions follow the same procedure.

> **Multiplying Rational Expressions**
>
> If $a, b, c, d \in R$ and $b, d \neq 0$, then
> $$\frac{a}{b} \cdot \frac{c}{d} = \frac{ac}{bd}$$

EXAMPLE 1

Multiply $\dfrac{5x}{3} \cdot \dfrac{2x^2}{7y}$

Solution: Multiply numerators and multiply denominators.

$$\frac{5x}{3} \cdot \frac{2x^2}{7y} = \frac{(5x)(2x^2)}{3(7y)}$$

$$= \frac{10x^3}{21y}$$

Rational expressions often contain factors common to both a numerator and a denominator. When this occurs we multiply according to this procedure:

> **To Multiply Rational Expressions**
>
> (1) Factor all numerators and denominators, completely.
> (2) Apply the fundamental principle of fractions by dividing out the factors common to both a numerator and a denominator.
> (3) Multiply remaining numerators and multiply remaining denominators.

MULTIPLICATION AND DIVISION OF RATIONAL EXPRESSIONS

EXAMPLE 2

Multiply $\dfrac{3x - 6}{y} \cdot \dfrac{2y^2}{5x - 10}$

Solution: Factor completely, then divide out the common factors.

$$\dfrac{3x - 6}{y} \cdot \dfrac{2y^2}{5x - 10} = \dfrac{3(x - 2)}{y} \cdot \dfrac{2 \cdot y \cdot y}{5(x - 2)}$$

$$= \dfrac{6y}{5}$$

EXAMPLE 3

Multiply $\dfrac{m - 3}{2m^2 + m - 6} \cdot \dfrac{6 - 4m}{m^2 - 9}$

Solution: Factor completely, then divide out opposites by introducing a factor of -1.

$$\dfrac{m - 3}{2m^2 + m - 6} \cdot \dfrac{6 - 4m}{m^2 - 9} = \dfrac{m - 3}{(2m - 3)(m + 2)} \cdot \dfrac{\overset{-1}{2(3 - 2m)}}{(m + 3)(m - 3)}$$

$$= \dfrac{-2}{(m + 2)(m + 3)}$$

Using previous properties we derive a method for dividing two fractions:

Let $a, b, c, d \in R$ and $b, c, d \neq 0$, then

$$\dfrac{a}{b} \div \dfrac{c}{d} = \dfrac{\dfrac{a}{b}}{\dfrac{c}{d}} \qquad \text{[definition of a fraction]}$$

$$= \dfrac{\dfrac{a}{b} \cdot \dfrac{d}{c}}{\dfrac{c}{d} \cdot \dfrac{d}{c}} \qquad \text{[fundamental principle]}$$

$$= \dfrac{\dfrac{a}{b} \cdot \dfrac{d}{c}}{1} \qquad \text{[multiplication of fractions]}$$

$$= \dfrac{a}{b} \cdot \dfrac{d}{c} \qquad \text{[dividing by 1]}$$

This derivation gives the following property for dividing fractions:

Dividing Rational Expressions

If $a, b, c, d \in R$ and $b, c, d \neq 0$, then

$$\frac{a}{b} \div \frac{c}{d} = \frac{a}{b} \cdot \frac{d}{c}$$

reciprocals

Thus, to divide rational expressions we multiply by the reciprocal (or multiplicative inverse) of the divisor. This is sometimes stated as "invert the divisor and multiply."

EXAMPLE 4 Divide $\dfrac{3t}{5} \div \dfrac{4t^2}{15}$

Solution: Multiply by the reciprocal of the second fraction.

$$\frac{3t}{5} \div \frac{4t^2}{15} = \frac{3t}{5} \cdot \frac{15}{4t^2}$$

$$= \frac{3\cancel{t}}{\cancel{5}} \cdot \frac{\cancel{5} \cdot 3}{4 \cdot \cancel{t} \cdot t}$$

$$= \frac{9}{4t}$$

EXAMPLE 5 Simplify $\dfrac{3m^2 - m - 2}{m^2 - 2m + 1} \div \dfrac{3m^2 + 8m + 4}{m^3 + 8} \cdot \dfrac{m^2 - 1}{m}$

Solution: Invert the second fraction. Then factor completely and divide out the common factors.

$$\frac{3m^2 - m - 2}{m^2 - 2m + 1} \div \frac{3m^2 + 8m + 4}{m^3 + 8} \cdot \frac{m^2 - 1}{m}$$

$$= \frac{3m^2 - m - 2}{m^2 - 2m + 1} \cdot \frac{m^3 + 8}{3m^2 + 8m + 4} \cdot \frac{m^2 - 1}{m}$$

$$= \frac{\cancel{(3m + 2)}\cancel{(m - 1)}}{\cancel{(m - 1)}\cancel{(m - 1)}} \cdot \frac{\cancel{(m + 2)}(m^2 - 2m + 4)}{\cancel{(3m + 2)}\cancel{(m + 2)}} \cdot \frac{(m + 1)\cancel{(m - 1)}}{m}$$

$$= \frac{(m + 1)(m^2 - 2m + 4)}{m}$$

It is also important to be able to raise a rational expression to a power. This is accomplished by using the distributive property of exponents:

$$\left(\frac{a}{b}\right)^n = \frac{a^n}{b^n}, \text{ where } b \neq 0$$

The numerator and denominator are both raised to the indicated power.

MULTIPLICATION AND DIVISION OF RATIONAL EXPRESSIONS

EXAMPLE 6

Perform the indicated operation: $\left(\dfrac{3x^2}{2y^3}\right)^4$

Solution: Raise both the numerator and the denominator to the fourth power.

$$\left(\dfrac{3x^2}{2y^3}\right)^4 = \dfrac{(3x^2)^4}{(2y^3)^4}$$

$$= \dfrac{81x^8}{16y^{12}}$$

EXAMPLE 7

Perform the indicated operation: $\left(\dfrac{3a - b}{a + 2b}\right)^2$

Solution: Square the numerator and denominator.

$$\left(\dfrac{3a - b}{a + 2b}\right)^2 = \dfrac{(3a - b)^2}{(a + 2b)^2}$$

$$= \dfrac{9a^2 - 6ab + b^2}{a^2 + 4ab + 4b^2}$$

Perform the indicated operations. Reduce all answers.

Exercise 4.2

1. $\dfrac{5x^3}{3y} \cdot \dfrac{2y^2}{3x}$

2. $\dfrac{-x^3}{5} \cdot \dfrac{25}{x}$

3. $\dfrac{5x^5y}{8z^2} \cdot \dfrac{4z}{15xy^4}$

4. $\dfrac{27x^3}{xy^5} \cdot \dfrac{4xy^2}{15x^2y}$

5. $\dfrac{7a}{12b^2} \cdot \dfrac{20b^3}{35a^3}$

6. $\dfrac{8x^4y^2}{5y^6} \cdot \dfrac{10x^2y^4}{x^3y^2}$

7. $15x \cdot \dfrac{3}{5x^4}$

8. $\dfrac{2x^2y^4}{9z^4} \cdot \dfrac{3z}{5x^3y^3}$

9. $\dfrac{xy^3}{yz} \div x^2z$

10. $\dfrac{mn^3}{18n^2} \div \dfrac{5m^4}{24m^3n}$

11. $\dfrac{4y}{15x^4} \cdot \dfrac{3x^3}{14} \cdot \dfrac{7x^2}{10y^2}$

12. $\dfrac{3x^6}{8} \div \dfrac{x^2}{2} \cdot \dfrac{1}{x}$

13. $x^3 \div \left(\dfrac{3}{x^2} \cdot \dfrac{9}{x}\right)$

14. $\dfrac{10a}{12b} \cdot \dfrac{3a^2c}{5a^3c} \div \dfrac{3bc}{6ab^2}$

15. $\dfrac{x + 1}{x + 3} \cdot \dfrac{x + 3}{x + 5}$

16. $\dfrac{a - 1}{a} \cdot \dfrac{1}{1 - a}$

17. $\dfrac{6x + 18}{5} \cdot \dfrac{25x}{7x + 21}$

18. $\dfrac{11y}{3y - 9} \cdot \dfrac{4y - 12}{22}$

RATIONAL EXPRESSIONS

19. $\dfrac{8x + 12}{5x} \cdot \dfrac{x}{2x + 3}$

20. $\dfrac{x^2 - 16}{x^2 - 25} \cdot \dfrac{x + 5}{x + 4}$

21. $\dfrac{3x - 12}{x^2 - 4} \cdot \dfrac{x - 2}{x - 4}$

22. $\dfrac{x^2 - 16}{x^2 - 25} \cdot \dfrac{x + 5}{x + 4}$

23. $\dfrac{8x - 8y}{16x - 16y} \cdot \dfrac{(x - y)^2}{2x - 2y}$

24. $\dfrac{x^3 y - y^3 x}{x^2 y - xy^2} \cdot \dfrac{1}{x + y}$

25. $\dfrac{x^2 - y^2}{y} \div \dfrac{x + y}{y^2}$

26. $\dfrac{x^2 - y^2}{x - y} \div \dfrac{x + y}{x}$

27. $\dfrac{x + 2xy}{3x^2} \div \dfrac{2y + 1}{6x}$

28. $x \div \dfrac{10x}{x + 5}$

29. $\dfrac{y}{x} \div \dfrac{xy}{xy - 1}$

30. $\dfrac{1}{x^2 - 9} \div \dfrac{1}{x^2 + 9}$

31. $\dfrac{x^2 - 4}{xy^2} \cdot \dfrac{2xy}{x^2 - 4x + 4}$

32. $\dfrac{(x - y)^2}{y + x} \cdot \dfrac{3y + 3x}{x^2 - y^2}$

33. $\dfrac{2y - 6}{y - 2x} \div \dfrac{4y - 12}{2y + 4x}$

34. $\dfrac{x^2}{x^2 - 1} \div \dfrac{x^3}{(1 - x)^2}$

35. $\dfrac{1 - x^2}{x + 1} \div \dfrac{(x - 1)^2}{x^2 - 1}$

36. $\dfrac{x - y}{x^2 + xy} \cdot \dfrac{x^2}{y^2 - xy}$

37. $\dfrac{6x - 12}{4xy + 4x} \cdot \dfrac{y^2 - 1}{2 - 3x + x^2}$

38. $\dfrac{x^3 - x}{x - 1} \div \dfrac{x^2 + x}{x^2 + 2x + 1}$

39. $(x^2 - 25) \cdot \dfrac{2x + 1}{x - 5}$

40. $\dfrac{x^2 - 36}{1 + x} \div \dfrac{6 - x}{x}$

41. $\dfrac{x^4 - 1}{x - 1} \cdot \dfrac{x + 1}{x^2 + 1}$

42. $\dfrac{12 - 6x}{3x + 9} \div \dfrac{4x - 8}{5x + 15}$

43. $\dfrac{(x - 1)^2}{(x + 1)^2} \cdot \dfrac{(x + 1)}{(1 - x)}$

44. $\dfrac{y^2 + 4y - 12}{y^2 + 2y - 8} \div \dfrac{y^2 + 7y + 6}{y^2 - 1}$

45. $\dfrac{x^4 + 5x^2 + 9}{x^2 - 9} \cdot \dfrac{x - 3}{x^2 - x + 3}$

46. $\dfrac{2y^2 + 9y + 10}{y^2 + 5y + 6} \cdot \dfrac{y^2 + 7y + 12}{2y^2 + 3y - 5}$

47. $\dfrac{x^3 - y^3}{3x + 3y} \div \dfrac{x^2 + xy + y^2}{x^2 - y^2}$

48. $\dfrac{15a^2 - ab - 2b^2}{15a^2 + 11ab + 2b^2} \div \dfrac{15a^2 + 4ab - 4b^2}{15a^2 + ab - 2b^2}$

49. $\dfrac{x + y}{x} \cdot \dfrac{(x - y)^2}{x} \cdot \dfrac{1}{x^2 - y^2}$

50. $\dfrac{1}{1 - x} \div \left(\dfrac{x}{1 - x} \cdot \dfrac{2}{x} \right)$

51. $\dfrac{x - x^2}{x + 1} \div \dfrac{2 - 2x}{6x} \cdot \dfrac{x^2 - 1}{9x^2}$

ADDITION AND SUBTRACTION OF RATIONAL EXPRESSIONS 149

52. $\dfrac{y^2 + 2y + 1}{y - 2} \cdot \dfrac{5y - 20}{y - 3} \cdot \dfrac{y^2 - 5y + 6}{y + 1}$

53. $\dfrac{2a^2 - a - 15}{a^2 - a - 20} \cdot \dfrac{a^2 + 12a + 35}{a^2 + 4a - 21} \cdot \dfrac{a - 5}{2a + 5}$

54. $\dfrac{3x - y}{y + x} \cdot \dfrac{4x^2 + 5xy + y^2}{y^2 + xy - 12x^2} \cdot \dfrac{10x^2 - 8xy}{8xy - 10x^2}$

55. $\dfrac{x^2 - x}{x - 3} \cdot \dfrac{x + 1}{x^2 + 4x} \div \dfrac{x^2 - 3x - 4}{x^2 - 16}$

56. $\dfrac{ac + bc + ad + bd}{ac - ad - bc + bd} \cdot \dfrac{c^2 - 2cd + d^2}{a^2 + 2ab + b^2} \cdot \dfrac{a - b}{c + d}$

57. $\dfrac{b^2 - a^2}{4ab - ab^2} \cdot \dfrac{2a^2 + ab - b^2}{2a - 2b} \div (a + b)$

58. $\dfrac{x - 1}{3 - x} \cdot \dfrac{3x^2 - 7x - 6}{2x^2 - x - 1} \cdot \dfrac{2x^2 - 9x - 5}{3x^2 - 13x - 10}$

ADDITION AND SUBTRACTION OF RATIONAL EXPRESSIONS

4.3

In arithmetic you learned that fractions can be combined by addition or subtraction only if their denominators are the same. For example,

$$\frac{2}{7} + \frac{4}{7} = \frac{2 + 4}{7} \text{ or } \frac{6}{7}$$

To combine fractions having different denominators, each must be written with a common denominator. The least common denominator (LCD) is the smallest number that all denominators will divide into exactly. Thus, the LCD of $\dfrac{5}{6}$ and $\dfrac{11}{15}$ is 30. To add these fractions, we first write each with a denominator of 30, then add as follows:

$$\frac{5}{6} + \frac{11}{15} = \frac{5 \cdot 5}{6 \cdot 5} + \frac{11 \cdot 2}{15 \cdot 2}$$

$$= \frac{25}{30} + \frac{22}{30}$$

$$= \frac{47}{30}$$

To Find the Least Common Denominator (LCD)

(1) Factor each denominator completely using exponential notation.
(2) Form a product of all different factors from each denominator, where each factor is raised to the highest power that occurs for it in any denominator.

EXAMPLE 1

Find the LCD for $\dfrac{7}{6m^2n}$ and $\dfrac{3}{4mn^3}$

Solution:

Step 1. Completely factor each denominator using exponents.
$6m^2n = 2 \cdot 3 \cdot m^2 \cdot n$
$4mn^3 = 2^2 \cdot m \cdot n^3$

Step 2. Form a product of all different factors. To each factor, apply the largest exponent that it has in any denominator.
LCD $= 2^2 \cdot 3 \cdot m^2 \cdot n^3$
$= 12m^2n^3$

EXAMPLE 2

Find the LCD for $\dfrac{x}{x^2 + 6x + 9}$ and $\dfrac{5}{2x^2 + 5x - 3}$

Solution:

Step 1. Factor completely using exponents.
$x^2 + 6x + 9 = (x + 3)^2$
$2x^2 + 5x - 3 = (2x - 1)(x + 3)$

Step 2. Form the product of different factors and apply the largest exponents.
LCD $= (x + 3)^2(2x - 1)$

Rational expressions are added or subtracted by first finding the least common denominator, then each fraction is rewritten using the LCD. These fractions are then combined using the following property:

Addition and Subtraction of Rational Expressions

If $a, b, c \in R$ and $c \neq 0$, then

$$\dfrac{a}{c} + \dfrac{b}{c} = \dfrac{a + b}{c} \quad \text{and} \quad \dfrac{a}{c} - \dfrac{b}{c} = \dfrac{a - b}{c}$$

This property states that rational expressions sharing a common denominator may be added or subtracted by simply combining the numerators. This result is then placed over the common denominator. The final answer should always be reduced to lowest terms.

EXAMPLE 3

Add $\dfrac{3x}{2y} + \dfrac{5x}{2y}$

ADDITION AND SUBTRACTION OF RATIONAL EXPRESSIONS

Solution: Since the denominators are the same, add the numerators, then reduce to lowest terms.

$$\frac{3x}{2y} + \frac{5x}{2y} = \frac{3x + 5x}{2y}$$

$$= \frac{8x}{2y}$$

$$= \frac{4x}{y}$$

EXAMPLE 4

Add $\dfrac{1}{6xy^2} + \dfrac{5}{3x^2y}$

Solution: The LCD is $6x^2y^2$. Rewrite each fraction using the LCD.

$$\frac{1}{6xy^2} + \frac{5}{3x^2y} = \frac{1 \cdot x}{6xy^2 \cdot x} + \frac{5 \cdot 2y}{3x^2y \cdot 2y}$$

$$= \frac{x}{6x^2y^2} + \frac{10y}{6x^2y^2}$$

$$= \frac{x + 10y}{6x^2y^2}$$

EXAMPLE 5

Subtract $\dfrac{p}{p + 3} - \dfrac{18}{p^2 - 9}$

Solution: Factor each denominator to obtain the LCD.

$$\frac{p}{p + 3} - \frac{18}{p^2 - 9} = \frac{p}{(p + 3)} - \frac{18}{(p + 3)(p - 3)}$$

The LCD is $(p + 3)(p - 3)$. Rewrite the first fraction:

$$= \frac{p \cdot (p - 3)}{(p + 3) \cdot (p - 3)} - \frac{18}{(p + 3)(p - 3)}$$

$$= \frac{p \cdot (p - 3) - 18}{(p + 3)(p - 3)}$$

$$= \frac{p^2 - 3p - 18}{(p + 3)(p - 3)}$$

Reduce to lowest terms.

$$= \frac{(p - 6)\cancel{(p + 3)}}{\cancel{(p + 3)}(p - 3)}$$

$$= \frac{p - 6}{p - 3}$$

EXAMPLE 6

Combine $\dfrac{3m - 1}{2m^2 - 5m - 3} - \dfrac{2m - 1}{2m + 1} + \dfrac{m}{3 - m}$

Solution: Factor each denominator.

$$\frac{3m-1}{2m^2-5m-3} - \frac{2m-1}{2m+1} + \frac{m}{3-m}$$

$$= \frac{3m-1}{(2m+1)(m-3)} - \frac{2m-1}{(2m+1)} + \frac{m}{(3-m)}$$

opposites

Multiply the numerator and denominator of the last fraction by -1.

$$= \frac{3m-1}{(2m+1)(m-3)} - \frac{2m-1}{(2m+1)} + \frac{-1 \cdot m}{-1 \cdot (3-m)}$$

$$= \frac{3m-1}{(2m+1)(m-3)} - \frac{2m-1}{(2m+1)} + \frac{-m}{(m-3)}$$

The LCD is $(2m+1)(m-3)$.

$$= \frac{3m-1}{(2m+1)(m-3)} - \frac{(2m-1)(m-3)}{(2m+1)(m-3)} + \frac{-m(2m+1)}{(m-3)(2m+1)}$$

Multiply out the numerator.

$$= \frac{3m-1-(2m-1)(m-3)-m(2m+1)}{(2m+1)(m-3)}$$

$$= \frac{3m-1-[2m^2-7m+3]-2m^2-m}{(2m+1)(m-3)}$$

$$= \frac{3m-1-2m^2+7m-3-2m^2-m}{(2m+1)(m-3)}$$

Combine like terms.

$$= \frac{-4m^2+9m-4}{(2m+1)(m-3)}$$

The numerator cannot be factored further. Thus, the fraction is in lowest terms.

Exercise 4.3

Add or subtract as indicated. Assume all denominators are not equal to zero.

1. $\dfrac{5x}{x+2} + \dfrac{10}{x+2}$

2. $\dfrac{x+1}{y} - \dfrac{x-1}{y}$

3. $\dfrac{5}{r} + \dfrac{3}{2r}$

4. $\dfrac{x}{3x} + \dfrac{1}{x^2}$

5. $\dfrac{x-2}{2} + \dfrac{4-x}{6}$

6. $\dfrac{3x-1}{10} + \dfrac{5-2x}{15}$

7. $\dfrac{x-2}{6} - \dfrac{x+1}{3}$

8. $\dfrac{1}{6xy^2} + \dfrac{3}{8x^2y}$

9. $\dfrac{5}{18x^2} + \dfrac{11}{12x^4}$

10. $\dfrac{9}{25a^3} + \dfrac{7}{15a}$

11. $\dfrac{1}{7y} - \dfrac{8}{y^2}$

12. $\dfrac{a}{b} - c$

13. $\dfrac{5x}{3y^2} + 3$

14. $\dfrac{-35x}{3-7x} - \dfrac{15}{7x-3}$

15. $\dfrac{3}{2(a-b)} - \dfrac{2}{5(a-b)}$

16. $\dfrac{5}{a-4} + \dfrac{2}{4-a}$

17. $\dfrac{2}{3-x} - \dfrac{1}{x-3}$

18. $\dfrac{1}{y-1} + \dfrac{1}{y}$

19. $\dfrac{5}{2x-6} - \dfrac{3}{x+3}$

20. $\dfrac{1}{3b-3} + \dfrac{b}{b-1}$

21. $\dfrac{5}{x^3-8} - \dfrac{3}{x^2+2x+4}$

22. $\dfrac{6}{y} + \dfrac{3y}{5-y}$

23. $\dfrac{a+b}{b^2} + \dfrac{1-a^2}{a-b}$

24. $\dfrac{x}{x+y} - \dfrac{y}{x-y}$

25. $\dfrac{x-2}{x-3} - \dfrac{x-4}{x+4}$

26. $\dfrac{1}{x^2-1} - \dfrac{1}{x^2+2x+1}$

27. $\dfrac{1}{2y+1} - \dfrac{3}{y-2}$

28. $\dfrac{3}{x^2-49} + \dfrac{2}{x^2+7x}$

29. $\dfrac{2}{b^2-b} + \dfrac{b-3}{b^3-b^2}$

30. $\dfrac{y-3}{y^2-y-2} - \dfrac{y-1}{y^2+2y+1}$

31. $\dfrac{x}{x^2-16} - \dfrac{x+1}{x^2-5x+4}$

32. $\dfrac{1}{x-2} - \dfrac{2x}{x^3-8}$

33. $\dfrac{2x}{x^2-16} + \dfrac{x+4}{x^2-3x-4}$

34. $\dfrac{x+1}{x+2} - \dfrac{x+2}{x+3}$

35. $a + 3 + \dfrac{a}{a-4}$

36. $\dfrac{1}{3b-3} - \dfrac{b^2+6}{b^2+3b-4}$

37. $\dfrac{x+1}{x^2-9} - \dfrac{x-2}{x^2+4x-21}$

38. $\dfrac{y+2}{y-3} - \dfrac{y+3}{y-2}$

39. $\dfrac{a-b}{a} - \dfrac{a}{a-b}$

40. $\dfrac{3}{4x} - \dfrac{2}{5y} + \dfrac{9}{20xy}$

41. $\dfrac{4}{b^3} - \dfrac{3}{8b^2} - \dfrac{5}{3b}$

42. $\dfrac{5}{y} - \dfrac{8}{y-6} - y$

43. $\dfrac{1}{t} + \dfrac{5}{t^2} - \dfrac{t+3}{t^2-1}$

44. $\dfrac{3}{x-2} - \dfrac{2}{x+2} - \dfrac{x}{x^2-4}$

45. $1 + \dfrac{1}{x-3} + \dfrac{3}{x^2+3x}$

46. $x + \dfrac{2}{x+2} + \dfrac{2}{x^2-4}$

47. $\dfrac{2}{y^2 - 1} + \dfrac{1}{1 + y} + \dfrac{1}{y^2 + y}$ 48. $\dfrac{3b}{b + 5} - \dfrac{3b}{b - 5} + \dfrac{150}{b^2 - 25}$

49. $\dfrac{6t + 18}{8t^2 + 6t - 5} + \dfrac{2}{1 - 2t} + \dfrac{3}{4t + 5}$

50. $\dfrac{15}{4x^2} - \dfrac{11x}{2x^2 - 50} + \dfrac{2x - 5}{x^2 - 10x + 25}$

51. $\dfrac{3}{y - 1} - \dfrac{y - 1}{y^2 + 2y + 1} - \dfrac{y + 1}{y^2 - 1}$

SIMPLIFYING COMPLEX FRACTIONS

4.4

A **complex fraction** is an expression containing at least one fraction in the numerator or denominator, or both.

EXAMPLE 1 Each of these expressions is a complex fraction.

(a) $\dfrac{-5}{\tfrac{2}{3}}$ (b) $\dfrac{x + \tfrac{1}{x}}{1 \cdot \tfrac{x}{2}}$ (c) $\dfrac{\tfrac{m^2 - 4}{m - 1}}{\tfrac{m + 2}{m^2 - 1}}$

To simplify a complex fraction means to write it as a common fraction reduced to lowest terms. This is accomplished by using the fundamental principle of fractions to multiply the numerator and denominator by the LCD of all fractions appearing within the complex fraction.

EXAMPLE 2 Simplify $\dfrac{-\tfrac{3}{5}}{\tfrac{7}{10}}$

Solution: Multiply the numerator and denominator by the LCD, 10.

$$\dfrac{-\tfrac{3}{5}}{\tfrac{7}{10}} = \dfrac{-\tfrac{3}{5} \cdot \overset{2}{\cancel{10}}}{\tfrac{7}{10} \cdot \cancel{10}}$$

$$= -\dfrac{6}{7}$$

EXAMPLE 3 Simplify $\dfrac{\tfrac{1}{m} - 1}{\tfrac{1}{m^2} - \tfrac{1}{m}}$

SIMPLIFYING COMPLEX FRACTIONS

Solution: Multiply the numerator and denominator by m^2.

$$\frac{\frac{1}{m} - 1}{\frac{1}{m^2} - \frac{1}{m}} = \frac{\left(\frac{1}{m} - 1\right)m^2}{\left(\frac{1}{m^2} - \frac{1}{m}\right)m^2}$$

Use the distributive property to multiply by m^2.

$$= \frac{\frac{1}{m} \cdot m^2 - 1 \cdot m^2}{\frac{1}{m^2} \cdot m^2 - \frac{1}{m} \cdot m^2}$$

$$= \frac{m - m^2}{1 - m}$$

Reduce to lowest terms.

$$= \frac{m\cancel{(1 - m)}}{\cancel{(1 - m)}}$$

$$= m$$

EXAMPLE 4

Simplify $\dfrac{\dfrac{4}{a^2 - b^2}}{\dfrac{1}{(a - b)}}$

Solution: Factor the denominator to obtain the LCD.

$$\frac{\dfrac{4}{a^2 - b^2}}{\dfrac{1}{a - b}} = \frac{\dfrac{4}{(a + b)(a - b)}}{\dfrac{1}{(a - b)}}$$

Multiply the numerator and denominator by $(a + b)(a - b)$.

$$= \frac{\dfrac{4}{\cancel{(a + b)(a - b)}} \cdot \cancel{(a + b)}\cancel{(a - b)}}{\dfrac{1}{\cancel{(a - b)}} \cdot (a + b)\cancel{(a - b)}}$$

$$= \frac{4}{a + b}$$

Sometimes a complex fraction will include other complex fractions. Such expressions are called continued complex fractions. Each included complex fraction must be simplified before simplifying the entire expression.

EXAMPLE 5

Simplify $\dfrac{\dfrac{1}{n}}{1 + \dfrac{1}{1 + \dfrac{1}{n}}}$

Solution: First we simplify the included complex fraction, $\dfrac{1}{1 + \dfrac{1}{n}}$.

$$\dfrac{\dfrac{1}{n}}{1 + \boxed{\dfrac{1}{1 + \dfrac{1}{n}}}} \leftarrow \left[\begin{array}{l}\text{simplify first by multiplying}\\ \text{numerator and denominator by } n.\end{array}\right]$$

$$\dfrac{\dfrac{1}{n}}{1 + \dfrac{1}{1 + \dfrac{1}{n}}} = \dfrac{\dfrac{1}{n}}{1 + \dfrac{1 \cdot n}{\left(1 + \dfrac{1}{n}\right) \cdot n}}$$

$$= \dfrac{\dfrac{1}{n}}{1 + \dfrac{n}{n + \dfrac{1}{\cancel{n}} \cdot \cancel{n}}}$$

$$= \dfrac{\dfrac{1}{n}}{1 + \dfrac{n}{n + 1}}$$

Now simplify the remaining complex fraction by multiplying the numerator and denominator by $n(n + 1)$.

$$= \dfrac{\dfrac{1}{\cancel{n}} \cdot \cancel{n}(n + 1)}{\left(1 + \dfrac{n}{n + 1}\right) \cdot n(n + 1)}$$

$$= \dfrac{n + 1}{1 \cdot n(n + 1) + \dfrac{n}{\cancel{(n + 1)}} \cdot \cancel{n(n+1)}}$$

$$= \dfrac{n + 1}{n(n + 1) + n^2}$$

SIMPLIFYING COMPLEX FRACTIONS

$$= \frac{n+1}{n^2 + n + n^2}$$

$$= \frac{n+1}{2n^2 + n} \text{ or } \frac{n+1}{n(2n+1)}$$

Simplify the following complex fractions.

Exercise 4.4

1. $\dfrac{\frac{3}{5}}{\frac{2}{3}}$

2. $\dfrac{1 + \frac{2}{3}}{3 - \frac{1}{3}}$

3. $\dfrac{\frac{1}{2} + \frac{1}{3}}{\frac{1}{6} - \frac{3}{2}}$

4. $\dfrac{1}{1 - \frac{4}{x}}$

5. $\dfrac{\frac{3xy}{4}}{\frac{3y}{8x^2}}$

6. $\dfrac{\frac{1}{x^2}}{1 - \frac{1}{x}}$

7. $\dfrac{x - \frac{x}{y}}{1 + \frac{1}{y}}$

8. $\dfrac{a - 1}{1 - \frac{2}{a^2}}$

9. $\dfrac{b}{\frac{b}{b-1}}$

10. $\dfrac{3 + \frac{2}{x}}{5 - \frac{1}{x}}$

11. $\dfrac{1 + \frac{1}{x}}{1 - \frac{1}{x}}$

12. $\dfrac{\frac{2}{a} - 3}{3 + \frac{2}{a}}$

13. $\dfrac{1 + \frac{2}{x}}{4 - \frac{4}{x}}$

14. $\dfrac{a - 3}{a - \frac{9}{a}}$

15. $\dfrac{\frac{x - 2y}{5x}}{\frac{8x - 16y}{10}}$

16. $\dfrac{\frac{2}{y} + 1}{1 + \frac{2}{y}}$

17. $\dfrac{\frac{x}{y} + 1}{\frac{x}{y} - 1}$

18. $\dfrac{x^2 - \frac{1}{x}}{x^2 - \frac{1}{x^2}}$

19. $\dfrac{\frac{x+y}{3x^2}}{\frac{x-y}{x}}$

20. $\dfrac{\frac{x^2 - 4}{2x + 1}}{\frac{x^2 + x - 2}{2x + 1}}$

21. $\dfrac{1 - \frac{7}{y+1}}{\frac{4}{y+1} + 1}$

22. $\dfrac{\frac{2}{x} + \frac{3}{2x}}{5 + \frac{1}{x}}$

23. $\dfrac{\frac{a}{b} + 2}{\frac{a^2}{b^2} - 4}$

24. $\dfrac{\frac{x}{4} - \frac{1}{x}}{1 + \frac{x+4}{x}}$

25. $\dfrac{x - \dfrac{x-3}{3}}{\dfrac{4}{9} + \dfrac{2}{3x}}$

26. $\dfrac{\dfrac{1}{y} - \dfrac{1}{1+y}}{\dfrac{1}{y} + 1}$

27. $\dfrac{3 - \dfrac{x}{y}}{9 - \dfrac{x^2}{y^2}}$

28. $\dfrac{\dfrac{9x^2}{5y^2} - 5}{\dfrac{3x}{y} + 5}$

29. $\dfrac{x + y}{\dfrac{1}{x} + \dfrac{1}{y}}$

30. $\dfrac{\dfrac{1}{a} + \dfrac{1}{b}}{\dfrac{1}{a} - \dfrac{1}{b}}$

31. $\dfrac{\dfrac{b}{a} - \dfrac{a}{b}}{\dfrac{1}{b} + \dfrac{1}{a}}$

32. $\dfrac{\dfrac{1}{2y} + \dfrac{3}{4y}}{\dfrac{1}{3y} + \dfrac{1}{6y}}$

33. $\dfrac{\dfrac{x}{x+3} - \dfrac{1}{x}}{3 - \dfrac{1}{x+3}}$

34. $\dfrac{\dfrac{1}{a} - \dfrac{1}{ab}}{1 - \dfrac{1}{b^2}}$

35. $\dfrac{\dfrac{1}{(x+a)^2} - \dfrac{1}{x^2}}{a}$

36. $\dfrac{\dfrac{a-b}{a} - \dfrac{a}{a+b}}{\dfrac{b^2}{a+b}}$

37. $\dfrac{\dfrac{1}{xy} + \dfrac{2}{yz} + \dfrac{3}{xz}}{\dfrac{2x + 3y + z}{xyz}}$

38. $\dfrac{x + 2 - \dfrac{12}{x+3}}{x - 5 + \dfrac{16}{x+3}}$

39. $x - \dfrac{x}{x + \dfrac{1}{2}}$

40. $1 + \dfrac{1}{1 + \dfrac{x}{y}}$

41. $\dfrac{\dfrac{1}{y}}{1 - \dfrac{y}{1 + \dfrac{1}{y}}}$

42. $\dfrac{1}{1 + \dfrac{1}{1 + \dfrac{1}{x}}}$

43. $\dfrac{1}{1 - \dfrac{1}{1 - \dfrac{1}{x}}}$

44. $\dfrac{\dfrac{1}{n}}{1 + \dfrac{1}{n - \dfrac{1}{n}}}$

EQUATIONS AND FORMULAS CONTAINING RATIONAL EXPRESSIONS

4.5

To solve an equation containing rational expressions, we clear the equation of fractions by multiplying both sides by the least common denominator. When applying this technique, you must be careful to use the distributive property and multiply the LCD times each term of the equation.

EXAMPLE 1 Solve $\dfrac{5y}{3} + 5 = \dfrac{y}{2} - 2$

EQUATIONS AND FORMULAS CONTAINING RATIONAL EXPRESSIONS 159

Solution: The LCD is 6; multiply it times both sides.

$$\frac{5y}{3} + 5 = \frac{y}{2} - 2$$

$$6\left[\frac{5y}{3} + 5\right] = 6\left[\frac{y}{2} - 2\right]$$

Use the distributive property and multiply each term by 6.

$$6 \cdot \frac{5y}{3} + 6 \cdot 5 = 6 \cdot \frac{y}{2} - 6 \cdot 2$$

$$10y + 30 = 3y - 12$$

The equation has been cleared of fractions. Solve for y.

$$7y = -42$$

$$y = -6$$

Check this solution by replacing it for y in the original equation.

$$\frac{5y}{3} + 5 = \frac{y}{2} - 2$$

$$\frac{5(-6)}{3} + 5 = \frac{-6}{2} - 2$$

$$-10 + 5 = -3 - 2$$

$$-5 = -5$$

As illustrated by the following examples, whenever both sides of an equation are multiplied by an expression containing the variable, the resulting answers may or may not be actual solutions. Answers that do not check in the original equation are called *extraneous solutions*.

EXAMPLE 2

Solve $\dfrac{2x}{x+1} + 1 = \dfrac{-2}{x+1}$

Solution: The LCD is $x + 1$; multiply it times both sides.

$$\frac{2x}{(x+1)} + 1 = \frac{-2}{(x+1)}$$

$$(x+1) \cdot \left[\frac{2x}{(x+1)} + 1\right] = (x+1) \cdot \left[\frac{-2}{(x+1)}\right]$$

Use the distributive property.

$$\cancel{(x+1)} \cdot \frac{2x}{\cancel{(x+1)}} + (x+1) \cdot 1 = \cancel{(x+1)} \cdot \frac{-2}{\cancel{(x+1)}}$$

$$2x + x + 1 = -2$$

$$3x = -3$$

$$x = -1$$

This solution will not check since it turns denominators into zero.

$$\frac{2x}{x+1} + 1 = \frac{-2}{x+1}$$

$$\frac{0}{-1+1} + 1 \neq \frac{-2}{-1+1}$$

$$\frac{0}{0} + 1 \neq \frac{-2}{0}$$

-1 is an extraneous solution. This equation has no solutions.

Remember, whenever the LCD contains a variable, the answers must be checked in the original equation to eliminate extraneous solutions.

EXAMPLE 3

Solve $\dfrac{2m}{m+2} = \dfrac{7m+17}{m^2+5m+6} + \dfrac{m-1}{m+3}$

Solution: Factor denominators to determine the LCD.

$$\frac{2m}{m+2} = \frac{7m+17}{m^2+5m+6} + \frac{m-1}{m+3}$$

$$\frac{2m}{(m+2)} = \frac{7m+17}{(m+3)(m+2)} + \frac{m-1}{(m+3)}$$

The LCD is $(m+3)(m+2)$; multiply it times both sides.

$$(m+3)(m+2) \cdot \left[\frac{2m}{(m+2)}\right] = (m+3)(m+2) \cdot \left[\frac{7m+17}{(m+3)(m+2)} + \frac{m-1}{(m+3)}\right]$$

$$(m+3)\cancel{(m+2)} \cdot \frac{2m}{\cancel{(m+2)}} = \cancel{(m+3)(m+2)} \cdot \frac{7m+17}{\cancel{(m+3)(m+2)}} + \cancel{(m+3)}(m+2) \cdot \frac{m-1}{\cancel{(m+3)}}$$

$$(m+3) \cdot 2m = 7m + 17 + (m+2)(m-1)$$

$$2m^2 + 6m = 7m + 17 + m^2 + m - 2$$

$$2m^2 + 6m = m^2 + 8m + 15$$

$$m^2 - 2m - 15 = 0$$

Solve this equation by factoring.

$$(m+3)(m-5) = 0$$

$m + 3 = 0$ or $m - 5 = 0$

$m = -3$ or $m = 5$

Checking these two answers in the original equation shows that -3 is extraneous since it turns the last two denominators into zero. Thus, the only solution is 5.

EXAMPLE 4

Solve $\dfrac{3}{p^2} + \dfrac{5}{p} = 2$

Solution: Multiply both sides by the LCD, p^2.

$$\frac{3}{p^2} + \frac{5}{p} = 2$$

$$p^2 \cdot \left[\frac{3}{p^2} + \frac{5}{p}\right] = p^2 \cdot [2]$$

$$p^2 \cdot \frac{3}{p^2} + p^2 \cdot \frac{5}{p} = 2p^2$$

$$3 + 5p = 2p^2$$

$$0 = 2p^2 - 5p - 3$$

Solve by factoring.

$$0 = (2p + 1)(p - 3)$$

$$2p + 1 = 0 \quad \text{or} \quad p - 3 = 0$$

$$p = -\frac{1}{2} \quad \text{or} \quad p = 3$$

Both of these answers check in the original equation. The solutions are $-\frac{1}{2}$ and 3. There are no extraneous solutions.

In section 2.5, Working with Formulas, you learned to rearrange formulas by solving for a specified variable. Many formulas contain rational expressions and can be rearranged using the techniques of this section.

EXAMPLE 5

A gas law in chemistry states that $\frac{P_1V_1}{T_1} = \frac{P_2V_2}{T_2}$. Solve this formula for T_2.

Solution: The LCD is T_1T_2; multiply it times both sides. (The subscripts 1 and 2 are used to indicate different variables.)

$$\frac{P_1V_1}{T_1} = \frac{P_2V_2}{T_2}$$

$$\cancel{T_1}T_2 \cdot \frac{P_1V_1}{\cancel{T_1}} = T_1\cancel{T_2} \cdot \frac{P_2V_2}{\cancel{T_2}}$$

$$T_2P_1V_1 = T_1P_2V_2$$

$$\frac{T_2P_1V_1}{P_1V_1} = \frac{T_1P_2V_2}{P_1V_1} \qquad \text{[divide both sides by } P_1V_1\text{]}$$

$$T_2 = \frac{T_1P_2V_2}{P_1V_1}$$

EXAMPLE 6

Solve $R = \frac{a}{a + r} - b$ for r.

162 RATIONAL EXPRESSIONS

Solution: Multiply both sides by the LCD, $a + r$.

$$R = \frac{a}{a+r} - b$$

$$(a+r)R = (a+r)\left[\frac{a}{a+r} - b\right]$$

$$(a+r)R = \cancel{(a+r)} \cdot \frac{a}{\cancel{(a+r)}} - (a+r)b$$

$$(a+r)R = a - (a+r)b$$

$$aR + rR = a - ab - rb$$

Obtain all terms containing an r on one side.

$$rR + rb = a - ab - aR$$

Remove the common factor, r.

$$r(R+b) = a - ab - aR$$

Divide both sides by $(R+b)$.

$$\frac{r(R+b)}{R+b} = \frac{a - ab - aR}{R+b}$$

$$r = \frac{a - ab - aR}{R+b}$$

Exercise 4.5   Solve.

1. $\dfrac{3}{x} = \dfrac{5}{x} + 1$

2. $x = 2 + \dfrac{x^2 - 4}{x}$

3. $\dfrac{3x - 6}{x - 2} = 3$

4. $\dfrac{y + 4}{3} - 7 = 3 - \dfrac{y + 2}{4}$

5. $\dfrac{a}{a - 3} - 4 = \dfrac{a}{a - 3}$

6. $\dfrac{5}{a + 1} = \dfrac{2}{11}$

7. $-\dfrac{2a - 11}{a - 2} = \dfrac{7}{a - 2}$

8. $\dfrac{1}{2y} = \dfrac{2}{y} - \dfrac{3}{8}$

9. $\dfrac{5y}{y + 2} = \dfrac{4y - 3}{2}$

10. $\dfrac{b}{b - 1} = \dfrac{b + 1}{b}$

11. $\dfrac{2x + 3}{2} + \dfrac{3x}{x - 1} = x$

12. $\dfrac{y + 1}{y - 2} = \dfrac{7}{4}$

13. $\dfrac{4}{a + 1} = \dfrac{3}{a} + \dfrac{1}{15}$

14. $\dfrac{1}{p + 1} = \dfrac{1}{2} + \dfrac{3}{p}$

15. $\dfrac{5}{t + 2} = \dfrac{3}{t - 2}$

16. $\dfrac{7}{b + 5} = \dfrac{3}{b - 1} - \dfrac{4}{b}$

17. $\dfrac{4}{p + 2} - \dfrac{5}{p - 2} = \dfrac{p}{p^2 - 4}$

18. $\dfrac{6}{y^2 - 9} = \dfrac{1}{y - 3} - \dfrac{1}{5}$

19. $\dfrac{t+2}{2t+8} = \dfrac{1}{2} - \dfrac{t}{t+4}$

20. $\dfrac{y-7}{y^2-2y} = \dfrac{y}{y-2} - \dfrac{y+4}{y}$

21. $\dfrac{x-1}{x^2-4} - \dfrac{2}{x+2} = \dfrac{4}{x-2}$

22. $\dfrac{1}{c-1} + \dfrac{2}{3c-3} = \dfrac{-5}{2}$

23. $\dfrac{12}{x^2-9} + \dfrac{3}{x+3} = \dfrac{2}{x-3}$

24. $\dfrac{4}{a} - \dfrac{6}{a-7} = \dfrac{2}{3a}$

25. $\dfrac{5y}{2y-12} = \dfrac{5}{12}$

26. $\dfrac{8}{x-4} + \dfrac{2x}{4-x} = 1$

27. $\dfrac{3}{x-4} = \dfrac{2}{4-x} + \dfrac{1}{x-2}$

28. $\dfrac{2}{x-2} - 3 = \dfrac{4}{2-x}$

29. $\dfrac{2}{a^2+a-12} = \dfrac{1}{a+4} + \dfrac{1}{a-3}$

30. $\dfrac{2}{y} + \dfrac{3}{2-y} = \dfrac{-3}{y^2-2y}$

31. $\dfrac{t^2-10}{t^2-t-20} = 1 + \dfrac{7}{t-5}$

32. $\dfrac{2x}{3x+3} - \dfrac{x-6}{8x+8} = \dfrac{x+2}{6x+6} + \dfrac{5}{12}$

33. $\dfrac{3y}{3y+15} + \dfrac{3y}{2y+10} = \dfrac{y+3}{4y+20} - \dfrac{7}{12}$

34. $\dfrac{1}{x} - \dfrac{5}{x^2+3x} + \dfrac{7-x}{x^2-x-12} = 0$

35. $\dfrac{x+11}{x^2+x-6} = \dfrac{5-3x}{x^2+4x+3} + \dfrac{4x-1}{x^2-x-2}$

Solve for the indicated letter. Assume denominators do not equal zero.

36. $\dfrac{1}{x} = \dfrac{1}{y}$ for x

37. $y = \dfrac{1}{x-1}$ for x

38. $C = \dfrac{5}{9}(F - 32)$ for F

39. $A = P(1 + rt)$ for r

40. $s = a + (n - 1)d$ for n

41. $\dfrac{1}{r} = \dfrac{1}{s} + \dfrac{2}{t}$ for r

42. $\dfrac{1}{r} = \dfrac{1}{s} + \dfrac{2}{t}$ for t

43. $R = \dfrac{a}{a+r} - b$ for a

44. $R = \dfrac{a}{a-r} - b$ for r

45. $\dfrac{a}{b-x} = \dfrac{b}{a-x}$ for x

46. $d = \dfrac{r}{1+rt}$ for r

47. $S = \dfrac{a-rt}{1-r}$ for r

48. $\dfrac{1}{r} = \dfrac{1}{r_1} + \dfrac{1}{r_2}$ for r

49. $\dfrac{t-a}{b} + \dfrac{t-b}{a} = 2$ for t

50. $\dfrac{P_1V_1}{T_1} = \dfrac{P_2V_2}{T_2}$ for T_1

51. $C = \dfrac{a}{1 + \dfrac{a}{\pi A}}$ for A

164   RATIONAL EXPRESSIONS

52. $\dfrac{1}{p} = \dfrac{1}{1 + \dfrac{1}{s}}$ for s **53.** $\dfrac{1}{p} = \dfrac{1}{1 + \dfrac{1}{s}}$ for p

54. $R = \dfrac{V_1 + V_2}{1 + \dfrac{V_1 V_2}{c^2}}$ for V_1

APPLIED PROBLEMS INVOLVING WORK AND MOTION

4.6

In this section we concentrate on the following three types of applications, each described by an equation containing rational expressions.

1. Problems using numerical relationships.
2. Applications involving work and time.
3. Problems involving distance, rate, and time.

The first example illustrates a word problem involving numbers.

EXAMPLE 1

One number is twice another. The sum of their reciprocals is $\dfrac{3}{10}$. Find the numbers.

Solution: Let x represent one of the numbers. The other number is twice x and is represented by $2x$. The corresponding reciprocals are $\dfrac{1}{x}$ and $\dfrac{1}{2x}$. Since their sum is $\dfrac{3}{10}$, we write this equation:

$$\frac{1}{x} + \frac{1}{2x} = \frac{3}{10}$$

Solving, we multiply both sides by the LCD, $10x$.

$$10x \cdot \left[\frac{1}{x} + \frac{1}{2x}\right] = 10x \cdot \left[\frac{3}{10}\right]$$

$$10x \cdot \frac{1}{x} + 10x \cdot \frac{1}{2x} = 10x \cdot \frac{3}{10}$$

$$10 + 5 = 3x$$

$$15 = 3x$$

$$5 = x$$

Since one number is 5, the other is $2 \cdot 5$ or 10. Checking, we add their reciprocals to obtain $\dfrac{3}{10}$.

$$\frac{1}{5} + \frac{1}{10} = \frac{2}{10} + \frac{1}{10} = \frac{3}{10}$$

APPLIED PROBLEMS INVOLVING WORK AND MOTION

The next example illustrates a problem requiring different amounts of time to complete a job.

EXAMPLE 2 Two pumps are used to fill an oil storage tank. One pump, working alone, takes 10 hours to fill the tank. The second pump, working alone, takes 15 hours to fill the tank. How long will it take both pumps working together to fill the tank?

Solution: Let x represent the number of hours to fill the tank when both pumps work together. In one hour both pumps together will fill $\frac{1}{x}$ of the tank. In one hour working alone, one pump will fill $\frac{1}{10}$ of the tank and the other will fill $\frac{1}{15}$ of the tank. The portion of the tank filled in one hour by both pumps $\left(\frac{1}{x}\right)$ must equal the sum of the portions filled by each pump in one hour $\left(\frac{1}{10} + \frac{1}{15}\right)$. Thus, we form the following equation:

$$\frac{1}{x} = \frac{1}{10} + \frac{1}{15}$$

Solving, we multiply both sides by the LCD, $30x$.

$$30x \cdot \frac{1}{x} = 30x \cdot \frac{1}{10} + 30x \cdot \frac{1}{15}$$

$$30 = 3x + 2x$$

$$30 = 5x$$

$$6 = x$$

Thus, to fill the storage tank the two pumps must work together for 6 hours.

The remaining examples utilize the distance formula, $d = rt$.

EXAMPLE 3 John made a trip at an average speed of 45 miles per hour. Sally, being somewhat of a speeder, made the same trip in one hour less time at an average speed of 60 miles per hour. What was the distance of the trip?

Solution: Let t represent the time necessary for John to make the trip. $(t - 1)$ would then represent the time necessary for Sally to make the same trip. According to the distance formula, distance equals rate multiplied by time. Thus, $50t$ would be the distance traveled by John while $60(t - 1)$ would be the distance traveled by Sally. This information is summarized in the following table:

RATIONAL EXPRESSIONS

	Rate	Time	Distance $d = rt$
John	45	t	$45t$
Sally	60	$(t-1)$	$60(t-1)$

Since the distance was the same for each person, we write and solve this equation:

$45t = 60(t - 1)$

$45t = 60t - 60$

$-15t = -60$

$t = 4$ hours

The distance traveled was represented by either $45t$ or $60(t - 1)$. Substituting 4 for t in either expression, we find that the distance was 180 miles.

EXAMPLE 4 A train travels 30 miles per hour faster than a car. The train travels 120 miles in the same time that the car travels 80 miles. Find the average speed for each.

Solution: Let x represent the rate of the car; then $x + 30$ represents the rate of the train. Since the time for the car and for the train is the same, we solve the distance formula, $d = rt$, for t.

$d = rt$

$\dfrac{d}{r} = \dfrac{rt}{r}$

$\dfrac{d}{r} = t$

The time for the car is $t = \dfrac{d}{r} = \dfrac{80}{x}$, while the time for the train is $t = \dfrac{d}{t} = \dfrac{120}{x + 30}$. This information is summarized in the following table:

	Distance	Rate	Time $t = \dfrac{d}{r}$
Car	80	x	$\dfrac{80}{x}$
Train	120	$x + 30$	$\dfrac{120}{x + 30}$

Since the car and train travel for the same time, we write and solve this equation:

$$\frac{80}{x} = \frac{120}{x + 30}$$

The LCD is $x(x + 30)$.

$$\cancel{x}(x + 30) \cdot \frac{80}{\cancel{x}} = x\cancel{(x + 30)} \cdot \frac{120}{\cancel{(x + 30)}}$$

$$(x + 30) \cdot 80 = x \cdot 120$$

$$80x + 2400 = 120x$$

$$2400 = 40x$$

$$60 = x$$

The rate of the car is 60 miles per hour, while the rate of the train is $x + 30$ or 90 miles per hour.

Exercise 4.6

Solve the following problems.

1. The denominator of a fraction is 10 more than the numerator. If 1 is added to both numerator and denominator, the value of the fraction is $\frac{2}{3}$. What is the fraction?

2. What number should be added to both the numerator and the denominator of $\frac{8}{11}$ to form the fraction $\frac{6}{7}$?

3. One number is twice another. The sum of their reciprocals is $\frac{5}{6}$. Find the two numbers.

4. A certain number, when divided by 3 more than the number, gives 4. What is the number?

5. Two numbers differ by 22. One-eighth the larger exceeds one-sixth the smaller by 2. Find the numbers.

6. The reciprocal of a number added to the reciprocal of twice the number is equal to the reciprocal of 2. Find the number.

7. What number must be added to the denominator of $\frac{15}{13}$ to make the result equal to $\frac{5}{3}$?

8. If one-half the reciprocal of a number is added to one-eighth, the result is the reciprocal of seven times the number. Find the number.

9. The sum of the reciprocals of two consecutive integers equals nine divided by the product of the two integers. Find the integers.

10. One train travels 30 miles per hour faster than a second train. The faster train travels 160 miles in the same time the slower train travels 100 miles. Find the rate of each.

11. A plane whose speed is 100 miles per hour faster than a second plane travels 1200 miles in the same time a second plane travels 900 miles. Find the rate of the faster plane.

12. Bob takes 30 minutes to drive to school in the morning but 45 minutes to return home in the afternoon because of the traffic. His rate in the morning is 10 miles per hour faster than his afternoon rate. How far is it from home to school?

13. A bus trip of 300 miles would have taken an hour less if the average speed had been increased by 10 miles per hour. Find the average speed.

14. Tom figures that if he increases his normal cycling speed by 3 miles per hour he will take 1 minute less to cycle 1 mile. Find his normal speed.

15. The difference in the rates of two trains is 13 miles per hour. The slower train takes 2 hours more to travel 168 miles than the faster train takes to travel 164 miles. Find the rate of the slower train.

16. An airplane with an air speed of 450 miles per hour requires 30 minutes longer to fly from city A to city B against a 30 miles per hour wind than from city B to city A with the wind. Find the distance between the two cities. Hint: The ground speed with the wind is 480 miles per hour and against the wind 420 miles per hour.

17. A jet airliner made a trip at an average speed of 400 miles per hour. A second jet airliner made the same trip in two hours less time at an average speed of 600 miles per hour. What was the distance of the trip?

18. Jack can go 15 miles downstream in his boat in the same time as he can go 10 miles upstream. If the river has a current of 4 miles per hour, find the speed of Jack's boat in still water.

19. Edith can address envelopes for a company in 4 hours. Lynda can do the same job in 3 hours. Working together, how long would it take them?

20. A certain man can paint a house in 4 days. A second man can paint the house in 5 days. Working together, how long would it take them to paint the house?

21. One pipe can fill a tank in 30 hours while a second pipe can fill the same tank in 45 hours. How long will it take to fill the tank if both pipes are running together?

22. Bob can do a job alone in 3 hours. If Art is allowed to assist him, the job can be finished in 1 hour 20 minutes. How long does it take Art to do the job alone?

23. Pipe A can fill a tank in 7 hours. Pipes A and B can fill the tank in 4 hours. How long does it take pipe B alone to fill the tank?

24. Joe can mow a lawn in 1 hour. Art can mow the lawn in 80 minutes. If they each use a lawnmower, how long will it take them together to mow the lawn?

25. One pipe can empty a tank in 2 hours and another pipe can empty the tank in 4 hours. How long will it take both pipes to empty the tank?

26. A storage tank has 2 inflow pipes and 1 outflow pipe. Each inflow pipe fills the tank in 3 hours. The ouflow pipe empties the tank in 5 hours. If all three pipes are open, how long will it take to fill the tank?

27. Two numbers are in the ratio of $\frac{6}{11}$. If the first is decreased by 4 and the second increased by 6, the resulting numbers are in the ratio of $\frac{4}{9}$. Find the two numbers.

Summary

KEY WORDS, PHRASES, AND PROPERTIES

Rational Expression—an expression made up of a quotient of two polynomials where the denominator is not zero.

Fundamental Principle of Fractions

If $a, b, c \in R$ and $b, c \neq 0$, then $\frac{a}{b} = \frac{ac}{bc}$

Multiplication of Rational Expressions

If $a, b, c, d \in R$ and $b, d \neq 0$, then $\frac{a}{b} \cdot \frac{c}{d} = \frac{ac}{bd}$

Division of Rational Expressions

If $a, b, c, d \in R$ and $b, c, d \neq 0$, then $\frac{a}{b} \div \frac{c}{d} = \frac{a}{b} \cdot \frac{d}{c}$

Addition and Subtraction of Rational Expressions

If $a, b, c \in R$ and $c \neq 0$, then $\frac{a}{c} + \frac{b}{c} = \frac{a + b}{c}$ and $\frac{a}{c} - \frac{b}{c} = \frac{a - b}{c}$

Finding the Least Common Denominator (LCD)
1. Factor completely each denominator using exponential notation.
2. Form a product of all different factors from each denominator and raise each factor to the highest power that occurs for it in any denominator.

Self-Checking Exercise

Give the domain of these rational expressions.

1. $\dfrac{x^2 + 9}{x^2 - 9}$
2. $\dfrac{2x - 1}{2x^2 + 5x - 3}$
3. $\dfrac{-6}{4x^2 - 25}$
4. $\dfrac{y^2 + 3y - 10}{6y^2 - 8y + 2}$

Reduce to lowest terms. Assume denominators are not equal to zero.

5. $\dfrac{6x + 12}{8x + 16}$
6. $\dfrac{7x - 21}{6 - 2x}$
7. $\dfrac{(x + y)^2}{x^2 - y^2}$
8. $\dfrac{y^2 - 1}{6 - 5y - y^2}$
9. $\dfrac{x^3 + 4x}{x^4 - 16}$
10. $\dfrac{6x^2 + 3x - 30}{3x^2 - 15x + 18}$
11. $\dfrac{a^2 + 3a - 10}{a^2 - 5a + 6}$
12. $\dfrac{6x^2 + 5x - 4}{4 - 3x - 10x^2}$

Multiply or divide as indicated. Assume denominators are not equal to zero.

13. $\dfrac{x^2 - 16}{x} \cdot \dfrac{2x}{4 - x}$
14. $\dfrac{6a^2b^3}{4a^3b} \cdot \dfrac{18a^2b^3}{12a^3b^2}$
15. $\dfrac{a^2 - 4}{16} \div \dfrac{2 - a}{8a}$
16. $\dfrac{x^3 + y^3}{x - y} \div \dfrac{(x + y)^2}{x^2 - y^2}$
17. $\dfrac{x^2 - 16}{5x - 15} \cdot \dfrac{10(x - 3)^2}{x^2 - 7x + 12}$
18. $\dfrac{y^2 - 5y + 6}{y - 1} \cdot \dfrac{y^2 - 2y + 1}{y - 2} \cdot \dfrac{5y - 20}{y - 3}$

Add or subtract as indicated. Assume denominators are not equal to zero.

19. $\dfrac{7}{24a^2b} + \dfrac{1}{6ab^3}$
20. $\dfrac{50}{9x - 63} - \dfrac{16}{3x}$
21. $\dfrac{x - y}{x} - \dfrac{x}{x + y}$
22. $\dfrac{x + 3}{x - 3} + \dfrac{3}{x^2 - 6x + 9}$
23. $\dfrac{4y - 16}{y^2 + y - 12} + \dfrac{3}{y^2 - 16}$
24. $\dfrac{2a}{a^2 - 16} - \dfrac{a + 1}{a^2 - 5a + 4}$
25. $\dfrac{3x}{x + 5} - \dfrac{3x}{x - 5} + \dfrac{3x}{x^2 - 25}$
26. $\dfrac{a^2}{9a^2 - 4} + \dfrac{9a}{3a + 2} + \dfrac{2a}{9a^2 + 12a + 4}$

Simplify. Assume denominators are not equal to zero.

27. $\dfrac{\dfrac{x}{y} + 3}{\dfrac{x^2}{y^2} - 9}$
28. $\left(\dfrac{2x^4}{3y^3}\right)^4$

29. $\dfrac{3}{x} \cdot \dfrac{4}{x^2} \div \dfrac{-2}{2-x}$

30. $\left(\dfrac{x^2 - y^2}{x - y}\right)^2$

31. $(2x - 4) \div \dfrac{4x - 8}{3x}$

32. $\dfrac{4x - 8}{3x} \div (2x - 4)$

Solve for the unknown.

33. $\dfrac{4}{x - 5} = \dfrac{3}{x - 3}$

34. $\dfrac{1}{5} = 5 - \dfrac{5x + 2}{x - 1}$

35. $\dfrac{1}{10x} - \dfrac{1}{10} = \dfrac{1}{x} - \dfrac{1}{5}$

36. $\dfrac{3}{x - 3} = 3 + \dfrac{x}{x - 3}$

37. $\dfrac{2}{x + 1} + \dfrac{1}{3x + 3} = \dfrac{1}{6}$

38. $\dfrac{a + 5}{2a - 2} - \dfrac{2a + 1}{a - 1} = \dfrac{a - 3}{2}$

Solve for the indicated letter.

39. $V = k + gt$; solve for t

40. $S = \dfrac{a}{1 - r}$; solve for r

41. $A = \dfrac{h}{2}(b + c)$; solve for c

42. $\dfrac{2}{x} + \dfrac{2}{a} = 2$; solve for x

Solve the following problems.

43. The difference of two numbers is 64 and their quotient is 17. Find the numbers.

44. Find two consecutive positive even integers such that the sum of their reciprocals is $\dfrac{7}{24}$.

45. If a number x is added to both the numerator and the denominator of $\dfrac{1}{5}$, the new fraction formed is $\dfrac{1}{2}$. Find the value of x.

46. One train travels 20 miles per hour faster than another train. The faster train travels 210 miles in the same time the slower train travels 140 miles. Find the rate of each train.

Solutions to Self-Checking Exercise

1. $\{x \mid x \in R \text{ and } x \neq \pm 3\}$

2. $\{x \mid x \in R \text{ and } x \neq \dfrac{1}{2}, x \neq -3\}$

3. $\left\{x \mid x \in R \text{ and } x \neq \pm \dfrac{5}{2}\right\}$

4. $\left\{y \mid y \in R \text{ and } y \neq 1, y \neq \dfrac{1}{3}\right\}$

5. $\dfrac{3}{4}$

6. $-\dfrac{7}{2}$

7. $\dfrac{x + y}{x - y}$

8. $\dfrac{-y - 1}{y + 6}$

9. $\dfrac{x}{x^2 - 4}$ or $\dfrac{x}{(x + 2)(x - 2)}$

10. $\dfrac{2x + 5}{x - 3}$

11. $\dfrac{a + 5}{a - 3}$

12. $\dfrac{-(3x + 4)}{5x + 4}$

13. $-2(x + 4)$

14. $\dfrac{9b^3}{4a^2}$

15. $\dfrac{-a(a + 2)}{2}$

16. $x^2 - xy + y^2$

17. $2(x + 4)$

18. $5(y - 1)(y - 4)$

19. $\dfrac{7b^2 + 4a}{24a^2b^3}$

20. $\dfrac{2(x + 168)}{9x(x - 7)}$

21. $\dfrac{-y^2}{x(x + y)}$

22. $\dfrac{x^2 - 6}{(x - 3)^2}$

23. $\dfrac{4y^2 - 29y + 55}{(y + 4)(y - 4)(y - 3)}$

24. $\dfrac{a^2 - 7a - 4}{(a + 4)(a - 4)(a - 1)}$

25. $\dfrac{-27x}{x^2 - 25}$

26. $\dfrac{4a(21a^2 + 2a - 10)}{(3a + 2)^2(3a - 2)}$

27. $\dfrac{y}{x - 3y}$

28. $\dfrac{16x^{16}}{81y^{12}}$

29. $\dfrac{6(x - 2)}{x^3}$

30. $x^2 + 2xy + y^2$

31. $\dfrac{3x}{2}$

32. $\dfrac{2}{3x}$

33. -3

34. -34

35. 9

36. no solution

37. 13

38. 0 and 1

39. $t = \dfrac{V - k}{g}$

40. $r = \dfrac{S - a}{S}$

41. $c = \dfrac{2A - bh}{h}$

42. $x = \dfrac{a}{a - 1}$

43. 4 and 68

44. 6 and 8

45. 3

46. 40 miles per hour and 60 miles per hour

EXPONENTS, ROOTS, AND RADICALS

5

5.1 Roots and Radicals
5.2 Negative Exponents and Scientific Notation
5.3 Rational Exponents
5.4 Simplifying Radicals
5.5 Operating with Radical Expressions
5.6 Rationalizing Denominators
5.7 Solving Equations Containing Radicals

ROOTS AND RADICALS

5.1

From section 1.6 we know that taking a square root is the opposite of squaring. That is, b is a square root of a providing $b^2 = a$. Every positive real number has two square roots, one positive and one negative. Zero has only one square root. And, negative numbers have no real square roots.

EXAMPLE 1   Consider these illustrations of square root:

(a) 5 and -5 are both square roots of 25, since $5^2 = 25$ and $(-5)^2 = 25$.
(b) 0 is the only square root of 0, since $0^2 = 0$.
(c) -9 has no real square roots since the square of a real number is never negative.

The non-negative square root is called the *principal square root*, and is indicated by the radical sign, $\sqrt{}$. Thus, for $a \geq 0$, the symbol $\sqrt{a}$ represents the non-negative square root of a, while the symbol $-\sqrt{a}$ represents the negative square root of a.

EXAMPLE 2   The following square roots have been evaluated:

(a) $\sqrt{36} = 6$
(b) $-\sqrt{36} = -6$
(c) $\sqrt{0} = 0$

Consider the symbol $\sqrt{a^2}$ where $a \neq 0$. a^2 has two square roots, $-a$ and a. One of these is positive and the other is negative. But, we cannot tell which is which. Since the principal square root is not negative, it must be written using the absolute value symbols, as $|a|$.

Definition

For $a \in R$, $\sqrt{a^2} = |a|$

EXAMPLE 3   The following have been simplified:

(a) $\sqrt{x^2} = |x|$
(b) $-\sqrt{x^2} = -|x|$
(c) $\sqrt{4m^2} = \sqrt{(2m)^2} = |2m|$ or $2|m|$

The radical symbol is also used to denote higher roots, such as cube roots, fourth roots, fifth roots, and so on.

> **The Radical Symbol**
> $$\text{If } \sqrt[n]{a} = b, \text{ then } b^n = a$$
> n is called the index, a is called the radicand, and the entire expression is called the radical.

Consider these radicals:
(a) $\sqrt[3]{8} = 2$ means $2^3 = 8$.
(b) $\sqrt[3]{-27} = -3$ means $(-3)^3 = -27$.
(c) $\sqrt[4]{81} = 3$ means $3^4 = 81$.
(d) $\sqrt[4]{-81}$ is not a real number.
(e) $\sqrt[5]{32} = 2$ means $2^5 = 32$.

EXAMPLE 4

Roots may be even or odd according to the index. Even roots are square roots, fourth roots, and so on, while odd roots are cube roots, fifth roots, and so on.

Even roots behave similiarly to square roots in that the radical sign represents the principal (non-negative) root.

These radicals have been simplified:
(a) $\sqrt[4]{x^4} = |x|$
(b) $\sqrt[4]{81a^4} = \sqrt[4]{(3a)^4} = |3a|$ or $3|a|$
(c) $\sqrt[6]{64m^6} = \sqrt[6]{(2m)^6} = |2m|$ or $2|m|$
(d) $-\sqrt[6]{m^6} = -|m|$

EXAMPLE 5

Odd roots behave quite differently than even roots. For each odd root, every real number has exactly one root. Thus, there is no need to define principal odd roots.

These radicals have been simplified:
(a) $\sqrt[3]{x^3} = x$
(b) $\sqrt[3]{8m^3} = \sqrt[3]{(2m)^3} = 2m$
(c) $\sqrt[3]{-27y^3} = \sqrt[3]{(-3y)^3} = -3y$
(d) $-\sqrt[3]{27y^3} = -\sqrt[3]{(3y)^3} = -3y$

EXAMPLE 6

> **Properties of Roots**
> If n is even, then $\sqrt[n]{a}$ represents the principal (non-negative) root of a and:
> 1. $\sqrt[n]{a^n} = |a|$
> 2. If $a \geq 0$ then $\sqrt[n]{a} \geq 0$
> 3. If $a < 0$ then $\sqrt[n]{a}$ does not exist as a real number.
>
> If n is odd then $\sqrt[n]{a}$ is not concerned with principal roots and:
> 1. $\sqrt[n]{a^n} = a$
> 2. If $a \geq 0$ then $\sqrt[n]{a} \geq 0$
> 3. If $a < 0$ then $\sqrt[n]{a} < 0$

EXAMPLE 7 Simplify $\sqrt[4]{x^8}$

Solution: Write x^8 as a fourth power.
$$\sqrt[4]{x^8} = \sqrt[4]{(x^2)^4}$$
$$= |x^2|$$
The absolute value symbol can be removed since x^2 cannot represent a negative number. Thus,
$$\sqrt[4]{x^8} = x^2$$

EXAMPLE 8 Simplify $\sqrt{a^6}$

Solution: Write a^6 as a square.
$$\sqrt{a^6} = \sqrt{(a^3)^2}$$
$$= |a^3| \text{ or } a^2 \cdot |a|$$
The absolute value symbol cannot be removed since a^3 can represent either a positive or a negative number.

EXAMPLE 9 Simplify $\sqrt{25m^2}$, where $m < 0$.

Solution: $\sqrt{25m^2} = |5m|$ or $5|m|$
Since $m < 0$, $|m| = -m$. Thus, $5|m| = 5(-m)$ or $-5m$.

EXAMPLE 10 Simplify $\sqrt[5]{-32x^{15}y^5}$

Solution: Write $-32x^{15}y^5$ as a fifth power.
$$\sqrt[5]{-32x^{15}y^5} = \sqrt[5]{(-2x^3y)^5} = -2x^3y$$
Do not include an absolute value symbol since this is an odd root.

As an aid to simplifying radicals, Table 5.1 lists commonly used powers.

TABLE 5.1 Commonly Used Powers

Number n	n^2	n^3	n^4	n^5	n^6
0	0	0	0	0	0
1	1	1	1	1	1
2	4	8	16	32	64
3	9	27	81	243	729
4	16	64	256	1024	4096
5	25	125	625	3125	
6	36	216	1296		
7	49	343	2401		
8	64	512			
9	81	729			
10	100	1000			
11	121				
12	144				
13	169				
14	196				
15	225				

Use Table 5.1 to help simplify $\sqrt[5]{-1024n^{10}}$.

EXAMPLE 11

Solution: According to the table $1024 = 4^5$. Thus, we write the radicand as a fifth power.
$$\sqrt[5]{-1024n^{10}} = \sqrt[5]{(-4n^2)^5}$$
$$= -4n^2$$

Radicals often contain numbers that are not perfect powers. As you learned in section 1.6, these are irrational numbers (non-terminating, non-repeating decimals). They can be approximated by using your calculator or by referring to the table at the back of the text entitled "Powers and Roots" (see page A-2).

Simplify.

Exercise 5.1

1. $-\sqrt{25}$
2. $\sqrt{4^2}$
3. $-\sqrt{256}$
4. $\sqrt{.09}$
5. $-\sqrt{-100}$
6. $\sqrt{0}$
7. $\sqrt[3]{-125}$
8. $-\sqrt[3]{-64}$
9. $-\sqrt[3]{64}$
10. $-\sqrt[3]{1000}$
11. $\sqrt[3]{3^3}$
12. $\sqrt[4]{-16}$
13. $\sqrt[3]{-64}$
14. $\sqrt{\dfrac{9}{16}}$
15. $-\sqrt{\dfrac{25}{36}}$

EXPONENTS, ROOTS, AND RADICALS

16. $\sqrt[4]{81}$ 17. $\sqrt[5]{2^5}$ 18. $\sqrt[3]{\dfrac{27}{8}}$

19. $-\sqrt{\dfrac{16}{25}}$ 20. $-\sqrt[6]{64}$ 21. $\sqrt[4]{(-9)^2}$

22. $\sqrt{x^4}$ 23. $\sqrt{y^6}$ 24. $\sqrt{49a^4}$

25. $\sqrt{x^2y^8}$ 26. $\sqrt{4x^2}$ 27. $-\sqrt{4x^2y^{18}}$

28. $\sqrt{36x^4}$ 29. $\sqrt{\dfrac{9}{16}x^2}$ 30. $\sqrt{3600a^6}$

31. $\sqrt[3]{a^9b^{18}c^{24}}$ 32. $\sqrt[3]{x^6y^9z^{12}}$ 33. $\sqrt{-9x^2}$

34. $\sqrt[3]{-64x^6y^3}$ 35. $\sqrt{x^4y^2}$ 36. $\sqrt[3]{27a^{12}}$

37. $\sqrt{81x^2}$, where $x < 0$ 38. $\sqrt{144y^4}$, where $y < 0$

39. $\sqrt{x^2y^6}$ 40. $\sqrt[3]{-8a^3y^6}$

41. $\sqrt[3]{216x^6y^3}$ 42. $-\sqrt[5]{-243x^{10}}$

43. $\sqrt{144(x+y)^2}$ 44. $\sqrt[3]{343(x+y)^6}$

45. $\sqrt[4]{625x^4y^8}$ 46. $\sqrt[5]{-32x^5y^{10}}$

47. $-\sqrt{169a^8b^{10}}$ 48. $\sqrt{(-4)^2 + (-3)^2}$

Use the table of powers to help simplify each expression.

49. $\sqrt[4]{1296x^4y^4}$ 50. $-\sqrt[6]{4096x^{12}y^6}$ 51. $\sqrt[3]{-729x^6y^9z^{12}}$

NEGATIVE EXPONENTS AND SCIENTIFIC NOTATION

5.2

Our initial discussion of exponents in section 1.6 was confined to natural (or positive integral) exponents. We now define negative integral exponents and the zero exponent. This is accomplished by assuming that the properties of natural number exponents from section 1.6 hold true for all integral exponents. Those properties will guide us to the proper definitions.

Consider zero as an exponent. We assume that a^0 exists and that the subtraction property of exponents holds true for all integral exponents. Then, for all non-zero real numbers a, the following argument is valid:

For $a \neq 0$, $\dfrac{a^n}{a^n} = 1$ [a non-zero real number divided by itself is 1]

But, $\dfrac{a^n}{a^n} = a^{n-n} = a^0$ [subtraction property of exponents]

Hence, $a^0 = 1$, where $a \neq 0$ [substitution]

NEGATIVE EXPONENTS AND SCIENTIFIC NOTATION

Definition

If $a \in R$ and $a \neq 0$, then
$$a^0 = 1$$
(Any non-zero real number raised to the zero power is 1.)

These equations illustrate the definition of a^0:

EXAMPLE 1

(a) $5^0 = 1$
(b) $(-6)^0 = 1$
(c) $-6^0 = -(6)^0 = -1$
(d) $2 \cdot 10^0 = 2 \cdot 1 = 2$
(e) $(2 \cdot 10)^0 = 20^0 = 1$

Next we wish to define negative integral exponents. We assume they exist and obey the previous properties. This argument then follows:

Let $a \in R$, $a \neq 0$ and let n be any integer

$$\frac{a^0}{a^n} = a^{0-n} = a^{-n} \qquad \text{[subtraction property of exponents]}$$

But, $\dfrac{a^0}{a^n} = \dfrac{1}{a^n}$ [definition of a^0]

Hence, $a^{-n} = \dfrac{1}{a^n}$, where $a \neq 0$ [substitution]

Definition

If $a \in R$, $a \neq 0$ and if n is any integer, then
$$a^{-n} = \frac{1}{a^n}$$

These equations illustrate the definition of a^{-n}.

EXAMPLE 2

(a) $10^{-1} = \dfrac{1}{10^1}$ or $\dfrac{1}{10}$
(b) $10^{-2} = \dfrac{1}{10^2}$ or $\dfrac{1}{100}$
(c) $(-2)^{-3} = \dfrac{1}{(-2)^3}$ or $-\dfrac{1}{8}$
(d) $x^{-4} = \dfrac{1}{x^4}$ where $x \neq 0$

Properties of Integral Exponents

For $a, b \in R$ and $m, n \in I$

(1) Definition of a^0
$$a^0 = 1, \text{ where } a \neq 0$$

(2) Definition of a^{-n}
$$a^{-n} = \frac{1}{a^n}, \text{ where } a \neq 0$$

(3) Addition Property of Exponents
$$a^m \cdot a^n = a^{m+n}$$

(4) Subtraction Property of Exponents
$$\frac{a^m}{a^n} = a^{m-n}, \text{ where } a \neq 0$$

(5) Power to a Power Property of Exponents
$$(a^m)^n = a^{mn}$$

(6) Distributive Property of Exponents
$$(ab)^n = a^n b^n$$
$$\left(\frac{a}{b}\right)^n = \frac{a^n}{b^n}, \text{ where } b \neq 0$$

EXAMPLE 3 The properties of integral exponents have been applied to the following. (Note that final answers are not left with negative exponents.)

(a) $(2x^{-4})(3x^6) = 6x^{-4+6} = 6x^2$

(b) $(5m^{-2})(3m^{-1}) = 15m^{-3} = 15 \cdot \frac{1}{m^3}$ or $\frac{15}{m^3}$

(c) $\frac{12a^5}{3a^7} = 4a^{5-7} = 4a^{-2}$ or $\frac{4}{a^2}$

(d) $(-4y^{-5})^{-2} = (-4)^{-2}(y^{-5})^{-2} = \frac{1}{(-4)^2} \cdot y^{10}$
$$= \frac{1}{16} \cdot y^{10}$$
$$= \frac{y^{10}}{16}$$

(e) $\frac{27x^5y^2}{9x^2y^2} = 3x^3y^0 = 3x^3 \cdot 1$
$$= 3x^3$$

Often we must simplify algebraic expressions containing negative exponents. Such expressions are simplified by converting negative exponents to positive exponents and simplifying the resulting complex fraction. Recall the definition of negative exponents:

$$x^{-n} = \frac{1}{x^n}, \text{ where } n \in I$$

EXAMPLE 4

Simplify $\dfrac{2m^{-3}}{7n^{-5}}$

Solution: Use the definition to convert negative exponents to positive exponents. Then, simplify the complex fraction.

$$\frac{2m^{-3}}{7n^{-5}} = \frac{2 \cdot \dfrac{1}{m^3}}{7 \cdot \dfrac{1}{n^5}}$$

Multiply the numerator and denominator by the LCD, $m^3 n^5$.

$$= \frac{2 \cdot \dfrac{1}{\cancel{m^3}} \cdot \cancel{m^3} n^5}{7 \cdot \dfrac{1}{\cancel{n^5}} \cdot m^3 \cancel{n^5}}$$

$$= \frac{2n^5}{7m^3}$$

EXAMPLE 5

Simplify $\dfrac{x^{-3} + y^{-3}}{2}$

Solution: Convert negative exponents to positive exponents.

$$\frac{x^{-3} + y^{-3}}{2} = \frac{\dfrac{1}{x^3} + \dfrac{1}{y^3}}{2}$$

Multiply the numerator and denominator by the LCD, $x^3 y^3$.

$$= \frac{\left[\dfrac{1}{x^3} + \dfrac{1}{y^3}\right] \cdot x^3 y^3}{2 \cdot x^3 y^3}$$

$$= \frac{\dfrac{1}{\cancel{x^3}} \cdot \cancel{x^3} y^3 + \dfrac{1}{\cancel{y^3}} \cdot x^3 \cancel{y^3}}{2x^3 y^3}$$

$$= \frac{y^3 + x^3}{2x^3 y^3}$$

Many important applications of integral exponents are found in science, where many numbers are very large or very small and involve an excessive number of zeros. For example, the sun is approximately 93,000,000 miles from earth, and light travels one foot in approximately 0.000000001 seconds. To simplify the writing of such numbers we use scientific notation.

> **Scientific Notation**
> A number is said to be in scientific notation when it is written as a product of a number between 1 and 10 and a power of 10.

EXAMPLE 6   In scientific notation, 93,000,000 is written as 9.3×10^7. The number between 1 and 10 is 9.3. The power of ten is 10^7.

In order to understand scientific notation you need to recall two rules from arithmetic.

(1) Each time a number is multiplied by 10, the decimal point is moved one place to the right.

Examples: $3.56 \times 10 = 35.6$
$3.56 \times 10^2 = 356$ [multiplied by 10 two times]
$3.56 \times 10^3 = 3,560$ [multiplied by 10 three times]

(2) Each time a number is divided by 10, the decimal point is moved one place to the left.

Examples: $7.6 \times 10^{-1} = \dfrac{7.6}{10} = 0.76$

$7.6 \times 10^{-2} = \dfrac{7.6}{100} = 0.076$ [divided by 10 two times]

$7.6 \times 10^{-3} = \dfrac{7.6}{1,000} = 0.0076$ [divided by 10 three times]

Remembering these rules, we formulate the following method for converting scientific notation to ordinary notation.

> **Converting to Ordinary Notation**
> Move the decimal point the same number of places as the exponent on 10. Move to the right if the exponent is positive; move to the left if the exponent is negative.

EXAMPLE 7   Convert 8.92×10^4 to ordinary notation.

Solution: The exponent of 4 tells us to move the decimal point 4 places to the right:

$8.92 \times 10^4 = 89200.$ or $89,200$

4 places

NEGATIVE EXPONENTS AND SCIENTIFIC NOTATION

Convert 5.14×10^{-6} to ordinary notation. **EXAMPLE 8**

Solution: The exponent of -6 tells us to move the decimal point 6 places to the left.

$$5.14 \times 10^{-6} = 0.00000514$$
$$\underbrace{\qquad}_{6 \text{ places}}$$

Note in Example 8 that a negative exponent in scientific notation indicates that the number is less than 1. We now formulate a method of converting ordinary notation to scientific notation.

> **Converting to Scientific Notation**
>
> Move the decimal point to create a number greater than or equal to 1 but less than 10. Then, attach the corresponding exponent to form the correct power of 10. The exponent will be negative if the original number is less than 1.

Convert 0.000285 to scientific notation. **EXAMPLE 9**

Solution: Move the decimal point four places to form 2.85. The original number is less than 1, so the exponent is -4.

$$0.000285 = 2.85 \times 10^{-4}$$

Convert 35,600,000 to scientific notation. **EXAMPLE 10**

Solution: Move the decimal point seven places to form 3.56. The original number is more than 1, so the exponent is 7.

$$35,600,000 = 3.56 \times 10^7$$

Some expressions can be easily evaluated by converting the numbers to scientific notation and applying the properties of exponents.

Evaluate $\dfrac{(0.0025)(63,000,000)}{(50,000)(0.021)}$ **EXAMPLE 11**

Solution: Convert to scientific notation. Then apply the addition and subtraction properties of exponents.

$$\frac{(0.0025)(63,000,000)}{(50,000)(0.021)} = \frac{(2.5 \times 10^{-3})(6.3 \times 10^7)}{(5 \times 10^4)(2.1 \times 10^{-2})}$$

$$= \frac{2.5 \times 6.3 \times 10^{-3} \times 10^7}{5 \times 2.1 \times 10^4 \times 10^{-2}}$$

$$= \frac{2.5 \times 6.3}{5 \times 2.1} \times \frac{10^4}{10^2}$$

$$= 1.5 \times 10^2$$

$$= 150$$

EXPONENTS, ROOTS, AND RADICALS

When performing lengthy and complex computations it can be helpful to estimate answers. This is done by rounding all numbers to one significant digit and converting to scientific notation. (We use the symbol ≈ to mean "approximately equal to.")

EXAMPLE 12

Estimate the value of $\dfrac{(0.000682)(513,000)}{0.00846}$.

Solution: Round numbers to one significant digit. Then convert to scientific notation and apply the addition and subtraction properties of exponents.

$$\dfrac{(0.000682)(513,000)}{0.00846} \approx \dfrac{(0.0007)(500,000)}{0.008}$$

$$\approx \dfrac{(7 \times 10^{-4})(5 \times 10^{5})}{8 \times 10^{-3}}$$

$$\approx \dfrac{35}{8} \times \dfrac{10^{1}}{10^{-3}}$$

$$\approx 4 \times 10^{1-(-3)}$$

$$\approx 4 \times 10^{4}$$

$$\approx 40,000$$

Calculator

With your instructor's approval, you may evaluate expressions containing negative exponents using your calculator. This can be done by combining the change sign key, (+/−) or (cs), with the exponential key, (y^x).

EXAMPLE 13

Evaluate $(3.62)^{-2}$. Round the answer to three significant digits.

Calculator Sequence: (3.62) (y^x) (2) (+/−) (=) .07631025
Rounded answer is 0.0763

Calculators can also perform operations in scientific notation by using the engineering notation key, (EE).

EXAMPLE 14

Evaluate $\dfrac{3.68 \times 10^{5}}{9.84 \times 10^{-2}}$. Round the answer to three significant digits.

Calculator Sequence: (3.68) (EE) (5) (÷) (9.84) (EE) (2) (+/−) (=)
3.739806
Rounded answer is 3.74×10^{6}

Exercise 5.2

Problems 1–20. Simplify the following by using the properties of integral exponents. Assume that no variables represent zero. Final answers should not contain negative exponents.

NEGATIVE EXPONENTS AND SCIENTIFIC NOTATION

1. $x^3 \cdot x^7$
2. $2x^4 \cdot x^2$
3. $x^6 \cdot x^{-2}$
4. $3x^{-4} \cdot x^{-6}$
5. $\dfrac{(4x)^2}{-4x^2}$
6. $\dfrac{x^{12}}{x^4}$
7. $(x^4)^{-3}$
8. $(3x^2)^{-2}$
9. $(4y^2)^{-3}$
10. $\dfrac{-3^0}{4^{-2}}$
11. $\dfrac{36x^6y^3}{9x^2y^2}$
12. $(-4y^{-4})^{-2}$
13. $(4x^{-3})(2x^{-1})$
14. $(xy^2z^3)^3$
15. $(2x^{-5})(7x^8)$
16. $(x^{-4})^2$
17. $(x^{-5})^0$
18. $(m^2n^{-1})^3$
19. $\dfrac{(3x^{-5})(4x^8)}{6x^3}$
20. $(-2m^{-2}n^3)^{-1}$

Evaluate the given expression.

21. $3^{-2} \cdot 3^4$
22. $3^{-1} \cdot 3^{-3}$
23. $4^{-2} \cdot 4^5$
24. $4 \cdot 4^{-2}$
25. $(2^2)^{-2}$
26. $(3^2)^{-2}$
27. $(3^{-3})^{-2}$
28. $(-2^{-2})^{-2}$

Convert each number to scientific notation.

29. 251,000,000
30. 10,000,000,000
31. 11,600,000,000
32. 985.16
33. 3,685.47
34. 0.000004
35. 0.000378
36. 0.00000000316

Write the following numbers in ordinary notation.

37. 3.12×10^{-4}
38. 3×10^{-5}
39. 7.14×10^3
40. 2.14×10^6
41. 5×10^{-3}
42. 4.86×10^4

Compute the following. Leave answers in scientific notation.

43. $\dfrac{(4 \times 10^4)(5 \times 10^5)}{2 \times 10^6}$
44. $\dfrac{(6 \times 10^{-4})(7 \times 10^8)}{3 \times 10^7}$
45. $\dfrac{15{,}000 \times .000002}{.005}$
46. $\dfrac{.000016 \times 500}{2{,}000{,}000}$
47. $\dfrac{.00348 \times .002}{.003}$
48. $\dfrac{.00018 \times 7500}{.000025 \times 9{,}000}$

Estimate the value of each of the following.

49. $\dfrac{(.00358)(.014)}{(.00081)(13)}$
50. $\dfrac{(.003)(.160)}{(.00004)(.02)}$

EXPONENTS, ROOTS, AND RADICALS

51. $\dfrac{(.00258)(.04)}{(.000062)(13)}$

52. $\dfrac{576.6 \times 548.2}{8{,}846{,}000}$

Simplify by first converting negative exponents to positive exponents.

53. $\dfrac{3x^{-4}}{4y^{-6}z^{-2}}$

54. $\dfrac{27y^{-1}}{36y^{-2}t^4}$

55. $x^{-1} + y^{-1}$

56. $\dfrac{a^{-2}}{a^{-2} - b^{-2}}$

57. $\dfrac{x^{-1} + y^{-1}}{x^{-1} - y^{-1}}$

58. $\dfrac{y^{-2} + z^{-2}}{z^{-2}}$

59. $(a^{-1} - b^{-1})^{-1}$

60. $\dfrac{2x^{-1} - y^{-1}}{x^{-1} - y^{-1}}$

61. $\dfrac{(a + b)^{-1}}{a^{-2} + b^{-2}}$

62. $\dfrac{5(x - y)^{-1}}{x^{-2} - y^{-2}}$

Use your calculator to evaluate the following. Express the answer in scientific notation and round to three significant digits.

63. $(4.83)^{-3}$

64. $(2.36 \times 10^3)^{-4}$

65. $2.65 \times 10^{-2} + 9.805 \times 10^{-1}$

66. $(8.72 \times 10^{-5})(9.604 \times 10^{-2})$

67. $\dfrac{5.607 \times 10^{18}}{9.683 \times 10^{20}}$

68. $\dfrac{(4.016 \times 10^4)(3.718 \times 10^3)}{(8.604 \times 10^2)(9.715 \times 10^1)}$

69. $\dfrac{(7.83 \times 10^4)(6281 + 723)}{5.2 \times 10^3}$

70. $\dfrac{(7.96 \times 10^{-3})(8.04 \times 10^{-2})}{5.05 \times 10^{-2}}$

71. $\sqrt{6.7051 \times 10^{-3}}$

72. $\sqrt[3]{7.615 \times 10^{-6}}$

73. $\sqrt{5.364 \times 10^5} \cdot \sqrt[3]{6.141 \times 10^{-3}}$

RATIONAL EXPONENTS

5.3

In this section we extend the properties of exponents to include *rational exponents*. These are exponents of the form m/n, where $m, n \in I$ and $n \neq 0$. Thus, we will study exponentials such as $16^{1/2}$, $4^{-(1/2)}$, $8^{2/3}$, and $36^{-(3/2)}$.

Assuming that the properties of exponents are true for rational numbers, we first define the exponent $1/n$, where $n \in I$ and $n \neq 0$. This gives meaning to such exponentials as $16^{1/2}$ and $8^{1/3}$.

RATIONAL EXPONENTS

Consider the exponential $a^{1/n}$, where $n \in I$ and $n \neq 0$. Multiplying n factors of $a^{1/n}$ produces the following:

$$\underbrace{a^{1/n} \cdot a^{1/n} \cdot a^{1/n} \ldots a^{1/n}}_{n \text{ factors}} = (a^{1/n})^n$$
$$= a^{n/n}$$
$$= a^1 \text{ or } a$$

This illustrates that $a^{1/n}$ is one of n equal factors of a. Therefore, it is reasonable to define $a^{1/n}$ as being the nth root of a. If n is an even integer, then $a^{1/n}$ is defined to be the principal n^{th} root of a.

Thus, we have formulated the following definition.

Definition of the Exponent 1/n.
For $n \in I$ and $n \geq 2$,
$$a^{1/n} = \sqrt[n]{a}$$
If n is even, then a must be non-negative.

These exponentials have been evaluated:

(a) $16^{1/2} = \sqrt{16} = 4$
(b) $8^{1/3} = \sqrt[3]{8} = 2$
(c) $(-27)^{1/3} = \sqrt[3]{-27} = -3$
(d) $32^{1/5} = \sqrt[5]{32} = 2$

EXAMPLE 1

We can now structure a definition for rational exponents of the form m/n, where $m, n \in I$ and $n \geq 2$. Assuming that the properties of exponents are true for rational numbers, we use the power to a power property and previous definitions as follows:

$$a^{m/n} = (a^m)^{1/n} \text{ or } (a^{1/n})^m$$
$$= \sqrt[n]{a^m} \text{ or } (\sqrt[n]{a})^m$$

To avoid even roots of negative numbers, the base a must be non-negative when n is even.

We now state the definition.

Definition of Rational Exponents
For $m, n \in I$ and $n \geq 2$,
$$a^{m/n} = \sqrt[n]{a^m} \text{ or } (\sqrt[n]{a})^m$$
If n is even, then $a \geq 0$.

EXPONENTS, ROOTS, AND RADICALS

This definition enables us to convert back and forth between exponential and radical form. Observe that the numerator of a rational exponent indicates the power, while the denominator gives the root.

$$a^{m/n} = \sqrt[n]{a^m} = (\sqrt[n]{a})^m$$

EXAMPLE 2

These exponentials have been converted to radical form:
(a) $m^{3/5} = \sqrt[5]{m^3}$ or $(\sqrt[5]{m})^3$
(b) $x^{3/4} = \sqrt[4]{x^3}$ or $(\sqrt[4]{x})^3$, where $x \geq 0$

EXAMPLE 3

These radicals have been converted to exponential form:
(a) $\sqrt[5]{k^2} = k^{2/5}$
(b) $\sqrt{x^3} = x^{3/2}$, where $x \geq 0$
(c) $(\sqrt[3]{b})^2 = b^{2/3}$

Exponentials involving rational exponents can be evaluated by applying the root and power in any order. However, numbers are usually kept smaller if the root is taken first.

EXAMPLE 4

Evaluate $16^{3/2}$

Solution: Convert to radical form, taking the root first.
$$16^{3/2} = (\sqrt{16})^3$$
$$= 4^3$$
$$= 64$$

EXAMPLE 5

The following have been evaluated.
(a) $(-27)^{2/3} = (\sqrt[3]{-27})^2 = (-3)^2 = 9$
(b) $32^{2/5} = (\sqrt[5]{32})^2 = 2^2 = 4$
(c) $\left(\dfrac{9}{100}\right)^{3/2} = \left(\sqrt{\dfrac{9}{100}}\right)^3 = \left(\dfrac{3}{10}\right)^3 = \dfrac{27}{1000}$

When evaluating exponentials with negative exponents, we recall the following definition from section 5.2:

$$x^{-n} = \dfrac{1}{x^n}, \text{ where } x \neq 0$$

Thus, we will change negative exponents to positive exponents, then evaluate.

EXAMPLE 6

These exponentials have been evaluated.
(a) $36^{-1/2} = \dfrac{1}{36^{1/2}} = \dfrac{1}{\sqrt{36}} = \dfrac{1}{6}$

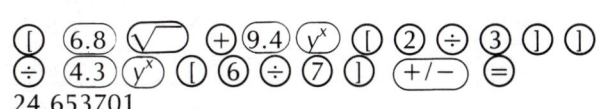

(b) $8^{-2/3} = \dfrac{1}{8^{2/3}} = \dfrac{1}{(\sqrt[3]{8})^2} = \dfrac{1}{2^2} = \dfrac{1}{4}$

(c) $\left(\dfrac{4}{9}\right)^{-3/2} = \dfrac{1}{\left(\dfrac{4}{9}\right)^{3/2}} = \dfrac{1}{\left(\sqrt{\dfrac{4}{9}}\right)^3} = \dfrac{1}{\left(\dfrac{2}{3}\right)^3} = \dfrac{1}{\dfrac{8}{27}} = \dfrac{27}{8}$

The calculator, with its exponential key (y^x) and grouping keys $(\,(\,)\,)$, can be used to evaluate exponentials involving rational exponents.

Evaluate $(26.7)^{9/13}$ and round the answer to three significant digits. **EXAMPLE 7**

Calculator Sequence: (26.7)(y^x)(()(9)(÷)(13)())(=) 9.7181751
The rounded answer is 9.72.

Evaluate $(127.3)^{-(4/3)}$ and round the answer to four significant digits. **EXAMPLE 8**

Calculator Sequence: (127.3)(y^x)(()(4)(÷)(3)())(+/−)(=)
0.00156157
The rounded answer is 0.001562.

Evaluate $\dfrac{(6.8)^{1/2} + (9.4)^{2/3}}{(4.3)^{-(6/7)}}$ and round the answer to two significant digits. **EXAMPLE 9**

Calculator Sequence: (()(6.8)(√)(+)(9.4)(y^x)(()(2)(÷)(3)())())
(÷)(4.3)(y^x)(()(6)(÷)(7)())(+/−)(=)
24.653701
The rounded answer is 25.

Evaluate these exponentials. **Exercise 5.3**

1. $64^{1/2}$
2. $32^{1/5}$
3. $-9^{1/2}$
4. $(-8)^{1/3}$
5. $\left(\dfrac{81}{25}\right)^{1/2}$
6. $\left(-\dfrac{64}{125}\right)^{1/3}$
7. $8^{1/3} \cdot 16^{1/2}$
8. $\left(\dfrac{25}{36}\right)^{1/2}$
9. $(9 + 16)^{1/2}$
10. $(-27)^{1/3}$
11. $(-27)^{-(1/3)}$
12. $\left(\dfrac{1}{4}\right)^{-(1/2)}$
13. $16^{-(1/2)} + 16^{1/2}$
14. $\left(\dfrac{4}{9}\right)^{-(1/2)}$
15. $\left(\dfrac{81}{25}\right)^{-(1/2)}$
16. $[8^{1/3} + (-8)^{1/3}]^2$
17. $27^{2/3}$
18. $\left(\dfrac{1}{4}\right)^{3/2}$

190 EXPONENTS, ROOTS, AND RADICALS

19. $8^{2/3} + 4^{-(1/2)}$
20. $64^{-(2/3)}$
21. $\left(\dfrac{27}{8}\right)^{-(2/3)}$
22. $\left(-\dfrac{8}{27}\right)^{4/3}$
23. $\left(\dfrac{1}{16}\right)^{-(3/4)}$
24. $4 - (5^2 - 3^2)^{1/2}$
25. $(-125)^{-(4/3)}$
26. $\dfrac{27^{2/3} \cdot 27^{5/3}}{27^{1/3}}$
27. $(-144)^{3/2}$

Change the following to exponential form.
28. $\sqrt[6]{x^5}$
29. $\sqrt{a^3}$
30. $\sqrt[3]{m+n}$
31. $\sqrt[3]{x} + \sqrt[3]{y}$
32. $\sqrt[3]{(m+n)^2}$
33. $\sqrt[5]{x^2} \cdot \sqrt[5]{y^8}$

Convert the following to radical form.
34. $x^{3/4}$
35. $a^{2/5}$
36. $(x^3 + y^3)^{1/3}$
37. $2x^{1/2}y^{1/3}$
38. $(-2a)^{2/3}$
39. $x^{2/3} \cdot y^{1/2}$
40. $(2x)^{-(3/2)}$
41. $(x^2 + y^2)^{-(1/2)}$
42. $(2a^4 + b^2)^{-(2/3)}$

Find decimal approximations using a calculator. Round results to the nearest hundredth.
43. $(34.7)^{2/3}$
44. $(34.7)^{7/13}$
45. $(1.76)^{-(5/3)}$
46. $(-28.4)^{-(2/3)}$
47. $(.824)^{7/5}$
48. $[(1.34)^{-(2/3)}]^{5/3}$
49. $\dfrac{6.9^{2/3} + 1.3^{3/4}}{6.2^{-(6/5)}}$
50. $\dfrac{22.8^{-(3/4)} + 6.4^{1/3}}{(.81)^{6/7} - 16.8^{-(2/5)}}$

SIMPLIFYING RADICALS

5.4

In this section we develop two important properties which will enable us to multiply, divide, and simplify radicals. These are the multiplication and division properties of radicals. The multiplication property is derived below.

$\sqrt[n]{a} \cdot \sqrt[n]{b} = a^{1/n} \cdot b^{1/n}$ [definition of rational exponents]

$\qquad\qquad = (a \cdot b)^{1/n}$ [distributive property of exponents]

$\qquad\qquad = \sqrt[n]{a \cdot b}$ [definition of rational exponents]

The Multiplication Property of Radicals

$\sqrt[n]{a} \cdot \sqrt[n]{b} = \sqrt[n]{a \cdot b}$, where

1. n is odd and $a, b \in R$
2. n is even and $a, b \geq 0$

This property states that the product of two radicals is equivalent to the radical of the product. The radicals, however, must have equal indexes.

SIMPLIFYING RADICALS

EXAMPLE 1

The multiplication property of radicals has been used to find the following products.
(a) $\sqrt{3} \cdot \sqrt{5} = \sqrt{3 \cdot 5} = \sqrt{15}$
(b) $\sqrt[3]{3} \cdot \sqrt[4]{7}$ cannot be multiplied using this property since the indexes (3 and 4) are different.
(c) $\sqrt[3]{7} \cdot \sqrt[3]{x} = \sqrt[3]{7x}$
(d) $\sqrt[5]{8m^2} \cdot \sqrt[5]{3m} = \sqrt[5]{24m^3}$
(e) $\sqrt{\dfrac{x}{3}} \cdot \sqrt{\dfrac{2y}{5}} \cdot \sqrt{\dfrac{1}{2}} = \sqrt{\dfrac{2xy}{30}} = \sqrt{\dfrac{xy}{15}}$

To simplify a radical means to write an equivalent form where the radicand contains no factor with a power greater than or equal to the index. Also, the radical must not contain a fraction. Radicals are simplified by reversing the multiplication property as shown in these examples. (You may find it helpful to refer to Table 5.1 in section 5.1.)

Simplify $\sqrt{32}$

EXAMPLE 2

Solution: Turn the radicand into a product by finding the largest perfect square that exactly divides into 32. Then, apply the multiplication property to form two radicals. 16 is the largest square that divides into 32 exactly. Thus,

$\sqrt{32} = \sqrt{16 \cdot 2}$
$\phantom{\sqrt{32}} = \sqrt{16} \cdot \sqrt{2}$
$\phantom{\sqrt{32}} = 4\sqrt{2}$

Simplify $3\sqrt{50}$

EXAMPLE 3

Solution: 50 is exactly divisible by the perfect square, 25.

$3 \cdot \sqrt{50} = 3 \cdot \sqrt{25 \cdot 2}$

Apply the multiplication property.

$\phantom{3 \cdot \sqrt{50}} = 3 \cdot \sqrt{25} \cdot \sqrt{2}$
$\phantom{3 \cdot \sqrt{50}} = 3 \cdot 5 \cdot \sqrt{2}$
$\phantom{3 \cdot \sqrt{50}} = 15\sqrt{2}$

Simplify $5\sqrt[3]{-24}$

EXAMPLE 4

Solution: -24 is exactly divisible by the perfect cube, -8.

$5\sqrt[3]{-24} = 5\sqrt[3]{(-8) \cdot 3}$

Use the multiplication property of radicals.

$\phantom{5\sqrt[3]{-24}} = 5 \cdot \sqrt[3]{-8} \cdot \sqrt[3]{3}$
$\phantom{5\sqrt[3]{-24}} = 5 \cdot (-2) \cdot \sqrt[3]{3}$
$\phantom{5\sqrt[3]{-24}} = -10\sqrt[3]{3}$

192   EXPONENTS, ROOTS, AND RADICALS

Radicals containing variables are simplified in a similar manner. We also recall from section 5.1 that:
$$\sqrt[n]{a^n} = a \text{ when } n \text{ is odd, and}$$
$$\sqrt[n]{a^n} = |a| \text{ when } n \text{ is even}$$

EXAMPLE 5 Simplify $\sqrt{25m^3}$, where $m \geq 0$.

Solution: $\sqrt{25m^3} = \sqrt{25m^2 \cdot m}$
$= \sqrt{25m^2} \cdot \sqrt{m}$
$= 5|m|\sqrt{m}$

Since $m \geq 0$, the absolute value symbol is not needed.
$= 5m\sqrt{m}$

EXAMPLE 6 Simplify $5\sqrt[3]{24x^5y^4}$

Solution: $5\sqrt[3]{24x^5y^4} = 5\sqrt[3]{(8x^3y^3)(3x^2y)}$
$= 5\sqrt[3]{8x^3y^3} \cdot \sqrt[3]{3x^2y}$
$= 5(2xy) \cdot \sqrt[3]{3x^2y}$
$= 10xy\sqrt[3]{3x^2y}$

The division property of radicals is derived similarly to the multiplication property.

$$\frac{\sqrt[n]{a}}{\sqrt[n]{b}} = \frac{a^{1/n}}{b^{1/n}} \qquad \text{[definition of rational exponents]}$$
$$= \left(\frac{a}{b}\right)^{1/n} \qquad \text{[distributive property of exponents]}$$
$$= \sqrt[n]{\frac{a}{b}} \qquad \text{[definition of rational exponents]}$$

The Division Property of Radicals
$$\frac{\sqrt[n]{a}}{\sqrt[n]{b}} = \sqrt[n]{\frac{a}{b}}, \text{ where } b \neq 0 \text{ and}$$
1. n is odd and $a, b \in R$
2. n is even, $a \geq 0$, and $b > 0$

The division property tells us that the quotient of two radicals is equivalent to the radical of the quotient. As with the multiplication property, the indexes of the radicals must be equal.

SIMPLIFYING RADICALS

EXAMPLE 7

The division property of radicals has been used to divide the following:

(a) $\dfrac{\sqrt{10}}{\sqrt{2}} = \sqrt{\dfrac{10}{2}} = \sqrt{5}$

(b) $\dfrac{\sqrt[3]{16x^4}}{\sqrt[3]{2x}} = \sqrt[3]{\dfrac{16x^4}{2x}} = \sqrt[3]{8x^3} = 2x$

The division property can also be reversed to simplify radicals containing quotients.

EXAMPLE 8

The division property has been used to simplify these radicals:

(a) $\sqrt{\dfrac{16p^4}{25}} = \dfrac{\sqrt{16p^4}}{\sqrt{25}} = \dfrac{4p^2}{5}$

(b) $\sqrt[3]{\dfrac{m}{125}} = \dfrac{\sqrt[3]{m}}{\sqrt[3]{125}} = \dfrac{\sqrt[3]{m}}{5}$

Exercise 5.4

Simplify each of the following.

1. $\sqrt{40}$
2. $\sqrt{32}$
3. $\sqrt[3]{40}$
4. $\sqrt{50}$
5. $\sqrt[3]{24}$
6. $2\sqrt{99}$
7. $3\sqrt{72}$
8. $\sqrt[3]{16}$
9. $5\sqrt{12}$
10. $3\sqrt{24}$
11. $-2\sqrt{80}$
12. $\sqrt{98}$
13. $\sqrt[3]{-250}$
14. $5\sqrt[3]{-24}$
15. $\sqrt{52}$
16. $\sqrt[4]{32}$
17. $-2\sqrt{-18}$
18. $3\sqrt[3]{54}$
19. $\sqrt{100x^2}$
20. $\sqrt{27x^4}$
21. $\sqrt{25m^3}$ where $m > 0$
22. $\sqrt{12x^5y^4}$, $x > 0$, $y > 0$
23. $\sqrt{20x^2}$, where $x < 0$
24. $\sqrt{50x^2}$
25. $\sqrt{18x^8}$
26. $-\sqrt[3]{162}$
27. $\sqrt{7 \cdot 10^3}$
28. $\sqrt{3 \cdot 10^5}$
29. $\sqrt[3]{40a^4b^7}$
30. $\sqrt{18a^8}$
31. $\sqrt{18x^3y^5}$, $x > 0$, $y > 0$
32. $\sqrt[3]{-16x^5y^7}$
33. $\sqrt{24m^5n^2}$, where $m > 0$
34. $-\sqrt[3]{8x^8}$
35. $\sqrt[3]{32x^5y^4}$
36. $\sqrt{200x^5y^8}$, where $x > 0$
37. $\sqrt{\dfrac{x}{y^2}}$, where $x > 0$, $y \neq 0$
38. $\sqrt[3]{\dfrac{m}{125}}$
39. $\sqrt{\dfrac{16x^4}{25}}$
40. $\dfrac{2}{\sqrt{4}} \cdot \sqrt{18}$
41. $\sqrt{\dfrac{72}{25}}$
42. $\sqrt[3]{\dfrac{27}{8}}$
43. $\sqrt{\dfrac{3}{16}}$
44. $\sqrt{\dfrac{3x^2}{4}}$

194 EXPONENTS, ROOTS, AND RADICALS

45. $\sqrt[3]{8x^4y^5}$
46. $\dfrac{\sqrt{90}}{\sqrt{10}}$
47. $\dfrac{\sqrt{75}}{\sqrt{3}}$

48. $\sqrt{\dfrac{2}{3}} \cdot \sqrt{\dfrac{1}{6}}$
49. $\sqrt{\dfrac{a}{b}} \cdot \sqrt{ab},\ a > 0,\ b > 0$

50. $\sqrt{3} \cdot \sqrt{27}$
51. $\sqrt{2} \cdot \sqrt{6}$

52. $\sqrt[3]{2} \cdot \sqrt[3]{4}$
53. $\dfrac{4}{\sqrt{2}} \cdot \dfrac{\sqrt{32}}{2}$

54. $\dfrac{\sqrt{48x^3y^3}}{\sqrt{3xy}},\ x > 0,\ y > 0$
55. $\sqrt[6]{x^2} \cdot \sqrt[6]{x^4}$

56. $\sqrt[4]{\dfrac{c^8 d^{12}}{81}}$
57. $\sqrt[3]{\dfrac{9}{8}} \cdot \sqrt[3]{\dfrac{24}{x^3}}$

OPERATING WITH RADICAL EXPRESSIONS

5.5

An algebraic expression containing radicals is called a radical expression. In this section you will learn to simplify radical expressions involving addition or subtraction.

Radical expressions can be added or subtracted by combining like terms. To be like terms, radicals must have the same index and the same radicand. Thus, $3\sqrt{5}$ and $2\sqrt{5}$ are like terms. However, $4\sqrt{6}$ and $5\sqrt[3]{6}$ are not like terms because their indexes differ. Also, $2\sqrt{5}$ and $4\sqrt{3}$ are not like terms because their radicands differ.

EXAMPLE 1 Simplify $3\sqrt{2} + 7\sqrt{2}$

Solution: Since these are like radicals, we use the distributive property to combine their numerical coefficients.

$$3\sqrt{2} + 7\sqrt{2} = (3 + 7)\sqrt{2}$$
$$= 10\sqrt{2}$$

EXAMPLE 2 Simplify $5\sqrt[3]{7} + 2\sqrt{10} - 3\sqrt[3]{7} - 9\sqrt{10}$

Solution: Combine coefficients of like terms.

$$5\sqrt[3]{7} + 2\sqrt{10} - 3\sqrt[3]{7} - 9\sqrt{10} = \underline{5\sqrt[3]{7} - 3\sqrt[3]{7}} + \underline{2\sqrt{10} - 9\sqrt{10}}$$
$$= 2\sqrt[3]{7} \qquad - \qquad 7\sqrt{10}$$

Radicals should always be simplified before attempting to add or subtract.

EXAMPLE 3 Combine $5\sqrt{27} + 4\sqrt{75} - \sqrt{3}$

Solution: Simplify each radical, then combine like terms.
$$5\sqrt{27} + 4\sqrt{75} - \sqrt{3} = 5\sqrt{9 \cdot 3} + 4\sqrt{25 \cdot 3} - \sqrt{3}$$
$$= 5 \cdot \sqrt{9} \cdot \sqrt{3} + 4 \cdot \sqrt{25} \cdot \sqrt{3} - \sqrt{3}$$
$$= 5 \cdot 3 \cdot \sqrt{3} + 4 \cdot 5 \cdot \sqrt{3} - \sqrt{3}$$
$$= 15\sqrt{3} + 20\sqrt{3} - 1\sqrt{3}$$
$$= 34\sqrt{3}$$

Combine $2\sqrt[3]{16x^4} - \sqrt[3]{54x^4} + 3\sqrt{8y^3} - \sqrt{18y^3}$, where $y \geq 0$. **EXAMPLE 4**

Solution: Simplify each radical.
$$2\sqrt[3]{16x^4} - \sqrt[3]{54x^4} + 3\sqrt{8y^3} - \sqrt{18y^3}$$
$$= 2\sqrt[3]{(8x^3)2x} - \sqrt[3]{(27x^3)2x} + 3\sqrt{(4y^2) \cdot 2y} - \sqrt{(9y^2) \cdot 2y}$$
$$= 2 \cdot 2x \cdot \sqrt[3]{2x} - 3x \cdot \sqrt[3]{2x} + 3 \cdot 2y \cdot \sqrt{2y} - 3y\sqrt{2y}$$
$$= 4x\sqrt[3]{2x} - 3x\sqrt[3]{2x} + 6y\sqrt{2y} - 3y\sqrt{2y}$$

Combine only the like terms.
$$= x\sqrt[3]{2x} + 3y\sqrt{2y}$$

A variety of radical expressions can be multiplied by using the multiplication property of radicals together with the special products studied in chapter 3.

Multiply $(3\sqrt{5}) \cdot (2\sqrt{7})$ **EXAMPLE 5**

Solution: Multiply the coefficients, then use the multiplication property to multiply the radicals.
$$(3\sqrt{5})(2\sqrt{7}) = 3 \cdot 2 \cdot \sqrt{5} \cdot \sqrt{7}$$
$$= 6 \cdot \sqrt{5 \cdot 7}$$
$$= 6\sqrt{35}$$

Multiply $(5\sqrt[3]{2a^2})(4\sqrt[3]{3a})$ **EXAMPLE 6**

Solution: Multiply the coefficients and multiply the radicals. Always simplify the answer.
$$(5\sqrt[3]{2a^2})(4\sqrt[3]{3a}) = 20\sqrt[3]{6a^3}$$
$$= 20a\sqrt[3]{6}$$

Multiply $\sqrt{3}(\sqrt{7} + \sqrt{5})$ **EXAMPLE 7**

Solution:
Use the distributive property and multiply $\sqrt{3}$ times each term of the binomial.
$$\sqrt{3}(\sqrt{7} + \sqrt{5}) = \sqrt{3} \cdot \sqrt{7} + \sqrt{3} \cdot \sqrt{5}$$
$$= \sqrt{21} + \sqrt{15}$$

Radicals having binomial form can be multiplied using the special products studied in section 3.3.

1. Product of Two Binomials (The FOIL Shortcut)

$$(a + b)(c + d) = \overset{F}{ab} + \overset{O}{ad} + \overset{I}{bc} + \overset{L}{bd}$$

2. Product of a Sum Times a Difference

$$(a + b)(a - b) = a^2 - b^2$$

3. Squaring a Binomial

$$(a + b)^2 = a^2 + 2ab + b^2$$
$$(a - b)^2 = a^2 - 2ab + b^2$$

EXAMPLE 8   Multiply $(2\sqrt{3} + 5)(\sqrt{7} - 1)$

Solution: Apply the FOIL Shortcut:

$$(2\sqrt{3} + 5)(\sqrt{7} - 1) = \overset{F}{2\sqrt{3} \cdot \sqrt{7}} - \overset{O}{2\sqrt{3} \cdot 1} + \overset{I}{5 \cdot \sqrt{7}} - \overset{L}{5 \cdot 1}$$
$$= 2\sqrt{21} - 2\sqrt{3} + 5\sqrt{7} - 5$$

This result cannot be simplified further.

EXAMPLE 9   Multiply $(3\sqrt{6} + \sqrt{2})(3\sqrt{6} - \sqrt{2})$

Solution: This is a sum times a difference, which produces a difference of two squares.

$$(3\sqrt{6} + \sqrt{2})(3\sqrt{6} - \sqrt{2}) = 3\sqrt{6} \cdot 3\sqrt{6} - \sqrt{2} \cdot \sqrt{2}$$
$$= 9 \cdot 6 - 2$$
$$= 54 - 2$$
$$= 52$$

EXAMPLE 10   Simplify $(3\sqrt{2} + \sqrt{5})^2$

Solution: Square the binomial:

$$(3\sqrt{2} + \sqrt{5})^2 = (3\sqrt{2})^2 + 2 \cdot 3\sqrt{2} \cdot \sqrt{5} + (\sqrt{5})^2$$
$$= 9 \cdot 2 + 6\sqrt{10} + 5$$
$$= 18 + 6\sqrt{10} + 5$$
$$= 23 + 6\sqrt{10}$$

This final example shows how to simplify a fraction containing a radical.

EXAMPLE 11   Simplify $\dfrac{16 + \sqrt{20}}{4}$

OPERATING WITH RADICAL EXPRESSIONS

Solution: Simplify the radical first:
$$\frac{16 + \sqrt{20}}{4} = \frac{16 + \sqrt{4 \cdot 5}}{4}$$
$$= \frac{16 + 2\sqrt{5}}{4}$$

Remove the common factor of 2 from the numerator. Then, reduce the fraction.
$$= \frac{2(8 + \sqrt{5})}{4}$$
$$= \frac{\cancel{2}(8 + \sqrt{5})}{\cancel{2} \cdot 2}$$
$$= \frac{8 + \sqrt{5}}{2}$$

Combine.

Exercise 5.5

1. $5\sqrt{12} + 3\sqrt{3}$
2. $3\sqrt{2} - 2\sqrt{18}$
3. $3\sqrt{27} + 4\sqrt{75}$
4. $3\sqrt{8} - 2\sqrt{72} - 3\sqrt{200}$
5. $7\sqrt[3]{54} - 3\sqrt[3]{16}$
6. $\sqrt{72} - 3\sqrt{12} + \sqrt{32}$
7. $\sqrt{45} - \sqrt{20} + \sqrt{500}$
8. $2\sqrt[3]{16} - 4\sqrt[3]{54}$
9. $2\sqrt{12} - 4\sqrt{27} + \sqrt{\dfrac{3}{4}}$
10. $\sqrt[3]{16} + 8\sqrt[3]{\dfrac{2}{27}}$
11. $4\sqrt{28} + 3\sqrt{63} - 2\sqrt{112}$
12. $3\sqrt[4]{32} + 2\sqrt[4]{162}$
13. $\sqrt[3]{81} - 3\sqrt[3]{24}$
14. $\sqrt{27x} - \sqrt{75x} + \sqrt{12x}$, $x \geq 0$
15. $\sqrt[3]{x^4y^2} + 2x\sqrt[3]{xy^2}$
16. $x\sqrt{xy^6} - y^3\sqrt{x^3} + xy^3\sqrt{x}$, $x \geq 0, y \geq 0$
17. $\sqrt{50x^2} - 8x\sqrt{18} - 2\sqrt{72x^2}$, $x \geq 0$
18. $2\sqrt[3]{54} - 3\sqrt[3]{16} + 4\sqrt{81}$
19. $10\sqrt[3]{40x} - 2\sqrt[3]{5x} + \sqrt[3]{625x}$
20. $5\sqrt[3]{a^4b^3c^2} - ab\sqrt[3]{ac^2}$

Multiply and simplify where possible.

21. $(\sqrt{3} \cdot \sqrt{6})^2$
22. $\sqrt{2}(\sqrt{18} - \sqrt{6})$
23. $\sqrt{2}(\sqrt{32} + \sqrt{6})$
24. $3\sqrt{5}(\sqrt{15} - 2)$
25. $\sqrt{5}(\sqrt{10} - \sqrt{5})$
26. $\sqrt{\dfrac{a}{b}} \cdot \sqrt{ab}$, $a > 0, b > 0$
27. $\sqrt{3x^2y^3} \cdot \sqrt{12x^5y}$, where $x \geq 0, y \geq 0$

28. $(\sqrt{6} + \sqrt{2})(\sqrt{6} - \sqrt{2})$
29. $(\sqrt{5} + 2\sqrt{3})(\sqrt{5} - 2\sqrt{3})$
30. $(\sqrt{x} + \sqrt{y})(\sqrt{x} - \sqrt{y})$, where $x \geq 0$, $y \geq 0$
31. $(1 + \sqrt{2})(2 - \sqrt{2})$
32. $(\sqrt{3} - \sqrt{5})(2\sqrt{3} + \sqrt{6})$
33. $(2\sqrt{3} + 1)^2$
34. $(5\sqrt{2} - 1)^2$
35. $(\sqrt{8} - \sqrt{2})^2$
36. $(\sqrt{3} + 2)^2 \cdot \sqrt{2}$
37. $(4\sqrt{x} + 1)^2$, where $x \geq 0$
38. $(\sqrt{7} + \sqrt{5})(2\sqrt{7} - 3\sqrt{5})$
39. $(3\sqrt{x} + 2\sqrt{y})^2$, where $x \geq 0$, $y \geq 0$
40. $(3\sqrt{2} - 2\sqrt{3})^2$
41. $(2\sqrt{7} - 3\sqrt{5})(4\sqrt{7} + 2\sqrt{5})$
42. $\sqrt{x}(\sqrt{x} + \sqrt{x^3})$, where $x \geq 0$
43. $(\sqrt[3]{2} + 6)(\sqrt[3]{2} - 6)$
44. $(2 + \sqrt[3]{3})(2 - \sqrt[3]{3})$

Simplify.

45. $\dfrac{8 + \sqrt{8}}{8}$
46. $\dfrac{10 + \sqrt{48}}{2}$
47. $\dfrac{20 + \sqrt{20}}{4}$
48. $\dfrac{8 + 2\sqrt{12}}{4}$
49. $\dfrac{15 + 10\sqrt{3}}{10}$
50. $\dfrac{-5 - 5\sqrt{2}}{5}$
51. $\dfrac{7 + \sqrt{49 - 4(10)}}{2}$

RATIONALIZING DENOMINATORS

5.6

A fraction is not considered to be in simplest form when it contains a radical in the denominator. The process of converting a denominator from a radical (irrational number) to a rational number is called *rationalizing the denominator*. In this section you will learn how to rationalize the following two types of denominators:

(1) Monomial denominators containing a radical, such as $\dfrac{6}{5\sqrt{3}}$

(2) Binomial denominators containing square roots, such as $\dfrac{6}{2 + \sqrt{5}}$

To rationalize a monomial denominator containing an nth root ($\sqrt[n]{}$), we multiply the numerator and denominator by a radical that will produce a perfect n power in the denominator. Such a radical is called a *rationalizing factor*.

EXAMPLE 1

Rationalize the denominator of $\dfrac{7}{2\sqrt{5}}$

Solution: Multiply the numerator and denominator by the rationalizing factor, $\sqrt{5}$. This will produce a perfect square of 25 in the denominator.

$$\frac{7}{2\sqrt{5}} = \frac{7 \cdot \sqrt{5}}{2\sqrt{5} \cdot \sqrt{5}}$$

$$= \frac{7\sqrt{5}}{2\sqrt{25}}$$

$$= \frac{7\sqrt{5}}{2 \cdot 5}$$

$$= \frac{7\sqrt{5}}{10}$$

EXAMPLE 2

Rationalize the denominator of $\sqrt{\frac{3}{5}}$

Solution: Use the division property of radicals to obtain a quotient of two radicals. You will then see that the rationalizing factor is $\sqrt{5}$.

$$\sqrt{\frac{3}{5}} = \frac{\sqrt{3}}{\sqrt{5}} \qquad \text{[division property of radicals]}$$

$$= \frac{\sqrt{3} \cdot \sqrt{5}}{\sqrt{5} \cdot \sqrt{5}}$$

$$= \frac{\sqrt{15}}{\sqrt{25}}$$

$$= \frac{\sqrt{15}}{5}$$

EXAMPLE 3

Simplify $\frac{6\sqrt{7}}{5\sqrt{3}}$

Solution: Multiply the numerator and denominator by $\sqrt{3}$. Always reduce the result to lowest terms.

$$\frac{6\sqrt{7}}{5\sqrt{3}} = \frac{6\sqrt{7} \cdot \sqrt{3}}{5\sqrt{3} \cdot \sqrt{3}}$$

$$= \frac{6\sqrt{21}}{5\sqrt{9}}$$

$$= \frac{6\sqrt{21}}{5 \cdot 3}$$

$$= \frac{2\sqrt{21}}{5}$$

The same process is used to rationalize monomial denominators containing higher roots, such as cube roots, fourth roots, and so on. Unlike square roots, you will not necessarily select a rationalizing factor equal to the radical appearing in the denominator.

EXAMPLE 4

Simplify $\dfrac{2\sqrt[3]{5}}{\sqrt[3]{4x}}$

Solution: Multiply the numerator and denominator by $\sqrt[3]{2x^2}$. This produces a perfect cube of $8x^3$ in the denominator.

$$\dfrac{2\sqrt[3]{5}}{\sqrt[3]{4x}} = \dfrac{2\sqrt[3]{5} \cdot \sqrt[3]{2x^2}}{\sqrt[3]{4x} \cdot \sqrt[3]{2x^2}}$$

$$= \dfrac{2\sqrt[3]{10x^2}}{\sqrt[3]{8x^3}}$$

$$= \dfrac{2\sqrt[3]{10x^2}}{2x} \quad \text{[reduce to lowest terms]}$$

$$= \dfrac{\sqrt[3]{10x^2}}{x}$$

Next, we discuss how to rationalize a binomial denominator containing a square root. You should recall that a sum times a difference produces a difference of two squares. Thus, our rationalizing factor will be a binomial identical to the denominator except for its middle sign. This rationalizing factor is called the conjugate.

EXAMPLE 5

Each of these binomials is followed by its conjugate.

Expression	Conjugate
$1 + \sqrt{3}$	$1 - \sqrt{3}$
$\sqrt{5} - 3$	$\sqrt{5} + 3$
$2\sqrt{7} + \sqrt{3}$	$2\sqrt{7} - \sqrt{3}$

The following examples illustrate how conjugates are used to rationalize binomial denominators containing square roots.

EXAMPLE 6

Simplify $\dfrac{3}{\sqrt{7} - \sqrt{3}}$

Solution: Multiply the numerator and denominator by the conjugate, $\sqrt{7} + \sqrt{3}$. This will produce a difference of two squares in the denominator.

$$\frac{3}{\sqrt{7}-\sqrt{3}} = \frac{3\cdot(\sqrt{7}+\sqrt{3})}{(\sqrt{7}-\sqrt{3})\cdot(\sqrt{7}+\sqrt{3})}$$
$$= \frac{3(\sqrt{7}+\sqrt{3})}{\sqrt{49}-\sqrt{9}}$$
$$= \frac{3(\sqrt{7}+\sqrt{3})}{7-3}$$
$$= \frac{3(\sqrt{7}+\sqrt{3})}{4} \text{ or } \frac{3\sqrt{7}+3\sqrt{3}}{4}$$

EXAMPLE 7

Simplify $\dfrac{\sqrt{3x}+2}{\sqrt{2x}+5}$, where $x \geq 0$.

Solution: Multiply the numerator and denominator by $\sqrt{2x}-5$.

$$\frac{\sqrt{3x}+2}{\sqrt{2x}+5} = \frac{(\sqrt{3x}+2)\cdot(\sqrt{2x}-5)}{(\sqrt{2x}+5)\cdot(\sqrt{2x}-5)}$$

$$\overset{\text{F} \qquad \text{O} \qquad \text{I} \qquad \text{L}}{= \frac{\sqrt{3x}\cdot\sqrt{2x} - 5\cdot\sqrt{3x} + 2\cdot\sqrt{2x} - 2\cdot 5}{\sqrt{4x^2}-25}}$$

$$= \frac{\sqrt{6x^2} - 5\sqrt{3x} + 2\sqrt{2x} - 10}{2x-25}$$

$$= \frac{x\sqrt{6} - 5\sqrt{3x} + 2\sqrt{2x} - 10}{2x-25}$$

Simplify.

Exercise 5.6

1. $\dfrac{3}{\sqrt{3}}$
2. $\dfrac{5}{\sqrt{2}}$
3. $\dfrac{1}{\sqrt{7}}$
4. $\dfrac{\sqrt{3}}{2\sqrt{2}}$
5. $\dfrac{2\sqrt{3}}{4\sqrt{5}}$
6. $\dfrac{\sqrt{x^6 y}}{\sqrt{3y}}$, $x>0, y>0$
7. $\dfrac{3}{\sqrt{18}}$
8. $\dfrac{2}{\sqrt[3]{9}}$
9. $\dfrac{\sqrt[3]{5}}{\sqrt[3]{4}}$
10. $\sqrt[3]{\dfrac{9}{32}}$
11. $\sqrt[3]{\dfrac{x^9}{y}}$, $y \neq 0$
12. $\sqrt{\dfrac{1}{6}}$
13. $\sqrt[3]{\dfrac{8b^3}{9c}}$, $c \neq 0$
14. $\sqrt{\dfrac{49x^2}{2}}$
15. $\dfrac{3}{\sqrt[3]{9}}$
16. $\dfrac{6}{\sqrt{2x}}$, $x>0$
17. $\dfrac{8+\sqrt{8}}{\sqrt{8}}$
18. $\dfrac{5-\sqrt{5}}{\sqrt{5}}$

EXPONENTS, ROOTS, AND RADICALS

19. $\dfrac{\sqrt{6} + \sqrt{5}}{\sqrt{2}}$

20. $\dfrac{-8\sqrt{5}}{\sqrt{12}}$

21. $\sqrt{\dfrac{72a^8}{b^3}},\ b > 0$

22. $\dfrac{24\sqrt[4]{64a}}{4\sqrt[4]{4a}},\ a > 0.$

23. $\dfrac{x}{\sqrt{xy}},\ x > 0,\ y > 0$

24. $\dfrac{3}{\sqrt{3} - \sqrt{2}}$

25. $\dfrac{3}{2 + \sqrt{3}}$

26. $\dfrac{\sqrt{7}}{\sqrt{7} - 2}$

27. $\dfrac{2\sqrt{3}}{2\sqrt{3} - 3}$

28. $\dfrac{2}{\sqrt{5} + \sqrt{2}}$

29. $\dfrac{4\sqrt{2}}{\sqrt{2} + \sqrt{6}}$

30. $\dfrac{3}{4 - \sqrt{5}}$

31. $\dfrac{\sqrt{3} - 1}{\sqrt{2} + \sqrt{3}}$

32. $\dfrac{3 + \sqrt{6}}{\sqrt{3} - \sqrt{2}}$

33. $\sqrt{\dfrac{1}{2} + \dfrac{1}{3}}$

34. $\dfrac{\sqrt{3} + \sqrt{2}}{\sqrt{3} - \sqrt{2}}$

35. $\dfrac{\sqrt{7} + \sqrt{5}}{\sqrt{7} - \sqrt{5}}$

36. $\dfrac{x}{x + \sqrt{y}},\ y > 0$

37. $\dfrac{\sqrt{x}}{\sqrt{x} - \sqrt{y}},\ x > 0,\ y > 0,\ x \ne y$

38. $\dfrac{3\sqrt{2} - \sqrt{5}}{3\sqrt{3} + \sqrt{5}}$

39. $\dfrac{1}{3\sqrt{x} + 2},\ x > 0$

40. $\dfrac{ab}{\sqrt{a^2 + b^2}}$

41. $\sqrt{48} - \sqrt{\dfrac{1}{3}}$

42. $\sqrt{50} + \sqrt{\dfrac{1}{2}}$

43. $\sqrt{90} - \sqrt{\dfrac{2}{5}}$

44. $\sqrt{\dfrac{1}{3}} + \dfrac{2\sqrt{3}}{3}$

45. $\sqrt[3]{72} + \dfrac{3}{\sqrt[3]{3}}$

46. $\dfrac{4}{\sqrt{6}} - \dfrac{2}{\sqrt{3}}$

47. $\sqrt{3} + \dfrac{2}{\sqrt{3}} - \sqrt{12}$

48. $\dfrac{x - 4}{\sqrt{x} + 2},\ x > 0$

49. $\dfrac{\sqrt{x}}{\sqrt{x} + \sqrt{y}},\ x > 0,\ y > 0$

50. $\dfrac{\sqrt{6} - 2}{3 - \sqrt{8}}$

Calculator problems. Use your calculator and round answers to three significant digits.

51. $\dfrac{5\sqrt{8} - 3\sqrt{5}}{\sqrt{7}}$

52. $\dfrac{4\sqrt{3}}{3\sqrt{2} - 5\sqrt{3}}$

53. $\dfrac{2\sqrt[3]{6} - 5\sqrt[3]{9}}{\sqrt{5}}$

54. $\dfrac{\sqrt[3]{17.62}}{\sqrt[5]{0.1648}}$

55. $\dfrac{6\sqrt[3]{17} + 3\sqrt[4]{23}}{\sqrt[5]{30}}$

56. $\dfrac{\sqrt[3]{32.4}}{5\sqrt{14.3} - 2\sqrt[3]{10}}$

SOLVING EQUATIONS CONTAINING RADICALS

5.7

In this section we develop a method for solving equations having radicals. We will eliminate radicals by raising both sides of the equation to a power which is the same as the index of the radical. However, this procedure can lead to extraneous solutions. Remember, an extraneous solution is caused by the solution process and will not check in the original equation. For example, if $x = 2$, we can square both sides to obtain $x^2 = 4$. The original equation, $x = 2$, has one solution—the number 2. The second equation has two solutions—the numbers 2 and -2. Observe that -2 will not check in the original equation $x = 2$. It is an extraneous solution brought about by the squaring process. This discussion gives rise to the following property.

The Power Property of Equations

Raising both sides of an equation to the same power will preserve every solution of the original equation, but may produce some extraneous solutions that will not check in the original equation.

To solve equations containing radicals, follow this procedure:

1. Isolate a single radical on one side of the equation.
2. Use the power property of equations by raising both sides to a power which is the same as the index of the radical.
3. Solve the resulting equation. If it still contains a radical, repeat the first two steps until all radicals are eliminated.
4. Check all proposed solutions in the original equation and discard the extraneous ones.

Solve $\sqrt{6m + 7} + 1 = 0$ **EXAMPLE 1**

Solution: Isolate the radical on one side.
$$\sqrt{6m + 7} + 1 = 0$$
$$\sqrt{6m + 7} = -1$$
Square both sides of the equation.
$$(\sqrt{6m + 7})^2 = (-1)^2$$
$$6m + 7 = 1$$
Solve the equation for m.
$$6m = -6$$
$$m = -1$$

EXPONENTS, ROOTS, AND RADICALS

Check the proposed answer in the original equation.
$$\sqrt{6m + 7} + 1 = 0$$
$$\sqrt{6(-1) + 7} + 1 = 0$$
$$\sqrt{-6 + 7} + 1 = 0$$
$$\sqrt{1} + 1 = 0$$
$$1 + 1 = 0$$
$$2 = 0 \quad \text{[false]}$$

Thus, -1 is an extraneous solution. The equation $\sqrt{6m + 7} + 1 = 0$ has no solution.

EXAMPLE 2 Solve $\sqrt{p + 5} - \sqrt{p} = 1$

Solution: Isolate a radical on one side.
$$\sqrt{p + 5} - \sqrt{p} = 1$$
$$\sqrt{p + 5} = \sqrt{p} + 1$$

Square both sides. Note that the right side is a binomial.
$$(\sqrt{p + 5})^2 = (\sqrt{p} + 1)^2$$
$$p + 5 = p + 2\sqrt{p} + 1$$

The equation still contains a radical. Repeat the first two steps.
$$4 = 2\sqrt{p}$$
$$4^2 = (2\sqrt{p})^2$$
$$16 = 4p$$
$$4 = p$$

Check for an extraneous solution.
$$\sqrt{p + 5} - \sqrt{p} = 1$$
$$\sqrt{4 + 5} - \sqrt{4} = 1$$
$$\sqrt{9} - \sqrt{4} = 1$$
$$3 - 2 = 1$$
$$1 = 1 \quad \text{[true]}$$

Thus, the solution is 4.

When solving equations with radicals, it is common to obtain a non-linear equation containing a polynomial. As you learned in section 3.6, such an equation is solved by applying the zero factor property, as the next example shows.

EXAMPLE 3 Solve $\sqrt{8x + 1} - \sqrt{4x - 3} = 2$

Solution: Isolate a radical on one side.
$$\sqrt{8x + 1} - \sqrt{4x - 3} = 2$$
$$\sqrt{8x + 1} = 2 + \sqrt{4x - 3}$$

Square both sides. Note that the right side is a binomial.
$$(\sqrt{8x + 1})^2 = (2 + \sqrt{4x - 3})^2$$
$$8x + 1 = 4 + 4\sqrt{4x - 3} + 4x - 3$$

Since the equation still contains a radical, repeat the first two steps.

$$4x = 4\sqrt{4x - 3} \quad \text{[divide both sides by 4]}$$
$$x = \sqrt{4x - 3}$$
$$x^2 = (\sqrt{4x - 3})^2$$
$$x^2 = 4x - 3$$
$$x^2 - 4x + 3 = 0 \quad \text{[use the zero factor property]}$$
$$(x - 3)(x - 1) = 0$$
$$x - 3 = 0 \quad \text{or} \quad x - 1 = 0$$
$$x = 3 \quad\quad\quad\quad x = 1$$

Check for extraneous solutions

$$\sqrt{8x + 1} - \sqrt{4x - 3} = 2 \quad\quad \sqrt{8x + 1} - \sqrt{4x - 3} = 2$$
$$\sqrt{8 \cdot 3 + 1} - \sqrt{4 \cdot 3 - 3} = 2 \quad\quad \sqrt{8 \cdot 1 + 1} - \sqrt{4 \cdot 1 - 3} = 2$$
$$\sqrt{25} - \sqrt{9} = 2 \quad\quad\quad\quad \sqrt{9} - \sqrt{1} = 2$$
$$5 - 3 = 2 \quad\quad\quad\quad\quad\quad 3 - 1 = 2$$

Both proposed solutions check. The solutions are 3 and 1.

As illustrated by the next example, the power property of equations is also used when radicals have an index greater than two.

Solve $\sqrt[4]{7y + 11} - \sqrt[4]{6y + 21} = 0$

EXAMPLE 4

Solution: Isolate the radicals and raise each side to the fourth power.
$$\sqrt[4]{7y + 11} - \sqrt[4]{6y + 21} = 0$$
$$\sqrt[4]{7y + 11} = \sqrt[4]{6y + 21}$$
$$(\sqrt[4]{7y + 11})^4 = (\sqrt[4]{6y + 21})^4$$
$$7y + 11 = 6y + 21$$
$$y = 10$$

Check for extraneous solutions.
$$\sqrt[4]{7y + 11} - \sqrt[4]{6y + 21} = 0$$
$$\sqrt[4]{7 \cdot 10 + 11} - \sqrt[4]{6 \cdot 10 + 21} = 0$$
$$\sqrt[4]{81} - \sqrt[4]{81} = 0$$
$$3 - 3 = 0$$
$$0 = 0 \quad \text{[true]}$$

The proposed answer checks. The solution is 10.

Exercise 5.7

Solve and check each equation.

1. $\sqrt{x} = 6$
2. $\sqrt{x - 1} = 7$
3. $\sqrt{2x - 3} = 5$
4. $\sqrt{x - 2} - 3 = 0$
5. $\sqrt{2x + 5} = 4$
6. $\sqrt{2x - 3} - 1 = 0$
7. $\sqrt{9 - 5x} = 3$
8. $2\sqrt{2x} = 12$
9. $\sqrt{x - 2} = 2 - x$
10. $\sqrt{3x + 1} + 1 = x$
11. $\sqrt{x^2 + 3x - 2} = 0$
12. $2x = \sqrt{7x - 3} + 3$
13. $\sqrt{3x + 10} = x + 4$
14. $\sqrt{x - 3} = x - 5$
15. $\sqrt{x + 3} = 2\sqrt{x}$
16. $\sqrt{x - 2} = 4 - x$
17. $\sqrt{x + 4} = x - 8$
18. $\sqrt{x + 5} - \sqrt{x} = 1$
19. $\sqrt{3x + 1} = \sqrt{x} + 3$
20. $\sqrt{5x - 1} - \sqrt{x} = 1$
21. $\sqrt{3x + 4} - \sqrt{x} = 2$
22. $\sqrt{4x + 1} = \sqrt{6x - 5}$
23. $\sqrt{2x + 6} - \sqrt{x + 4} = 1$
24. $\sqrt{x} + \sqrt{2} = \sqrt{x + 2}$
25. $\sqrt{2x + 3} + 3 = 4 + \sqrt{x + 1}$
26. $\sqrt{2x + 7} - \sqrt{x + 3} = 5$
27. $\sqrt{x + 6} - 2 = \sqrt{x}$
28. $\sqrt{x - 6} = \sqrt{x + 4} - \sqrt{2}$
29. $\sqrt[3]{x} = -2$
30. $\sqrt[3]{x} = -3$
31. $\sqrt[3]{4x + 4} = 3$
32. $\sqrt[4]{x - 1} = 1$
33. $\sqrt[4]{3x + 1} - 2 = 0$
34. $\sqrt[4]{5x - 9} + \sqrt[4]{4x - 1} = 0$
35. $\sqrt[4]{4x - 1} - \sqrt[4]{2x + 1} = 0$
36. $\sqrt[3]{2x - 11} - \sqrt[3]{5x + 1} = 0$

Summary

DEFINITIONS

$a^0 = 1$, where $a \neq 0$

$a^{-n} = \dfrac{1}{a^n}$, where $a \neq 0$

$\sqrt{a^2} = |a|$, $a \in R$

If $\sqrt[n]{a} = b$, then $b^n = a$

$a^{1/n} = \sqrt[n]{a}$, $n \in I$ and $n \geq 2$

$a^{m/n} = \sqrt[n]{a^m}$ or $(\sqrt[n]{a})^m$, where $m, n \in I$ and $n \geq 2$. If n is even, then $a \geq 0$.

PROPERTIES

n even

$\sqrt[n]{a}$ represents the principal root of a

1. $\sqrt[n]{a^n} = |a|$
2. $\sqrt[n]{a} \geq 0$ if $a \geq 0$
3. $\sqrt[n]{a}$ does not exist if $a < 0$

n odd

$\sqrt[n]{a}$ is not concerned with the principal root.
1. $\sqrt[n]{a} \geq 0$ if $a \geq 0$
2. $\sqrt[n]{a} < 0$ if $a < 0$

Multiplication Property

$\sqrt[n]{a} \cdot \sqrt[n]{b} = \sqrt[n]{ab}$ where *n* is odd and $a, b \in R$ or *n* is even and $a, b, \geq 0$

Division Property

$\dfrac{\sqrt[n]{a}}{\sqrt[n]{b}} = \sqrt[n]{\dfrac{a}{b}}$ where $b \neq 0$ and *n* is odd and $a, b \in R$ or *n* is even and $a \geq 0, b > 0$.

Power Property of Equations

Raising both sides of an equation to the same power will preserve every solution of the original equation but may produce some extraneous solutions that will not check in the original.

Self-Checking Exercise

Simplify.

1. $\sqrt{0}$
2. $-\sqrt{169}$
3. $\sqrt{-36}$
4. $\sqrt{121x^2}$
5. $\sqrt[3]{-1000}$
6. $\sqrt[5]{32x^{10}}$
7. $\sqrt[3]{8x^3y^6}$
8. $\sqrt[3]{\dfrac{a^3}{b^3}}, b \neq 0$
9. $\sqrt{9x^4}, x \geq 0$
10. $\sqrt{63}$
11. $\sqrt{60x^2}$
12. $\sqrt{147}$
13. $\sqrt[3]{48}$
14. $\sqrt[3]{40}$
15. $\sqrt{\dfrac{63}{4x^2}}, x \neq 0$
16. $\sqrt{2x^4} \cdot \sqrt{8x^3}, x > 0$
17. $\sqrt[3]{\dfrac{x^5y^3}{x^2y}}, x, y \neq 0$
18. $\sqrt[4]{32x^4}, x \geq 0$

Evaluate.

19. $-(16)^{3/4}$
20. $-32^{3/5}$
21. $81^{-(3/2)}$
22. $\left(\dfrac{121}{9}\right)^{-(3/2)}$

Write in radical form.

23. $3x^{3/4}$
24. $(3x)^{2/5}$
25. $(1 - x^2)^{1/3}$
26. $(2 + x)^{-(3/5)}$

Simplify by rationalizing denominators. Assume all variables represent positive numbers and denominators are not equal to zero.

27. $\dfrac{5}{\sqrt{3}}$
28. $\dfrac{44}{\sqrt{2x}}$
29. $\dfrac{-7}{\sqrt{27}}$
30. $\dfrac{2 + \sqrt{3}}{\sqrt{3}}$
31. $\dfrac{1}{2 - \sqrt{2}}$
32. $\dfrac{-2}{2 + \sqrt{5}}$
33. $\dfrac{x}{\sqrt{x} - 3}$
34. $\dfrac{\sqrt{2} + \sqrt{3}}{\sqrt{2} - \sqrt{3}}$
35. $\dfrac{\sqrt{x} - \sqrt{2y}}{\sqrt{x} + \sqrt{2y}}$

Combine and simplify.

36. $\sqrt{12} - 2\sqrt{27} + 3\sqrt{75}$
37. $\sqrt{5}(\sqrt{15} - \sqrt{3})$
38. $4\sqrt{18} + 4\sqrt{50} - 2\sqrt{8}$
39. $\sqrt[3]{54} - 5\sqrt[3]{2}$
40. $\sqrt[3]{4x^2} \cdot \sqrt[3]{2xy^3}$
41. $\dfrac{\sqrt{ab}}{\sqrt{ab^2} \cdot \sqrt{a^2b}}; a \neq 0, b \neq 0$
42. $(3\sqrt{3} - \sqrt{2})^2$
43. $(\sqrt{3} + 2\sqrt{2})^2$
44. $(\sqrt{7} + \sqrt{11})(\sqrt{7} - \sqrt{11})$
45. $(\sqrt{x} - 2)(2\sqrt{x} + 1), x > 0$
46. $6\sqrt[3]{128} - 3\sqrt[3]{16}$
47. $2\sqrt{3x^3} + x\sqrt{12x}, x > 0$

Solve.
48. $\sqrt{y} - 6 = 0$
49. $x = \sqrt{x^2 - 4x - 8}$
50. $2\sqrt{x} = \sqrt{3x + 25}$
51. $\sqrt{x - 3} + 5 = x$
52. $\sqrt[3]{a + 1} + 1 = 0$

Miscellaneous.
53. Convert 21,500,000 to scientific notation.
54. Convert 0.000324 to scientific notation.
55. Convert 1.13×10^{-3} to ordinary notation.
56. Simplify $\dfrac{2x^{-3}y^2}{z^{-4}}$, where $z \neq 0$.
57. Simplify $\dfrac{x^{-1} + y^{-1}}{2x^{-1}}$, where $x, y \neq 0$.

Solutions to Self-Checking Exercise

1. 0
2. -13
3. not a real number
4. $11 \cdot |x|$
5. -10
6. $2x^2$
7. $2xy^2$
8. $\dfrac{a}{b}$
9. $3x^2$
10. $3\sqrt{7}$
11. $2|x|\sqrt{15}$
12. $7\sqrt{3}$
13. $2\sqrt[3]{6}$
14. $2\sqrt[3]{5}$
15. $\dfrac{3\sqrt{7}}{2|x|}$
16. $4x^3\sqrt{x}$
17. $x\sqrt[3]{y^2}$
18. $2x\sqrt[4]{2}$
19. -8
20. -8
21. $\dfrac{1}{729}$
22. $\dfrac{27}{1331}$
23. $3\sqrt[4]{x^3}$
24. $\sqrt[5]{(3x)^2}$
25. $\sqrt[3]{1 - x^2}$
26. $\dfrac{1}{\sqrt[5]{(2 + x)^3}}$
27. $\dfrac{5\sqrt{3}}{3}$
28. $\dfrac{22\sqrt{2x}}{x}$
29. $\dfrac{-7\sqrt{3}}{9}$
30. $\dfrac{2\sqrt{3} + 3}{3}$
31. $\dfrac{2 + \sqrt{2}}{2}$
32. $2(2 - \sqrt{5})$
33. $\dfrac{x(\sqrt{x} + 3)}{x - 9}$
34. $-5 - 2\sqrt{6}$
35. $\dfrac{x - 2\sqrt{2xy} + 2y}{x - 2y}$
36. $11\sqrt{3}$
37. $5\sqrt{3} - \sqrt{15}$
38. $28\sqrt{2}$
39. $-2\sqrt[3]{2}$
40. $2xy$
41. $\dfrac{1}{ab}$
42. $29 - 6\sqrt{6}$

43. $11 + 4\sqrt{6}$
44. -4
45. $2x - 3\sqrt{x} - 2$
46. $18\sqrt[3]{2}$
47. $4x\sqrt{3x}$
48. $y = 36$
49. no solution
50. $x = 25$
51. $x = 7$
52. $a = -2$
53. 2.15×10^7
54. 3.24×10^{-4}
55. 0.00113
56. $\dfrac{2y^2z^4}{x^3}$
57. $\dfrac{y + x}{2y}$

SOLVING QUADRATIC EQUATIONS AND INEQUALITIES

6

6.1 Solving Quadratic Equations by Factoring
6.2 Solving Quadratic Equations by Completing The Square
6.3 The Quadratic Formula
6.4 Solving Quadratic Inequalities
6.5 The Set of Complex Numbers
6.6 Operating with Complex Numbers

SOLVING QUADRATIC EQUATIONS BY FACTORING

6.1

A *quadratic equation* is any second degree equation which can be written in this form:

$$ax^2 + bx + c = 0, \text{ where } a, b, c \in R \text{ and } a \neq 0$$

EXAMPLE 1

Each of these is a quadratic equation:
(a) $2x^2 + 3x + 5 = 0$, where $a = 2$, $b = 3$, and $c = 5$.
(b) $4m^2 + m = 0$, where $a = 4$, $b = 1$, and $c = 0$.
(c) $y^2 - 16 = 0$, where $a = 1$, $b = 0$, and $c = -16$.

A quadratic equation is said to be in *standard form* when written as $ax^2 + bx + c = 0$.

EXAMPLE 2

Place the quadratic equation, $3x^2 = 1 - 2x$, in standard form and identify the values of a, b, and c.

Solution: Obtain all terms on one side, setting the expression equal to zero.

$$3x^2 = 1 - 2x$$
$$3x^2 + 2x - 1 = 0$$
$$3x^2 + 2x + (-1) = 0$$
$$\downarrow \quad \downarrow \quad \downarrow$$
$$a \quad b \quad c$$

Thus, $a = 3$, $b = 2$, and $c = -1$.

Sometimes equations that do not look quadratic can be simplified and placed in standard quadratic form.

EXAMPLE 3

Simplify the equation $\dfrac{10}{m + 2} + 12 = \dfrac{3}{m}$ to see if it fits the quadratic form.

Solution: Clear the equation of fractions by multiplying both sides by the LCD, $m(m + 2)$.

$$\frac{10}{m + 2} + 12 = \frac{3}{m}$$

$$m(m + 2) \cdot \left[\frac{10}{(m + 2)}\right] + m(m + 2) \cdot [12] = m(m + 2) \cdot \left[\frac{3}{m}\right]$$

$$10m + 12m(m + 2) = (m + 2) \cdot 3$$

Multiply and combine like terms.

$$10m + 12m^2 + 24m = 3m + 6$$
$$12m^2 + 31m - 6 = 0$$

Thus, the original equation can be simplified to a quadratic form where $a = 12$, $b = 31$, and $c = -6$.

SOLVING QUADRATIC EQUATIONS BY FACTORING

Many quadratic equations are factorable and can be solved using the zero factor property from section 3.7.

> **Zero Factor Property**
> If $m, n \in R$ and if $m \cdot n = 0$, then $m = 0$ or $n = 0$.

To use this property, the quadratic equation is simplified and placed in standard form. The polynomial is factored and each variable factor is set equal to zero.

Solve $x^2 - x = 12$ **EXAMPLE 4**

Solution: Place the equation in standard form and factor.
$$x^2 - x = 12$$
$$x^2 - x - 12 = 0$$
$$(x - 4)(x + 3) = 0$$
Use the zero factor property to set both factors equal to zero.
$$x - 4 = 0 \quad \text{or} \quad x + 3 = 0$$
$$x = 4 \qquad\qquad x = -3$$
The solutions are 4 and -3. Remember, you should always check solutions by substituting into the original equation.

Solve $(p + 1)(3p + 1) = 2p(p + 1) + 1$ **EXAMPLE 5**

Solution: Multiply, combine like terms, and place in standard form.
$$(p + 1)(3p + 1) = 2p(p + 1) + 1$$
$$3p^2 + 4p + 1 = 2p^2 + 2p + 1$$
$$p^2 + 2p = 0$$
Factor, then apply the zero factor property.
$$p(p + 2) = 0$$
$$p = 0 \quad \text{or} \quad p + 2 = 0$$
$$p = 0 \qquad\qquad p = -2$$
The solutions are 0 and -2. Both check in the original equation.

Solve $\dfrac{5}{6} - \dfrac{1}{y} = \dfrac{1}{y + 1}$ **EXAMPLE 6**

Solution: Clear the equation of fractions by multiplying both sides by the LCD, $6y(y + 1)$.
$$\frac{5}{6} - \frac{1}{y} = \frac{1}{y + 1}$$
$$6y(y + 1) \cdot \frac{5}{6} - 6y(y + 1) \cdot \frac{1}{y} = 6y(y + 1) \cdot \frac{1}{(y + 1)}$$
$$5y(y + 1) - 6(y + 1) = 6y$$

Simplify and write in standard form.
$$5y^2 + 5y - 6y - 6 = 6y$$
$$5y^2 - 7y - 6 = 0$$
Factor, then apply the zero factor property.
$$(5y + 3)(y - 2) = 0$$
$$5y + 3 = 0 \quad \text{or} \quad y - 2 = 0$$
$$5y = -3 \qquad\qquad y = 2$$
$$y = -\frac{3}{5}$$

The solutions are $-\frac{3}{5}$ and 2. Both check in the original equation. Neither is extraneous.

Some non-quadratic equations can be placed in quadratic form and factored. Such equations are said to be quadratic in form.

EXAMPLE 7 Solve $y^4 - 7y^2 = 18$

Solution: Place in quadratic form and use the zero factor property.
$$y^4 - 7y^2 - 18 = 0$$
$$(y^2 - 9)(y^2 + 2) = 0$$
$$(y + 3)(y - 3)(y^2 + 2) = 0$$
$$y + 3 = 0 \quad \text{or} \quad y - 3 = 0 \quad \text{or} \quad y^2 + 2 = 0$$
$$y = -3 \quad \text{or} \quad y = 3 \quad \text{or} \quad y^2 = -2$$
Since the square of y cannot be negative, the only solutions are -3 and 3. Both check in the original equation.

EXAMPLE 8 Solve $x^{-2} - 2x^{-1} = 8$

Solution: Place in quadratic form and use the zero factor property.
$$x^{-2} - 2x^{-1} - 8 = 0$$
$$(x^{-1} - 4)(x^{-1} + 2) = 0$$
$$x^{-1} - 4 = 0 \quad \text{or} \quad x^{-1} + 2 = 0$$
$$x^{-1} = 4 \qquad\qquad x^{-1} = -2$$
$$\frac{1}{x} = 4 \qquad\qquad \frac{1}{x} = -2$$
$$x = \frac{1}{4} \qquad\qquad x = -\frac{1}{2}$$

The solutions are $\frac{1}{4}$ and $-\frac{1}{2}$. Both check in the original equation.

EXAMPLE 9 Solve $2n = \sqrt{6 - 5n}$

Solution: This equation contains a radical. Square both sides to produce a quadratic equation.

$$2n = \sqrt{6 - 5n}$$
$$(2n)^2 = (\sqrt{6 - 5n})^2$$
$$4n^2 = 6 - 5n$$
$$4n^2 + 5n - 6 = 0$$
$$(4n - 3)(n + 2) = 0$$
$$4n - 3 = 0 \quad \text{or} \quad n + 2 = 0$$
$$n = \frac{3}{4} \qquad\qquad n = -2$$

Both numbers must be checked in the original equation.

$$2n = \sqrt{6 - 5n} \qquad\qquad 2n = \sqrt{6 - 5n}$$
$$2 \cdot \frac{3}{4} = \sqrt{6 - 5 \cdot \frac{3}{4}} \qquad 2(-2) = \sqrt{6 - 5(-2)}$$
$$\qquad\qquad\qquad\qquad\qquad -4 = \sqrt{16}$$
$$\frac{3}{2} = \sqrt{\frac{9}{4}} \qquad\qquad\qquad -4 = 4 \quad \text{[false]}$$
$$\frac{3}{2} = \frac{3}{2} \quad \text{[true]}$$

-2 is extraneous. The only solution is $\frac{3}{4}$.

Exercise 6.1

Place the following in standard form and identify *a*, *b*, and *c*.
1. $x^2 - 3x = 5$
2. $4x^2 = 5x$
3. $2x^2 = 1 - 3x$
4. $\frac{3}{x} - 1 = \frac{4}{x^2}$

Solve the following equations.
5. $3x^2 = 5x$
6. $4c^2 = 8c$
7. $3y^2 + 2y = 1$
8. $2x^2 + 1 = 3x$
9. $5a^2 - 6 = -13a$
10. $5y^2 = 3(3 - 4y)$
11. $15x^2 = 8 - 14x$
12. $5y^2 = 42y + 27$
13. $3(x + 1)(x + 2) = x - (2x - 3)(x + 4)$
14. $(3y + 5)(2 - y) = (2y + 1)(3y + 2)$
15. $(x + 1)^2 - (x + 1) - 6 = 0$
16. $(y - 7)^2 - 10(y - 7) + 25 = 0$

17. $(2x - 1)^2 - 25 = 0$

18. $\dfrac{x}{2} - \dfrac{4}{x} = -\dfrac{7}{2}$

19. $\dfrac{x}{4} - \dfrac{3}{4} = \dfrac{1}{x}$

20. $3 = \dfrac{10}{y^2} - \dfrac{7}{y}$

21. $\dfrac{1}{y} + \dfrac{2}{y^2} = 3$

22. $7 + \dfrac{5}{y} + 2y = 0$

23. $\dfrac{4}{x + 1} + \dfrac{x - 1}{2} = 2$

24. $\dfrac{y}{y + 2} + \dfrac{4}{y - 1} = 2$

25. $\dfrac{y}{y + 1} - \dfrac{3}{y - 3} = 2$

26. $\dfrac{1}{x - 1} + \dfrac{1}{x - 4} = \dfrac{5}{4}$

27. $\dfrac{4}{3x} + \dfrac{3}{3x + 1} + 2 = 0$

28. $x^4 - 17x^2 + 16 = 0$

29. $x^4 - 8x^2 + 16 = 0$

30. $y^4 - 29y^2 + 100 = 0$

31. $y^4 - 10x^2 + 9 = 0$

32. $x^4 - 5x^2 + 4 = 0$

33. $x^4 - 3x^2 - 4 = 0$

34. $x^{-2} - x^{-1} - 12 = 0$

35. $x^{-2} + 9x^{-1} - 10 = 0$

36. $x^{-2} + 5x^{-1} + 6 = 0$

37. $x^{-4} - 9x^{-2} = 0$

38. $x^{-2} + 9x^{-1} - 10 = 0$

39. $y + y^{-1} = \left(\dfrac{6}{13}\right)^{-1}$

40. $6(y - 1)^{-2} + (y - 1)^{-1} = 2$

41. $7(y + 5)^{-1} = 6(y + 5)^{-2} + 1$

42. $\sqrt{x + 1} = 1 - \sqrt{2x}$

43. $\sqrt{2x + 4} = \sqrt{x + 3} + 1$

44. $\sqrt{x + 4} = 2 - \sqrt{2x}$

45. $3y - 14\sqrt{y} + 8 = 0$

46. $5y - 7\sqrt{y} + 2 = 0$

47. $4\sqrt{y - 4} = y$

Solve the following applied problems.

48. The sum of the squares of two consecutive odd numbers is 74. Find the numbers.

49. If two times an integer is added to four times the square of the integer, the result is 110. Find the integer(s).

50. A number less two times its reciprocal is $\dfrac{17}{3}$. Find the number(s).

SOLVING QUADRATIC EQUATIONS BY COMPLETING THE SQUARE

6.2

In the previous section quadratic equations were solved by factoring. However, many trinomials are not factorable using integers. Thus, it is not possible to solve all quadratic equations by factoring. In this section we develop a procedure called completing the square, which can be used to solve any quadratic equation. Completing the square will also be used in the following section to derive the quadratic formula. The quadratic formula enables us to solve any quadratic equation by simply evaluating a formula. To help us in completing the square, we must review the concept of perfect square trinom-

SOLVING QUADRATIC EQUATIONS BY COMPLETING THE SQUARE

ials first studied in section 3.6. Then, we will develop the square root property of equations.

A *perfect square trinomial* is a trinomial that is the square of a binomial. For example, $x^2 + 6x + 9$ is a perfect square trinomial because $x^2 + 6x + 9 = (x + 3)^2$. We will find it helpful to see how the coefficient of the middle term is related to the constant term.

EXAMPLE 1

$x^2 + 6x + 9$ is a perfect square trinomial. The constant terms is 9. It is equal to the square of one-half the coefficient of x.

Solution: $x^2 + 6x + 9$

$$\left(\frac{1}{2} \cdot 6\right)^2$$

To prove this relationship, we square a general binomial, then note the relationship between the second and third terms. We assume that x is the variable and k is a real constant.

$$(x + k)^2 = x^2 + 2kx + k^2$$

$$\left(\frac{1}{2} \cdot 2k\right)^2$$

This relationship also holds true for a difference.

$$(x - k)^2 = x^2 - 2kx + k^2$$

$$\left[\frac{1}{2}(-2k)\right]^2$$

The only restriction is that the coefficient of the first term must be 1.

> **Rule for Perfect Square Trinomials**
>
> The constant term of a perfect square trinomial is equal to the square of one-half the coefficient of the middle term, provided the coefficient of the first term is 1.

Fill in the blank so that $x^2 - 12x + \underline{}$ is a perfect square trinomial.

EXAMPLE 2

Solution: The constant term is found by squaring one-half the coefficient of x.

$$x^2 - 12x + \underline{} = x^2 - 12x + 36$$

$$\left[\frac{1}{2}(-12)\right]^2 = (-6)^2 = 36$$

217

EXAMPLE 3 Determine the constant so that $m^2 + 5m + \underline{}$ is a perfect square trinomial.

Solution: Square one-half the coefficient of m.

$$m^2 + 5m + \underline{} = m^2 + 5m + \frac{25}{4}$$

$$\left(\frac{1}{2} \cdot 5\right)^2 = \left(\frac{5}{2}\right)^2 = \frac{25}{4}$$

Next we develop the square root property of equations by solving quadratic equations of the form $x^2 = k$, where k is a non-negative real constant. Consider the following argument:

If $x^2 = k$, then from the definition of square root, x must be a square root of k. Since each positive real number has two square roots, the solutions are $\sqrt{k}$ and $-\sqrt{k}$.

Square Root Property of Equations

If $x^2 = k$, where x is the variable and k is a nonnegative real constant, then $x = \sqrt{k}$ or $x = -\sqrt{k}$. (For convenience, the two equations can be written $x = \pm\sqrt{k}$.)

EXAMPLE 4 Solve $x^2 = 49$

Solution: Apply the square root property of equations. If $x^2 = 49$, then $x = \pm\sqrt{49}$. Thus, $x = \pm 7$. The solutions are 7 and -7.

EXAMPLE 5 Solve $(y + 6)^2 = 16$

Solution:
$$(y + 6)^2 = 16$$
$$y + 6 = \pm\sqrt{16} \qquad \text{[square root property]}$$
$$y = -6 \pm \sqrt{16}$$

Using the plus sign: $y = -6 + \sqrt{16} = -6 + 4$ or -2.
Using the minus sign: $y = -6 - \sqrt{16} = -6 - 4$ or -10.
Thus, the two solutions are -2 and -10.

EXAMPLE 6 Solve $(x + 3)^2 = 15$

Solution:
$$(x + 3)^2 = 15$$
$$x + 3 = \pm\sqrt{15} \qquad \text{[square root property]}$$
$$x = -3 \pm \sqrt{15}$$

The solutions are $-3 + \sqrt{15}$ and $-3 - \sqrt{15}$.

EXAMPLE 7

Solve $p^2 - 8p + 16 = 5$

Solution: $p^2 - 8p + 16$ is a perfect square trinomial and is equal to $(p - 4)^2$.

$$p^2 - 8p + 16 = 5$$
$$(p - 4)^2 = 5$$
$$p - 4 = \pm\sqrt{5} \quad \text{[square root property]}$$
$$p = 4 \pm \sqrt{5}$$

The solutions are $4 + \sqrt{5}$ and $4 - \sqrt{5}$.

Using the square root property of equations together with our knowledge of perfect square trinomials, we illustrate how to solve quadratic equations by completing the square.

EXAMPLE 8

Solve $x^2 + 6x + 7 = 0$ by completing the square.

Solution: Rearrange the equation so that the constant term is isolated on one side.

$$x^2 + 6x + 7 = 0$$
$$x^2 + 6x = -7$$

Determine what constant is needed to form a perfect square trinomial. Then add it to both sides. This procedure is called completing the square.

$$x^2 + 6x = -7$$
$$x^2 + 6x + 9 = -7 + 9 \quad \text{[add the constant 9 to both sides]}$$

$\left(\frac{1}{2} \cdot 6\right)^2$

$$x^2 + 6x + 9 = 2$$

Write the perfect square trinomial as the square of a binomial. Then, apply the square root property of equations.

$$x^2 + 6x + 9 = 2$$
$$(x + 3)^2 = 2$$
$$x + 3 = \pm\sqrt{2} \quad \text{[square root property]}$$
$$x = -3 \pm \sqrt{2}$$

The solutions are $-3 + \sqrt{2}$ and $-3 - \sqrt{2}$.

EXAMPLE 9

Solve $2x^2 + 5x - 1 = 0$ by completing the square.

Solution: Rearrange the equation to isolate the constant term on one side.

$$2x^2 + 5x - 1 = 0$$
$$2x^2 + 5x = 1$$

Divide both sides of the equation by 2. Remember, the coefficient of the squared term must be 1 before you complete the square.

$2x^2 + 5x = 1$

$x^2 + \dfrac{5}{2}x = \dfrac{1}{2}$

Complete the square.

$x^2 + \dfrac{5}{2}x = \dfrac{1}{2}$

$x^2 + \dfrac{5}{2}x + \dfrac{25}{16} = \dfrac{1}{2} + \dfrac{25}{16}$ $\left[\text{add } \dfrac{25}{16} \text{ to both sides}\right]$

$\left(\dfrac{1}{2} \cdot \dfrac{5}{2}\right)^2$

$x^2 + \dfrac{5}{2}x + \dfrac{25}{16} = \dfrac{8 + 25}{16}$

$x^2 + \dfrac{5}{2}x + \dfrac{25}{16} = \dfrac{33}{16}$

Write the perfect square trinomial as the square of a binomial. Then, apply the square root property of equations.

$x^2 + \dfrac{5}{2}x + \dfrac{25}{16} = \dfrac{33}{16}$

$\left(x + \dfrac{5}{4}\right)^2 = \dfrac{33}{16}$

$x + \dfrac{5}{4} = \pm\sqrt{\dfrac{33}{16}}$

$x + \dfrac{5}{4} = \dfrac{\pm\sqrt{33}}{4}$

$x = -\dfrac{5}{4} \pm \dfrac{\sqrt{33}}{4}$

$x = \dfrac{-5 \pm \sqrt{33}}{4}$

The solutions are $\dfrac{-5 + \sqrt{33}}{4}$ and $\dfrac{-5 - \sqrt{33}}{4}$.

As shown in this final example, completing the square may produce a negative number under the radical. If this occurs, the equation has no real solutions. In a later section we will expand our number system to include complex numbers so that solutions will exist.

EXAMPLE 10   Solve $3y^2 + 3y + 5 = 0$ by completing the square.

Solution:
$$3y^2 + 3y = -5$$
$$y^2 + 1y = -\frac{5}{3}$$
$$y^2 + 1y + \frac{1}{4} = -\frac{5}{3} + \frac{1}{4} \quad \left[\text{add } \frac{1}{4} \text{ to both sides}\right]$$
$$\left(\frac{1}{2}\cdot 1\right)^2$$
$$\left(y + \frac{1}{2}\right)^2 = \frac{-20 + 3}{12}$$
$$\left(y + \frac{1}{2}\right)^2 = -\frac{17}{12}$$
$$y + \frac{1}{2} = \pm\sqrt{-\frac{17}{12}}$$

The square root of a negative number does not exist in the real number system. Consequently, this equation has no real solutions.

Summary for Completing the Square

Step (1) Isolate the constant term on one side of the equation.
Step (2) Divide both sides of the equation by the coefficient of the squared variable. This makes the leading coefficient 1.
Step (3) Complete the square by squaring one-half the coefficient of the second term and adding it to both sides of the equation.
Step (4) Write the perfect square trinomial as the square of a binomial. Solve the resulting equation by applying the square root property of equations.

Find the term that must be added to make the expression a perfect square trinomial.

Exercise 6.2

1. $x^2 - 6x$
2. $x^2 + 3x$
3. $x^2 + \frac{1}{2}x$
4. $x^2 - \frac{1}{4}x$

Solve the following by the square root property of equations.

5. $x^2 = 36$
6. $x^2 = 12$
7. $2y^2 = 36$
8. $(y - 5)^2 = 9$
9. $(x + 3)^2 = 16$
10. $(x - 4)^2 = 5$

11. $x^2 - 8x + 16 = 3$
12. $x^2 + 2x + 1 = 8$
13. $(3a + 4)^2 = 50$
14. $(2a - 3)^2 = 75$
15. $(x - 2)^2 = -2$
16. $(1 - 2y)^2 = 49$
17. $(3 - 4t)^2 = 48$
18. $3(1 - 3x)^2 = 36$

Solve by completing the square
19. $x^2 + 2x = 3$
20. $x^2 - 8x = 5$
21. $y^2 - 6y - 7 = 0$
22. $y^2 + 4y + 1 = 0$
23. $y^2 - 6y - 6 = 0$
24. $a^2 - 2a + 5 = 0$
25. $x^2 - x - \dfrac{1}{2} = 0$
26. $a^2 + 3a = 1$
27. $x^2 + x - 1 = 0$
28. $5 = x^2 - 6x$
29. $t^2 - 6t + 1 = 0$
30. $2t^2 - 4t = 7$
31. $2a^2 = 4 - 3a$
32. $2x^2 + 3x - 2 = 0$
33. $9x^2 - 12x + 5 = 0$
34. $2y^2 - 8y + 3 = 0$
35. $2x^2 - 2x = -1$
36. $4y^2 - 12y - 81 = 0$
37. $4t^2 - 12t - 13 = 0$
38. $4t^2 - 4t - 3 = 0$

Solve the following applied problems.

39. The sum of the squares of two consecutive odd numbers is 34. Find the numbers.

40. The sum of a number and its reciprocal is $\dfrac{13}{6}$. Find the number.

41. If two times an integer is added to four times the square of the integer, the result is 72. Find the integer.

42. Two pipes can fill a tank in 2 hours. One pipe takes 3 hours longer than the other to fill the tank alone. How long does it take each to fill the tank?

Calculator problems. Solve the following by completing the square. Then, find the solution to the nearest hundredth.

43. $x^2 + 5x = 7$
44. $2x^2 - 5x + \dfrac{1}{2} = 0$
45. $.2x^2 - .01x = 4$
46. $4 - 3x^2 = 7x$
47. $5 + 4x - 3x^2 = 0$
48. $\dfrac{1 - x}{2} = \dfrac{2}{4 + x}$

THE QUADRATIC FORMULA

6.3

In the previous section you learned to solve quadratic equations by completing the square. While this method works for all quadratic equations, it is sometimes complicated and time consuming. Thus, in this section we will

THE QUADRATIC FORMULA

derive the quadratic formula, which enables us to solve any quadratic equation by simply evaluating a formula.

To derive the quadratic formula we complete the square on the general form of a quadratic equation, $ax^2 + bx + c = 0$, where $a, b, c \in R$ and $a \neq 0$.

Step 1. Isolate the constant term on one side of the equation.
$$ax^2 + bx + c = 0$$
$$ax^2 + bx = -c$$

Step 2. Divide both sides by a
$$ax^2 + bx = -c$$
$$x^2 + \frac{b}{a}x = -\frac{c}{a}$$

Step 3. Complete the square.
$$x^2 + \frac{b}{a}x = -\frac{c}{a}$$
$$x^2 + \frac{b}{a}x + \frac{b^2}{4a^2} = -\frac{c}{a} + \frac{b^2}{4a^2} \qquad \left[\text{add } \frac{b^2}{4a^2} \text{ to both sides}\right]$$
$$\left(\frac{1}{2}\cdot\frac{b}{a}\right)^2$$

Step 4. Write the perfect square trinomial as the square of a binomial. Then, apply the square root property of equations.
$$x^2 + \frac{b}{a}x + \frac{b^2}{4a^2} = -\frac{c}{a} + \frac{b^2}{4a^2}$$
$$\left(x + \frac{b}{2a}\right)^2 = \frac{-4ac + b^2}{4a^2}$$
$$x + \frac{b}{2a} = \pm\sqrt{\frac{b^2 - 4ac}{4a^2}}$$
$$x + \frac{b}{2a} = \pm\frac{\sqrt{b^2 - 4ac}}{|2a|}$$

Note : a may be either positive or negative, but due to the $\pm$ sign, there is no resultant effect on the absolute value.
$$x + \frac{b}{2a} = \pm\frac{\sqrt{b^2 - 4ac}}{2a}$$
$$x = -\frac{b}{2a} \pm \frac{\sqrt{b^2 - 4ac}}{2a}$$
$$x = \frac{-b \pm \sqrt{b^2 - 4ac}}{2a}$$

224 SOLVING QUADRATIC EQUATIONS AND INEQUALITIES

This derivation provides us with the quadratic formula. We will show how it is used to solve any quadratic equation quickly and efficiently.

The Quadratic Formula

For quadratic equations of the form $ax^2 + bx + c = 0$, where a, b, $c \in R$ and $a \neq 0$, the solutions are:

$$x = \frac{-b \pm \sqrt{b^2 - 4ac}}{2a}$$

(This formula must be memorized.)

To use the quadratic formula we identify the values of a, b, and c. These are then substituted into the formula and the result is simplified, producing the solutions.

EXAMPLE 1   Solve $2x^2 - 5x = 3$

Solution: Rewrite the equation so that all terms are on one side. Then, identify the values of a, b, and c.

$$2x^2 - 5x = 3$$

$$\underbrace{2x^2}_{a} \underbrace{- 5x}_{b} \underbrace{- 3}_{c} = 0$$

$$a = 2, \; b = -5, \; c = -3$$

Substitute these values into the quadratic formula and simplify.

$$\begin{aligned}
x &= \frac{-b \pm \sqrt{b^2 - 4ac}}{2a} \\
&= \frac{-(-5) \pm \sqrt{(-5)^2 - 4 \cdot 2 \cdot (-3)}}{2 \cdot 2} \\
&= \frac{5 \pm \sqrt{25 + 24}}{2 \cdot 2} \\
&= \frac{5 \pm \sqrt{49}}{4} \\
&= \frac{5 \pm 7}{4}
\end{aligned}$$

To separate the two solutions, we first use the plus sign and then the minus sign.

$$x = \frac{5 + 7}{4} = \frac{12}{4} = 3$$

$$x = \frac{5 - 7}{4} = -\frac{2}{4} = -\frac{1}{2}$$

The solutions are 3 and $-\frac{1}{2}$.

EXAMPLE 2

Solve $3m^2 = 4m + 3$

Solution: Rewrite the equation so that all terms are on one side and place the equation in quadratic form. Then, identify the values of a, b, and c.

$$3m^2 = 4m + 3$$
$$3m^2 - 4m - 3 = 0$$
$$a = 3, \, b = -4, \, c = -3$$

Substitute into the quadratic formula and simplify.

$$m = \frac{-b \pm \sqrt{b^2 - 4ac}}{2a}$$

$$= \frac{4 \pm \sqrt{16 - 4 \cdot 3 \cdot (-3)}}{2 \cdot 3}$$

$$= \frac{4 \pm \sqrt{16 + 36}}{6}$$

$$= \frac{4 \pm \sqrt{52}}{6}$$

Simplify the radical and reduce the fraction.

$$m = \frac{4 \pm \sqrt{4 \cdot 13}}{6}$$

$$= \frac{4 \pm 2\sqrt{13}}{6}$$

$$= \frac{\cancel{2}(2 \pm \sqrt{13})}{\cancel{2} \cdot 3}$$

$$= \frac{2 \pm \sqrt{13}}{3}$$

The solutions are $\frac{2 + \sqrt{13}}{3}$ and $\frac{2 - \sqrt{13}}{3}$.

In the previous example the solutions were $\dfrac{2 + \sqrt{13}}{3}$ and $\dfrac{2 - \sqrt{13}}{3}$. Both involve radicals and are exact solutions. Sometimes it is necessary to approximate such solutions with decimals. This is done either by using the Powers and Roots table on page A-2 or by using a calculator.

Using the table we find that the approximate value of $\sqrt{13}$ is 3.606. The two solutions are converted to decimal form as follows:

$$\dfrac{2 + \sqrt{13}}{3} \approx \dfrac{2 + 3.606}{3} \qquad \dfrac{2 - \sqrt{13}}{3} \approx \dfrac{2 - 3.606}{3}$$

$$\approx \dfrac{5.606}{3} \qquad\qquad \approx \dfrac{-1.606}{3}$$

$$\approx 1.869 \qquad\qquad\qquad \approx -0.535$$

Using a calculator with a storage (or memory) key (STO) and a recall key (RCL) enables us to perform the calculations quickly using the following sequences:

$$\dfrac{2 + \sqrt{13}}{3} \Rightarrow \text{① ② ⊕ ⑬ √ STO ① ÷ ③ = } 1.869$$

Go directly to the next calculation. Do not turn the calculator off.

$$\dfrac{2 - \sqrt{13}}{3} \Rightarrow \text{① ② ⊖ RCL ① ÷ ③ = } -0.535$$

Thus, the approximate decimal solutions of $3m^2 = 4m + 3$ are 1.869 and -0.535.

Sometimes quadratic equations contain fractions. When this occurs, it is easier to use the quadratic formula after the equation has been cleared of fractions.

EXAMPLE 3 Solve $\dfrac{1}{6}x^2 = \dfrac{5}{6}x - \dfrac{1}{2}$

Solution: Clear the equation of fractions by multiplying each side by the LCD, 6.

$$\dfrac{1}{6}x^2 = \dfrac{5}{6}x - \dfrac{1}{2}$$

$$6 \cdot \dfrac{1}{6}x^2 = 6 \cdot \dfrac{5}{6}x - 6 \cdot \dfrac{1}{2}$$

$$x^2 = 5x - 3$$

Obtain all terms on one side. Then, identify a, b, and c.

$$x^2 - 5x + 3 = 0$$
$$a = 1, b = -5, c = 3$$

THE QUADRATIC FORMULA

Substitute into the quadratic formula and simplify.

$$x = \frac{-b \pm \sqrt{b^2 - 4ac}}{2a}$$

$$x = \frac{5 \pm \sqrt{25 - 4 \cdot 1 \cdot 3}}{2 \cdot 1}$$

$$x = \frac{5 \pm \sqrt{25 - 12}}{2}$$

$$x = \frac{5 \pm \sqrt{13}}{2}$$

The solutions are $\dfrac{5 + \sqrt{13}}{2}$ and $\dfrac{5 - \sqrt{13}}{2}$.

As with completing the square, the quadratic formula may produce a negative radicand. When this happens, there are no real solutions to the original equation.

Solve $k^2 + k + 5 = 0$ **EXAMPLE 4**

Solution: Identify a, b, and c. Then use the quadratic formula.

$$k^2 + k + 5 = 0$$
$$a = 1, b = 1, c = 5$$
$$k = \frac{-b \pm \sqrt{b^2 - 4ac}}{2a}$$
$$= \frac{-1 \pm \sqrt{1 - 4 \cdot 1 \cdot 5}}{2 \cdot 1}$$
$$= \frac{-1 \pm \sqrt{1 - 20}}{2}$$
$$= \frac{-1 \pm \sqrt{-19}}{2}$$

The square root of a negative number does not exist in the real number system. Thus, the equation has no real solutions.

We have seen that the solution of a quadratic equation is given by the formula, $\dfrac{-b \pm \sqrt{b^2 - 4ac}}{2a}$. Observe in the following three solutions how the value of the radicand ($b^2 - 4ac$) determines both the number and the type of solutions:

Solution	Radicand	Comment
$\dfrac{3 \pm \sqrt{-7}}{2}$	-7	No real solutions
$\dfrac{3 \pm \sqrt{7}}{2}$	$+7$	Two real solutions $\dfrac{3 + \sqrt{7}}{2}$ and $\dfrac{3 - \sqrt{7}}{2}$
$\dfrac{3 \pm \sqrt{0}}{2}$	0	One real solution $\dfrac{3}{2}$

Since the radicand of the quadratic formula, $b^2 - 4ac$, determines the nature of the solutions, it is called the discriminant (symbolized D).

> **Discriminant of the Quadratic Equation**
> If $ax^2 + bx + c = 0$, where $a, b, c \in R$ and $a \neq 0$, then $D = b^2 - 4ac$ is called the discriminant, and:
> (1) If $b^2 - 4ac < 0$, there are no real solutions.
> (2) If $b^2 - 4ac > 0$, there are two real solutions.
> (3) If $b^2 - 4ac = 0$, there is one real solution.

As shown by the following examples, the nature of the solutions can be determined by simply finding the value of the discriminant instead of actually solving the equation.

EXAMPLE 5 Without solving, classify the solutions of $2x^2 - x + 3 = 0$.

Solution: Find the value of the discriminant.
$$D = b^2 - 4ac$$
$$= (-1)^2 - 4 \cdot 2 \cdot 3$$
$$= 1 - 24$$
$$= -23$$
Since $D < 0$, there are no real solutions.

EXAMPLE 6 Classify the solutions of $y^2 - 3y + 1 = 0$.

Solution: Find the value of the discriminant.
$$D = b^2 - 4ac$$
$$= (-3)^2 - 4 \cdot 1 \cdot 1$$
$$= 9 - 4$$
$$= 5$$
Since $D > 0$, there are two real solutions.

THE QUADRATIC FORMULA

EXAMPLE 7 Find the value of k so that the equation, $x^2 + kx + 25 = 0$, will have only one real solution.

Solution: For only one real solution the discriminant must be zero.
$$b^2 - 4ac = 0$$
$$k^2 - 4 \cdot 1 \cdot 25 = 0$$
Solve this equation for the values of k.
$$k^2 - 100 = 0$$
$$k^2 = 100$$
$$k = \pm\sqrt{100} \quad \text{[square root property]}$$
$$k = \pm 10$$

Thus, the equation $x^2 + kx + 25 = 0$ will have only one solution when k is either 10 or -10. In other words, the equations, $x^2 + 10x + 25 = 0$ and $x^2 - 10x + 25 = 0$, will each have only one solution.

Exercise 6.3

Solve by using the quadratic formula.

1. $x^2 - 5x - 6 = 0$
2. $x^2 - 3x + 1 = 0$
3. $x^2 - x + 2 = 0$
4. $3y^2 - 2y = 2$
5. $3y^2 - 5y = 1$
6. $y^2 + 4y = -4$
7. $2t^2 - 2t = 5$
8. $4t^2 + 7t - 1 = 0$
9. $y^2 - 7y + 5 = 0$
10. $2c(c + 1) = 7$
11. $(c + 2)(c - 3) = 1$
12. $3x^2 - 2x = 2$
13. $5x^2 - 2x = 1$
14. $4y^2 + 12y + 9 = 0$
15. $\dfrac{3x^2}{2} - x - \dfrac{1}{3} = 0$
16. $\dfrac{x^2}{2} - 3x - 1 = 0$
17. $\dfrac{y^2}{6} - \dfrac{y}{3} - 1 = 0$
18. $\dfrac{x^2 - 3}{4} = \dfrac{2x}{5}$
19. $\dfrac{a^2}{4} - \dfrac{a}{6} + 1 = 0$
20. $\dfrac{2}{a} - 1 = \dfrac{2}{a^2}$
21. $25t^2 + 10t - 12 = 0$
22. $18b^2 + 3b = 10$
23. $4x^2 + 12x + 9 = 0$
24. $5x^2 + 2x + 1 = 0$
25. $(x + 4)(x - 4) = 4x$
26. $(a + 3)(a - 3) = 6(a + 1)$
27. $3t(t - 1) = 1$
28. $(y - 3)(y + 1) = 2y$
29. $(t - 4)(2 - 3t) = 5$
30. $(2x - 3)(x + 2) = (3x - 1)(x + 1)$

Without solving, classify the solutions. Use the discriminant.

31. $x^2 + x - 3 = 0$
32. $2x^2 - 4x + 3 = 0$
33. $12x^2 - 5x - 2 = 0$
34. $2x^2 - x + 2 = 0$
35. $4x^2 + 4x + 1 = 0$
36. $3x^2 - 5 = 0$

Miscellaneous.
37. Find k so that $2x^2 + kx + 8 = 0$ will have one real solution.
38. Find k so that $x^2 + kx + 16 = 0$ will have one real solution.
39. Find k so that $x^2 - 5x + k = 0$ will have one real solution.
40. Find k so that $k^2x^2 - 40x + 25 = 0$ will have one real solution.
41. Solve $x^{-2} + 5 + 7x^{-1} = 0$.
42. Solve $\dfrac{1}{x + 2} + \dfrac{1}{x - 3} = 2$.
43. The sum of the squares of two consecutive even natural numbers is 100. Find the numbers.
44. One number is 1 less than twice another. The sum of their squares is 34. Find the two numbers.

Calculator problems. Find the solution to the nearest hundredth.
45. $7x^2 - 15x - 23 = 0$
46. $21x^2 + 11x - 18 = 0$
47. $x^{-2} - 7x^{-1} - 18 = 0$
48. $17x^2 - 5x - 23 = 0$
49. $2.3x^2 - 4.2x - 1.9 = 0$
50. $13.7x^2 + 7.2x - 6.3 = 0$

SOLVING QUADRATIC INEQUALITIES

6.4

Throughout this chapter we have studied several methods for solving quadratic equations. Now we develop a procedure for solving quadratic inequalities. Quadratic inequalities are defined to have one of these forms:

$$ax^2 + bx + c < 0 \quad \text{or} \quad ax^2 + bx + c > 0,$$

where $a, b, c \in R$ and $a \neq 0$

To develop a technique for solving these inequalities we consider $x^2 - x - 6 > 0$. First we factor the left side, obtaining:

$$(x - 3)(x + 2) > 0$$

For the product of $(x - 3)(x + 2)$ to be positive, both factors must be positive or both factors must be negative. So, we next determine the values which make each factor positive, negative, or zero.

(1) $x - 3$ is positive when $x - 3 > 0$ or when $x > 3$.

(2) $x - 3$ is negative when $x - 3 < 0$ or when $x < 3$.

(3) $x - 3$ is zero when $x - 3 = 0$ or when $x = 3$.
3 is called the critical value of the factor $x - 3$ because it separates the positive values of the factor from the negative values.

(4) $x + 2$ is positive when $x + 2 > 0$ or when $x > -2$.

(5) $x + 2$ is negative when $x + 2 < 0$ or when $x < -2$.

(6) $x + 2$ is zero when $x + 2 = 0$ or when $x = -2$.
Thus, the critical value of the factor $x + 2$ is $x = -2$.

SOLVING QUADRATIC INEQUALITIES   **231**

These values are then recorded on a graph of the standard line in Figure 6.1.

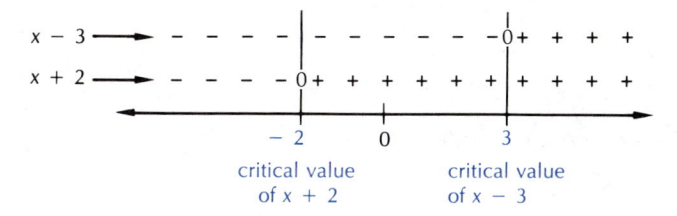

$x - 3$ ⟶ — — — — | — — — — — — 0 + + + +
$x + 2$ ⟶ — — — — 0 + + + + + + | + + + +

    −2   0   3
critical value   critical value
 of $x + 2$ of $x - 3$

FIGURE 6.1

Recall that the original inequality, $x^2 - x - 6 > 0$, was factored, producing:
$$(x - 3)(x + 2) > 0$$
And, as stated earlier, the product of $(x - 3)(x + 2)$ is positive when both factors are positive or when both factors are negative. So, we proceed to our graph in Figure 6.2 and locate the regions where both factors are recorded positive or where both factors are recorded negative.

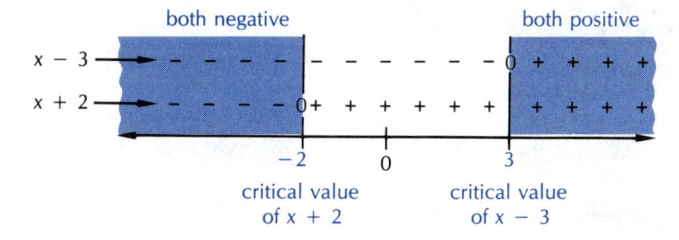

                 both negative   both positive

$x - 3$ ⟶ — — — — | — — — — — — 0 + + + +
$x + 2$ ⟶ — — — — 0 + + + + + + | + + + +

    −2   0   3
critical value   critical value
 of $x + 2$ of $x - 3$

FIGURE 6.2

As seen, both factors are negative when $x < -2$ and both factors are positive when $x > 3$. The solution includes both intervals and is written in set notation as:
$$\{x \mid x < -2 \quad \text{or} \quad x > 3\}$$
This solution is graphed in Figure 6.3.

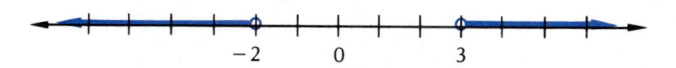

    −2   0   3

FIGURE 6.3

Next, we use the same technique to solve the quadratic inequality $x^2 + 2x - 8 < 0$. First, we factor the left side, obtaining:
$$(x + 4)(x - 2) < 0$$
Next, we find the values that make each factor positive, negative, or zero.

   (1) $x + 4 > 0$ when $x > -4$.

   (2) $x + 4 < 0$ when $x < -4$.

(3) $x + 4 = 0$ when $x = -4$. So, -4 is the critical value of $x + 4$.

(4) $x - 2 > 0$ when $x > 2$.

(5) $x - 2 < 0$ when $x < 2$.

(6) $x - 2 = 0$ when $x = 2$. So, 2 is the critical value of $x - 2$.

These values are recorded on the number line in Figure 6.4.

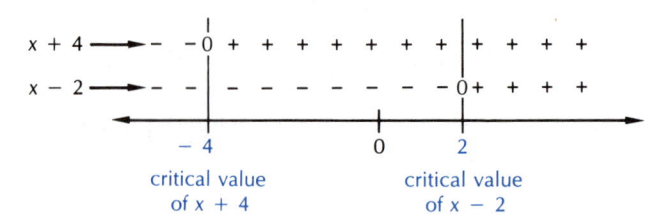

FIGURE 6.4

The original inequality, $x^2 + 2x - 8 < 0$, was factored producing:

$$(x + 4)(x - 2) < 0$$

For the product of $(x + 4)(x - 2)$ to be negative, the factors must have opposite signs. That is, one factor must be positive while the other is negative. Proceeding to our graph in Figure 6.5, we locate the region where one factor is recorded positive and the other is recorded negative.

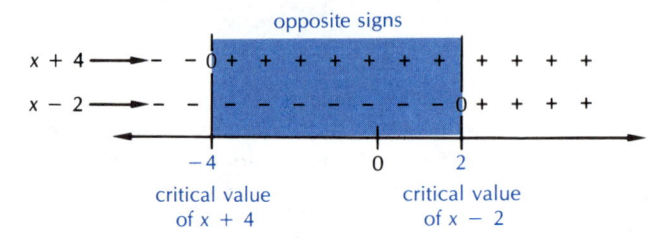

FIGURE 6.5

As seen in Figure 6.5, the factors have opposite signs when $x > -4$ and $x < 2$. In set notation the solution is written as:

$$\{x \mid -4 < x < 2\}$$

This solution is graphed in Figure 6.6.

FIGURE 6.6

This procedure for solving quadratic inequalities is summarized in the following chart.

SOLVING QUADRATIC INEQUALITIES

To Solve a Quadratic Inequality
(1) Write the inequality in either the form
$$ax^2 + bx + c < 0 \quad \text{or} \quad ax^2 + bx + c > 0.$$
(2) Factor the trinomial $ax^2 + bx + c$. (If the trinomial will not factor, the critical values can be found using the quadratic formula.)
(3) Find the critical value for each factor. Also, determine the values for which each factor is positive and the values for which each factor is negative. Record these values on a graph of the number line.
(4) From the graph, find the interval that forms the correct product, either positive or negative.

Solve $2x^2 - x > 15$

EXAMPLE 1

Solution: Write the inequality in the form $ax^2 + bx + c > 0$.
$$2x^2 - x > 15$$
$$2x^2 - x - 15 > 0$$
Factor the trinomial.
$$(2x + 5)(x - 3) > 0$$
Note that the product will be positive only when both factors are positive or when both factors are negative.

Find the critical value of each factor. Also, determine the values for which each factor is positive and the values for which each factor is negative.

(1) $2x + 5 > 0$ when $2x > -5$ or when $x > -\frac{5}{2}$.

(2) $2x + 5 < 0$ when $2x < -5$ or when $x < -\frac{5}{2}$.

(3) critical value of $2x + 5$ is $x = -\frac{5}{2}$.

(4) $x - 3 > 0$ when $x > 3$.
(5) $x - 3 < 0$ when $x < 3$.
(6) critical value of $x - 3$ is $x = 3$.

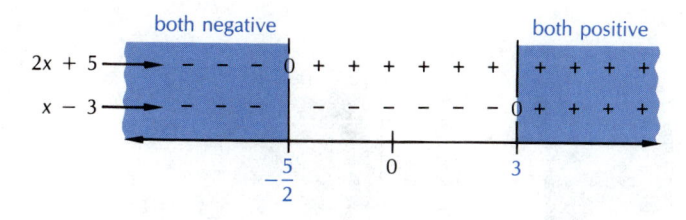

The solution set is $\{x \mid x < -\frac{5}{2} \text{ or } x > 3\}$.

SOLVING QUADRATIC EQUATIONS AND INEQUALITIES

EXAMPLE 2 Solve $x^2 - 3x \leq 0$

Solution: Factor the expression $x^2 - 3x$.

$x^2 - 3x \leq 0$

$x(x - 3) \leq 0$

Note that the product will be negative only if the factors have opposite signs. It will equal zero when $x = 0$ or when $x = 3$.

Find the critical values as well as the values which make the factors positive or negative.

(1) x is positive when $x > 0$.
(2) x is negative when $x < 0$.
(3) critical value of x is $x = 0$.
(4) $x - 3 > 0$ when $x > 3$.
(5) $x - 3 < 0$ when $x < 3$.
(6) critical value of $x - 3$ is $x = 3$.

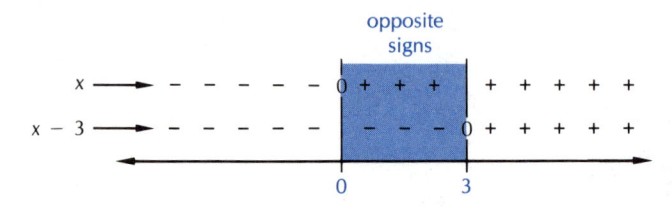

The solution set is $\{x \mid 0 \leq x \leq 3\}$.

As illustrated by the next two examples, this procedure can be used to solve certain inequalities that are not quadratic.

EXAMPLE 3 Solve $(x + 4)(x + 1)(x - 3) > 0$

Solution: This inequality is third degree rather than quadratic. Since it is in factored form, our procedure will work. The following graph shows the critical values and the regions where each factor is positive or negative. The solution consists of the intervals where the product of three factors is positive. This occurs when all factors are positive or when one factor is positive and the other two are negative.

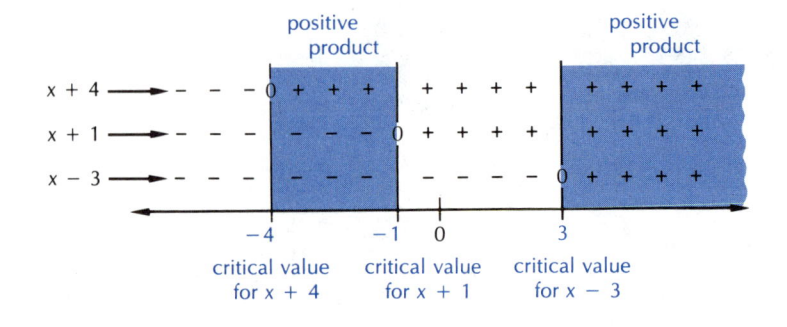

SOLVING QUADRATIC INEQUALITIES

The solution set is $\{x \mid -4 < x < -1 \text{ or } x > 3\}$.

EXAMPLE 4

Solve $\dfrac{2x + 3}{x - 4} < 0$

Solution: This inequality is fractional rather than quadratic. However, the same procedure will work because the quotient of two factors, like their product, is negative when the factors have opposite signs. The following graph shows the critical values and the regions where each factor is positive or negative. The solution consists of the region where the factors have opposite signs.

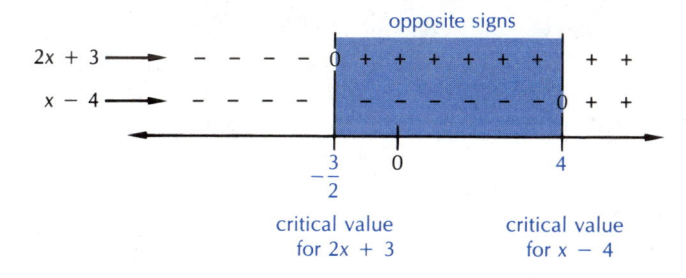

The solution set is $\left\{ x \mid -\dfrac{3}{2} < x < 4 \right\}$.

Solve.

Exercise 6.4

1. $x^2 - x - 6 < 0$
2. $x^2 + 4x - 5 < 0$
3. $x^2 - 5x + 6 \leq 0$
4. $x^2 + 3x - 4 \leq 0$
5. $y^2 - y - 6 > 0$
6. $y^2 - y - 12 \geq 0$
7. $y^2 - 3y - 4 \geq 0$
8. $x^2 - 5x + 4 > 0$
9. $-t^2 + 2t + 8 > 0$
10. $t^2 > 4$
11. $a^2 \leq 5a$
12. $a^2 + 6a \geq 0$
13. $x^2 - 4x \geq -4$
14. $x^2 < 4x$
15. $2y^2 + 5y > 12$
16. $2x^2 + 9x < 5$
17. $-x^2 + 3x - 2 > 0$
18. $3y^2 + 14y - 5 < 0$
19. $3x^2 + 2x - 1 > 0$
20. $2s^2 + 5s > 12$
21. $3t^2 - 13t - 10 \leq 0$
22. $2x^2 - 2x > 4$
23. $2y^2 - 9y + 4 > 0$
24. $6x^2 - x > 2$
25. $5y^2 - 3y - 2 \geq 0$
26. $6t^2 < 5t - 1$
27. $4t^2 + 7t + 3 > 0$
28. $4x^2 > 0$
29. $(x - 2)(x - 4)(x - 2) < 0$
30. $(y - 4)(3y - 1)(y + 2) < 0$
31. $(a - 2)(a + 3)(a - 5) > 0$
32. $(x + 1)(x + 4)(2 - x) \leq 0$
33. $(x + 2)(x - 2)^2(x - 3) \geq 0$
34. $x^4 - 9x^2 < 0$

35. $\dfrac{x}{x-1} \geq 0$ 36. $\dfrac{x-4}{x-1} \leq 0$

37. $\dfrac{6}{x-5} \leq 0$ 38. $\dfrac{x+1}{x-2} > 0$

39. $\dfrac{x}{x+1} + 2 < 0$ 40. $\dfrac{x-1}{x} - 2 > 0$

THE SET OF COMPLEX NUMBERS

6.5

In previous sections we studied solution techniques for quadratic equations. We also learned that some quadratic equations, such as $x^2 = -4$, have no real solutions. To resolve this problem, mathematicians created a new number system called the set of complex numbers. Complex numbers are built upon the "imaginary unit" i, where i^2 is defined to be -1. Since $i^2 = -1$, we agree that i is a square root of -1 and write $i = \sqrt{-1}$.

> **Definition of the Imaginary Unit**
> The imaginary unit is i, where $i^2 = -1$. It is agreed that $i = \sqrt{-1}$.

Any real number, except zero, can be multiplied times the imaginary unit to produce an imaginary number.

> **Definition of Imaginary Numbers**
> If $b \in R$, $b \neq 0$, and i is the imaginary unit, then any number of the form bi is an imaginary number.
> ($0 \cdot i$ is not considered to be an imaginary number because $0 \cdot i = 0$.)

Imaginary numbers are inappropriately named. They are not a figment of imagination. They do exist and form an important and useful number system. However, the term "imaginary" is still commonly used.

EXAMPLE 1   Each of these is an imaginary number.

(a) $2i$

(b) $-i$ [means $-1 \cdot i$]

(c) $-3i$

(d) $\dfrac{2}{3}i$

(e) $\sqrt{3} \cdot i$ or $i\sqrt{3}$

Although the imaginary unit is not a real number, we assume that it behaves according to the algebraic laws that we have already developed. Thus, we are able to simplify higher powers of i.

$$i^3 = i^2 \cdot i = (-1) \cdot i = -i$$
$$i^4 = i^2 \cdot i^2 = (-1) \cdot (-1) = 1$$

Since $i^4 = 1$, we are able to simplify expressions of the form i^n, where $n \in N$. The expression is written as two factors where the exponent of one is the highest multiple of 4 less than or equal to n.

Simplify i^{37}

EXAMPLE 2

Solution: 36 is the highest multiple of 4 contained in 37.

$$i^{37} = i^{36} \cdot i$$
$$= (i^4)^9 \cdot i$$
$$= 1^9 \cdot i$$
$$= 1 \cdot i$$
$$= i$$

Thus, i^{37} simplifies to i.

Simplify i^{51}

EXAMPLE 3

Solution: 48 is the highest multiple of 4 contained in 51.

$$i^{51} = i^{48} \cdot i^3$$
$$= (i^4)^{12} \cdot i^3$$
$$= 1 \cdot i^3$$
$$= i^3$$
$$= -i$$

As shown in the following argument, the imaginary unit can be used to simplify square roots of negative numbers. Let a represent a positive real number, then:

$\sqrt{-a} = i\sqrt{a}$, since by definition of square root, $(i\sqrt{a})^2 = i^2 a$ or $-a$

This provides us with a property for simplifying square roots of negative numbers.

Square Roots of Negative Numbers
For $a \in R, a > 0$
$$\sqrt{-a} = i\sqrt{a}$$

These square roots have been simplified.
(a) $\sqrt{-7} = i\sqrt{7}$
(b) $\sqrt{-36} = i\sqrt{36} = 6i$
(c) $\sqrt{-18} = i\sqrt{18} = i\sqrt{9 \cdot 2} = 3i\sqrt{2}$

The multiplication and division properties of radicals do not hold for square roots of negative numbers. In such cases the radicals must first be written using the imaginary unit. For example,

$$(\sqrt{-2})(\sqrt{-8}) = (i\sqrt{2})(i\sqrt{8})$$
$$= i^2\sqrt{16}$$
$$= -1 \cdot 4$$
$$= -4$$

If we incorrectly use the multiplication property of radicals, we would obtain the following wrong answer:

$$(\sqrt{-2})(\sqrt{-8}) = \sqrt{(-2)(-8)} \quad \text{[incorrect]}$$
$$= \sqrt{16}$$
$$= 4 \quad \text{[a wrong answer]}$$

EXAMPLE 5 Multiply $(2\sqrt{-3})(5\sqrt{-6})$

Solution: Write the radicals using the imaginary unit, then multiply.

$$(2\sqrt{-3})(5\sqrt{-6}) = (2i\sqrt{3})(5i\sqrt{6})$$
$$= 10i^2\sqrt{18}$$
$$= 10(-1)\sqrt{9 \cdot 2}$$
$$= 10(-1)3\sqrt{2}$$
$$= -30\sqrt{2}$$

EXAMPLE 6 Divide $\dfrac{\sqrt{-50}}{\sqrt{2}}$

Solution: Use the imaginary unit, then divide.

$$\frac{\sqrt{-50}}{\sqrt{2}} = \frac{i\sqrt{50}}{\sqrt{2}}$$
$$= i\sqrt{\frac{50}{2}}$$
$$= i\sqrt{25}$$
$$= 5i$$

More complicated expressions involving square roots of negative numbers can also be simplified by first converting to the imaginary unit. Examples similar to the following arise when the quadratic formula is used with a negative discriminant.

EXAMPLE 7 Simplify $\dfrac{6 + \sqrt{-32}}{2}$

Solution: Use the imaginary unit, then factor and reduce.

$$\frac{6 + \sqrt{-32}}{2} = \frac{6 + i\sqrt{32}}{2}$$
$$= \frac{6 + i\sqrt{16 \cdot 2}}{2}$$
$$= \frac{6 + 4i\sqrt{2}}{2}$$
$$= \frac{\cancel{2}(3 + 2i\sqrt{2})}{\cancel{2}}$$
$$= 3 + 2i\sqrt{2}$$

Now that you have had experience using the imaginary unit, we will define the set of complex numbers and relate it to the familiar set of real numbers. Complex numbers were created by simply combining the real numbers and the imaginary numbers to form such numbers as, $2 + 3i$, or, $-5 + i\sqrt{3}$. This suggests the following definition.

> **The Set of Complex Numbers**
>
> If $a, b \in R$ and if i is the imaginary unit, then any number of the form $a + bi$ is a complex number.

Each of the following is a complex number.

(a) $-5 + 3i$

(b) $2 - 4i$ since $2 - 4i = 2 + (-4)i$

(c) 6 since $6 = 6 + 0i$

(d) $5i$ since $5i = 0 + 5i$

EXAMPLE 8

The standard form of a complex number is $a + bi$, where $a, b \in R$. For any complex number in standard form, a is called the real part and b is called the imaginary part. The following table illustrates the standard form and parts of several complex numbers.

Complex Number	Standard Form	Real Part	Imaginary Part
$6 - 2i$	$6 + (-2)i$	6	-2
-12	$-12 + 0i$	-12	0
$7i$	$0 + 7i$	0	7
0	$0 + 0i$	0	0

By writing the real number a as, $a + 0i$, it is evident that every real number is also a complex number. Similarly, writing the imaginary number bi as,

SOLVING QUADRATIC EQUATIONS AND INEQUALITIES

0 + bi, shows that every imaginary number is also a complex number. In other words, the set of real numbers and the set of imaginary numbers are subsets of the complex numbers. Other subsets and their interrelationships are illustrated by the chart in Figure 6.7.

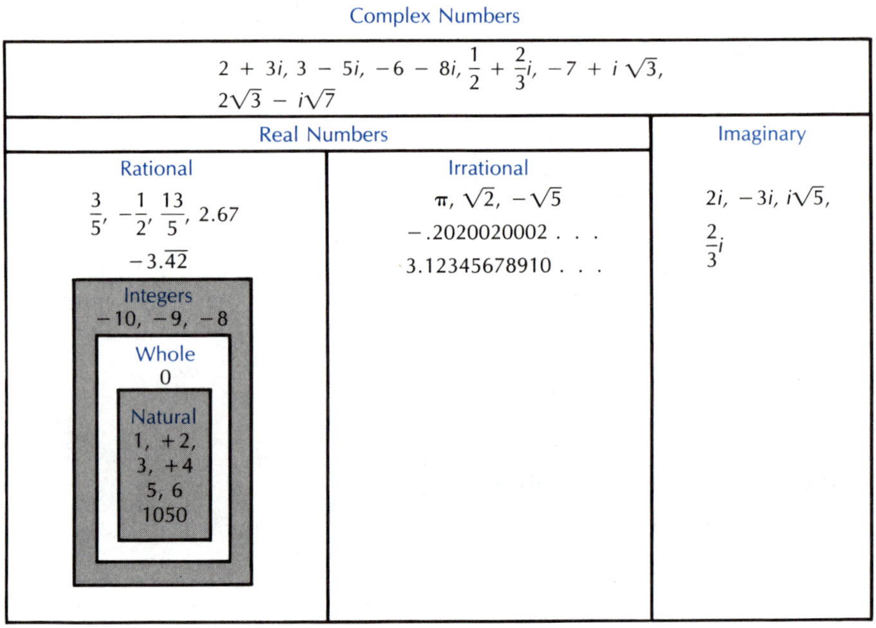

Complex Numbers

$2 + 3i$, $3 - 5i$, $-6 - 8i$, $\frac{1}{2} + \frac{2}{3}i$, $-7 + i\sqrt{3}$, $2\sqrt{3} - i\sqrt{7}$

Real Numbers		Imaginary
Rational $\frac{3}{5}$, $-\frac{1}{2}$, $\frac{13}{5}$, 2.67, $-3.\overline{42}$ **Integers** $-10, -9, -8$ **Whole** 0 **Natural** 1, +2, 3, +4 5, 6 1050	**Irrational** π, $\sqrt{2}$, $-\sqrt{5}$ $-.2020020002\ldots$ $3.12345678910\ldots$	$2i$, $-3i$, $i\sqrt{5}$, $\frac{2}{3}i$

FIGURE 6.7

Exercise 6.5

True or false.
1. $\sqrt{-4}$ is a real number.
2. Every real number is also a complex number.
3. All imaginary numbers are real numbers.
4. $-i^2 = -1$
5. $\sqrt[3]{-64} = 4i$
6. $i^{50} = 1$

Write each number in standard form $a + bi$, and identify the real part and the imaginary part.

7. $2 + 5i$
8. $6 - 9i$
9. $-\frac{1}{2} - i$
10. $-\frac{3}{4}i$
11. $-1 + \sqrt{-2}$
12. $4 + \sqrt{-4}$

Find each of the following powers of i.

13. i^{10}
14. i^{15}
15. i^{100}
16. i^{28}
17. i^{30}
18. i^{203}

Simplify.

19. $\sqrt{-25}$
20. $\sqrt{-144}$
21. $-2i^2$
22. $-1 - \sqrt{-16}$
23. $4\sqrt{-49}$
24. $\sqrt{-3} \cdot \sqrt{-4}$
25. $\sqrt{5} \cdot \sqrt{-8}$
26. $\sqrt{-4} \cdot \sqrt{-9}$
27. $2\sqrt{-12}$
28. $6\sqrt{-18}$
29. $-\sqrt{-4} \cdot \sqrt{9}$
30. $-\sqrt{-1}$
31. $\sqrt{-2} \cdot \sqrt{-75}$
32. $\sqrt{-5} \cdot \sqrt{-5}$
33. $\dfrac{\sqrt{-8}}{2}$
34. $\dfrac{\sqrt{-90}}{\sqrt{10}}$
35. $\dfrac{\sqrt{-50}}{\sqrt{-2}}$
36. $\dfrac{\sqrt{-400}}{\sqrt{-100}}$
37. $\dfrac{\sqrt{-9}}{\sqrt{-4}}$
38. $\dfrac{\sqrt{-3} \cdot \sqrt{-27}}{\sqrt{-121} \cdot \sqrt{3}}$
39. $\dfrac{2 + \sqrt{-8}}{2}$
40. $\dfrac{3 + \sqrt{-12}}{3}$
41. $\dfrac{-2 - \sqrt{-20}}{2}$
42. $\dfrac{-6 + \sqrt{-4}}{2}$
43. $\dfrac{4\sqrt{-50}}{\sqrt{2}}$
44. $\dfrac{\sqrt{-48}}{\sqrt{-75}}$

OPERATING WITH COMPLEX NUMBERS

6.6

Recall that a complex number was defined in the previous section as having the form $a + bi$, where $a, b \in R$ and where i is the imaginary unit. We defined i^2 to be -1 and agreed that $i = \sqrt{-1}$. In this section we continue our study of complex numbers in order to learn how to perform the operations of addition, subtraction, multiplication, and division.

The set of complex numbers is designated by C and we assume the following properties:

1. The operations of addition and multiplication obey the properties of real numbers.
2. The additive identity is unique and is equal to $0 + 0i$ or simply 0.

3. The multiplicative identity is unique and is equal to $1 + 0i$ or simply 1.
4. Each complex number, $a + bi$, has a unique additive inverse, $-a - bi$, such that $(a + bi) + (-a - bi) = 0 + 0i$ or 0.
5. Zero, or $0 + 0i$, times any complex number is zero.

Two complex numbers are equal provided their real parts are equal and their imaginary parts are equal.

Definition of Equality
$a + bi = c + di$ if and only if $a = c$ and $b = d$

EXAMPLE 1

If $x + 5i = -3 + yi$, find x and y.

Solution: Use the definition of equality to set the real parts equal and the imaginary parts equals.
Since $x + 5i = -3 + yi$, then $x = -3$ and $y = 5$.

Complex numbers are added and subtracted by simply combining like terms. That is, the real parts are combined and the imaginary parts are combined.

Definition of Addition and Subtraction
$(a + bi) + (c + di) = (a + c) + (b + d)i$
$(a + bi) - (c + di) = (a - c) + (b - d)i$

EXAMPLE 2

The indicated operations have been performed.
(a) $(3 + 2i) + (-8 + 7i) = (3 - 8) + (2 + 7)i$
$= -5 + 9i$
(b) $(-2 + 7i) - (4 + 5i) = (-2 - 4) + (7 - 5)i$
$= -6 + 2i$
(c) $5i + (-3 + 2i) \quad = (0 + 5i) + (-3 + 2i)$
$= -3 + 7i$
(d) $8 - (7 - 4i) \quad = (8 + 0i) - (7 - 4i)$
$= 1 + 4i$

Complex numbers are multiplied in the same manner as polynomials. The distributive property can be used as well as the corresponding shortcuts. Products are simplified using the fact that $i^2 = -1$.

OPERATING WITH COMPLEX NUMBERS

EXAMPLE 3

The distributive property has been used to multiply the following complex numbers.

(a) $3(4 + 2i) = 12 + 6i$

(b) $-5i(3 - 4i) = -15i + 20i^2$
$= -15i + 20(-1)$
$= -15i - 20$
$= -20 - 15i$

Complex numbers of the form $a + bi$ can be multiplied as binomials using the FOIL method (first, outer, inner, last).

EXAMPLE 4

Multiply $(3 + 5i)(2 - 7i)$

Solution: Use the FOIL method, remembering that $i^2 = -1$.

$$\begin{array}{rl} & \quad\ \ \text{F}\qquad\ \ \ \text{O}\qquad\ \ \ \ \text{I}\qquad\ \ \ \ \text{L} \\ (3 + 5i)(2 - 7i) =& 3 \cdot 2 + 3(-7i) + (5i)2 + (5i)(-7i) \\ =& 6 - 21i + 10i - 35i^2 \\ =& 6 - 11i - 35(-1) \\ =& 6 - 11i + 35 \\ =& 41 - 11i \end{array}$$

This procedure suggests the following definition.

Multiplication of Complex Numbers

$$\begin{array}{rl} & \qquad\qquad\ \ \text{F}\ \quad\ \ \text{O}\quad\ \ \ \ \text{I}\quad\ \ \ \text{L} \\ (a + bi)(c + di) =& ac + a(di) + (bi)c + (bi)(di) \\ =& ac + adi + bci + bdi^2 \\ =& ac + adi + bci - bd \\ =& (ac - bd) + (ad + bc)i \end{array}$$

EXAMPLE 5

The following complex numbers have been multiplied.

(a) $-3(5 - 4i) = -15 + 12i$

(b) $i(1 - i) = i - i^2$
$= i + 1$
$= 1 + i$

(c) $(-3 + 5i)(2 - i) = -6 + 3i + 10i - 5i^2$
$= -6 + 13i + 5$
$= -1 + 13i$

The complex number $a - bi$ is called the **conjugate** of the complex number $a + bi$. As shown below, the product of a complex number and its conjugate is always a real number.

$$(a + bi)(a - bi) = a^2 - abi + abi - b^2i^2$$
$$= a^2 - b^2i^2$$
$$= a^2 - b^2(-1)$$
$$= a^2 + b^2$$

> **Conjugate of a Complex Number**
> The product of a complex number and its conjugate is a real number.
> $$(a + bi)(a - bi) = a^2 + b^2$$

EXAMPLE 6 Multiply $3 + 5i$ times its conjugate.

Solution: The conjugate of $3 + 5i$ is $3 - 5i$.
$$(3 + 5i)(3 - 5i) = 3^2 + 5^2$$
$$= 9 + 25$$
$$= 34$$

EXAMPLE 7 Multiply $-3i$ times its conjugate.

Solution: Since $-3i = 0 - 3i$, its conjugate is $0 + 3i$.
$$(0 - 3i)(0 + 3i) = 0^2 + 3^2$$
$$= 0 + 9$$
$$= 9$$

Division of complex numbers is similar to rationalizing a binomial denominator. The quotient of two complex numbers is found by multiplying the numerator and denominator by the conjugate of the denominator.

EXAMPLE 8 Find the quotient of $\dfrac{3 - 2i}{6 + 5i}$.

Solution: Multiply the numerator and denominator by $6 - 5i$, which is the conjugate of the denominator.

$$\frac{3 - 2i}{6 + 5i} = \frac{(3 - 2i)(6 - 5i)}{(6 + 5i)(6 - 5i)}$$
$$= \frac{18 - 15i - 12i + 10i^2}{6^2 + 5^2}$$
$$= \frac{18 - 27i - 10}{36 + 25}$$
$$= \frac{8 - 27i}{61} \text{ or } \frac{8}{61} - \frac{27}{61}i$$

OPERATING WITH COMPLEX NUMBERS

EXAMPLE 9

Find the quotient of $\dfrac{1-i}{i}$.

Solution: Since $i = 0 + i$, its conjugate is $0 - i$ or simply $-i$. Thus, we multiply the numerator and denominator by $-i$.

$$\dfrac{1-i}{i} = \dfrac{(1-i)(-i)}{i(-i)}$$

$$= \dfrac{-i + i^2}{-i^2}$$

$$= \dfrac{-i - 1}{-(-1)}$$

$$= \dfrac{-i - 1}{1}$$

$$= -1 - i$$

The reciprocal (or multiplicative inverse) of a complex number $a + bi$ is defined to be $\dfrac{1}{a+bi}$, where $a \neq 0$ and $b \neq 0$. A reciprocal can be written in standard form by multiplying the numerator and denominator by the conjugate of the denominator.

Write the reciprocal of $2 - 3i$ in standard form.

EXAMPLE 10

Solution: The reciprocal of $2 - 3i$ is $\dfrac{1}{2-3i}$.

Multiply the numerator and denominator by $2 + 3i$.

$$\dfrac{1}{2-3i} = \dfrac{1(2+3i)}{(2-3i)(2+3i)}$$

$$= \dfrac{2+3i}{2^2 + 3^2}$$

$$= \dfrac{2+3i}{4+9}$$

$$= \dfrac{2-3i}{13} \text{ or } \dfrac{2}{13} + \dfrac{3}{13}i$$

Therefore, the reciprocal of $2 - 3i$ is $\dfrac{2}{13} + \dfrac{3}{13}i$.

In section 6.3 we learned that a quadratic equation has no real solutions when its discriminant ($b^2 - 4ac$) is negative. As illustrated by the next example, such equations do have complex solutions.

Find the complex solutions of $2x^2 - 6x + 5 = 0$.

EXAMPLE 11

Solution: Use the quadratic formula:

$$2x^2 - 6x + 5 = 0$$
$$a = 2, b = -6, c = 5$$
$$x = \frac{-b \pm \sqrt{b^2 - 4ac}}{2a}$$
$$= \frac{6 \pm \sqrt{36 - 4 \cdot 2 \cdot 5}}{2 \cdot 2}$$
$$= \frac{6 \pm \sqrt{36 - 40}}{4}$$
$$= \frac{6 \pm \sqrt{-4}}{4}$$
$$= \frac{6 \pm 2i}{4}$$

Factor a 2 from the numerator and denominator. Then, reduce the fraction.

$$= \frac{\cancel{2}(3 \pm i)}{\cancel{2} \cdot 2}$$
$$= \frac{3 \pm i}{2}$$

The complex solutions are $\frac{3 + i}{2}$ and $\frac{3 - i}{2}$.

Exercise 6.6

Add or subtract as indicated.

1. $(1 + 4i) + (3 + 5i)$
2. $(2 + 5i) + (2 - 6i)$
3. $(3 + 2i) + (5 + i)$
4. $(5 + 3i) + (-6 + i)$
5. $(-1 - 2i) - (3 - i)$
6. $(-2 + 5i) - (-4 - 2i)$
7. $(4 + i) - 7$
8. $(3 + 4i) - (5i - 3)$

Multiply as indicated.

9. $i(3 + 5i)$
10. $(2 + 3i)(6 + 4i)$
11. $(3 - i)(4 - 2i)$
12. $(5 + 3i)(-6 + i)$
13. $(3 - 2i)(-1 + 5i)$
14. $(3 + i)(3 - i)$
15. $i^3(i + 2)$
16. $3i(4 - i)$
17. $\sqrt{-3}(2 - \sqrt{-12})$
18. $(2 - 3i)^2$
19. $(3 + i)^2$
20. $(1 - i)^2$
21. $(6 - 2i)^2$
22. $(\sqrt{2} + \sqrt{-1})^2$

Divide as indicated.

23. $\dfrac{3}{2i}$
24. $\dfrac{3 - 4i}{i}$
25. $\dfrac{1}{3 + i}$
26. $\dfrac{2}{i - 5}$
27. $\dfrac{6}{3 - i}$
28. $\dfrac{2i}{1 - 2i}$
29. $\dfrac{i - 2}{1 + 2i}$
30. $\dfrac{5 + i}{2 + 3i}$
31. $\dfrac{-1 + 2i}{4 + i}$
32. $\dfrac{\sqrt{2} + \sqrt{3}i}{\sqrt{2} - \sqrt{3}i}$

Perform the indicated operations and simplify.

33. $(2i\sqrt{-5})^2 + (2i\sqrt{5})^2$
34. $[(3 + 4i) - (6 + i)] - (5 - i)$
35. $[(3 - i) - (2 + 5i)] - (3 - 4i)$
36. $(2 - i)^2 + (6 - 2i) - (2 + i)^2$

Solve using the complex number system.

37. $x^2 = -25$
38. $x^2 + 8 = 0$
39. $(x - 3)^2 = -4$
40. $(y - 1)^2 + 4 = 0$
41. $y^2 - y + 3 = 0$
42. $2a^2 + a + 3 = 0$
43. $3x^2 + 5x + 3 = 0$
44. $3y^2 + 2y + 4 = 0$
45. $2t^2 - 4t + 5 = 0$
46. $2t^2 + 6 = 0$
47. $\dfrac{1}{2}x^2 - 2x + 4 = 0$
48. $\dfrac{1}{2}y^2 + 2 = y$

Miscellaneous.

49. Find the value of $x^2 + x - 1$ if $x = 2 - 3i$.
50. Find the value of $3x^2 - 4x - 5$ if $x = 4 + 3i$.
51. Find the reciprocal of $-2 + 4i$.
52. Multiply the conjugate of $4 - 3i$ by the conjugate of $3 - 4i$.

Summary

KEY WORDS AND PHRASES

Quadratic Equation—any second degree equation which can be written in the form $ax^2 + bx + c = 0$, where $a, b, c \in R$, $a \neq 0$.

Standard Form of the Quadratic Equation—the form $ax^2 + bx + c = 0$.

Quadratic Formula—$x = \dfrac{-b \pm \sqrt{b^2 - 4ac}}{2a}$

Discriminant—$D = b^2 - 4ac$

Critical Value—a number that separates the positive values of a factor from the negative values.

Imaginary Unit—i is the imaginary unit, where $i^2 = -1$.

Imaginary Number—any number of the form bi, where $b \in R$, $b \neq 0$ and i is the imaginary unit.

Complex Number—any number of the form $a + bi$, where $a, b \in R$ and i is the imaginary unit.

Standard Form of a Complex Number—$a + bi$, where a is the real part and b is the imaginary part.

Equal Complex Numbers—$a + bi = c + di$ if and only if $a = c$ and $b = d$.

Addition and Subtraction of Complex Numbers—

$$(a + bi) + (c + di) = (a + c) + (b + d)i$$
$$(a + bi) - (c + di) = (a - c) + (b - d)i$$

Multiplication of Complex Numbers—

$$(a + bi)(c + di) = (ac - bd) + (ad + bc)i$$

Conjugate—the conjugate of $a + bi$ is $a - bi$.

Division of Complex Numbers—The quotient of two complex numbers is found by multiplying the numerator and denominator by the conjugate of the denominator.

PROPERTIES

Zero Factor Property—If $m \cdot n = 0$ and $m, n \in R$, then $m = 0$ or $n = 0$.

Square Root Property—If $x^2 = k$, where k is a non-negative real constant, then $x = \sqrt{k}$ or $x = -\sqrt{k}$.

Completing the Square—
  (1) Isolate the constant term on one side of the equation.
  (2) Make leading coefficient of the squared variable 1.
  (3) Complete the square by squaring one-half the coefficient of the second term and adding it to both sides of the equation.
  (4) Solve the resulting equation by applying the square root property of equations.

Solving a Quadratic Inequality—
  (1) Write in the form $ax^2 + bx + c < 0$ or $ax^2 + bx + c > 0$.
  (2) Factor the quadratic $ax^2 + bx + c$.
  (3) Find the critical values for each factor. Also, graph the values where each factor is positive and negative.
  (4) From the graph, find the interval that forms the correct product, either positive or negative.

Self-Checking Exercise

Solve by factoring.
1. $3a^2 = 5a$
2. $3a^2 = 8a + 3$
3. $10m^2 - 29m = -21$
4. $9x^2 + 30x + 25 = 0$
5. $2(x - 2)^2 = 2 - (x - 3)^2$
6. $\dfrac{4}{x + 1} - \dfrac{1}{x} = 1$
7. $\dfrac{4}{x} - 1 = \dfrac{1}{x - 1}$
8. $3x - 2 = \sqrt{2x - 1}$
9. $\sqrt{2x + 5} = x + 1$
10. $x^4 - 5x^2 + 4 = 0$
11. $x^{-2} + 4x^{-1} - 5 = 0$
12. $x^{-2} - 6x^{-1} = 16$

Solve by completing the square.
13. $y^2 + 2y = 5$
14. $(x + 1)^2 = 49$
15. $x^2 - 2x = 14$
16. $x^2 - 4x = -2$
17. $2x^2 + 3x - 4 = 0$
18. $5x^2 = x + 2$

Solve by using the quadratic formula.
19. $3t^2 + 2t - 1 = 0$
20. $6x^2 = 5 - 7x$
21. $5x^2 + 2x - 4 = 0$
22. $x - 1 = \dfrac{12}{x} + \dfrac{x}{2}$
23. $x^2 + \dfrac{7}{2}x = 3$
24. $\dfrac{2}{x - 1} - \dfrac{x - 4}{x - 3} = 1$

Solve the inequalities.
25. $x^2 - x - 20 < 0$
26. $2x^2 + 11x + 5 \leq 0$
27. $-x^2 - 7x - 6 < 0$
28. $3x^2 - 7x + 2 \geq 0$
29. $x^2 - 3x - 28 \leq 0$
30. $x^2 + 4x + 4 > 0$
31. $(x - 3)(x + 4)(x + 1) < 0$
32. $x^4 - x^2 > 0$
33. $\dfrac{x}{x - 3} < 0$
34. $\dfrac{x}{x + 1} \geq 0$

Simplify and write the result in standard form.
35. $-3i^2$
36. i^{43}
37. i^{86}
38. $\sqrt{-4} \cdot \sqrt{-4}$
39. $(i\sqrt{5})^2 + \sqrt{-9} \cdot \sqrt{-4}$
40. $(6 + 6i) - (-3 - 6i)$
41. $(4 - 3i) + (-1 + 6i)$
42. $(3 - 2i)(1 + 2i)$
43. $(5 + 3i)(2 - 4i) - (3 - 8i)$
44. $\dfrac{3 + 4i}{3 - 2i}$
45. $\dfrac{2 - i}{3 + i}$
46. $\dfrac{1 + i}{2 + 3i}$

Solve using the complex numbers.
47. $x^2 - 3x + 3 = 0$
48. $4x^2 + 6x + 5 = 0$
49. $2x^2 - 5x + 4 = 0$
50. $2x(3 - x) = 5$

Classify the roots without solving. Use the discriminant.

51. $5x^2 - x + 2 = 0$
52. $2x^2 - x - 6 = 0$
53. $10x - 4x^2 = \dfrac{25}{4}$
54. $x^2 - 7x + 16 = 0$

Miscellaneous.

55. Find k so that $x^2 - 12x + k = 0$ has one real root.
56. Find k so that $x^2 + kx + \dfrac{1}{2}k = 0$ has one real root.
57. Find k so that $x^2 + kx + 1 = 0$ has no real roots.

Solutions to Self-Checking Exercise

1. 0 and $\dfrac{5}{3}$
2. $-\dfrac{1}{3}$ and 3
3. $\dfrac{7}{5}$ and $\dfrac{3}{2}$
4. $-\dfrac{5}{3}$
5. 3 and $\dfrac{5}{3}$
6. 1
7. 2
8. $\dfrac{5}{9}$ and 1
9. 2
10. $-2, -1, 1,$ and 2
11. $-\dfrac{1}{5}$ and 1
12. $-\dfrac{1}{2}$ and $\dfrac{1}{8}$
13. $-1 \pm \sqrt{6}$
14. -8 and 6
15. $1 \pm \sqrt{15}$
16. $2 \pm \sqrt{2}$
17. $\dfrac{-3 \pm \sqrt{41}}{4}$
18. $\dfrac{1 \pm \sqrt{41}}{10}$
19. -1 and $\dfrac{1}{3}$
20. $-\dfrac{5}{3}$ and $\dfrac{1}{2}$
21. $\dfrac{-1 \pm \sqrt{21}}{5}$
22. -4 and 6
23. $\dfrac{-7 \pm \sqrt{97}}{4}$
24. $\dfrac{11 \pm \sqrt{17}}{4}$
25. $\{x \mid -4 < x < 5\}$
26. $\left\{x \mid -5 \leq x \leq -\dfrac{1}{2}\right\}$
27. $\{x \mid x < -6 \text{ or } x > -1\}$
28. $\left\{x \mid x \leq \dfrac{1}{3} \text{ or } x \geq 2\right\}$
29. $\{x \mid -4 \leq x \leq 7\}$
30. $\{x \mid x \neq -2\}$
31. $\{x \mid x < -4 \text{ or } -1 < x < 3\}$
32. $\{x \mid x < -1 \text{ or } x > 1\}$
33. $\{x \mid 0 < x < 3\}$
34. $\{x \mid x < -1 \text{ or } x \geq 0\}$
35. 3
36. $-i$

SOLUTIONS TO SELF-CHECKING EXERCISE

37. -1
38. -4
39. -11
40. $9 + 12i$
41. $3 + 3i$
42. $7 + 4i$
43. $19 - 6i$
44. $\dfrac{1 + 18i}{13}$
45. $\dfrac{1 - i}{2}$
46. $\dfrac{5 - i}{13}$
47. $\dfrac{3 \pm i\sqrt{3}}{2}$
48. $\dfrac{-3 \pm i\sqrt{11}}{4}$
49. $\dfrac{5 + i\sqrt{7}}{4}$
50. $\dfrac{3 \pm i}{2}$
51. $D = -39$; no real solutions
52. $D = 49$; two real solutions
53. $D = 0$; one real solution
54. $D = -15$; no real solutions
55. $k = 36$
56. $k = 0$ or $k = 2$
57. $\{k \mid -2 < k < 2\}$

RELATIONS, FUNCTIONS, AND THEIR GRAPHS

7

- 7.1 Ordered Pairs, Relations, and Graphs
- 7.2 Functions
- 7.3 Inverse Relations and Functions
- 7.4 The Linear Function
- 7.5 Equations of the Straight Line
- 7.6 Linear Inequalities
- 7.7 Variation

ORDERED PAIRS, RELATIONS, AND GRAPHS

7.1

An *ordered pair* is an arrangement of two numbers within parentheses to show that one number is first and the other is second. In the ordered pair (a, b), a is the first component and b is the second component. Given two ordered pairs (a, b) and (c, d), they are equal if and only if $a = c$ and $b = d$. Ordered pairs of the form (x, y) are commonly used to write solutions for equations and inequalities having two variables. For example, the equation $y = 3x$ produces infinitely many ordered pairs when values are arbitrarily substituted for x and the equation is solved for y. The following table of values illustrates some of these ordered pairs.

Computation using $y = 3x$	Table of Values for $y = 3x$	Corresponding Ordered Pairs
	x | y	
If $x = -3$, $y = 3(-3)$ or -9	-3 | -9	$(-3, -9)$
If $x = -2$, $y = 3(-2)$ or -6	-2 | -6	$(-2, -6)$
If $x = 0$, $y = 3 \cdot 0$ or 0	0 | 0	$(0, 0)$
If $x = \frac{2}{3}$, $y = 3 \cdot \frac{2}{3}$ or 2	$\frac{2}{3}$ | 2	$\left(\frac{2}{3}, 2\right)$
If $x = 1$, $y = 3 \cdot 1$ or 3	1 | 3	$(1, 3)$
If $x = \sqrt{2}$, $y = 3\sqrt{2}$	$\sqrt{2}$ | $3\sqrt{2}$	$(\sqrt{2}, 3\sqrt{2})$

Ordered pairs are used to denote points in a plane, which will will enable us to graph solutions of equations and inequalities having two variables. The French philosopher and mathematician René Descartes (1596–1650) found, as shown in Figure 7.1, that a plane can be partitioned into a coordinate grid by constructing two perpendicular real number lines called *coordinate axes*. The horizontal number number line is called the *x-axis* while the vertical number line is called the *y-axis*. Their point of intersection is called the *origin*. This configuration is called the *rectangular coordinate system* or, in honor of its inventor, the *Cartesian coordinate system*.

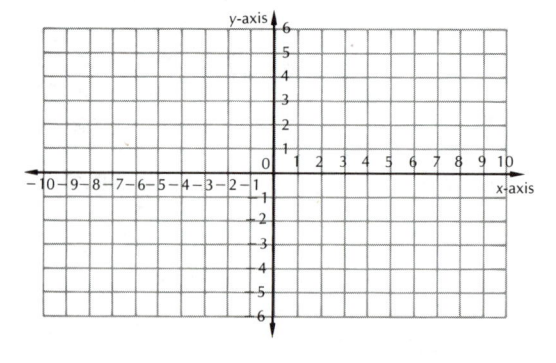

FIGURE 7.1

ORDERED PAIRS, RELATIONS, AND GRAPHS   255

To name a point using the rectangular coordinate system, we use an ordered pair of the form (x, y). The x-coordinate is called the abscissa and measures the directed distance to the point along the x-axis. The y-coordinate is called the ordinate and measures the directed distance to the point along the y-axis. To illustrate, study Figure 7.2, where we have located two points, (−5, −3) and (7, 4).

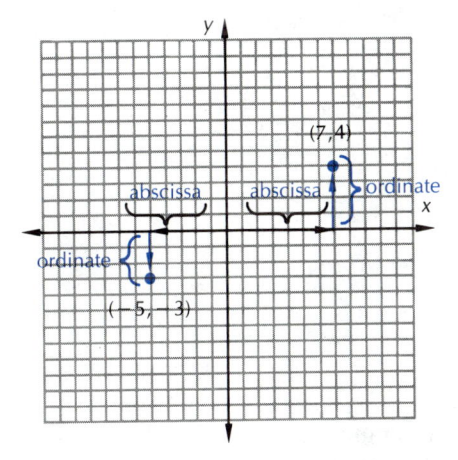

FIGURE 7.2

The x and y-axes separate the plane into regions called quadrants. As shown in Figure 7.3, the quadrants are numbered I through IV, reading clockwise from upper right. The x and y-axes are considered to be boundary lines and are not part of any quadrant.

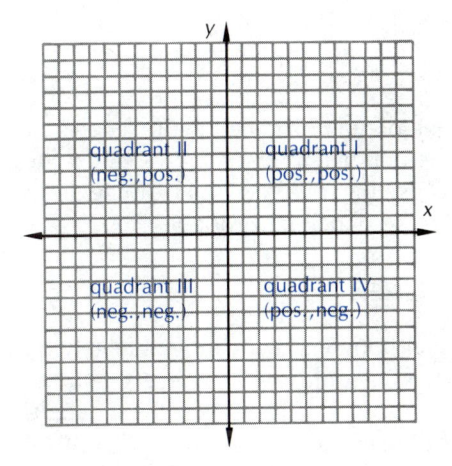

FIGURE 7.3

Using ordered pairs, name each of the points in Figure 7.4. Also, give the associated quadrant.

EXAMPLE 1

RELATIONS, FUNCTIONS, AND THEIR GRAPHS

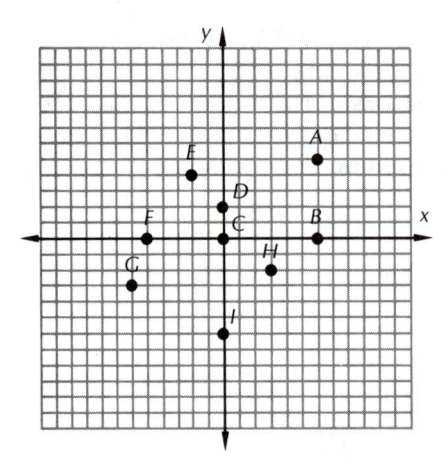

FIGURE 7.4

Solution: A (6, 5) is in quadrant I.
E (−2, 4) is in quadrant II.
G (−6, −3) is in quadrant III.
H (3, −2) is in quadrant IV.
These other points are located on the x or y axis and thus are not within any quadrant: B(6, 0); C(0, 0); D(0, 2); F(−5, 0); and I(0, −6).

A non-empty set of ordered pairs is said to be a relation. The set of first components (x-coordinates) of the ordered pairs is called the domain of the relation, while the set of second components (y-coordinates) is called the range.

EXAMPLE 2 State the domain and range of the relation {(−6, 2), (3, 5), (0, −3), (3, −7)}.

Solution: The domain is {−6, 3, 0}, the range is {2, 5, −3, −7}.

A variable representing the numbers in the domain is called the independent variable, while a variable representing the numbers of the range is called the dependent variable. Thus, for relations having ordered pairs of the form (x, y), the independent variable is x while the dependent variable is y.

EXAMPLE 3 Find the domain and range of the relation $\{(x, y) \mid y = x^2\}$.

Solution: The domain consists of all real numbers that can legitimately be substituted for the independent variable, x. Since every number can be squared, there are no restrictions placed on the independent variable, and thus the domain consists of the entire set of real numbers, symbolized $\{x \mid x \in R\}$. On the other hand, a restriction has been placed on the dependent variable, y. We know that the square of a real number cannot be negative. Since y is equal to the square of x, it is never negative. Thus, the range is $\{y \mid y \geq 0\}$.

ORDERED PAIRS, RELATIONS, AND GRAPHS 257

Ordered pairs and the rectangular coordinate system provide a means of drawing the graph of a relation. A graph of a relation is the set of points whose coordinates satisfy the condition defining the given relation. This condition is generally an equation or inequality having two variables. To construct a graph we follow this procedure:

Constructing a Graph
(1) Make a table of values by arbitrarily selecting values for one of the variables. This provides ordered pairs. Obtain as many as necessary to determine a pattern of points.
(2) Plot on a rectangular coordinate system the points corresponding to the ordered pairs. Remember, if you need more points, go back and expand the table of values.
(3) Look for a pattern of points and connect them with a "smooth" line or curve.

Graph the relation $\{(x, y) \mid y = 3x - 1\}$.

EXAMPLE 4

Solution: Form the following table by assigning arbitrary values to x and calculating the corresponding values of y.

Table of Values   Ordered Pairs

x	y
−2	−7
−1	−4
0	−1
1	2
2	5

$(-2, -7)$
$(-1, -4)$
$(0, -1)$
$(1, 2)$
$(2, 5)$

The points are plotted in Figure 7.5 and connected with a "smooth" straight line.

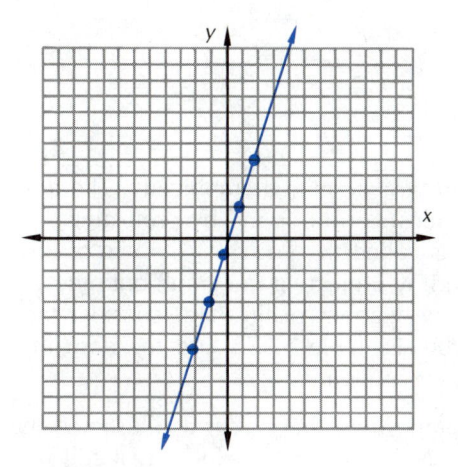

FIGURE 7.5

EXAMPLE 5

For the relation $A = \{(x, y) \mid y = \sqrt{x + 1}\}$, give the domain, range, and graph.

Solution: The domain consists of all real numbers that can legitimately be substituted for x. Since the expression $x + 1$ is contained under a radical, we know that $x + 1 \geq 0$. Solving for x produces the domain $\{x \mid x \geq -1\}$. The range is the set of all y-values given by the equation $y = \sqrt{x + 1}$. Thus, y is the principal square root of $x + 1$. Since principal roots are never negative, we deduce that the range is $\{y \mid y \geq 0\}$. The graph is found by forming a table of values and plotting the ordered pairs, as seen in Figure 7.6.

Table of Values   Ordered Pairs

x	y
−1	0
0	1
3	2
8	3

(−1, 0)
(0, 1)
(3, 2)
(8, 3)

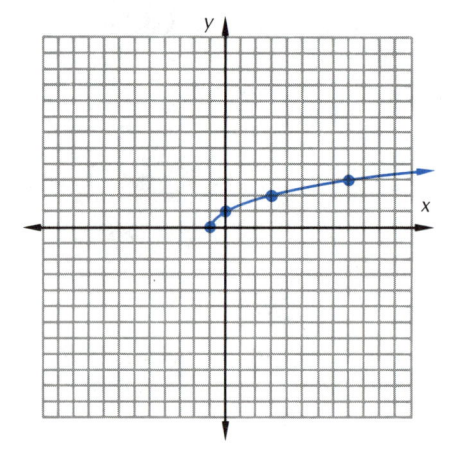

FIGURE 7.6

The domain and range of a relation can be determined graphically. The domain, being the set of all possible x values, can be found by projecting the graph onto the x-axis. This projection is called the horizontal sweep of the graph. Similarly, the range can be found by projecting the graph onto the y-axis, producing the vertical sweep of the graph. To illustrate, consider the graph of the relation $y = \sqrt{x + 1}$ from Example 5. To find the domain, project the graph onto the x-axis as shown in Figure 7.7.

The range is found by projecting the graph onto the y-axis, as illustrated in Figure 7.8. Note that even though the graph seems to level out, it still increases vertically without bound.

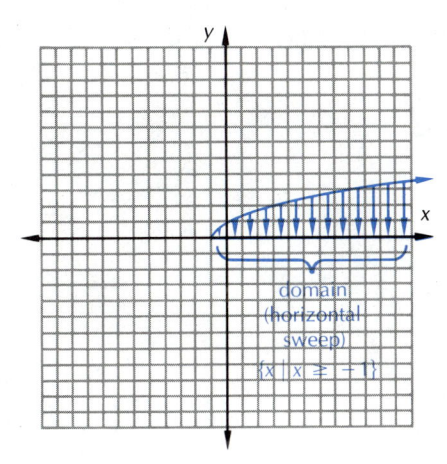

FIGURE 7.7

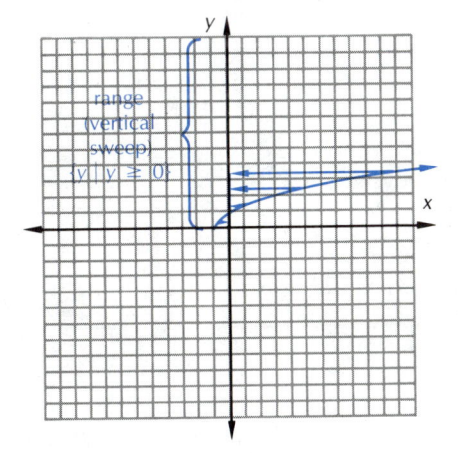

FIGURE 7.8

Graph the relation defined by the equation $y = |x - 2|$. Also, determine the domain and range graphically.

EXAMPLE 6

Solution: Form the following table by assigning values to x. Remember to take the absolute value when computing the corresponding values of y.

Table of Values   Ordered Pairs

x	y
−4	6
−2	4
0	2
2	0
4	2
8	6

(−4, 6)
(−2, 4)
(0, 2)
(2, 0)
(4, 2)
(8, 6)

The points are plotted in Figure 7.9, producing a "V-shaped" graph.

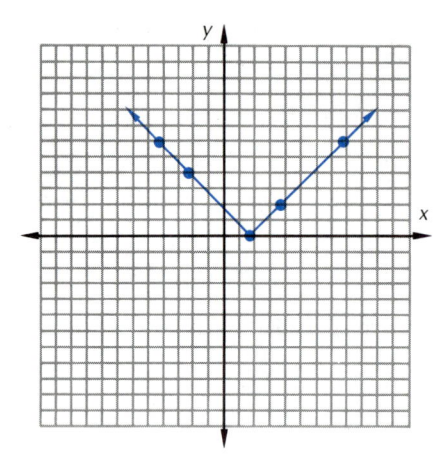

FIGURE 7.9

The projection of the graph onto the x-axis gives the domain, $\{x \mid x \in R\}$. The projection onto the y-axis provides us with the range, $\{y \mid y \geq 0\}$.

EXAMPLE 7   Graph the equation $x = y^2 - 2y - 8$ and determine the domain and range graphically.

Solution: Since this equation is solved for x, it is easier to select values for y and compute the corresponding values of x.

Table of Values

x	y
7	−3
0	−2
−5	−1
−8	0
−9	1
−8	2
−5	3
0	4
7	5

Ordered Pairs

(7, −3)
(0, −2)
(−5, −1)
(−8, 0)
(−9, 1)
(−8, 2)
(−5, 3)
(0, 4)
(7, 5)

The points are plotted in Figure 7.10, producing a "U-shaped" curve called a *parabola*.

ORDERED PAIRS, RELATIONS, AND GRAPHS

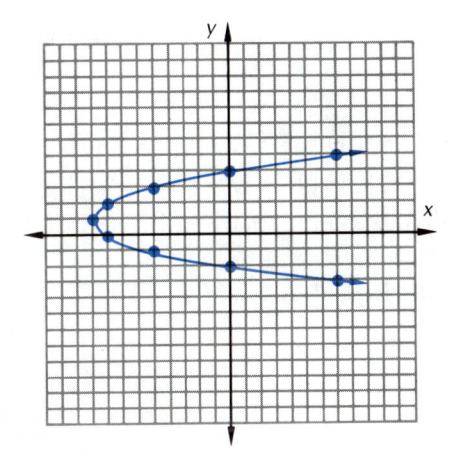

FIGURE 7.10

By projecting the graph onto the x and y-axes, you should see that the domain is $\{x \mid x \geq -9\}$ and the range is $\{y \mid y \in R\}$.

Give the domain and range of the following relations.

Exercise 7.1

1. $\{(-4, 3), (-1, 2), (4, 3), (0, 0)\}$
2. $\{(2, 0), (0, 2), (2, -1), (3, 4)\}$
3. $\{(-3, 0), (0, 0), (3, -2), (1, 3), (-3, -3)\}$
4. $\{(1, 1), (3, -2), (5, 1), (8, 3)\}$
5. $\{(2, 4), (3, 4), (4, 4)\}$
6. $\{(-1, -1), (2, 2), (3, 3)\}$

Give the domain and range of the relations.

7. $y = x^2 + 1$
8. $y = x$
9. $y = x + 2$
10. $y = x^2 - 2$
11. $y = \sqrt{x}$
12. $y = \sqrt{x - 1}$
13. $y = \sqrt{x - 4}$
14. $y = |x|$
15. $x = y^2$
16. $x = \sqrt{y - 2}$
17. $x = |y|$
18. $y = \dfrac{1}{x}$

Graph the following relations.

19. $\{(x, y) \mid y = 3x - 1\}$
20. $\{(x, y) \mid x = \sqrt{y}\}$
21. $\{(x, y) \mid x = |y|\}$
22. $\{(x, y) \mid y = 4x\}$
23. $\{(x, y) \mid x = y^2 - 1\}$
24. $\{(x, y) \mid y = 2x^2 + x - 2\}$
25. $\{(x, y) \mid x = 2y - 1\}$
26. $\{(x, y) \mid x = 4y\}$

RELATIONS, FUNCTIONS, AND THEIR GRAPHS

Graph the following relations and determine the domain and range graphically.

27. $\{(x, y) \mid x - 2y = 6\}$
28. $\{(x, y) \mid x^2 + 2 = y\}$
29. $\{(x, y) \mid y = \sqrt{x + 5}\}$
30. $\{(x, y) \mid y = \sqrt{1 - 2x}\}$
31. $\{(x, y) \mid y = |x + 2|\}$
32. $\{(x, y) \mid x = \sqrt{y - 2}\}$
33. $\{(x, y) \mid x = y^2 - 2y - 6\}$
34. $\{(x, y) \mid |x| + |y| = 4\}$
35. $\{(x, y) \mid y = x^2 + 3x - 4\}$
36. $\{(x, y) \mid x = -y^2 + 3\}$
37. $\{(x, y) \mid y = \sqrt{4 - x}\}$
38. $\{(x, y) \mid y = \sqrt{4 - x^2}\}$

FUNCTIONS

7.2

A special type of relation is called a function and is defined as follows:

> **Definition of a Function**
>
> A function is a relation where for each value of the first component of the ordered pairs, there is exactly one corresponding value of the second component.

In other words, we can say that a function is a set of ordered pairs where for each x value in the domain, there corresponds only one y value in the range. To illustrate, the following sets, S and T, are functions:

$$S = \{(-2, -1), (0, 3), (4, 5)\}$$
$$T = \{(1, 3), (2, 3), (0, -6)\}$$

In set T the first two ordered pairs have the same y value, 3. This does not violate the definition of a function because each x value has only one corresponding y value. The definition is violated only when two or more ordered pairs share the same x value. For this reason, the sets M and N are not functions.

$$M = \{(5, -1), (5, 0), (5, 4)\}$$
$$N = \{(-4, 0), (-1, 1), (0, 3), (0, 4)\}$$

In set M all three ordered pairs share the same x value, 5. In set N the last two ordered pairs share the same x value, 0.

Many functional relationships are defined by equations. To define a function, each x value of the domain must produce exactly one corresponding y value.

FUNCTIONS

Each of these equations defines a function because each value of x in the domain will produce exactly one y value. This has been illustrated for each equation by selecting a single value of x. You should verify further by substituting more values for x.

EXAMPLE 1

(a) $y = 3x - 1$; if $x = 2$, then y is only 5.
(b) $y = \dfrac{6}{x}$; if $x = -3$, then y is only -2.
(c) $y = x^2 + 1$; if $x = 0$, then y is only 1.
(d) $y = \sqrt{5 - x}$; if $x = 1$, then y is only 2.

None of these equations or inequalities defines a function. In each case, some values of x will produce more than one corresponding value of y.

EXAMPLE 2

(a) $y < x + 1$; if $x = 2$, then y represents every real number less than 3.
(b) $|y| = x$; if $x = 1$, then $y = 1$ and $y = -1$.
(c) $x^2 + y^2 = 4$; if $x = 0$, then $y = 2$ and $y = -2$.

As we have learned, a functional relationship exists only if each x value in the domain produces exactly one corresponding y value. Graphically, this means that each x value of the domain produces only one point on the curve. This is shown by the arrows in Figure 7.11, where we have graphed the function $y = x^2 + 2$.

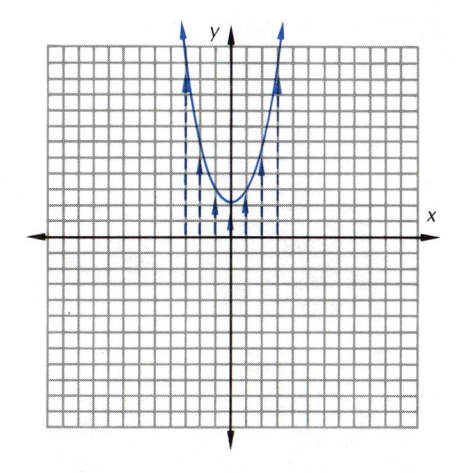

FIGURE 7.11

On the other hand, if a relation is not a function, then for at least one x value of the domain there are two or more different points on the graph. This is illustrated by the arrows in Figure 7.12, which shows the graph of the relation $|y| = x$.

RELATIONS, FUNCTIONS, AND THEIR GRAPHS

Note that one x-value yields two points, and these points lie on the same vertical line.

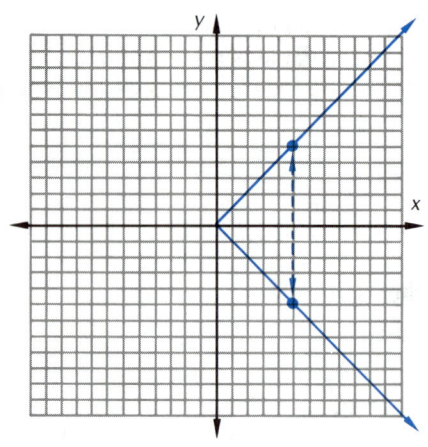

FIGURE 7.12

These illustrations suggest the following test for the graph of a function.

> **Vertical Line Test for a Function**
> Imagine a set of vertical lines passing through the graph. If any one of these lines cuts the graph more than once, then the graph does not represent a function.

EXAMPLE 3 Use the vertical line test to determine whether the curves in Figure 7.13 represent graphs of functions.

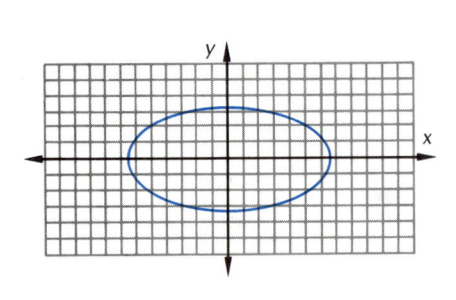

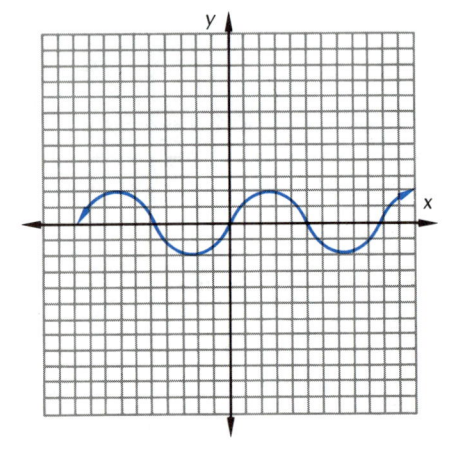

FIGURE 7.13

Solution: As shown in Figure 7.14, imagine a set of vertical lines passing through each graph. See whether any of these lines cut the graph more than once.

FUNCTIONS   265

not a function   a function

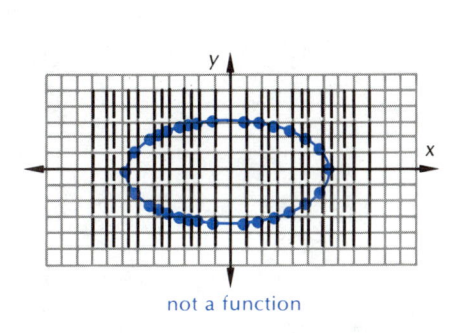

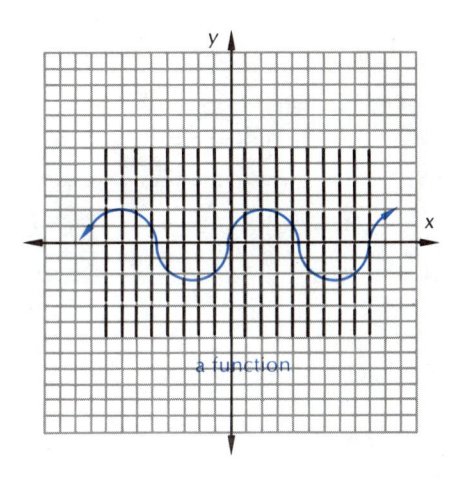

FIGURE 7.14

It is common to use lowercase letters such as *f*, *g*, and *h* to represent functions. For example, using the letter *f*, the function described by $y = 3x$ can be written as:

$$f = \{(x, y) \mid y = 3x\}$$

Given a function represented by *f*, another way to symbolize the variable *y* is to write *f(x)*, read "*f* of *x*." This symbolism is called functional notation and does not mean "*f* times *x*." Since *f(x)* symbolizes *y*, the preceding function can be written as:

$$f = \{(x, f(x)) \mid f(x) = 3x\}$$

Often you will find that the symbol *f(x)* has *x* replaced by a number, such as *f*(2). This is read "*f* of 2" and instructs us to refer to the function represented by *f* and determine the *y* value when *x* is 2. Since $f(x) = 3x$, we substitute 2 for *x*, obtaining $f(2) = 3 \cdot 2$ or 6. The notation $f(2) = 6$ simply means that for the function *f*, when $x = 2$ then $y = 6$. This also produces the ordered pair (2, 6).

We will use the following functions to illustrate functional notation in Examples 4 through 7.

$$f = \{(x, f(x)) \mid f(x) = 4x - 3\}$$
$$g = \{(x, g(x)) \mid g(x) = 2x^2 - 3\}$$

Find $f\left(\dfrac{3}{4}\right)$

EXAMPLE 4

Solution: $f(x) = 4x - 3$

$$f\left(\dfrac{3}{4}\right) = 4 \cdot \dfrac{3}{4} - 3$$
$$= 3 - 3$$
$$= 0$$

EXAMPLE 5 Find $f(3) - f(0)$

Solution: $f(x) = 4x - 3$
$f(3) = 4 \cdot 3 - 3 = 9$
$f(0) = 4 \cdot 0 - 3 = -3$
$f(3) - f(0) = 9 - (-3)$
$= 9 + 3$
$= 12$

EXAMPLE 6 Find $g(3a)$

Solution: $g(x) = 2x^2 - 3$
$g(3a) = 2(3a)^2 - 3$
$= 2(9a^2) - 3$
$= 18a^2 - 3$

EXAMPLE 7 Find $g(a + h)$

Solution: $g(x) = 2x^2 - 3$
$g(a + h) = 2(a + h)^2 - 3$
$= 2(a^2 + 2ah + h^2) - 3$
$= 2a^2 + 4ah + 2h^2 - 3$

In algebra we often refer to the zeros of a function. They are defined to be the values of x that make the functional value zero. To find the zeros of a function described by an equation, set y equal to zero and solve for x. Graphically, the zeros of a function tell where the graph crosses the x-axis. That is, they locate the x-intercepts.

EXAMPLE 8 Find the zeros and graph the function described by $f(x) = x^2 - 2x - 8$.

Solution: To find the zeros, let $f(x) = 0$ and solve for x. Then, form a table of values and graph the function. Remember, the zeros name the x-intercepts.

$f(x) = x^2 - 2x - 8$
$0 = x^2 - 2x - 8$
$0 = (x - 4)(x + 2)$
$x - 4 = 0$ or $x + 2 = 0$
$x = 4$ or $x = -2$

The zeros are 4 and -2.

FUNCTIONS **267**

	Table of Values		Ordered Pairs
	x	f(x)	
	−3	7	(−3, 7)
	−2	0	(−2, 0)
	−1	−5	(−1, −5)
	0	−8	(0, −8)
zeros are	1	−9	(1, −9)
−2 and 4	2	−8	(2, −8)
	3	−5	(3, −5)
	4	0	(4, 0)
	5	7	(5, 7)

The graph is shown in Figure 7.15. Note that the graph crosses the x-axis at $x = 4$ and $x = -2$.

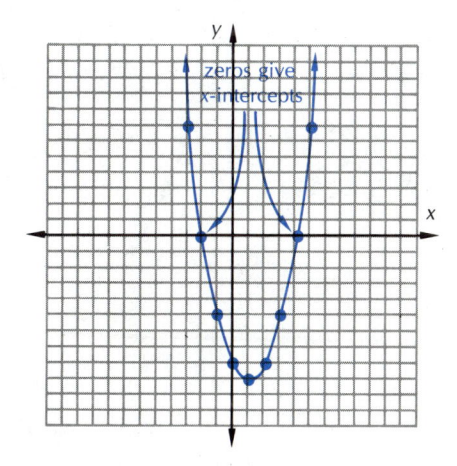

FIGURE 7.15

Often it is necessary to find the domain and range of a function. This can generally be accomplished either algebraically or graphically. Algebraically, we identify the restrictions placed on the variables. Graphically, we project the graph of the function onto the x-axis and onto the y-axis.

Find the domain and range of $f(x) = \sqrt{3x - 6}$.   **EXAMPLE 9**

Solution: To find the domain, we note that the expression $3x - 6$ cannot be negative. Thus,

$$3x - 6 \geq 0$$
$$3x \geq 6$$
$$x \geq 2$$

The domain is $\{x \mid x \geq 2\}$.

To find the range, we see that f(x) represents a principal square root and therefore cannot be negative. The radical is zero when x is 2, and can be increased without bound by selecting larger and larger values of x. Thus, the range values begin with zero and increase indefinitely.

The range if $\{f(x) \mid f(x) \geq 0\}$.

EXAMPLE 10 Determine the domain and range of the functions whose graphs appear in Figures 7.16 and 7.17.

Solution: Find the horizontal and vertical sweeps by projecting the graphs onto the x and y-axes, respectively.

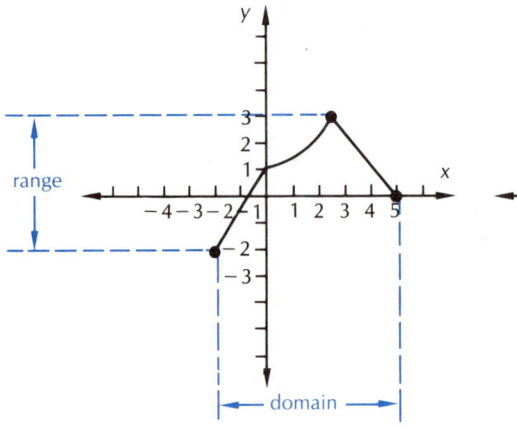

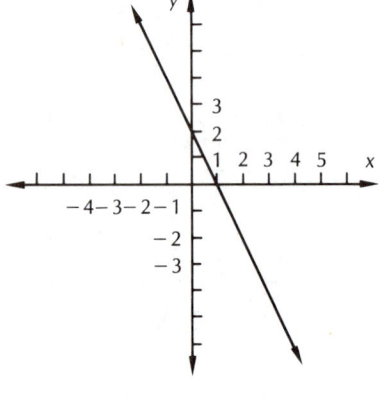

Domain is $\{x \mid -2 \leq x \leq 5\}$
Range is $\{y \mid -2 \leq y \leq 3\}$

FIGURE 7.16

Domain is $\{x \mid x \in R\}$
Range is $\{y \mid y \in R\}$

FIGURE 7.17

Exercise 7.2 Determine if the following curves represent graphs of functions. If a function, give the domain and range.

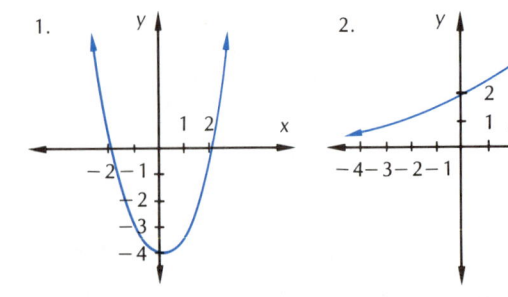

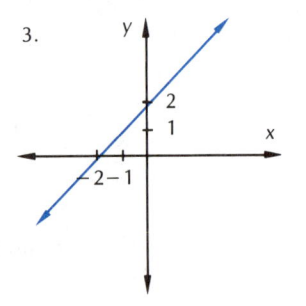

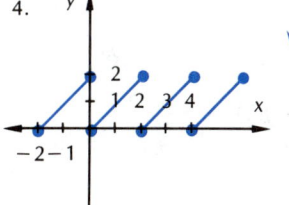

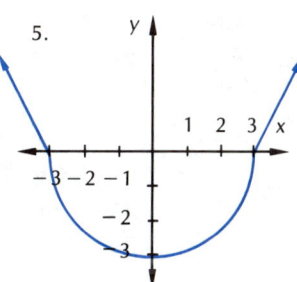

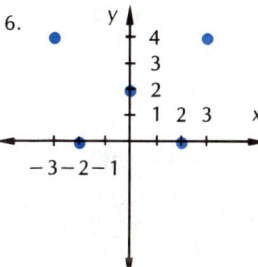

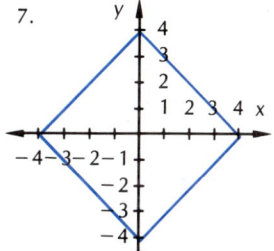

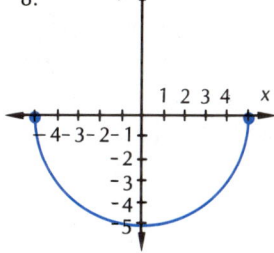

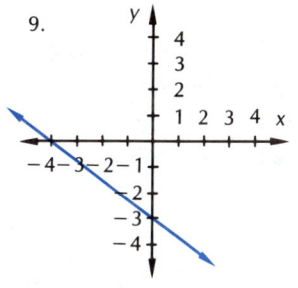

FUNCTIONS   **269**

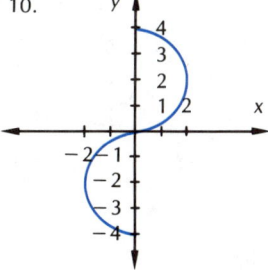

4., 5., 6., 7., 8., 9., 10. (graphs)

State whether these relations represent functions.
11. $y = 3x$
12. $y > x - 2$
13. $x + y = 6$
14. $x = y^2$
15. $|x| = y$
16. $x = \sqrt{y}$
17. $y^2 = x^2 - 1$
18. $y = x^2 - 1$

Find an expression for $f(x)$ and then find $f(3)$ and $f(-2)$.
19. $3x + y = 8$
20. $4x - y = 1$
21. $x^2 - y = 2$
22. $x^2 + y = 5$
23. $x = y^3$
24. $\sqrt{x+3} - y = 0$

Given these functions: $f = \{(x, f(x)) \mid f(x) = 4x - 1\}$
$g = \{(x, g,(x)) \mid g(x) = 3x^2 - 4x - 3\}$
$h = \{(x, h(x)) \mid h(x) = \sqrt{x+2}\}$

evaluate the following, if possible.

25. $f(0)$
26. $g(-1)$
27. $h(-3)$
28. $f(3)$
29. $g(2) + h(1)$
30. $f(-1) + f(1)$
31. $g(1) - h(0)$
32. $f(a + h)$
33. $g(a + h)$
34. $h(2a - 1)$
35. $f\left(\frac{3}{4}\right) + f\left(\frac{1}{4}\right)$
36. $g(-2) - g(2)$

Find any real zeros of the following functions.

37. $y = 5x - 2$
38. $f(x) = x^2 - x - 6$
39. $g(x) = x^2 - 25$
40. $h(x) = x^2$
41. $f(x) = x^2 + 4$
42. $y = x^2 + 3x - 2$
43. $y = 2x^2 + 5x - 3$
44. $f(x) = x^2 - 3$

Graph the following functions and give the domain and range.

45. $f(x) = 3x - 1$
46. $h(x) = |x - 2|$
47. $g(x) = \sqrt{x - 3}$
48. $f(x) = 3x^2 - 1$
49. $g(x) = \sqrt{4 - x}$
50. $h(x) = 5 - 2x$
51. $f(x) = x^2 + 6x + 9$
52. $g(x) = -x^2 + 4$

Miscellaneous.

53. If $f(x) = x^2 - 3x - 1$, find $f\left(\frac{1}{2}\right) - f\left(-\frac{1}{2}\right)$
54. If $g(x) = |x - 3|$, find $g(3) - g(-3)$
55. If $f(x) = 2x^2 - 1$, find $f(a + h)$
56. If $f(x) = 3x^2 + x - 1$, find $f(a + h)$
57. If $f(x) = x^2 - 3x + 2$, find $f(a + h) - f(a)$
58. If $f(x) = -x^2 + 2x - 3$, find $f(a + h) - f(a)$

INVERSE RELATIONS AND FUNCTIONS

7.3

Suppose G represents the relation $\{(-1, 2), (0, 0), (3, 5)\}$. Another relation can be formed by interchanging the x and y values of each ordered pair as follows:

$$\{(2, -1), (0, 0), (5, 3)\}$$

INVERSE RELATIONS AND FUNCTIONS

This new relation is called the *inverse* of G and is symbolized G^{-1}, read "the inverse of G." It does not mean $\frac{1}{G}$.

Since the inverse is formed by interchanging the x and y values, the domain of G becomes the range of G^{-1} and the range of G becomes the domain of G^{-1}, as shown in the following chart.

	Relation G	Inverse G^{-1}
	{(2, −1), (0, 0), (5, 3)}	{(−1, 2), (0, 0), (3, 5)}
Domain	{2, 0, 5}	{−1, 0, 3}
Range	{−1, 0, 3}	{2, 0, 5}

EXAMPLE 1

Find the inverse of the relation $F = \{(-1, 6), (-2, -3), (4, 0), (6, 8)\}$, and give the domain and range of F^{-1}.

Solution: To form the inverse relation, interchange the x and y values of each ordered pair.

$F^{-1} = \{(6, -1), (-3, -2), (0, 4), (8, 6)\}$

The domain of F^{-1} is {6, −3, 0, 8}.

The range of F^{-1} is {−1, −2, 4, 6}.

Notice that the inverse F^{-1} is a function because each x value corresponds with exactly one y value.

EXAMPLE 2

Find the inverse of the relation $G = \{(-3, 1), (-1, 1), (0, 7), (2, 4)\}$, and give the domain and range of G^{-1}.

Solution: Interchange the components of each ordered pair.

$G^{-1} = \{(1, -3), (1, -1), (7, 0), (4, 2)\}$

The domain of G^{-1} is {1, 7, 4}.

The range of G^{-1} is {−3, −1, 0, 2}.

Observe that the inverse G^{-1} is not a function because two ordered pairs share the same x value, 1.

As you have seen, an inverse relation may or may not be a function. We are particularly interested in functions whose inverses are also functions. This occurs only when the original function has no two ordered pairs with the same y values. Consider these examples:

Function	Inverse Relation
{(2, 5), (4, 5), (6, 8)}	{(5, 2), (5, 4), (8, 6)}
same y value	not a function
{(1, 4), (2, 6), (3, 8)}	{(4, 1), (6, 2), (8, 3)}
y values differ	is a function

We are now able to define an inverse function.

Inverse Function

Given a function f having no two ordered pairs with the same y value, the inverse function, symbolized f^{-1}, is such that if (a, b) belongs to f, then (b, a) belongs to f^{-1}.

If a function f is defined by an equation, we can find the equation of the inverse function f^{-1} (if it exists) by interchanging x and y, then solving for y.

The domain and range of the inverse function should always be obtained by interchanging the domain and range of the original function. This suggests the following procedure:

Finding the Inverse Function Algebraically

(1) Write the original function and give its domain and range.
(2) Interchange the domain and range to obtain the domain and range of the inverse.
(3) Interchange x and y in the original function, then solve for y. Remember, the $f^{-1}(x)$ is another name for y in the inverse function.

EXAMPLE 3   If $f(x) = \sqrt{x + 2}$, find the equation of the inverse function and determine its domain and range.

Solution:

Step 1.   The original function is given by the equation $f(x) = \sqrt{x + 2}$.

The domain is $\{x \mid x \geq -2\}$.

The range if $\{y \mid y \geq 0\}$.

Step 2.   We interchange the domain and range of the original function to obtain the domain and range of the inverse.

INVERSE RELATIONS AND FUNCTIONS

The domain of the inverse is $\{x \mid x \geq 0\}$.

The range of the inverse is $\{y \mid y \geq -2\}$.

Step 3. To find the equation of the inverse, interchange x and y in the original function and solve for y.

$$f(x) = \sqrt{x + 2} \quad \text{[original equation]}$$
$$y = \sqrt{x + 2}$$
$$x = \sqrt{y + 2} \quad \text{[interchange } x \text{ and } y\text{]}$$
$$x^2 = (\sqrt{y + 2})^2 \quad \text{[square both sides]}$$
$$x^2 = y + 2$$
$$y + 2 = x^2$$
$$y = x^2 - 2 \text{ or } f^{-1}(x) = x^2 - 2$$

Thus, the inverse function is defined by $f^{-1}(x) = x^2 - 2$. Its domain is $\{x \mid x \geq 0\}$ and its range is $\{f^{-1}(x) \mid f^{-1}(x) \geq -2\}$.

Functional notation, first discussed in section 7.2, can also be used with inverse functions, as shown in the next example.

EXAMPLE 4

Find the inverse of $f(x) = \dfrac{3}{x - 2}$. Also, evaluate $f(5)$ and $f^{-1}(1)$.

Solution: To find the inverse, interchange x and y, then solve for y.

$$y = \frac{3}{x - 2} \quad \text{[original function]}$$
$$x = \frac{3}{y - 2} \quad \text{[interchange } x \text{ and } y\text{]}$$
$$x(y - 2) = 3 \quad \text{[multiply both sides by } y - 2\text{]}$$
$$xy - 2x = 3$$
$$xy = 3 + 2x$$
$$y = \frac{3 + 2x}{x} \text{ or } f^{-1}(x) = \frac{3 + 2x}{x}$$

To find $f(5)$ and $f^{-1}(1)$, substitute the values of x into the proper function and evaluate.

(a) $f(5) = \dfrac{3}{5 - 2} = \dfrac{3}{3} = 1$

(b) $f^{-1}(1) = \dfrac{3 + 2 \cdot 1}{1} = \dfrac{5}{1} = 5$

Thus, $f(5) = 1$ and $f^{-1}(1) = 5$. This shows that $(5, 1)$ is a member of the function f, and that $(1, 5)$ is a member of the inverse function, f^{-1}.

To develop a procedure for graphing an inverse function, we refer to Figure 7.18 and assume that (a, b) belongs to the function f. Then, (b, a) would belong to the inverse function, f^{-1}. The line y = x is the perpendicular bisector of the line segment joining the points (a, b) and (b, a). This means that the points (a, b) and (b, a) are symmetrical to each other with respect to the line y = x.

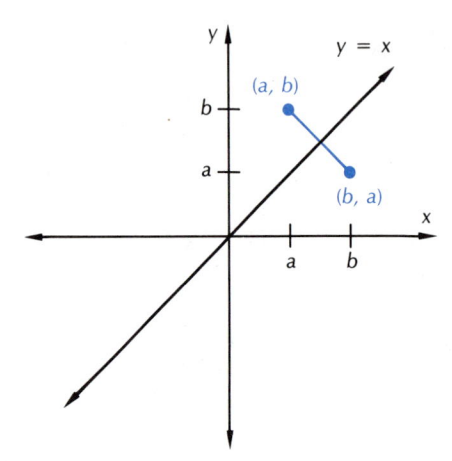

FIGURE 7.18

Thus, as illustrated in Figure 7.19, the graph of the inverse can be found by simply reflecting the graph of the function across the line y = x, giving its mirror image.

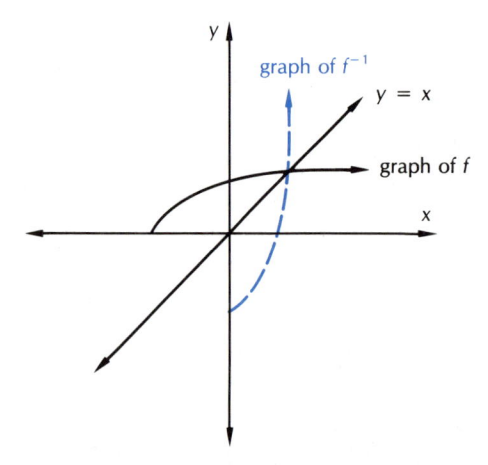

FIGURE 7.19

INVERSE RELATIONS AND FUNCTIONS

Graph the inverse of the function g shown in Figure 7.20 **EXAMPLE 5**

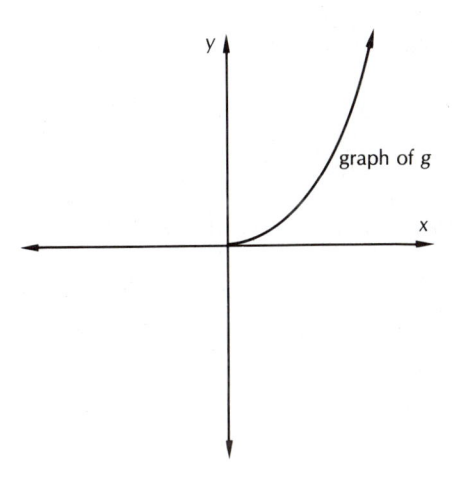

FIGURE 7.20

Solution: As illustrated in Figure 7.21, first draw in the line $y = x$. Then, reflect the graph of g across this line to obtain the mirror image.

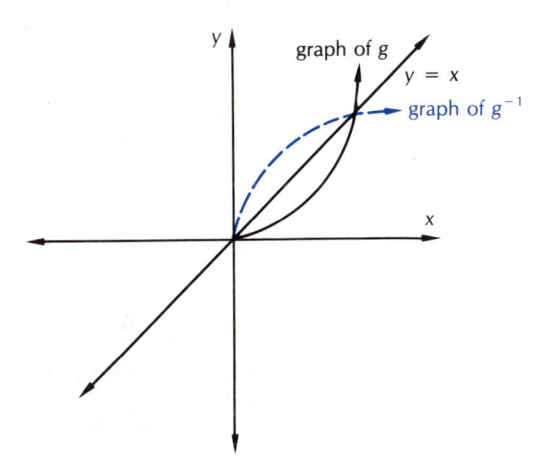

FIGURE 7.21

Find the inverse for each of the following relations. Determine if the inverse is a function. **Exercise 7.3**
   1. {(3, 0) (1, 1) (0, 2) (4, 3)}
   2. {(1, 0) (2, 0) (3, 0)}

3. {(4, 6) (3, 9) (−2, 3)}
4. {(1, 1) (2, 2) (3, 3) (4, 4) (5, 5)}
5. {(1, −1) (2, −2) (3, −3) (4, −4)}
6. {(0, 0) (1, 1) (2, 4) (3, 1)}

Graph each equation with a solid curve or line. Graph the inverse on the same coordinate system as a dashed curve or line.

7. $y = 3x$
8. $y = 2x + 3$
9. $y = x^2 + 2$
10. $y = \sqrt{x}$
11. $y = \sqrt{x - 2}$
12. $y = |x|$

Which graphs below have inverses that are functions?

13.

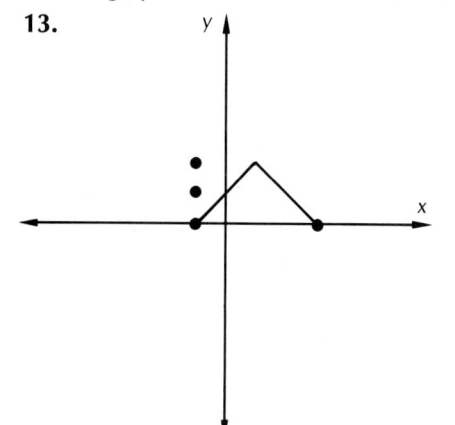

14.

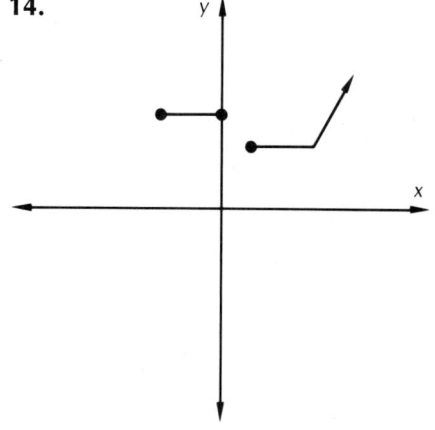

15.

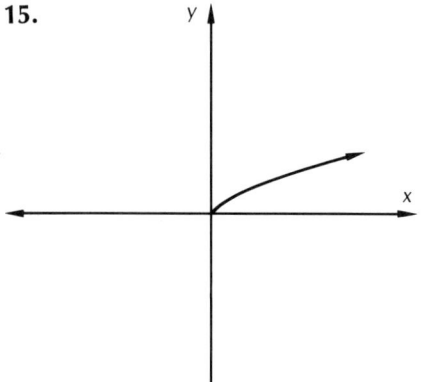

16.

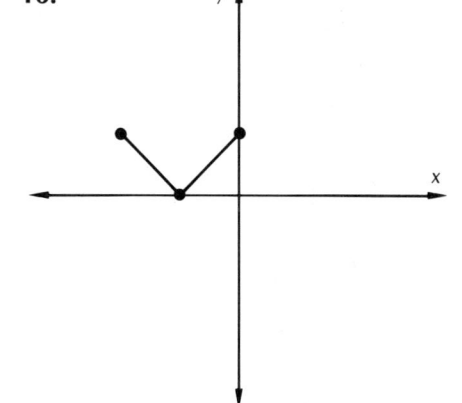

17.

18.

19.

20.

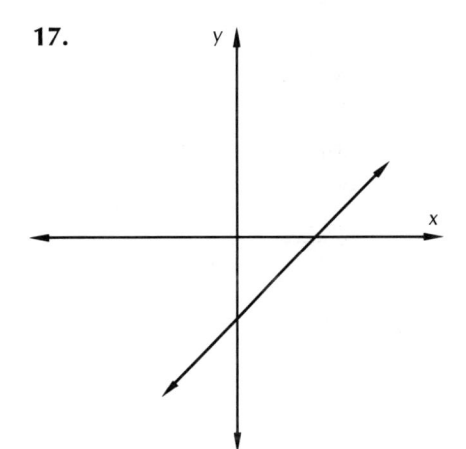

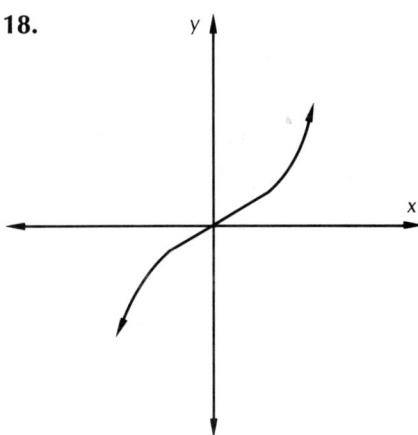

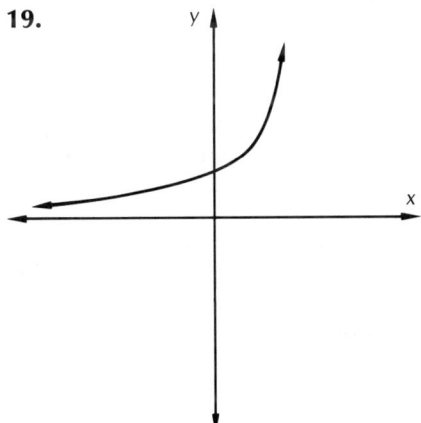

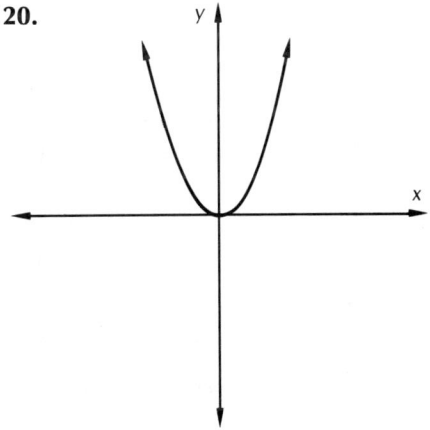

Find the inverse (f^{-1}) and its domain and range for each of the following.

21. $f(x) = 3x - 2$
22. $f(x) = 5x - 2$
23. $f(x) = 2x$
24. $y = 4x + 1$
25. $3y = x + 1$
26. $y - 3x = 2$
27. $3x + 2y = 12$
28. $f(x) = 1 - 3x$
29. $f(x) = \frac{1}{2}x - 2$
30. $f(x) = \frac{3x - 5}{4}$
31. $f(x) = \sqrt{x - 1}$
32. $f(x) = x^2, x \geq 0$
33. $y = \sqrt{x + 2}$
34. $f(x) = \frac{x - 1}{3}, -1 \leq x \leq 2$
35. $y = (x - 1)^2, x \geq 1$
36. $y = x^3$

RELATIONS, FUNCTIONS, AND THEIR GRAPHS

37. $y = \dfrac{4x - 1}{3}, 1 \leq x \leq 3$

38. $y = \dfrac{3}{x + 2}$

Find the inverse of $f(x)$. Then evaluate $f^{-1}(3)$, $f^{-1}(0)$ and $f^{-1}(-2)$.

39. $f(x) = 3x + 1$

40. $f(x) = 4x - 2$

41. $f(x) = \sqrt{x - 1}$

42. $f(x) = \dfrac{4}{x - 2}$

Evaluate the following if $f(x) = 3x - 1$ and $g(x) = x^2 + 2, x \geq 0$.

43. $f[g^{-1}(2)]$

44. $g[f^{-1}(4)]$

45. $f[f^{-1}(-3)]$

46. $g^{-1}[g^{-1}(2)]$

THE LINEAR FUNCTION

7.4

A *linear function* is any function whose equation can be written in the form,

$$f(x) = ax + b, \text{ where } a, b \in R$$

As examples in this section will illustrate, the graph of a linear function is a straight line.

EXAMPLE 1   Each of these equations describes a linear function.

(a) $y = 2x + 3$
(b) $h(x) = 3$, since $h(x) = 0x + 3$
(c) $x + y = 5$, since $y = -x + 5$
(d) $3x + 2y - 6 = 0$, since $y = -\dfrac{3}{2}x + 3$

As shown by Example 1(d), a linear function may also be described by a first degree equation in two variables fitting the form

$$Ax + By + C = 0, \text{ where } A, B, C \in R \text{ and } B \neq 0$$

EXAMPLE 2   Graph the linear function $f(x) = 2x + 1$.

Solution: Form a table of values and plot the corresponding points to produce the straight line in Figure 7.22.

THE LINEAR FUNCTION

Table of Values

x	f(x)
−4	−7
−2	−3
0	1
2	5
3	7

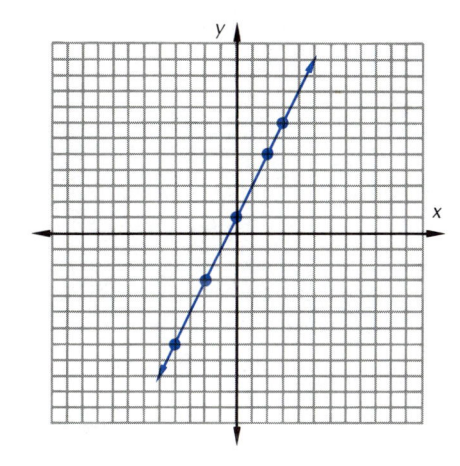

FIGURE 7.22

When graphing a linear function where both variables appear on the same side of the equation, such as $2x - 3y = 0$, it is convenient to obtain two special points called the x and y-intercepts. The x-intercept is the x value (or abscissa) of the point where the graph intersects the x-axis. Similarly, the y-intercept is the y value (or ordinate) of the point where the graph intersects the y-axis.

Finding x- and y-Intercepts
(1) To find the x-intercept, substitute 0 for y and solve the equation for x.
(2) To find the y-intercept, substitute 0 for x and solve the equation for y.

The following example illustrates how to graph a straight line by finding the x and y-intercepts.

Graph $2x - 3y = 6$ by finding the x and y-intercepts.

EXAMPLE 3

Solution: To find the x intercept, let $y = 0$.

$$2x - 3y = 6$$
$$2x - 3 \cdot 0 = 6$$
$$2x = 6$$
$$x = 3$$

The x intercept is 3.

To find the y-intercept, let $x = 0$.

$$2x - 3y = 6$$
$$2 \cdot 0 - 3y = 6$$
$$-3y = 6$$
$$y = -2$$

The y-intercept is -2.

As a check, we obtain a third point by selecting a convenient x value, such as -3.

$$2x - 3y = 6$$
$$2(-3) - 3y = 6$$
$$-6 - 3y = 6$$
$$-3y = 12$$
$$y = -4$$

The check point is $(-3, -4)$.

The graph is shown in Figure 7.23.

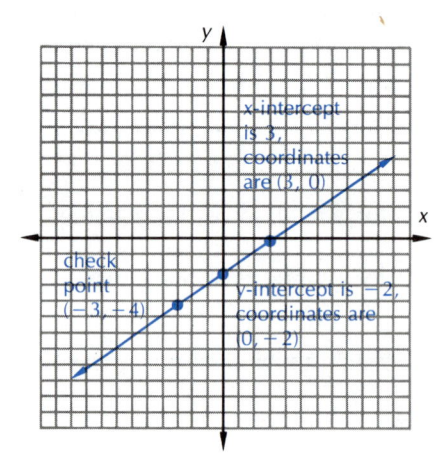

FIGURE 7.23

There are two special types of equations whose graphs are straight lines. These equations are of the form $y = k$ and $x = k$, where k is a real constant. To illustrate the first, consider the equation $y = 5$. This equation states that y is always 5 no matter what values x represents. As shown by the following table of values, the equation $y = 5$ produces the set of ordered pairs where every ordinate is 5. The graph, in Figure 7.24, is a horizontal straight line 5 units above the x-axis and is seen to represent a function. Since y is constant, always representing the number 5, the equation $y = 5$ or $f(x) = 5$ is said to define a constant function.

THE LINEAR FUNCTION **281**

Table of Values
for y = 5

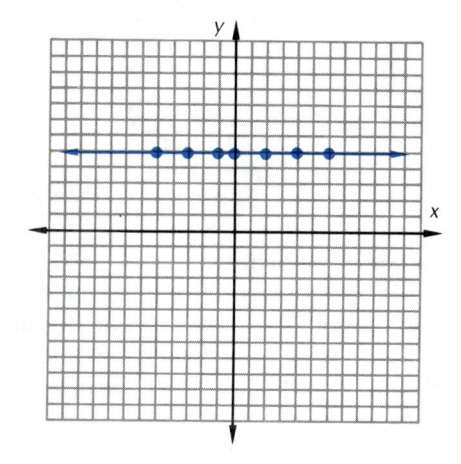

x	y
−5	5
−3	5
−1	5
0	5
2	5
4	5
6	5

FIGURE 7.24

The Constant Function

A function defined by the equation $f(x) = k$, where k is a real constant, is called a constant function. Its graph is a horizontal straight line k units from the x-axis.

To illustrate the second special form, we consider the equation $x = 4$. This equation states that x is always 4 regardless of the y value. Thus, in the table of values the abscissa is always 4. The graph, pictured in Figure 7.25, is a vertical straight line 4 units to the right of the x-axis. This is not the graph of a function.

Table of Values
for x = 4

x	y
4	6
4	4
4	2
4	0
4	−1
4	−3
4	−5

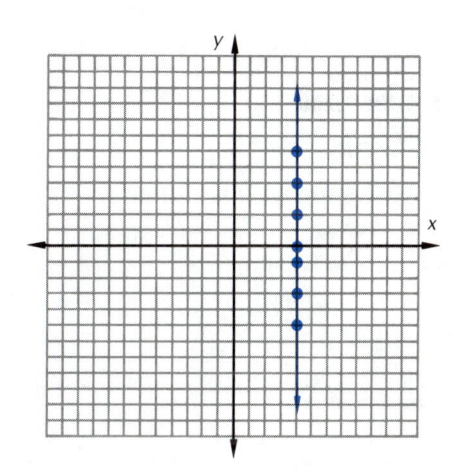

FIGURE 7.25

> **Equations of the Form $x = k$**
>
> An equation of the form $x = k$, where k is a real constant, does not define a function. Its graph is a vertical straight line k units from the y-axis.

As we have seen, two distinct points determine a straight line. A straight line is also determined by a point on a line and the slope of the line. Slope is a measure of "slant" or "incline." Using the graph in Figure 7.26, we define slope by selecting two different points on the line, L. They are labeled P_1 and P_2 with coordinates (x_1, y_1) and (x_2, y_2), respectively. Moving along the line from P_1 to P_2, the x value changes from x_1 to x_2 (a directed distance of $x_2 - x_1$). The corresponding y values change from y_1 to y_2 (a directed distance of $y_2 - y_1$). The ratio of the change in y (called the rise) to the corresponding change in x (called the run) is defined to be the slope of the line and is symbolized by the letter m.

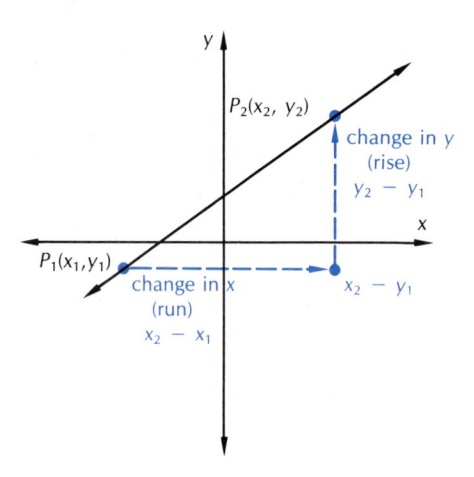

FIGURE 7.26

> **Definition of the Slope of a Line**
>
> Given two distinct points on a straight line, (x_1, y_1) and (x_2, y_2), the slope m is defined as:
>
> $$m = \frac{\text{change in } y}{\text{change in } x} \quad \text{or} \quad \frac{\text{rise}}{\text{run}}$$
>
> In terms of the given points, we have the formula:
>
> $$m = \frac{y_2 - y_1}{x_2 - x_1}$$

THE LINEAR FUNCTION

It can be shown using similar triangles that no matter which points are chosen from the line, the slope remains the same. Thus, to use the slope formula, obtain any two distinct points on the line. Either point may be represented by (x_1, y_1). Be sure, however, to subtract the x values in the same order as the y values.

Find the slope of the line passing through the points $(-2, -3)$ and $(6, -1)$, as illustrated in Figure 7.27.

EXAMPLE 4

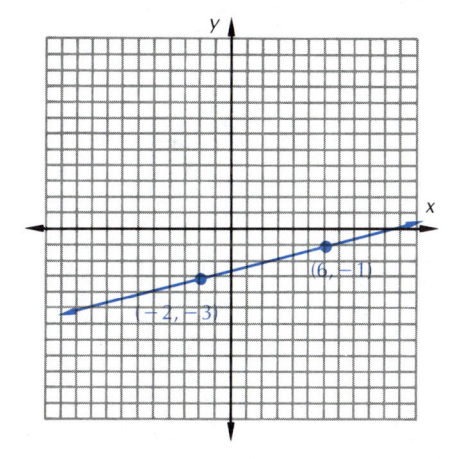

FIGURE 7.27

Solution: Letting $(x_1, y_1) = (-2, -3)$ and $(x_2, y_2) = (6, -1)$, use the slope formula:

$$m = \frac{y_2 - y_1}{x_2 - x_1} = \frac{-1 - (-3)}{6 - (-2)}$$

$$= \frac{-1 + 3}{6 + 2}$$

$$= \frac{2}{8}$$

$$= \frac{1}{4}$$

The slope is $\frac{1}{4}$. This means the line rises 1 unit for every 4 units it goes to the right. Note how reversing the points in the formula does not change the slope. Let $(x_1, y_1) = (6, -1)$ and $(x_2, y_2) = (-2, -3)$.

$$m = \frac{y_2 - y_1}{x_2 - x_1} = \frac{-3 - (-1)}{-2 - 6}$$

$$= \frac{-3 + 1}{-8}$$

$$= \frac{-2}{-8}$$

$$= \frac{1}{4}$$

This example also illustrates that a line having a positive slope will rise as it goes to the right.

EXAMPLE 5 Graph and find the slope of the line whose equation is $2x + y = 6$.

Solution: We find two different points on the line by obtaining the x and y-intercepts.

To find the x-intercept, let $y = 0$.

$2x + y = 6$
$2x + 0 = 6$
$x = 3$

This gives the point (3, 0).

To find the y intercept, let $x = 0$.

$2x + y = 6$
$2 \cdot 0 + y = 6$
$y = 6$

This gives the point (0, 6).

The line is graphed in Figure 7.28.

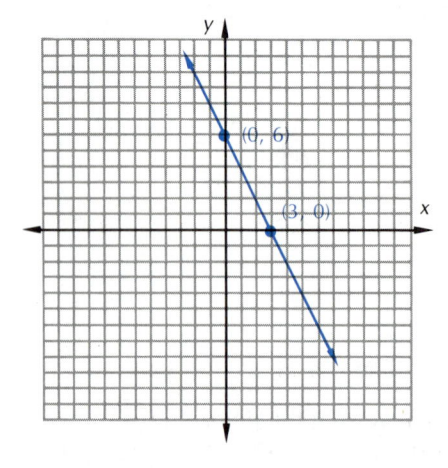

FIGURE 7.28

THE LINEAR FUNCTION   285

To find the slope, we use the slope formula for the points (0, 6) and (3, 0).

$$m = \frac{y_2 - y_1}{x_2 - x_1} = \frac{6 - 0}{0 - 3} = -2$$

The slope is -2, which means the line will fall 2 units for each unit it goes to the right. This example shows that a line having a negative slope will fall as it goes to the right.

The next two examples examine the concept of slope as it applies to horizontal and vertical lines.

Find the slope of the horizontal line whose equation is $y = 5$.   **EXAMPLE 6**

Solution: Select any two points from the graph in Figure 7.29, such as $(-2, 5)$ and $(4, 5)$. Then, use the slope formula.

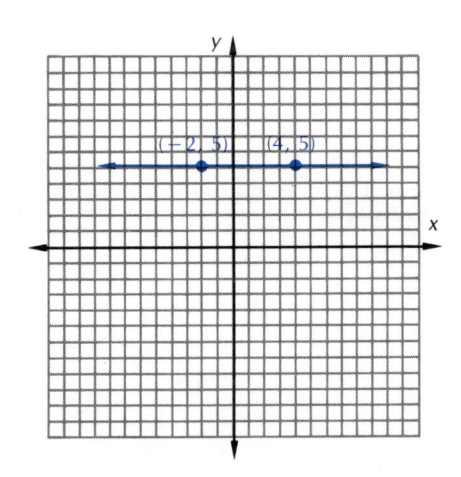

FIGURE 7.29

$$m = \frac{y_2 - y_1}{x_2 - x_1} = \frac{5 - 5}{-2 - 4} = \frac{0}{-6} = 0$$

The slope of a horizontal line is 0. It neither rises nor falls as it moves to the right.

Find the slope of a vertical line whose equation is $x = 3$.   **EXAMPLE 7**

Solution: Select any two points from the graph in Figure 7.30, such as $(3, 6)$ and $(3, -2)$. Then, use the slope formula.

RELATIONS, FUNCTIONS, AND THEIR GRAPHS

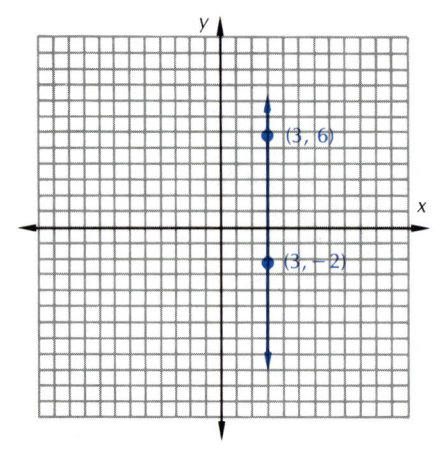

FIGURE 7.30

$$m = \frac{y_2 - y_1}{x_2 - x_1} = \frac{-2 - 6}{3 - 3} = \frac{-8}{0}$$

Since division by zero is not possible, the slope of a vertical line is undefined.

Slopes of Straight Lines
1. A positive slope means the line rises as it goes to the right.
2. A negative slope means the line falls as it goes to the right.
3. The slope of a horizontal line is 0.
4. The slope of a vertical line is undefined.

Slope can be used to tell whether two straight lines are parallel or perpendicular. Parallel lines never intersect. Thus, as illustrated in Figure 7.31, they must have the same "inclination" or slope.

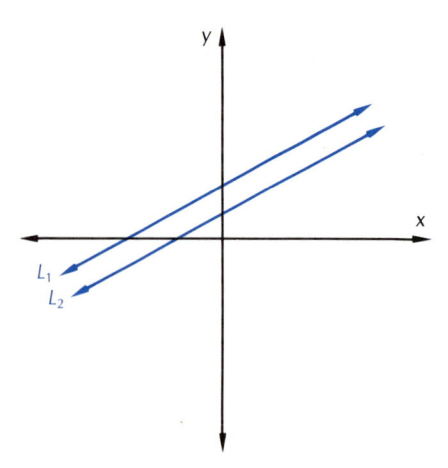

FIGURE 7.31

> **Slopes of Parallel Lines**
>
> Given two distinct lines L_1 and L_2 with respective slopes of m_1 and m_2, then:
> (1) If L_1 is parallel to L_2, then $m_1 = m_2$.
> (2) If $m_1 = m_2$, then L_1 is parallel to L_2.

Are lines L_1 and L_2 parallel, if L_1 passes through $(-1, 1)$ and $(2, 7)$ while L_2 passes through $(2, -1)$ and $(5, 5)$?

EXAMPLE 8

Solution: Determine the slope of each line:

$$m_1 = \frac{1 - 7}{-1 - 2} = \frac{-6}{-3} = 2$$

$$m_2 = \frac{-1 - 5}{2 - 5} = \frac{-6}{-3} = 2$$

Since the slopes are equal, the lines are parallel.

Perpendicular lines, as shown in Figure 7.32, meet at right angles (90°). It can be shown that perpendicular lines, neither of which is vertical, have slopes that are negative reciprocals. Negative reciprocals are two numbers whose product is -1, such as $\frac{1}{2}$ and -2.

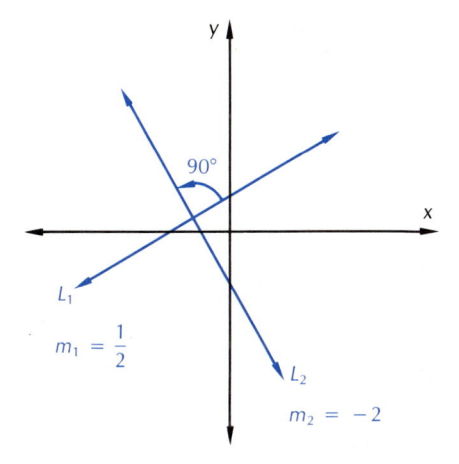

FIGURE 7.32

> **Slopes of Perpendicular Lines**
>
> Given two lines L_1 and L_2 with respective slopes of m_1 and m_2, then:
> (1) If L_1 is perpendicular to L_2, then $m_1 \cdot m_2 = -1$.
> (2) If $m_1 \cdot m_2 = -1$, then L_1 is perpendicular to L_2.

288 RELATIONS, FUNCTIONS, AND THEIR GRAPHS

EXAMPLE 9 Is L_1 perpendicular to L_2 if L_1 passes through $(-2, 4)$ and $(3, -11)$ while L_2 passes through $(-6, -1)$ and $(0, 1)$?

Solution: Determine the slope of each line:

$$m_1 = \frac{4 - (-11)}{-2 - 3} = \frac{15}{-5} = -3$$

$$m_2 = \frac{-1 - 1}{-6 - 0} = \frac{-2}{-6} = \frac{1}{3}$$

Find the product of $m_1 \cdot m_2$.

$$m_1 \cdot m_2 = (-3) \cdot \frac{1}{3} = -1$$

Since the product of slopes is -1, the lines are perpendicular to each other.

Exercise 7.4

Find the x and y-intercepts of the following equations.

1. $2x - y = 8$
2. $x - y = 2$
3. $3x = 4 + y$
4. $2x + 5y = 20$
5. $5x + 3y = 8$
6. $f(x) = 4$
7. $\frac{1}{3}x - \frac{1}{2}y = 2$
8. $y = \frac{3}{5}x - 2$

Find the slope of the line passing through the two points listed.

9. $(2, 5)$ $(6, 2)$
10. $(-3, 8)$ $(3, -1)$
11. $(0, 0)$ $(7, 1)$
12. $(4, -6)$ $(-4, 4)$
13. $(-4, 2)$ $(3, -2)$
14. $(3, 5)$ $(3, 8)$
15. $(-2, 3)$ $(3, 3)$
16. $\left(-2, \frac{1}{2}\right)$ $\left(-\frac{1}{2}, 3\right)$
17. $\left(\frac{1}{2}, 5\right)$ $\left(3, -\frac{1}{4}\right)$
18. $\left(1, -\frac{1}{2}\right)$ $\left(-\frac{1}{3}, -2\right)$

Graph the following and find the slope.

19. $f(x) = x + 2$
20. $2x + 3y = 6$
21. $y = -4x + 3$
22. $x - 2y = 8$
23. $f(x) = -1$
24. $f(x) = -x - 4$
25. $3x - 2y = 4$
26. $x - y + 4 = 0$
27. $x = 4$
28. $2y = x + 4$

Find the slope of each $Line_1$ and $Line_2$.

29. $Line_1$: $2x - 3y = 6$
    $Line_2$: $x + y = 2$
30. $Line_1$: $x = 5$
    $Line_2$: $y = 2$
31. $Line_1$: $3x + y - 2 = 0$
    $Line_2$: $x = y + 4$
32. $Line_1$: $5x + 2y = 14$
    $Line_2$: $y = \frac{1}{2}x + 1$

Determine if the following pairs of lines are parallel, perpendicular, or neither.

33. $\begin{cases} 4x - 2y = 8 \\ x - y = 6 \end{cases}$

34. $\begin{cases} 2x + y = 3 \\ x - y = 2 \end{cases}$

35. $\begin{cases} 3x - y = 6 \\ 3x = 4 + y \end{cases}$

36. $\begin{cases} x - \frac{1}{2}y = 3 \\ 2x - y = 5 \end{cases}$

37. $\begin{cases} x = 3 \\ y = 3 \end{cases}$

38. $\begin{cases} 2x + y = 5 \\ 2x - y = 5 \end{cases}$

39. $\begin{cases} \text{Line}_1 \text{ passes through } (4, 6) \text{ and } (3, -3) \\ \text{Line}_2 \text{ passes through } (7, -5) \text{ and } (8, 4) \end{cases}$

40. $\begin{cases} \text{Line}_1 \text{ passes through } (5, -6) \text{ and } (0, 2) \\ \text{Line}_2 \text{ passes through } (-2, 1) \text{ and } (6, 6) \end{cases}$

41. $\begin{cases} \text{Line}_1 \text{ passes through } (-11, 4) \text{ and } (-3, -2) \\ \text{Line}_2 \text{ passes through } (12, -7) \text{ and } (8, 9) \end{cases}$

42. $\begin{cases} \text{Line}_1 \text{ passes through } (1, 3) \text{ and } (1, 4) \\ \text{Line}_2 \text{ passes through } (1, 3) \text{ and } (2, 3) \end{cases}$

EQUATIONS OF THE STRAIGHT LINE

7.5

The concept of slope can be used to develop two important forms of the equation for a linear function. These are called the point-slope form and the slope-intercept form.

To derive the point-slope form, we are given a line L shown in Figure 7.33 with slope m passing through the fixed point P_1, whose coordinates are (x_1, y_1). Point P with coordinates (x, y) represents any other point on line L.

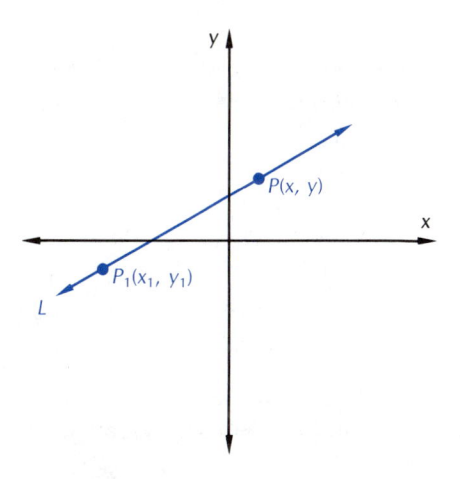

FIGURE 7.33

RELATIONS, FUNCTIONS, AND THEIR GRAPHS

Next, using P and P_1, we apply the slope formula to compute m:

$$m = \frac{y - y_1}{x - x_1}$$

Multiplying both sides of this equation by the LCD, $x - x_1$, produces the form:

$$y - y_1 = m(x - x_1)$$

Since this form gives the equation of a straight line with a given slope passing through a fixed point, it is called the point-slope form.

Point-Slope Form of a Straight Line

The equation of a line with slope m that passes through the fixed point (x_1, y_1) is:

$$y - y_1 = m(x - x_1)$$

The next two examples illustrate the use of the point-slope form to find equations of straight lines. Unless instructed otherwise, we will write answers in the form $Ax + By + C = 0$.

EXAMPLE 1 Find the equation of the line whose slope is -3 that passes through the point $(2, -4)$.

Solution: Use the point-slope form, where $m = -3$, $x_1 = 2$, and $y_1 = -4$.

$$y - y_1 = m(x - x_1)$$
$$y - (-4) = -3(x - 2)$$
$$y + 4 = -3x + 6$$
$$3x + y - 2 = 0$$

As illustrated by the next example, the point-slope form can be used in conjunction with the slope formula to find the equation of a line passing through two fixed points.

EXAMPLE 2 Find the equation of a line passing through $(-2, 3)$ and $(6, -1)$.

Solution: Use the slope formula to determine the slope. Then, apply the point-slope form.

$$m = \frac{3 - (-1)}{-2 - 6} = \frac{4}{-8} = -\frac{1}{2}$$

Either point can be used as (x_1, y_1) in the point-slope form. Letting $(x_1, y_1) = (-2, 3)$, we have:

EQUATIONS OF THE STRAIGHT LINE

$$y - y_1 = m(x - x_1)$$
$$y - 3 = -\frac{1}{2}[x - (-2)]$$
$$2y - 6 = -1[x + 2]$$
$$2y - 6 = -x - 2$$
$$x + 2y - 4 = 0$$

On the other hand, letting $(x_1, y_1) = (6, -1)$ produces the same equation.

$$y - y_1 = m(x - x_1)$$
$$y - (-1) = -\frac{1}{2}(x - 6)$$
$$2y + 2 = -x + 6$$
$$x + 2y - 4 = 0$$

To derive the second form of the linear function called the slope-intercept form, we are given a line L in Figure 7.34, with slope m and y-intercept b.

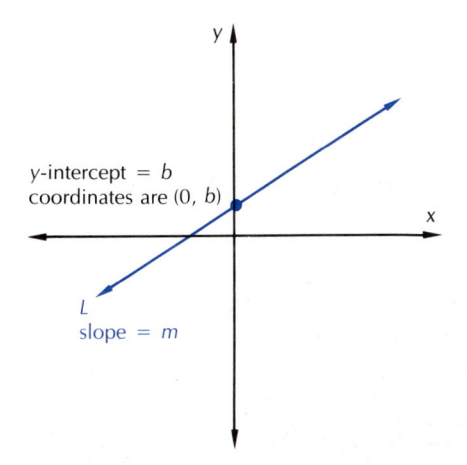

FIGURE 7.34

We use the point-slope form with $(x_1, y_1) = (0, b)$ to obtain the following equation solved for y.

$$y - y_1 = m(x - x_1)$$
$$y - b = m(x - 0)$$
$$y - b = mx$$
$$y = mx + b$$

Since this form is an equation of a straight line with a given slope and y-intercept, it is called the slope-intercept form.

RELATIONS, FUNCTIONS, AND THEIR GRAPHS

> **Slope-Intercept Form of a Straight Line**
> The equation of a line with slope m and y-intercept b is:
> $$y = mx + b$$

The slope-intercept form enables us to find the slope and y-intercept of a line by simply solving its equation for y. The coefficient of x is the slope while the constant term is the y-intercept.

$$y = \underset{\text{slope}}{m}x + \underset{y\text{-intercept}}{b}$$

EXAMPLE 3 Find the slope and y-intercept of the straight line described by $3x + 4y - 12 = 0$.

Solution: Write the equation in the slope-intercept form by solving for y.
$$3x + 4y - 12 = 0$$
$$4y = -3x + 12$$
$$y = -\frac{3}{4}x + 3$$

The slope is $-\frac{3}{4}$ and the y-intercept is 3.

When graphing a straight line, it is often helpful to write the linear equation in the slope-intercept form. The slope and y-intercept are quickly determined and indicated on a graph.

EXAMPLE 4 Graph $3x - 2y - 6 = 0$ by determining its slope and y-intercept.

Solution: Write the equation in the slope-intercept form by solving it for y.
$$3x - 2y - 6 = 0$$
$$-2y = -3x + 6$$
$$y = \frac{3}{2}x - 3$$

The slope is $\frac{3}{2}$ and the y-intercept is -3. These values determine the graph in Figure 7.35. First, we plot the y-intercept, -3. Then, a second point is found by using the slope. Recall the slope is the ratio of the rise to the run. Thus, a slope of $\frac{3}{2}$ means that the line rises 3 units for every 2 units it goes to the right. Thus, we determine the point (2, 0).

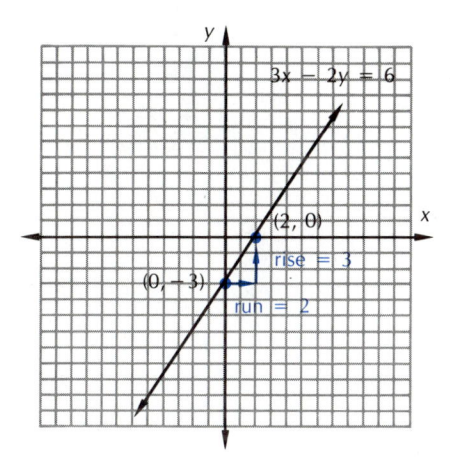

FIGURE 7.35

As shown by the following examples, the slope-intercept and point-slope forms of a straight line can be used together to find the equation of a line which passes through a fixed point and is either parallel or perpendicular to another given line.

EXAMPLE 5

Find the equation of the line which passes through the point (4, 2) and is parallel to the line $3x + y - 4 = 0$.

Solution: The slope of the given line is found by solving for y to place it in the slope-intercept form.

$$3x + y - 4 = 0$$
$$y = -3x + 4$$

The slope is -3. Thus, the line parallel to $3x + y - 4 = 0$ also has a slope of -3. Using the point-slope form where $m = -3$ and $(x_1, y_1) = (4, 2)$, we find the equation of the desired line.

$$y - y_1 = m(x - x_1)$$
$$y - 2 = -3(x - 4)$$
$$y - 2 = -3x + 12$$
$$3x + y - 14 = 0$$

294 RELATIONS, FUNCTIONS, AND THEIR GRAPHS

Figure 7.36 illustrates this example.

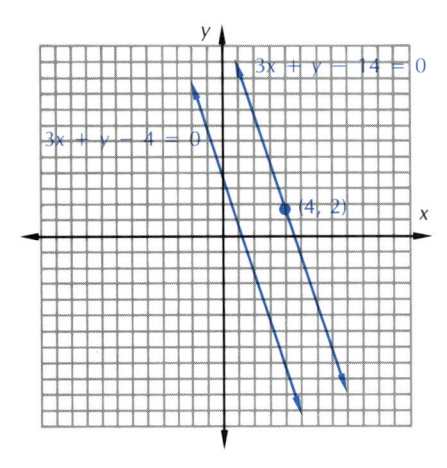

FIGURE 7.36

EXAMPLE 6 Find the equation of a line passing through the point $(-2, 3)$ that is perpendicular to the line $-2x + y + 3 = 0$.

Solution: The slope of the given line is found by solving for y.

$$-2x + y + 3 = 0$$
$$y = 2x - 3$$

Since the slope of the given line is 2, the slope of the perpendicular line is $-\frac{1}{2}$ (the negative reciprocal of 2). Using the point-slope form where $m = -\frac{1}{2}$ and $(x_1, y_1) = (-2, 3)$, we find the equation of the perpendicular line.

$$y - y_1 = m(x - x_1)$$
$$y - 3 = -\frac{1}{2}(x + 2)$$
$$2y - 6 = -x - 2$$
$$x + 2y - 4 = 0$$

This example is illustrated in Figure 7.37.

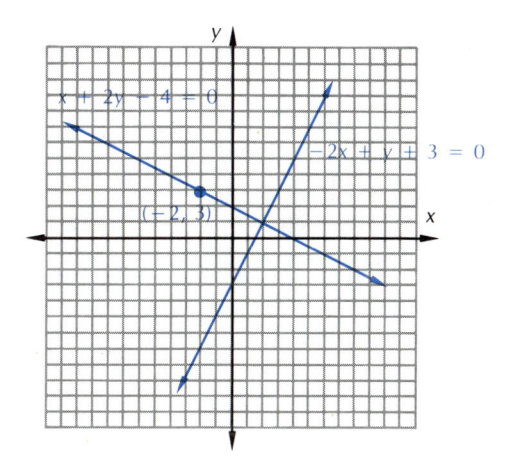

FIGURE 7.37

(Graph showing lines $x + 2y - 4 = 0$ and $-2x + y + 3 = 0$ intersecting at $(-2, 3)$)

Exercise 7.5

Write the equation of the line with the given slope and passing through the given point.

1. $m = 3$, $(-2, 3)$
2. $m = 0$, $(-4, 4)$
3. $m = -1$, $(-1, -2)$
4. $m = \dfrac{1}{2}$, $(-2, 3)$
5. $m = \dfrac{3}{4}$, $(2, 1)$
6. undefined slope, $(-3, 2)$
7. $m = -\dfrac{4}{5}$, $(5, -1)$
8. $m = \dfrac{3}{5}$, $(-5, 2)$

Write the equation of the line passing through the two given points.

9. $(3, -2)$ and $(1, 6)$
10. $(4, -5)$ $(-4, 4)$
11. $(-2, 4)$ and $(3, 4)$
12. $(3, 0)$ $(1, -1)$
13. $\left(\dfrac{1}{2}, 4\right)$ and $\left(\dfrac{3}{4}, 1\right)$
14. $(3, 5)$ and $(1, 5)$

Use the slope-intercept form to find the slope and y-intercept.

15. $y = -2x - 1$
16. $y = -x$
17. $x + y = 3$
18. $2x - 3y = 0$
19. $y = 3$
20. $2y = x + 2$
21. $5x + \dfrac{1}{2}y = 3$
22. $-y = 2x - 1$

Write the equation of the line meeting the conditions stated.
23. Parallel to $3x - y = 5$ and passing through $(-2, 4)$
24. Parallel to $2x + y = 4$ and passing through $(2, 3)$
25. Parallel to $2x - y = 8$ and passing through $(-6, 2)$
26. Parallel to $2x + 5y = 20$ and passing through $(2, 5)$
27. Perpendicular to $2x - y = 2$ and passing through $(6, 5)$
28. Perpendicular to $y = -4x + 6$ and passing through $(-4, 7)$
29. Perpendicular to $y = 3x + 2$ and passing through $(4, 5)$
30. Perpendicular to $3x + 2y = 8$ and passing through $(5, 0)$
31. x-intercept is 5 and y-intercept is 3
32. x-intercept is 2 and y-intercept is -3
33. x-intercept is $\frac{1}{2}$ and y-intercept is $\frac{1}{3}$
34. x-intercept is 4 and parallel to $y = x$

Graph by determining the slope and y-intercept.
35. $2x + 3y = 6$
36. $3y = x - 2$
37. $5x - 2y = 10$
38. $3x - y = 4$
39. $x - 4y - 6 = 0$
40. $2x + 3y = 0$

LINEAR INEQUALITIES

7.6

We have studied linear relations of the form $Ax + By + C = 0$. If the equal sign is replaced by any of the inequality symbols, $<$, $>$, $\leq$, or $\geq$, we have a linear inequality in two variables. For example, $2x + y - 6 < 0$ is a linear inequality in the variables x and y.

To graph a linear inequality, we need to recall that the graph of a linear equality is a straight line. As shown in Figure 7.38, a non-vertical line divides the plane into an upper region and a lower region called half-planes. The upper half-plane contains all the points above the line, while the lower half-plane contains all points below the line. The line itself is called the boundary

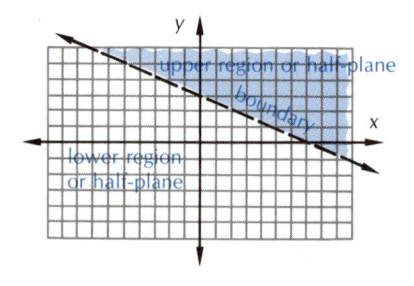

FIGURE 7.38

and is not contained within either half-plane. Thus, the boundary is illustrated by a dashed line.

The graph of a linear inequality will consist of either the upper or the lower half-plane, but not both. The boundary is included in the graph only if the inequality contains an equal sign: $\leq$ or $\geq$.

Graphing a Linear Inequality

To construct the graph of a linear inequality:
(1) Plot the boundary by replacing the inequality with an equal sign. If the boundary is to be included in the graph, draw it solid; otherwise it is dashed.
(2) Select any point not on the boundary as a test point. This point will be used to determine which half-plane is included in the graph. (If the boundary does not pass through the origin (0, 0), then the most convenient test point is the origin.)
(3) Substitute the coordinates of the test point into the inequality. The resulting statement will be either true or false. If it is true, the graph includes the half-plane containing the test point. If false, the graph includes the opposing half-plane.

Graph $2x + y \leq 8$

EXAMPLE 1

Solution:

Step 1. The boundary line given by the equation $2x + y = 8$ is graphed by plotting its x and y-intercepts. When $x = 0$, $y = 8$, and when $y = 0$, $x = 4$. The inequality contains an equal sign, so the boundary is drawn solid to show that it is included in the graph (see Figure 7.39).

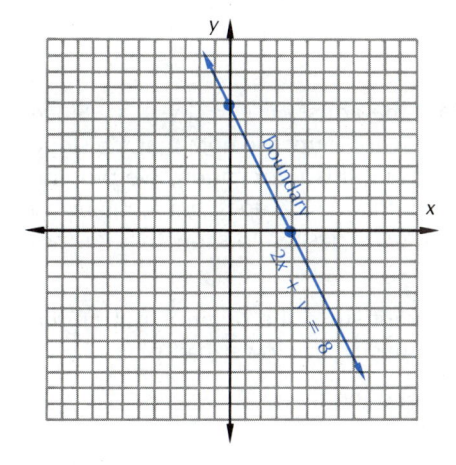

FIGURE 7.39

Step 2. Since the origin is not on the boundary, we choose (0, 0) as the test point. Note that it is contained in the lower half-plane.

Step 3. We substitute the coordinates into the inequality.

$$2x + y \leq 8$$
$$2 \cdot 0 + 0 \leq 8$$
$$0 \leq 8 \quad \text{[true]}$$

Since the statement is true, the graph includes the lower half-plane containing the test point. The graph of the inequality is the shaded region together with the boundary shown in Figure 7.40.

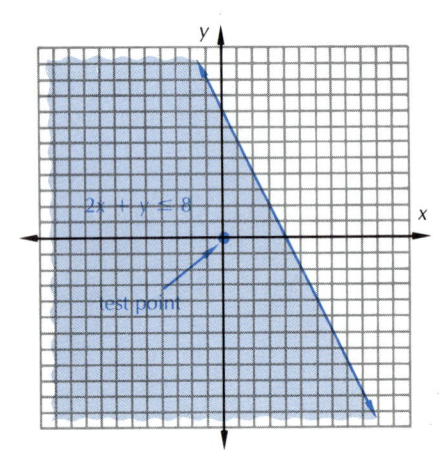

FIGURE 7.40

EXAMPLE 2 Graph $y > 2x + 6$

Solution:

Step 1. The boundary line is given by the equation $y = 2x + 6$. Since the equation is in the slope-intercept form, we note that the slope is 2 and the y-intercept is 6. Using this information, we plot the boundary in Figure 7.41. The inequality does not contain an equal sign, and thus the boundary is dashed to show that it is not included in the graph.

Step 2. We choose (0, 0) as the test point since the origin is not on the boundary. Note that it is contained in the lower half-plane.

Step 3. We substitute the coordinates into the inequality.

$$y > 2x + 6$$
$$0 > 2 \cdot 0 + 6$$
$$0 > 6 \quad \text{[false]}$$

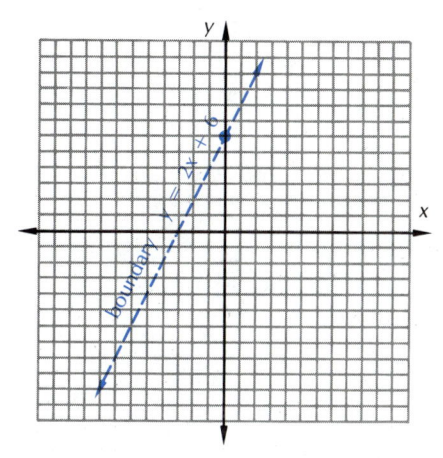

FIGURE 7.41

Since the statement is false the graph does not contain the lower half-plane containing the test point. Instead, it includes the opposing half-plane. Thus, as shown in Figure 7.42, the graph contains the upper half-plane, but not the boundary.

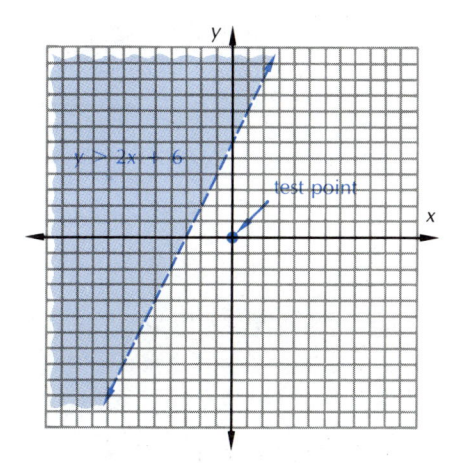

FIGURE 7.42

With practice you can shorten this graphing technique by combining steps as shown in the next example.

Graph $3y \leq x$ **EXAMPLE 3**

Solution: We first graph the boundary $3y = x$ by plotting two points, (0, 0) and (6, 2). The boundary is drawn solid to show that it is part of the graph. Since (0, 0) is on the boundary, we select another test point, say (1, 8).

$3y \leq x$
$3 \cdot 8 \leq 1$
$24 \leq 1$ [false]

Since this statement is false, the graph lies in the opposing half-plane. It also includes the boundary (see Figure 7.43).

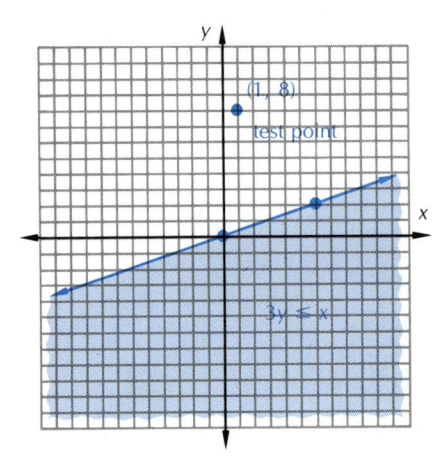

FIGURE 7.43

EXAMPLE 4   Graph $x > -3$

Solution: The boundary is the vertical line, $x = -3$. Using (0, 0) as the test point, we arrive at a true statement.

$x > -3$
$0 > -3$ [true]

Thus, the graph lies to the right of $x = -3$ and does not include the boundary (see Figure 7.44).

Sometimes we are given two or more linear inequalities and are asked to find their intersection. This is called solving a system of linear inequalities, and means to find that region of the plane where the coordinates of every point satisfy each inequality. This is accomplished by placing the graph of each inequality on the same coordinate system and locating the region common to all graphs.

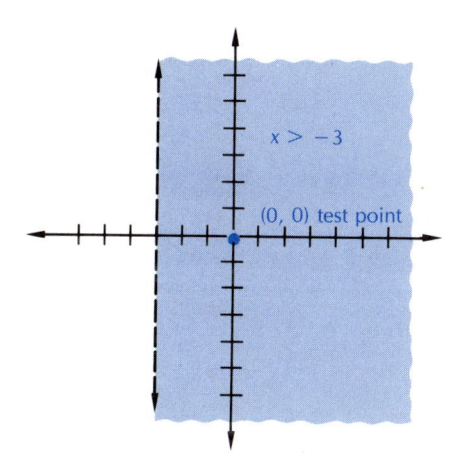

FIGURE 7.44

Solve this system:

$x + y \geq 5$

$x - y \leq 2$

EXAMPLE 5

Solution: First we graph each inequality separately. The graph of $x + y \geq 5$ is shown in Figure 7.45, while the graph of $x - y \leq 2$ is shown in Figure 7.46.

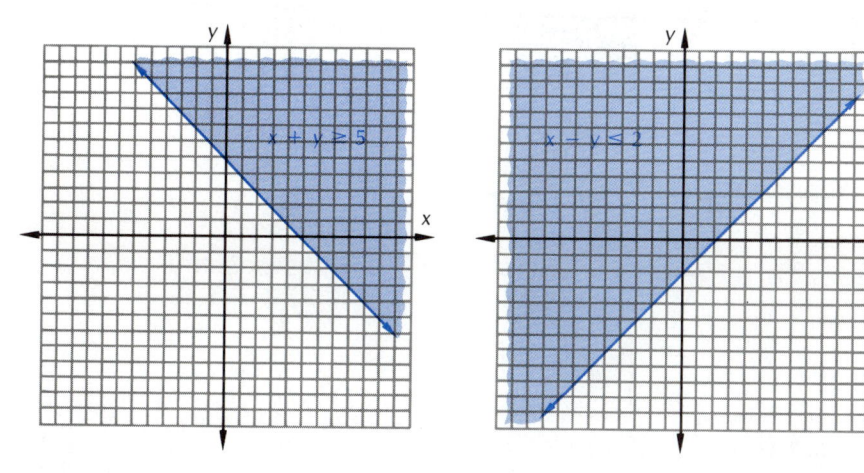

FIGURE 7.45

FIGURE 7.46

Then, to solve the system we find the intersection by placing both graphs on the same coordinate system and locating the common region, shown in Figure 7.47.

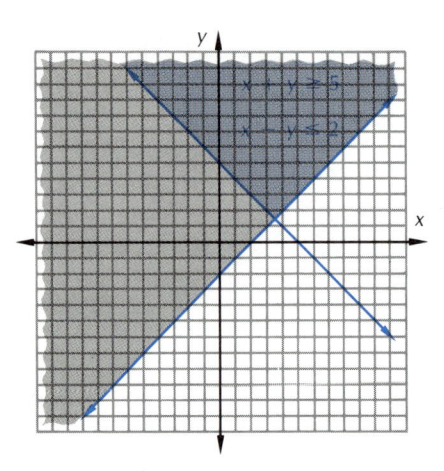

FIGURE 7.47

EXAMPLE 6 Solve this system:

$x > 2$
$y > 4$

Solution: The graph of $x > 2$ is the half-plane to the right of the vertical line $x = 2$. The graph of $y > 4$ is the half-plane above the horizontal line $y = 4$. The solution of this system is the region common to both graphs, seen in Figure 7.48.

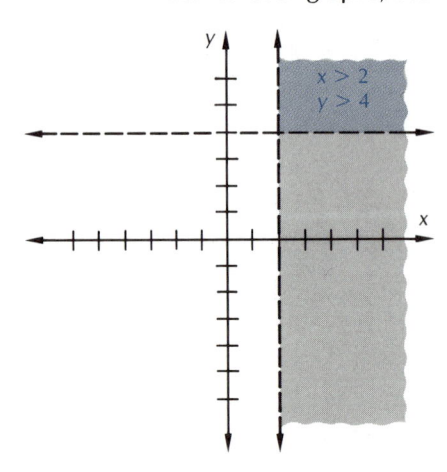

FIGURE 7.48

Exercise 7.6 Determine if the following points belong to the region defined by the given inequality.

1. $(2, -1); x - 2y < 3$
2. $(0, -3); 2x + y > 1$
3. $(4, 3); 2x - 3y \leq -1$
4. $(-2, -1); y < -2x + 3$

Draw the graphs of each inequality.

5. $y > x + 2$
6. $y \leq x + 2$
7. $y < 2x - 1$
8. $x + y \leq 3$
9. $x + 2y \geq -4$
10. $y < 2x - 4$
11. $x + y \leq 5$
12. $x > 2$
13. $x > -2$
14. $y \leq 1$
15. $y > -3$
16. $2x + y > 4$
17. $x + 2y \leq 4$
18. $y \geq -2x - 4$
19. $2x - 3y < -6$
20. $4x + 8y > 0$
21. $y \geq x$
22. $-6 > 2x + y$

Solve the following systems by graphing.

23. $x + y > 5$
    $x - y < 2$
24. $x > 3$
    $y > 2$
25. $y > x + 3$
    $y \geq -2x + 1$
26. $y \geq x + 2$
    $x + y \geq 0$
27. $x \leq 3$
    $y \leq x$
28. $y > x - 1$
    $x + 2 < y$
29. $x + y < 2$
    $x \geq 0$
30. $x \leq 1$
    $x + y \leq 4$
31. $3x + 2y < 12$
    $x - 3y < 6$
32. $3x - 4y > 12$
    $2x + 4y < 0$

VARIATION

7.7

In this section we will examine two types of functions that are widely applied in science and industry. They are called direct variation and inverse variation.

> **Direct Variation**
>
> Any function defined by the equation $y = kx$ (where k is a positive constant) is called a direct variation.

In a direct variation, the variable y is said to vary directly as the variable x, or more simply, *y varies as x*. y is also said to be proportional to x. The positive number k is called the constant of variation or the constant of proportionality.

The circumference (C) of a circle is given by the equation,

$$C = \pi d$$

where d is the diameter. Thus, the circumference of a circle varies directly as the diameter. The constant of variation is π.

In a direct variation, increasing values of one variable result in increasing values of the other variable. Referring to Example 1, it is obvious that increasing the diameter of a circle produces a proportionate increase in the circumference. Direct variations, as shown in the following, may also involve powers or roots.

Direct Variation of Powers or Roots

Assuming that k is a positive constant:
1. If $y = kx^n$, we say that y varies directly as the nth power of x.
2. If $y = k\sqrt[n]{x}$, we say that y varies directly as the nth root of x.

EXAMPLE 2 The area (A) of a circle is given by the equation,
$$A = \pi r^2$$
where r is the radius of the circle. Thus, the area of a circle varies directly as the square of the radius. The constant of variation is π.

EXAMPLE 3 Write the variation statement showing that the period (T) of the pendulum varies directly as the square root of its length (l).

Solution: Letting k represent a positive constant, we write:
$$T = k\sqrt{l}$$

A second type of function describes a relationship where one variable increases as the other decreases. This is called an inverse variation and is defined as follows:

Inverse Variation

Any function defined by the equation $xy = k$ or $y = \dfrac{k}{x}$ (where k is a positive constant) is called an inverse variation or an inverse proportion.

For an inverse variation the variable y is said to vary inversely as the variable x. As with direct variations, the positive number k is called the constant of variation or the constant of proportionality.

EXAMPLE 4 Write a variation statement showing that the volume of gas (V) varies inversely as the pressure (P).

Solution: Letting k represent a positive constant, we write:
$$V = \frac{k}{P}$$

As with direct variations, inverse variations may involve powers or roots.

VARIATION 305

> **Inverse Variations of Powers or Roots**
>
> Assuming that k is a positive constant:
>
> 1. If $y = \dfrac{k}{x^n}$, we say that y varies inversely as the nth power of x.
> 2. If $y = \dfrac{k}{\sqrt[n]{x}}$, we say that y varies inversely as the nth root of x.

It is also possible to have combinations of inverse and direct variation, as illustrated by the following example.

EXAMPLE 5 Write a variation statement showing that the resistance (R) of an electrical wire varies directly as its length (l) and inversely as the square of its diameter (d).

Solution: Letting k represent a positive constant, we write:

$$R = \frac{k \cdot l}{d^2}$$

Often we find that one variable varies as the product of two or more variables. This is called a joint variation.

EXAMPLE 6 The area (A) of a triangle is given by the equation

$$A = \frac{1}{2}bh$$

where b is the length of its base and h is the length of its altitude. Thus, the area of a triangle varies jointly as the length of its base and altitude.

Given a general variation statement, the constant of variation (k) can be determined if we know one set of values for the respective variables. The value of k can then be substituted into the variation statement to produce a specific equation that fully describes the problem.

EXAMPLE 7 Suppose y varies directly as the square of x, and $y = 12$ when $x = 6$. Find k and write the specific equation.

Solution: Since y varies directly as the square of x, we write:

$$y = kx^2$$

We know that $y = 12$ when $x = 6$. Thus, we substitute these values into the variation and solve for k.

$y = kx^2$

$12 = k \cdot 36$

$\dfrac{12}{36} = k$ or $k = \dfrac{1}{3}$

Replacing $\frac{1}{3}$ for k in the variation $y = kx^2$ produces the specific equation, $y = \frac{1}{3}x^2$.

As illustrated by the following examples, many applied problems can be solved by finding the constant of variation (k) to form the specific equation.

EXAMPLE 8   The volume (V) of a gas varies inversely as its pressure (P). A gas in a cylinder occupies 60 cubic inches when its pressure is 10 pounds per square inch. What will be the volume if the pressure is increased to 15 pounds per square inch?

Solution: Form the specific equation by finding k. Since V varies inversely as P, we write the general variation statement.

$$V = \frac{k}{P}$$

Substitute 60 cubic inches for V and 10 pounds per square inch for P.

$$V = \frac{k}{P}$$

$$60 = \frac{k}{10}$$

$$k = 600$$

Write the specific equation by replacing 600 for k in the general variation.

$$V = \frac{k}{P}$$

$$V = \frac{600}{P}$$

Use the specific equation to find the volume when the pressure is increased to 15 pounds per square inch.

$$V = \frac{600}{P}$$

$$= \frac{600}{15}$$

$$= 40$$

Thus, the volume decreases to 40 cubic inches when the pressure is increased to 15 pounds per square inch.

The preceding example suggests the following procedure for solving applied variation problems.

> **To Solve Variation Problems:**
> (1) Write the general variation statement.
> (2) Substitute the known set of values.
> (3) Solve for the constant of variation (k).
> (4) Form the specific equation by substituting the value of k into the general variation statement.
> (5) Solve the problem by substituting the remaining values into the specific equation.

EXAMPLE 9

The strength (S) of a rectangular beam varies jointly as its width (w) and the square of its depth (d). If the strength of a beam 2 inches wide and 8 inches deep is 896 pounds per square inch, what is the strength of a beam 4 inches wide and 10 inches deep?

Solution: Write the general variation statement.

$S = kwd^2$

Substitute the known set of values. $S = 896$ when $w = 2$ and $d = 8$.

$896 = k \cdot 2 \cdot 64$

Solve for k.

$896 = k \cdot 128$

$\dfrac{896}{128} = k$ or $k = 7$

Form the specific equation.

$S = kwd^2$

$S = 7wd^2$

Solve the problem by using the specific equation where $w = 4$ and $d = 10$.

$S = 7wd^2$
$ = 7 \cdot 4 \cdot 10^2$
$ = 7 \cdot 4 \cdot 100$
$ = 2800$

Thus, the strength of a beam 4 inches wide and 10 inches deep is 2800 pounds per square inch.

Exercise 7.7

Write the general variation statement.

1. The period P of a pendulum varies directly as the square root of the length L of the pendulum.

2. The tension T of a string is proportional to the distance s that is stretched.
3. The speed s of a pulley is inversely proportional to its diameter d.
4. The heat H developed in a wire carrying current varies jointly as the resistance R and the square of the current I.

Solve by determining the constant of variation and substituting known values.

5. y varies directly as x. y is 20 when x is 5. Find y when x is 7.
6. y varies directly as x. $y = 36$ when $x = 4$. Find x when $y = 40$.
7. y is inversely proportional to x. If y is 8 when x is 5, find x when y is 48.
8. If f varies directly as t^2 and $f = 8$ when $t = 3$, find f when $t = 6$.
9. If x varies as y and $x = 7$ when $y = 4$, find x when $y = 10$.
10. z varies inversely as w. If $z = 6$ when $w = \frac{1}{2}$, find z when $w = 6$.
11. If y varies inversely as the square of x and $y = 8$ when $x = 6$, find y when $x = 20$.
12. z varies jointly as x and the square of y. If $z = 64$ when $x = 2$ and $y = 6$, find x when $z = 20$ and $y = 2$.
13. If C varies directly as D and $C = 12\pi$ when $D = 12$, find C when $D = 3$.
14. If y varies jointly as r, s, and t, and $y = 30$ when $r = 3$, $s = 4$, and $t = 5$, what is r when $y = 40$, $s = 5$ and $t = 2$?
15. The distance in feet an object will fall from rest varies directly as the square of the time in seconds. A body will fall 16 feet the first second. How far will it fall in 4 seconds?
16. The weight of a certain material varies directly as the surface area of the material. If 10 square feet weighs 1.5 pounds, how much will 14 square feet weigh?
17. The distance d a person can see to the horizon from a point h feet above the surface of the earth varies as the square root of the height h. If at a height of 500 feet, the horizon is 28 miles, how far is the horizon from a height of 1000 feet?
18. The length a spring stretches is directly proportional to the force applied. If a force of 5 pounds stretches a spring 3 inches, how much force is needed to stretch the same spring 8 inches?
19. Interest earned varies jointly as the principal and time. If $2000 is invested for 3 years and earns $480, how much interest would be earned on $5000 for 4 years?
20. The surface area of a sphere varies directly as the square of the radius. If the surface area is 36π square inches when the radius is 3 inches, what is the surface area when the radius is 10 inches?
21. At constant temperature, the resistance of a wire varies directly as its length and inversely as the square of its diameter. If a piece of wire 0.1 inch in diameter and 50 feet long has a resistance of 0.1 ohm, what is the resistance of a piece of wire of the same material that is 2,000 feet long and 0.2 inch in diameter?

22. The current I in amperes in an electric circuit varies inversely as the resistance R in ohms when the electromotive force is constant. In a certain circuit I is 15 amperes where R is 2 ohms. Find I when R is 0.2 ohms.

23. The force p of the wind on a plane surface varies directly as the square of the speed s of the wind. If the force is 2 pounds per square foot when the speed of the wind is 20 miles per hour, find the force when the speed of the wind is 50 miles per hour.

24. When a ball is thrown vertically upward, the height reached varies directly as the square of the speed with which the ball is thrown. A ball thrown with a speed of 40 feet per second reaches a height of 25 feet. Find the height reached when the ball is thrown with a speed of 60 feet per second.

25. The weight of a piece of copper wire varies jointly as the length and as the square of the diameter. If 50 feet of copper wire with a diameter of $\frac{1}{2}$ inch weighs $1\frac{1}{4}$ pounds, find the weight of 175 feet of copper wire having a diameter 80% of that of the first wire.

26. When a small steel ball drops freely from rest, the distance traveled varies as the square of the time of falling. If the ball falls 64 feet in 2 seconds, find how far it falls in 3 seconds.

27. The friction force required to prevent a car from skidding in making a certain turn varies directly as the square of the speed of the car. If the friction force required at 20 miles per hour is 240 pounds, find the friction force at 30 miles per hour.

28. The volume V of a gas varies directly as the absolute temperature T and inversely as the pressure P. If a certain amount of gas occupies 100 cubic feet at a pressure of 20 pounds per square inch and at a temperature of 200°, find its volume when the pressure is 25 pounds per square inch and the temperature is 400°.

29. In silver plating, without varying the number of amperes of current, the number of grams n of silver deposited on a given surface is directly proportional to the number of hours h. If 1.4 grams of silver is deposited in one-half of an hour, how many grams would be deposited in three hours?

30. The force required to compress a spring is proportional to the change in the length of the spring. If a force of 16 kilograms is required to compress a certain spring 2 centimeters, how much force is required to compress the spring from 20 centimeters to 16 centimeters?

31. The time required for an elevator to lift a weight varies jointly as the weight and the distance through which the weight is lifted and inversely as the power of the motor. If 30 seconds is required for a 4-horsepower motor to lift 600 pounds through 40 feet, what power is necessary to lift 1000 pounds 100 feet in 50 seconds?

32. The intensity of illumination from a given light source varies inversely as the square of the distance from the source. If a book is 24 inches from the light, how far from the light must it be placed if it is to receive one-

fourth as much illumination? (Hint: assume the illumination is some fixed number.)

33. The volume of a cone varies jointly as the altitude and square of the radius. When the radius is 4 and the altitude is 6, the volume is 32π. What must the altitude be if the volume is 12π when the radius is 2?
34. If y varies directly as the cube of x, what is the effect on y when x is doubled?
35. If y varies inversely as x, what is the effect on y when x is doubled?
36. If y varies jointly as x and z, what is the effect on y when x is tripled and z is doubled?
37. If S varies directly as the square of x and inversely as y, what change in S results when x is doubled and y is tripled?
38. If y varies directly as x^2 and inversely as z, what is the effect on y if both x and z are halved?
39. y varies directly as x and inversely as the square root of t. If $y = 2.48$ when $x = .64$ and $t = 14.23$, find y when $x = .96$ and $t = 8.14$. Give the answer to the nearest hundredth.
40. x varies jointly as y and z^2 and inversely as q. If $x = 43.913$ when $y = 12.424$, $z = 5.401$ and $q = 41.216$, find x when $y = 6.823$, $z = 3.517$, and $q = 6.236$. Give the answer to the nearest thousandth.

Summary

KEY WORDS AND PHRASES

Ordered Pair—an arrangement of two numbers within parentheses to show one number is first and the other is second.

Rectangular Coordinate System—a partitioning of the plane into a coordinate grid by constructing two perpendicular real number lines.

Coordinate Axes—the two perpendicular number lines forming the rectangular coordinate system (the horizontal and vertical lines are called the x-axis and y-axis, respectively).

Origin—the point of intersection of the coordinate axes.

Abscissa—the first coordinate of an ordered pair.

Ordinate—the second coordinate of an ordered pair.

Quadrant—one of four regions that the plane is separated into by the coordinate axes.

Relation—a non-empty set of ordered pairs.

Domain of a Relation—the set of first components of a relation.

Range of a Relation—the set of second components of a relation.

Independent Variable—a variable representing the numbers in the domain.

Dependent Variable—a variable representing the numbers in the range.

Graph of a Relation—set of points whose coordinates satisfy the condition defining the given relation.

Function—a relation where for each value of the first component of the ordered pairs, there is exactly one corresponding value of the second component.

Functional Notation—a symbolism (f(x)) for the letter y in a function.

Zeros of a Function—the values of x that make the functional value zero.

Inverse of Relation G—the relation formed by interchanging the x and y values of each ordered pair of G.

Inverse Function—if a function f has no two ordered pairs with the same y value, the inverse function, f^{-1}, is such that if (a, b) belongs to f, then (b, a) belongs to f^{-1}.

Linear Function—any function whose equation can be written in the form $f(x) = ax + b$, where $a, b \in R$.

x-Intercept—the absicssa (first coordinate) of the point where the graph intersects the x-axis.

y-Intercept—the ordinate (second coordinate) of the point where the graph intersects the y-axis.

Constant Function—the function defined by $f(x) = k$, where k is a real constant.

Slope of a Line—given two distinct points (x_1, y_1) and (x_2, y_2) on a straight line,

$$\text{the slope } m = \frac{\text{change in } y}{\text{change in } x} \text{ or } \frac{\text{rise}}{\text{run}}$$

Linear Inequality—a linear relation of the form $Ax + By + C = 0$, with the equal sign replaced by any of the symbols, $<, >, \leq$, or $\geq$.

FORMULAS

Slope: $m = \dfrac{y_2 - y_1}{x_2 - x_1}$

Slope-Intercept Form: $y = mx + b$

Point-Slope Form: [given m and (x_1, y_1)]

$$y - y_1 = m(x - x_1)$$

Vertical Line Equation: $x = k$

Horizontal Line Equation: $y = k$

Slope of Vertical Line: undefined

Slope of Horizontal Line: $m = 0$

Parallel Lines (non-vertical): $m_1 = m_2$

Perpendicular Lines (neither vertical): $m_1 \cdot m_2 = -1$

Direct Variation: $y = kx$

Inverse Variation: $y = \dfrac{k}{x}$

Joint Variation: $y = kxz$

Self-Checking Exercise

True or false.
1. The abscissa of the ordered pair $(-2, 3)$ is 3.
2. $(-3, -2)$ names a point in quadrant IV.
3. The ordered pair $(0, 4)$ is not located in a quadrant.
4. A non-empty set of ordered pairs is a relation.

Graph these relations and give the domain and range.
5. $\{(-3, 2) (0, 4) (1, 2) (3, -1)\}$
6. $\{(x, y) \mid y = -3x + 2\}$
7. $\{(x, y) \mid y = x^2 + 2\}$
8. $\{(x, y) \mid y = \sqrt{x - 1}\}$
9. $\{(x, y) \mid x = |y + 2|\}$
10. $\{(x, y) \mid x = y^2 - y + 4\}$

Determine which of the following relations represents a function. For each function give the domain and range.
11. $x - y = 8$
12. $y = |x|$
13. $x^2 + y^2 = 9$
14. $y = x^2 + 4$
15. $y = \sqrt{x - 4}$
16. $x = y^2$

Find an expression for $f(x)$ and evaluate $f(4)$ and $f(-3)$.
17. $x + 3y = 8$
18. $x^2 - y = 4$
19. $x = y^3 + 1$
20. $\sqrt{x + 2} - y = 0$

Find any real zeros of the following functions.
21. $y = 2x + 7$
22. $3x + 5y = 8$
23. $f(x) = x^2 - x + 1$
24. $f(x) = 3x^2 - 5x - 2$

Graph the following functions and give the domain and range.
25. $f(x) = |2x + 1|$
26. $g(x) = \sqrt{x + 3}$
27. $h(x) = x^2 + 4x + 4$
28. $f(x) = 3 - x^2$

Graph each equation with a solid curve and on the same coordinate system graph the inverse with a dashed curve or line.
29. $y = x + 4$
30. $y = \sqrt{x - 3}$
31. $y = x^2 + 1$
32. $y = |x| + 2$

Find the inverse (f^{-1}) and its domain and range.
33. $f(x) = \frac{1}{3}x - 1$
34. $y = \sqrt{x + 1}$
35. $y = \frac{x - 2}{3}, -1 \leq x \leq 5$
36. $3y = 2x + 1$

If $f(x) = 2x + 4$ and $g(x) = \sqrt{x - 1}$, evaluate the following after finding f^{-1} and g^{-1}.
37. $f[g^{-1}(4)]$
38. $g[f^{-1}(2)]$
39. $f[f^{-1}(1)]$
40. $f^{-1}[g^{-1}(1)]$

Find the x and y-intercepts and the slope of each of the following lines.

41. $2x = 4 + y$

42. $y = \frac{2}{5}x - 3$

43. $\frac{1}{2}x - \frac{1}{3}y = 1$

44. $x - y = 0$

Find the slope of the line passing through the two given points.

45. $(4, 4)$ and $(-2, 2)$

46. $(0, 6)$ and $(3, 5)$

47. $(-2, 4)$ and $(-6, 4)$

48. $(3, 5)$ and $(3, -2)$

Write the equation of the line meeting the conditions stated. Put your answer in the form $Ax + By + C = 0$.

49. passing through $(2, -1)$ and $(4, -5)$

50. passing through $(0, -2)$ and $(-2, 3)$

51. slope of $-\frac{3}{5}$ and containing $(3, -2)$

52. x-intercept 4 and y-intercept -2

53. parallel to $3x - 2y = 5$ and passing through $(6, 1)$

54. perpendicular to $x + 3y = 4$ and passing through $(1, 1)$

55. a vertical line passing through $(-2, 3)$

56. slope 6 and y-intercept 6

Draw the graphs of the following inequalities.

57. $y \leq 3$

58. $2y > x - 4$

59. $2x - 3y \geq 6$

60. $x > -1$

Solve these systems graphically.

61. $\begin{cases} x + 2y < 1 \\ x > 2y \end{cases}$

62. $\begin{cases} x - y < 2 \\ x < 2 \end{cases}$

Solve by determining the constant of variation and substituting known values.

63. z varies jointly as x and y and inversely as $\sqrt{w}$, and $x = 10$ when $y = 2$, $z = 8$, and $w = 25$. Find x when $y = 4$, $z = 4$, and $w = 36$.

64. The current in an electrical circuit is inversely proportional to the resistance. If the current is 60 amps when the resistance is 20 ohms, find the current when the resistance is 10 ohms.

65. The volume of a gas in a container is inversely proportional to the pressure. If a pressure of 60 pounds per square inch corresponds to a volume of 60 cubic feet, what pressure is needed to produce a volume of 100 cubic feet?

66. If T varies directly as V^3, what is the corresponding change in T if V is tripled?

Miscellaneous

67. If $f(x) = x^2 - 2x + 3$, find $f(a + h)$.

68. If $f(x) = -x^2 + 3x - 2$, find $f(a + h)$.

314 RELATIONS, FUNCTIONS, AND THEIR GRAPHS

69. If $f(x) = |x^2 - 4x - 2|$, find $f\left(\dfrac{1}{2}\right) - f\left(-\dfrac{1}{2}\right)$.

70. If $f(x) = \sqrt{x - 3}$ and $g(x) = 4 - x - x^2$, evaluate $f(7) - g(3)$.

Solutions to Self-Checking Exercise

1. False
2. False
3. True
4. True

5.

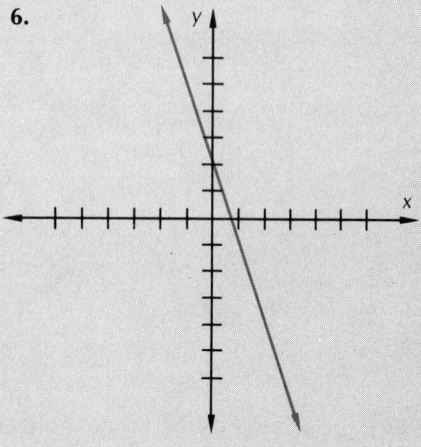

domain: $\{-3, 0, 1, 3\}$
range: $\{2, 4, -1\}$

6.

domain: $\{x \mid x \in R\}$
range: $\{y \mid y \in R\}$

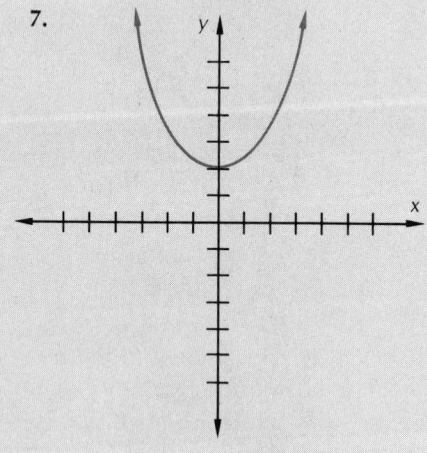

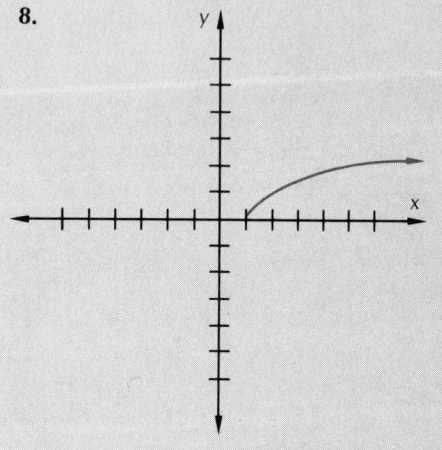

7.

domain: $\{x \mid x \in R\}$
range: $\{y \mid y \geq 2\}$

8.

domain: $\{x \mid x \geq 1\}$
range: $\{y \mid y \geq 0\}$

SOLUTIONS TO SELF-CHECKING EXERCISE

9.

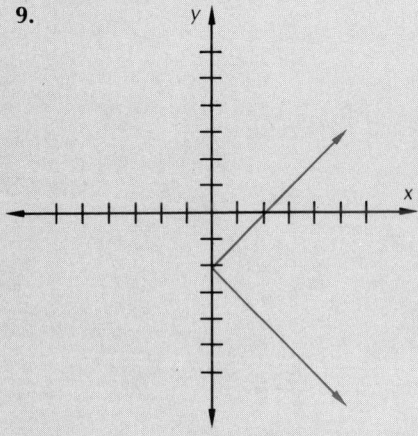

domain: $\{x \mid x \geq 0\}$
range: $\{y \mid y \in R\}$

10.

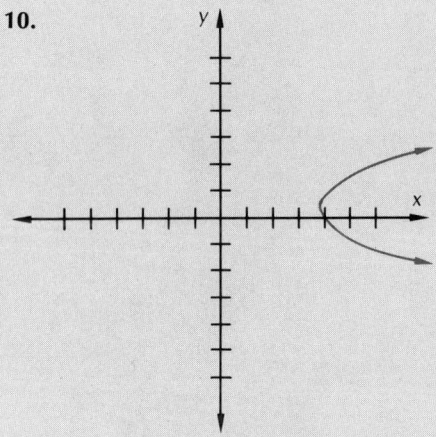

domain: $\{x \mid x \geq \frac{15}{4}\}$
range: $\{y \mid y \in R\}$

11. function
domain: $\{x \mid x \in R\}$
range: $\{f(x) \mid f(x) \in R\}$

12. function
domain: $\{x \mid x \in R\}$
range: $\{f(x) \mid f(x) \geq 0\}$

13. not a function

14. function
domain: $\{x \mid x \in R\}$
range: $\{f(x) \mid f(x) \geq 4\}$

15. function
domain: $\{x \mid x \geq 4\}$
range: $\{f(x) \mid f(x) \geq 0\}$

16. not a function

17. $f(x) = -\frac{1}{3}x + \frac{8}{3}$

$f(4) = \frac{4}{3}$

$f(-3) = \frac{11}{3}$

18. $f(x) = x^2 - 4$

$f(4) = 12$

$f(-3) = 5$

19. $f(x) = \sqrt[3]{x - 1}$

$f(4) = \sqrt[3]{3}$

$f(-3) = \sqrt[3]{-4}$

20. $f(x) = \sqrt{x + 2}$

$f(4) = \sqrt{6}$

$f(-3) = $ does not exist

21. $-\frac{7}{2}$

22. $\frac{8}{3}$

23. none exist

24. 2 and $-\frac{1}{3}$

25.

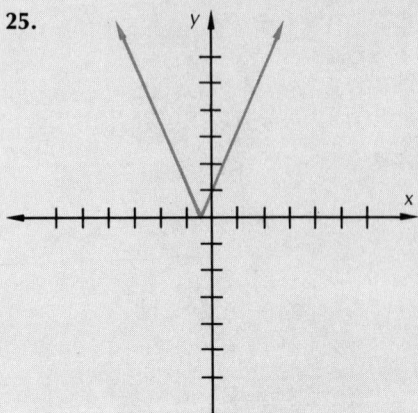

domain: $\{x \mid x \in R\}$
range: $\{f(x) \mid f(x) \geq 0\}$

26.

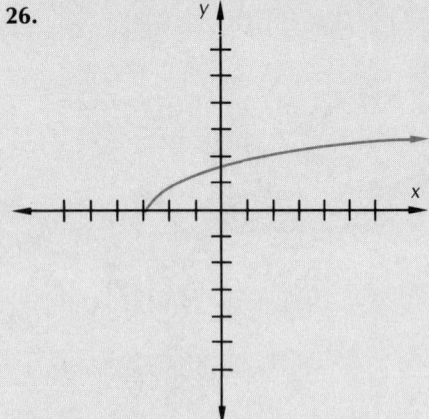

domain: $\{x \mid x \geq -3\}$
range: $\{f(x) \mid f(x) \geq 0\}$

27.

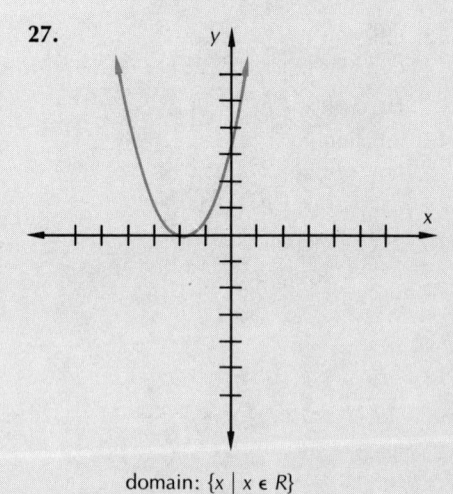

domain: $\{x \mid x \in R\}$
range: $\{f(x) \mid f(x) \geq 0\}$

28.

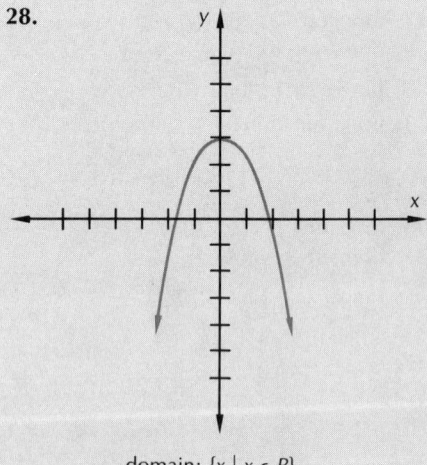

domain: $\{x \mid x \in R\}$
range: $\{f(x) \mid f(x) \leq 3\}$

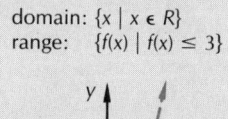

29.

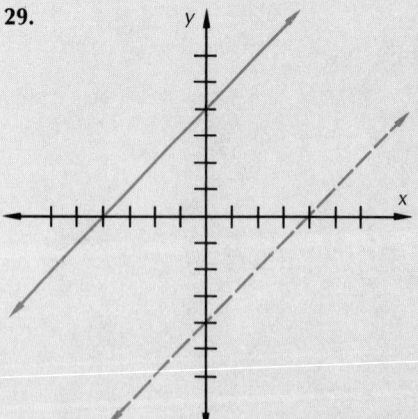

30.

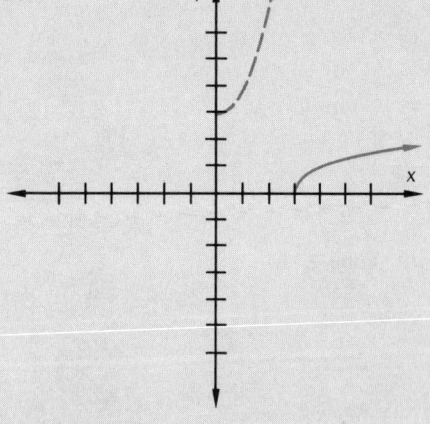

31.

32.

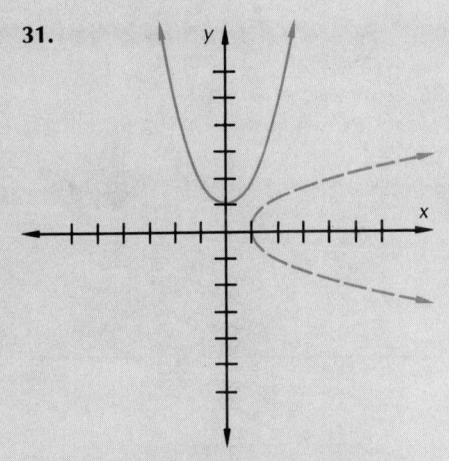

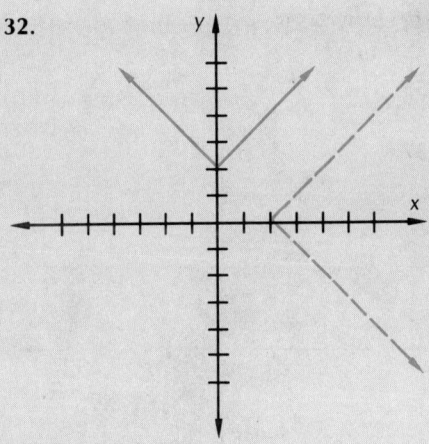

33. $f^{-1}(x) = 3x + 3$
    domain of f^{-1}: $\{x \mid x \in R\}$
    range of f^{-1}: $\{f^{-1}(x) \mid f^{-1}(x) \in R\}$
34. $f^{-1}(x) = x^2 - 1$
    domain of f^{-1}: $\{x \mid x \geq 0\}$
    range of f^{-1}: $\{f^{-1}(x) \mid f^{-1}(x) \geq -1\}$
35. $f^{-1}(x) = 3x + 2$
    domain of f^{-1}: $\{x \mid -1 \leq x \leq 1\}$
    range of f^{-1}: $\{f^{-1}(x) \mid -1 \leq f^{-1}(x) \leq 5\}$
36. $f^{-1}(x) = \dfrac{3x - 1}{2}$
    domain of f^{-1}: $\{x \mid x \in R\}$
    range of f^{-1}: $\{f^{-1}(x) \mid f^{-1}(x) \in R\}$
37. 38
38. does not exist
39. 1
40. -1
41. x-intercept is 2
    y-intercept is -4
    $m = 2$
42. x-intercept is $\dfrac{15}{2}$
    y-intercept is -3
    $m = \dfrac{2}{5}$
43. x-intercept is 2
    y-intercept is -3
    $m = \dfrac{3}{2}$
44. x-intercept is 0
    y-intercept is 0
    $m = 1$
45. $m = \dfrac{1}{3}$
46. $m = -\dfrac{1}{3}$
47. $m = 0$
48. undefined
49. $2x + y - 3 = 0$
50. $5x + 2y + 4 = 0$
51. $3x + 5y + 1 = 0$
52. $-x + 2y + 4 = 0$
    (or $x - 2y - 4 = 0$)

53. $-3x + 2y + 16 = 0$
    (or $3x - 2y - 16 = 0$)
54. $-3x + y + 2 = 0$
    (or $3x - y - 2 = 0$)
55. $x + 2 = 0$
56. $-6x + y - 6 = 0$
    (or $6x - y + 6 = 0$)

57. 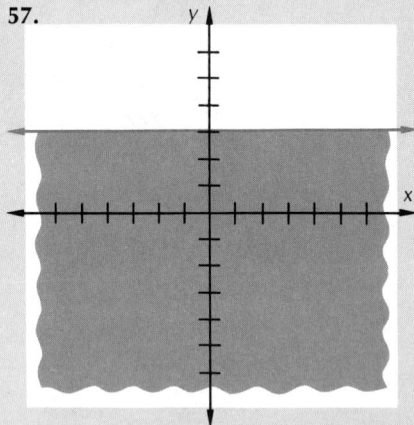
58. 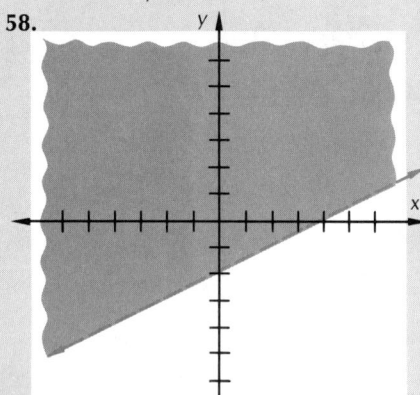
59. 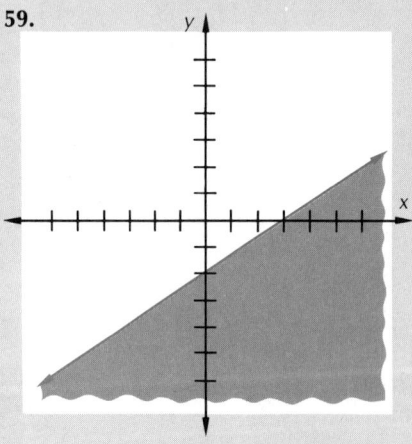
60. 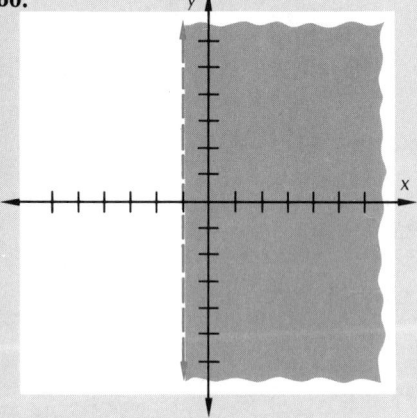
61. 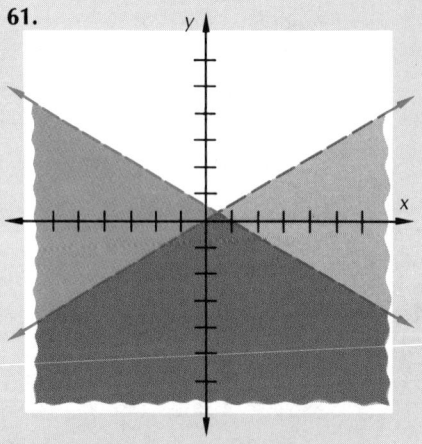
62.

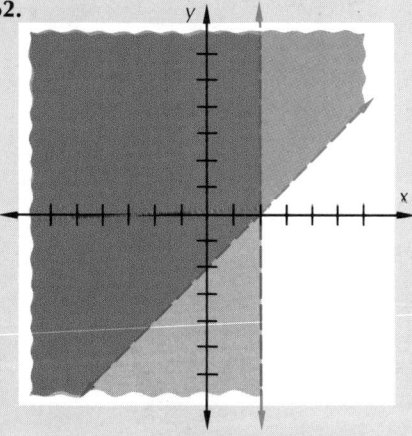

SOLUTIONS TO SELF-CHECKING EXERCISE

63. 3
64. 120 amperes
65. 36 pounds per square inch
66. 27 times as great
67. $a^2 + 2ah + h^2 - 2a - 2h + 3$
68. $-a^2 - 2ah - h^2 + 3a + 3h - 2$
69. $\dfrac{7}{2}$
70. 10

SYSTEMS OF LINEAR EQUATIONS

8

8.1 Systems of Linear Equations in Two Variables
8.2 Systems of Linear Equations in Three Variables
8.3 Second and Third Order Determinants
8.4 Cramer's Rule for Solving Systems of Linear Equations
8.5 Using Systems of Linear Equations to Solve Applied Problems

SYSTEMS OF LINEAR EQUATIONS IN TWO VARIABLES

8.1

A *system of linear equations in two variables* consists of two or more linear equations, each containing the same two variables. The solution of such a system is composed of all ordered pairs that satisfy every equation of the system simultaneously.

EXAMPLE 1

Determine whether the ordered pair $(-3, 2)$ is a solution of this linear system:

$2x + 5y = 4$
$3x - y = -11$

Solution: Substitute -3 for x and 2 for y in both equations.

$2x + 5y = 4$	$3x - y = -11$
$2(-3) + 5(2) = 4$	$3(-3) - 2 = -11$
$-6 + 10 = 4$	$-9 - 2 = -11$
true	true

Since $(-3, 2)$ satisfies both equations, it is a solution of the system.

The graph of a linear equation in two variables is a straight line. Thus, the graph of a system of two linear equations must contain two straight lines. A solution, since it satisfies both equations simultaneously, must represent a point of intersection. Consequently, as illustrated in the following, there are three possibilities for the solution of a system of two linear equations.

Consistent-Inconsistent-Dependent

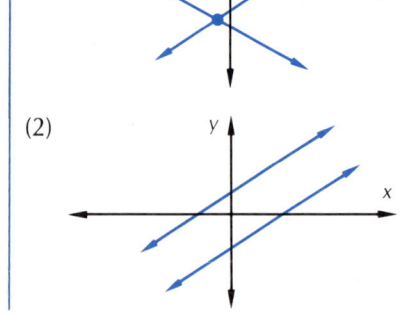

(1) The two lines intersect at a single point. The coordinates of this point give the solution of the system. When this occurs the system is said to be *consistent and independent*. We will simply say *consistent*.

(2) The two lines are parallel. There is no solution common to both equations. When this occurs the system is said to be *inconsistent*.

(3)

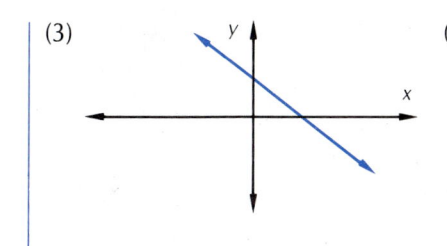

(3) The two equations are equivalent and produce the same line. Such a system has infinitely many solutions representing every point of the common line. When this occurs the system is said to be dependent.

A system of two linear equations in two variables can be solved by graphing both straight lines on the same coordinate system and finding the coordinates of the point of intersection. However, this method can be inaccurate since it is difficult to read exact coordinates from a graph. Thus, we will examine two algebraic methods called elimination and substitution.

To use the elimination method, we rewrite one or both of the equations in equivalent forms so that the coefficients of the same variable (either x or y) will be opposites of each other. Then, the equations are added, thereby eliminating one variable. The elimination method is explained in the following examples.

EXAMPLE 2

Solve the system:

$3x - 2y = 12$
$5x + 2y = 4$

Solution: Since the coefficients of y are opposites, adding the two equations will eliminate the variable y.

$$3x - 2y = 12$$
$$5x + 2y = 4$$
$$\overline{8x + 0 = 16}$$

The resulting equation contains only one variable and is solved for x.

$8x + 0 = 16$
$8x = 16$
$x = 2$

The x value of the solution is 2. To find the y value, replace 2 for x in either of the original two equations. Choosing the first equation produces the following:

$3x - 2y = 12$
$3(2) - 2y = 12$
$6 - 2y = 12$
$-2y = 6$
$y = -3$

SYSTEMS OF LINEAR EQUATIONS

The solution of this linear system is $x = 2$, $y = 3$, or more simply, the ordered pair $(2, -3)$. This solution is checked, as follows, by substituting 2 for x and -3 for y in both equations of the original system.

$$3x - 2y = 12 \qquad\qquad 5x + 2y = 4$$
$$3(2) - 2(-3) = 12 \qquad\qquad 5(2) + 2(-3) = 4$$
$$6 + 6 = 12 \qquad\qquad 10 - 6 = 4$$
$$\text{true} \qquad\qquad\qquad\qquad \text{true}$$

EXAMPLE 3

Solve the system:
$$7x + 5y = 11$$
$$3x - 2y = 13$$

Solution: Adding these two equations produces $10x + 3y = 24$. This does not help solve the system because neither variable was eliminated. Before adding we must multiply both sides of each equation by appropriate numbers so that the coefficients of either x or y will be opposites. Electing to eliminate y, we multiply the first equation by 2 and the second equation by 5, then add as follows:

$$7x + 5y = 11 \quad\xrightarrow{\text{multiply by 2}}\quad 14x + 10y = 22$$
$$3x - 2y = 13 \quad\xrightarrow{\text{multiply by 5}}\quad 15x - 10y = 65$$
$$\overline{\qquad\qquad\qquad\qquad\qquad 29x + 0 \;\;= 87}$$

The resulting equation contains only one variable and is solved for x.

$$29x + 0 = 87$$
$$29x = 87$$
$$x = 3$$

To find y, replace 3 for x in either equation of the original system. Choosing the first equation gives:

$$7x + 5y = 11$$
$$7(3) + 5y = 11$$
$$21 + 5y = 11$$
$$5y = -10$$
$$y = -2$$

The solution of this system is $(3, -2.)$. Remember to check this answer in both of the original equations.

The next two examples show what happens when we use an algebraic method to solve an inconsistent or dependent system of linear equations.

EXAMPLE 4

Solve the system:
$2x - 6y = 7$
$x = 2 + 3y$

Solution: Both equations are placed in the same form, $ax + by = c$. Then, -2 is multiplied times the second equation and we add as shown:

$$\begin{array}{l} 2x - 6y = 7 \\ x - 3y = 2 \end{array} \xrightarrow{\text{multiply by } -2} \begin{array}{l} 2x - 6y = 7 \\ -2x + 6y = -4 \\ \hline 0 + 0 = 3 \\ \text{false} \end{array}$$

Note that both variables were eliminated, producing a false statement. This indicates that the linear system is inconsistent. The graph consists of two parallel lines, so the system has no solution.

EXAMPLE 5

Solve the system:
$4x - y = 3$
$-8x + 2y = 6$

Solution: We multiply the first equation by 2, then add.

$$\begin{array}{l} 4x - y = -3 \\ -8x + 2y = 6 \end{array} \xrightarrow{\text{multiply by } 2} \begin{array}{l} 8x - 2y = -6 \\ -8x + 2y = 6 \\ \hline 0 + 0 = 0 \\ \text{true} \end{array}$$

As in the previous example, both variables were eliminated. However, in this case a true statement resulted. This indicates that the two equations are equivalent. There are infinitely many solutions for all points on the common line, $4x - y = -3$. We simply state that the system is *dependent*.

Systems of linear equations are also solved using a second algebraic procedure called the substitution method. This method uses the fact that any quantity may be substituted for its equal, and works particularly well when one of the equations has been solved for one of its variables. The remaining examples illustrate the substitution method.

EXAMPLE 6

Solve the system:
$y = 2x + 1$
$x + 3y = 17$

Solution: The first equation is solved for y. Substitute its equal, 2x + 1, for y in the second equation.

$$x + 3y = 17$$
$$x + 3(2x + 1) = 17$$

Solve this equation for x.

$$x + 6x + 3 = 17$$
$$7x + 3 = 17$$
$$7x = 14$$
$$x = 2$$

To find y, replace 2 for x in the equation y = 2x + 1.

$$y = 2x + 1$$
$$y = 2(2) + 1$$
$$y = 4 + 1$$
$$y = 5$$

The solution is (2, 5). Check this answer in the other equation, x + 3y = 17.

EXAMPLE 7

Solve the system:
$$2x - 3y = 6$$
$$3x - 4y = 7$$

Solution: To use the substitution method, solve one of the equations for either x or y. We elect to solve the first equation for x.

$$2x - 3y = 6$$
$$2x = 6 + 3y$$
$$x = \frac{6 + 3y}{2}$$

Substitute $\frac{6 + 3y}{2}$ for x in the second equation.

$$3x - 4y = 7$$
$$3\left(\frac{6 + 3y}{2}\right) - 4y = 7$$

Multiply both sides of the equation by 2 to eliminate fractions, then solve for y.

$$3(6 + 3y) - 8y = 14$$
$$18 + 9y - 8y = 14$$
$$18 + y = 14$$
$$y = -4$$

To find x, replace -4 for y in the equation $x = \dfrac{6 + 3y}{2}$.

$$x = \dfrac{6 + 3y}{2}$$

$$x = \dfrac{6 + 3(-4)}{2} = \dfrac{6 - 12}{3} = -\dfrac{6}{3} = -2$$

The solution is $(-2, -4)$. Check this answer.

The substitution method can be used to solve a system of two linear equations in two variables by simply solving either equation for one of its variables and substituting this result into the other equation. This method is summarized in the following.

To Solve a Linear System by Substitution:
(1) Solve either equation for one of its variables.
(2) Substitute that result into the other equation.
(3) Solve the equation obtained in step 2 to find the value of one variable.
(4) To find the other value, substitute the solution from step 3 into the equation formed in step 1.
(5) Check the solution by substituting into both equations of the original system.

Some linear systems are complicated by the presence of fractional equations. When this occurs, clear each equation of fractions by multiplying both sides by the lowest common denominator (LCD). The resulting system can then be solved using either elimination or substitution.

Solve the system:

$$\dfrac{x}{3} + \dfrac{y}{15} = \dfrac{1}{6}$$

$$\dfrac{x}{4} - y = 1$$

EXAMPLE 8

Solution: Clear the equations of fractions by multiplying the first equation by 30 and the second equation by 4.

$$30\left[\dfrac{x}{3} + \dfrac{y}{15}\right] = \dfrac{1}{6} \cdot 30 \Rightarrow 10x + 2y = 5$$

$$4\left[\dfrac{x}{4} - y\right] = 1 \cdot 4 \Rightarrow x - 4y = 4$$

SYSTEMS OF LINEAR EQUATIONS

To solve by substitution, solve the second equation for x.

$$x - 4y = 4$$
$$x = 4 + 4y$$

Substitute into the first equation.

$$10x + 2y = 5$$
$$10(4 + 4y) + 2y = 5$$
$$40 + 40y + 2y = 5$$
$$42y = -35$$
$$y = -\frac{35}{42} = -\frac{5}{6}$$

To find x replace $-\frac{5}{6}$ for y in the equation $x = 4 + 4y$.

$$x = 4 + 4y$$
$$x = 4 + 4\left(-\frac{5}{6}\right) = 4 - \frac{10}{3} = \frac{12 - 10}{3} = \frac{2}{3}$$

The solution is $\left(\frac{2}{3}, -\frac{5}{6}\right)$. Check this answer.

Exercise 8.1

Determine if the ordered pair listed is a solution of the given linear system.

1. $(-2, -3)$
$$\begin{cases} 2x - 5y = 11 \\ x + 2y = -8 \end{cases}$$

2. $(-1, 2)$
$$\begin{cases} -3x + 2y = 7 \\ x - y = 3 \end{cases}$$

Determine whether the following systems are consistent, inconsistent, or dependent.

3. $\begin{cases} 2x - y = 4 \\ 2x + y = 4 \end{cases}$

4. $\begin{cases} 2x + y = 6 \\ x - y = -6 \end{cases}$

5. $\begin{cases} x + 2y = 0 \\ 2x + 4y = 0 \end{cases}$

6. $\begin{cases} 2x - 3y = 5 \\ 2x - 3y = 1 \end{cases}$

Solve the systems by elimination.

7. $x + y = 8$
   $x - y = 4$

8. $3x - y = 7$
   $x + 2y = 7$

9. $4x - y = 2$
   $8x - 2y = 1$

10. $5x + 3y = 19$
    $7x - y = 11$

11. $4x - y = 11$
    $2x + 3y = -5$
12. $2x + y = 3$
    $4x + 2y = 5$
13. $x - 2y = 0$
    $2x - y = 0$
14. $2x - 3y = 7$
    $6x - 9y = 21$
15. $2x + 3y = 5$
    $3x + 2y = 1$
16. $3x + 7y = 2$
    $x - 3y = 4$
17. $7x + 5y = 11$
    $3x - 2y = 13$
18. $2x - 5y = 3$
    $4x - 10y = 6$

Solve the systems by substitution.

19. $x = 4 - 2y$
    $3x + y = 12$
20. $2x + y = 6$
    $-x + 4y = 15$
21. $x + 4y = 16$
    $2x + 3y = 17$
22. $3x - 5y = 2$
    $x + 5y = 6$
23. $3x - y = 2$
    $5x + y = 6$
24. $7x - y = 24$
    $x = 2y + 9$
25. $3x + y = 5$
    $2x - 5y = -8$
26. $5x - 4y = -1$
    $3x + y = -38$

Solve these systems. Use any method you wish.

27. $2x + 6y = 5$
    $4x = 3y$
28. $15x - 10y = 30$
    $18x - 12y = 30$
29. $7x + 6y = -11$
    $8x - 5y = -60$
30. $\frac{1}{2}x = 5y + 1$
    $2y = 2x - 8$
31. $\frac{1}{2}x - \frac{1}{4}y = 2$
    $x + y = 1$
32. $\frac{1}{2}x + \frac{1}{3}y = \frac{2}{3}$
    $\frac{1}{3}x + \frac{1}{5}y = \frac{7}{15}$
33. $12x - 9y = 28$
    $20x - 15y = 35$
34. $6y = 2 + x$
    $\frac{1}{2}x - 3y = -1$
35. $\frac{x}{2} + \frac{y}{6} = \frac{1}{12}$
    $\frac{x}{4} - y = 1$
36. $\frac{x}{3} - \frac{y}{4} = 8$
    $\frac{2x}{5} - \frac{y}{2} = 2$

Calculator problems. Express your answers to the nearest thousandth.

37. $1.324x - 3.854y = 1.530$
    $-4.621x + 4.536y = 5.850$
38. $2.641x - 4.923y = 36.401$
    $5.132x - 3.020y = -24.800$
39. $.512x - .076y = .364$
    $.391x + .210y = .966$
40. $.512x = 3.640 + .760y$
    $3.911x + 2.111y = .966$

SYSTEMS OF LINEAR EQUATIONS IN THREE VARIABLES

8.2

A linear equation in three variables has the form

$$ax + by + cz = d$$

where a, b, c, d are real constants.

A solution of a linear equation in three variables is an ordered triple, written (x, y, z), whose coordinates satisfy the equation. A linear equation in three variables has infinitely many solutions, and when graphed in three dimensions produces a plane.

EXAMPLE 1 Show that the ordered triple $(1, -2, 3)$ is a solution of $2x - 3y + z = 11$.

Solution: Substitute the coordinates for their respective variables to see if the equation is satisfied.

$$2x - 3y + z = 11$$
$$2(1) - 3(-2) + 3 = 11$$
$$2 + 6 + 3 = 11$$
$$\text{true}$$

The ordered triple $(1, -2, 3)$ is a solution of $2x - 3y + z = 11$.

In this section we discuss how to solve linear systems in three variables, such as:

$$x + 2y - z = 6$$
$$2x - y + 3z = -13$$
$$3x - 2y + 3z = -16$$

The solution of such a system is composed of all ordered triples that satisfy each equation simultaneously. Graphically, the solution represents the point(s) of intersection of three planes. Thus, as illustrated in the following, there are four possibilities for the solution of a linear system in three variables.

Possible Solutions for a Linear System in Three Variables

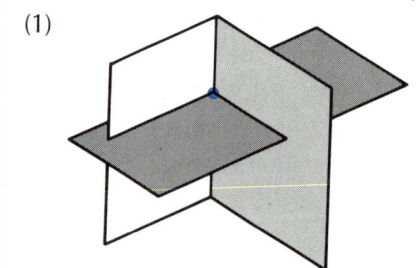

(1)

(1) The three planes intersect at exactly one common point. The coordinates of this point give the solution of the system.

SYSTEMS OF LINEAR EQUATIONS IN THREE VARIABLES   **331**

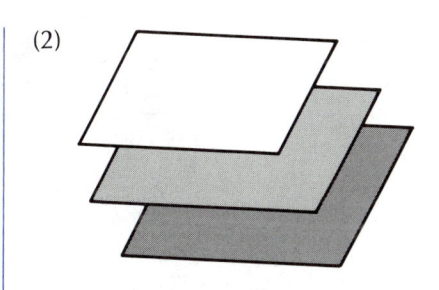

(2) Planes may be parallel or intersect in such a manner that there are no points in common. The system has no solution.

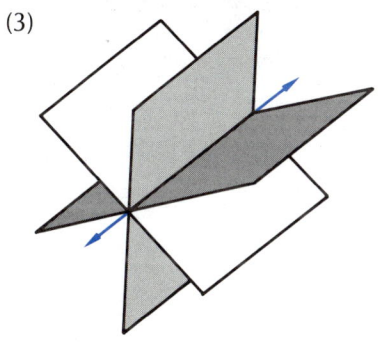

(3) The three planes may intersect in a straight line. The solution consists of the ordered pairs that satisfy the equation of the intersection line.

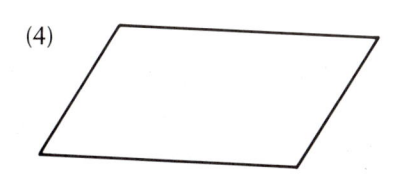

(4) The three equations may be equivalent, producing the same plane. The solution is represented by every point on the plane.

Solving systems of linear equations in three variables by three-dimensional graphing is difficult, inaccurate, and obviously impractical. Thus, we solve by the algebraic method of elimination.

Solve the system:

$$x + 2y - z = 6 \quad (1)$$
$$2x - y + 3z = -13 \quad (2)$$
$$3x - 2y + 3z = -16 \quad (3)$$

EXAMPLE 2

Solution: First we select a pair of equations and eliminate one of the variables. Adding equations (1) and (3) eliminates y.

$$x + 2y - z = 6$$
$$3x - 2y + 3z = -16$$
$$\overline{4x + 2z = -10} \quad (4)$$

This produced an equation having only two variables, x and z. To obtain another equation in x and z we select a different pair of equations and eliminate the same variable, y. Multiply equation (2) by -2 and add it to equation (3).

SYSTEMS OF LINEAR EQUATIONS

$$2x - y + 3z = -13 \xrightarrow{\text{multiply by } -2} -4x + 2y - 6z = 26$$
$$3x - 2y + 3z = -16 \quad (3)$$
$$\overline{}$$
$$-x - 3z = 10 \quad (5)$$

Equations (4) and (5) now comprise a linear system in two variables. Solve for x and z by elimination.

$$4x + 2z = -10 4x + 2z = -10$$
$$-x - 3z = 10 \xrightarrow{\text{multiply by } 4} -4x - 12z = 40$$
$$\overline{}$$
$$-10z = 30$$
$$z = -3$$

To find x, replace -3 for z in the equation $-x - 3z = 10$.

$$-x - 3z = 10$$
$$-x - 3(-3) = 10$$
$$-x + 9 = 10$$
$$-x = 1$$
$$x = -1$$

We have found x and z. To find y, select any equation of the original system and replace -1 for x and -3 for z. Selecting equation (1), we solve for y as follows:

$$x + 2y - z = 6$$
$$-1 + 2y - (-3) = 6$$
$$2y + 2 = 6$$
$$2y = 4$$
$$y = 2$$

The solution of the system is $(-1, 2, -3)$. It can be checked by substituting into all three equations of the system.

EXAMPLE 3 Solve the system:
$$4x - 6y + 10z = -7 \quad (1)$$
$$3x + 4y = -1 \quad (2)$$
$$14y - 6z = 7 \quad (3)$$

Solution: We begin by using equations (1) and (2) to eliminate x. Multiply equation (1) by 3 and equation (2) by -4. Then add.

$$4x - 6y + 10z = -7 \xrightarrow{\text{multiply by } 3} 12x - 18y + 30z = -21 \quad (4)$$
$$3x + 4y = -1 \xrightarrow{\text{multiply by } -4} -12x - 16y = 4$$
$$\overline{}$$
$$-34y + 30z = -17$$

This produced an equation having only the variables y and z. Equation (3) in the original system contains exactly the same variables. Thus, equations (3) and (4) comprise a linear system in two variables. Solve for y and z by elimination.

$$14y - 6z = 7 \xrightarrow{\text{multiply by 5}} 70y - 30z = 35$$
$$-34y + 30z = -17 \qquad\qquad\qquad -34y + 30z = -17$$
$$\qquad\qquad\qquad\qquad\qquad\qquad\qquad 36y = 18$$
$$\qquad\qquad\qquad\qquad\qquad\qquad\qquad y = \frac{18}{36} \text{ or } \frac{1}{2}$$

To find z, replace $\frac{1}{2}$ for y in the equation $14y - 6z = 7$.

$$14y - 6z = 7$$
$$14\left(\frac{1}{2}\right) - 6z = 7$$
$$7 - 6z = 7$$
$$-6z = 0$$
$$z = 0$$

To find x, select equation (2) and replace $\frac{1}{2}$ for y.

$$3x + 4y = -1$$
$$3x + 4\left(\frac{1}{2}\right) = -1$$
$$3x + 2 = -1$$
$$3x = -3$$
$$x = -1$$

The solution of the system is $\left(-1, \frac{1}{2}, 0\right)$.

As discussed earlier, there are several possibilities for solutions of linear systems in three variables. The following example illustrates a system having no solution.

Solve the system:

EXAMPLE 4

$$3x - y + 4z = 7 \qquad (1)$$
$$6x - 2y + 8z = 1 \qquad (2)$$
$$4y - z = 5 \qquad (3)$$

Solution: Eliminate x by multiplying equation (1) by -2 and adding it to equation (2).

$$3x - y + 4z = 7 \xrightarrow{\text{multiply by } -2} -6x + 2y - 8z = -14$$
$$6x - 2y + 8z = 1 \qquad\qquad\qquad\qquad 6x - 2y + 8z = 1$$
$$\qquad\qquad\qquad\qquad\qquad\qquad\qquad 0 = -13 \quad \text{(false)}$$

The false statement indicates that the system has no solution. If the system were graphed, at least two planes would be parallel.

> **Summary**
>
> To solve a system of three linear equations in three variables, follow these steps:
> (1) Select a pair of equations and eliminate one variable. This gives an equation in two variables.
> (2) Select a different pair of equations and eliminate the same variable. This gives another equation in two variables.
> (3) The equations from steps (1) and (2) produce a linear system in two variables. Solve for the two variables using either elimination or substitution.
> (4) Find the value of the third variable by substituting the two values from step (3) into any of the original three equations.
> (5) Check the solution by substituting into all three of the original equations.

Exercise 8.2

Solve each system of equations.

1. $x - y - z = 2$
    $x - y + z = 2$
    $x + y + z = 0$

2. $x - 3y + z = 1$
    $x - y + z = 3$
    $2x - 3y - 2z = 5$

3. $x + y + 2z = 0$
    $-2x - 2y + z = -10$
    $3x + 2y + z = 5$

4. $x + y + z = -1$
    $x - y + z = -1$
    $3x + 2y - z = -7$

5. $2x - 3y + z = 0$
    $x + y - z = 0$
    $-3x + 2y - 3z = 0$

6. $3x + 2y + z = 3$
    $x - 3y - 5z = 5$
    $2x + y - z = 0$

7. $2x + y - z = -5$
    $-5x - 3y + 2z = 7$
    $x + 4y - 3z = 0$

8. $2x + y + 2z = 1$
    $x + 2y - 3z = 4$
    $3x - y + z = 0$

9. $x - 2y - 3z = -4$
    $2x + 3y + 4z = 8$
    $5x + 4y - z = -6$

10. $2x + 2y + z = 4$
    $x - 2y + z = -1$
    $3x + y - 3z = 4$

11. $x - 2y + 3z = 2$
    $2x - 3y + z = 1$
    $3x - y + 2z = 9$

12. $x + 5y - z = 2$
    $2x + y + z = 7$
    $x - y + 2z = 11$

13. $x + y + 2z = 1$
    $2x + y + z = 4$
    $x - y + 3z = -2$

14. $x + 2y = 8$
    $2x + y + z = 11$
    $x + y + 2z = 13$

15. $x + y + z = 3$
    $x - z = 1$
    $y - z = -4$

16. $2x + y = 1$
    $2y - 3z = 4$
    $x + y = 3$

17. $2x - z = 1$
    $y + 3z = 8$
    $x - 2y = 4$

18. $3x + y - z = 5$
    $x + 2y + z = 3$
    $2x - y + 2z = 6$

19. $x - 3y + z = 4$
    $3x + 2y + z = 3$
    $-6x - 4y - 2z = 1$

20. $x - 5y + 2z = 10$
    $3x - y + 2z = 0$
    $x - 2y - 3z = 16$

21. $x + y = 2$
    $2x + z = 3$
    $-x + z = 0$

22. $x + y = 2$
    $4y + z = 0$
    $x - z = 5$

23. $2x + 3y + z = 1$
    $x - y - 2z = -2$
    $3x - 2y - z = 3$

24. $4x - 2y + 3z = 5$
    $2x - 3y + 5z = 4$
    $3x + y + 2z = 3$

25. $4x - 7y - 7z = 6$
    $x + 2y + \frac{1}{2}z = 0$
    $5x + 3y - 2z = 1$

26. $y - 2z = 1$
    $2x - y = 5$
    $4x - 2z = 9$

27. $3x + y = 6$
    $y - 2z = -7$
    $x - 2y + z = 0$

28. $3x + 3y - 2z = 10$
    $x - 2y + z = -1$
    $-2x - y + 3z = -3$

SECOND AND THIRD ORDER DETERMINANTS

8.3

A rectangular array of real numbers arranged in horizontal rows and vertical columns is called a matrix. The numbers are elements or entries and are enclosed within brackets. Thus,

$$A = \begin{bmatrix} 4 & 5 \\ -1 & 2 \\ 0 & 4 \end{bmatrix}$$

is a matrix containing three rows and two columns. In general, a matrix of m rows and n columns is said to have an order or dimension of $m \times n$ (read m by n). Hence, matrix A has an order of 3×2 (read 3 by 2). If a matrix has n rows and n columns, it is called a square matrix of order n. Thus,

$$B = \begin{bmatrix} 2 & 0 & 3 \\ -1 & 4 & 1 \\ 4 & 5 & -2 \end{bmatrix}$$

is a square matrix of order 3.

Associated with each square matrix A is a real number called the determinant of A, symbolized |A|. The determinant of A is displayed in the same form as the matrix, except vertical lines are used in place of brackets. In this section we will be concerned with determinants of 2 × 2 and 3 × 3 matrices. These are called second order and third order determinants, respectively.

Definition of Second Order Determinants

If $A = \begin{bmatrix} a_1 & b_1 \\ a_2 & b_2 \end{bmatrix}$ then $|A| = \begin{vmatrix} a_1 & b_1 \\ a_2 & b_2 \end{vmatrix}$

is the second order determinant and $|A|$ has a numeric value determined as follows:

$$\begin{vmatrix} a_1 & b_1 \\ a_2 & b_2 \end{vmatrix} = a_1 b_2 - a_2 b_1$$

This definition states that the value is obtained by multiplying the elements on the two diagonals, then subtracting the second product from the first. Schematically we have the following:

first   second
$\begin{vmatrix} 3 & -5 \\ 2 & 4 \end{vmatrix}$ first   second
$= 3 \cdot 4 - (-5) \cdot 2 = 12 + 10 = 22$

EXAMPLE 1 These second order determinants have been evaluated:

(a) $\begin{vmatrix} 0 & 4 \\ -5 & 2 \end{vmatrix} = 0 \cdot 2 - (4)(-5) = 0 + 20 = 20$

(b) $\begin{vmatrix} -1 & -2 \\ -3 & -5 \end{vmatrix} = 5 - 6 = -1$

(c) $\begin{vmatrix} 0 & 6 \\ 0 & 4 \end{vmatrix} = 0 - 0 = 0$

Definition of Third Order Determinants

If $A = \begin{bmatrix} a_1 & b_1 & c_1 \\ a_2 & b_2 & c_2 \\ a_3 & b_3 & c_3 \end{bmatrix}$ then $|A| = \begin{vmatrix} a_1 & b_1 & c_1 \\ a_2 & b_2 & c_2 \\ a_3 & b_3 & c_3 \end{vmatrix}$

is the third order determinant and $|A|$ has a numeric value determined as follows:

$$\begin{vmatrix} a_1 & b_1 & c_1 \\ a_2 & b_2 & c_2 \\ a_3 & b_3 & c_3 \end{vmatrix} = a_1 b_2 c_3 - a_1 b_3 c_2 + a_3 b_1 c_2 - a_2 b_1 c_3 + a_2 b_3 c_1 - a_3 b_2 c_1$$

This definition gives a cumbersome method of evaluating a third order determinant. Fortunately, this rule need not be memorized since we shall learn how a third order determinant can be evaluated using a series of second order determinants. To accomplish this we must study two important concepts called minors and cofactors.

The minor of an element in a determinant is the determinant that remains after deleting the row and column in which the element appears. For example, given the third order determinant

$$\begin{vmatrix} a_1 & b_1 & c_1 \\ a_2 & b_2 & c_2 \\ a_3 & b_3 & c_3 \end{vmatrix}$$

the minor of the element a_1 is found by deleting the first row and the first column as follows:

$$\begin{vmatrix} a_1 & b_1 & c_1 \\ a_2 & b_2 & c_2 \\ a_3 & b_3 & c_3 \end{vmatrix} \text{ to obtain } \begin{vmatrix} b_2 & c_2 \\ b_3 & c_3 \end{vmatrix}$$

Similarly, the minor of c_2 is found by deleting the second row and the third column as follows:

$$\begin{vmatrix} a_1 & b_1 & c_1 \\ a_2 & b_2 & c_2 \\ a_3 & b_3 & c_3 \end{vmatrix} \text{ to obtain } \begin{vmatrix} a_1 & b_1 \\ a_3 & b_3 \end{vmatrix}$$

Associated with each minor is either a plus sign or a minus sign. To determine the correct sign, we refer to the following checkerboard pattern called the array of signs:

$$\begin{array}{|ccc|} \hline + & - & + \\ - & + & - \\ + & - & + \\ \hline \end{array}$$

The cofactor of any element of a determinant is the associated minor prefixed by the corresponding sign. For example, given the determinant,

$$\begin{vmatrix} a_1 & b_1 & c_1 \\ a_2 & b_2 & c_2 \\ a_3 & b_3 & c_3 \end{vmatrix}$$

the cofactor of the element a_2 is found by attaching a minus sign (from row two, column one of the array of signs) to the minor obtained by deleting the first row and second column, as illustrated in the following:

$$-\begin{vmatrix} a_1 & b_1 & c_1 \\ a_2 & b_2 & c_2 \\ a_3 & b_3 & c_3 \end{vmatrix} \text{ to obtain } -\begin{vmatrix} b_1 & c_1 \\ b_3 & c_3 \end{vmatrix} = -(b_1c_3 - c_1b_3)$$

For the given determinant, evaluate the cofactor of the element in row three, column two.

$$\begin{vmatrix} 2 & 3 & -1 \\ 6 & 4 & 1 \\ -3 & 5 & 2 \end{vmatrix}$$

EXAMPLE 2

Solution: 5 is the element in row three, column two. Determine its minor and attach a minus sign by referring to the array of signs.

$$\begin{vmatrix} 2 & 3 & -1 \\ 6 & 4 & 1 \\ -3 & 5 & 2 \end{vmatrix} \text{ to obtain } - \begin{vmatrix} 2 & -1 \\ 6 & 1 \end{vmatrix} = -(2 + 6) = -8$$

EXAMPLE 3 Find the cofactor of each element in the first column of this determinant.

$$\begin{vmatrix} -3 & 4 & 6 \\ 2 & 3 & 0 \\ 5 & -1 & -2 \end{vmatrix}$$

Solution: The cofactor of -3 is:

$$+ \begin{vmatrix} -3 & 4 & 6 \\ 2 & 3 & 0 \\ 5 & -1 & -2 \end{vmatrix} \text{ or } + \begin{vmatrix} 3 & 0 \\ -1 & -2 \end{vmatrix} = +(-6 + 0) = -6$$

The cofactor of 2 is:

$$- \begin{vmatrix} -3 & 4 & 6 \\ 2 & 3 & 0 \\ 5 & -1 & -2 \end{vmatrix} \text{ or } - \begin{vmatrix} 4 & 6 \\ -1 & -2 \end{vmatrix} = -(-8 + 6) = 2$$

The cofactor of 5 is:

$$+ \begin{vmatrix} -3 & 4 & 6 \\ 2 & 3 & 0 \\ 5 & -1 & -2 \end{vmatrix} \text{ or } + \begin{vmatrix} 4 & 6 \\ 3 & 0 \end{vmatrix} = +(0 - 18) = -18$$

We are now able to provide a method for evaluating third order determinants using cofactors. This method will reduce a third order determinant to a series of second order determinants, and can also be used to evaluate higher order determinants.

> **Evaluating Determinants by Cofactors**
>
> To evaluate a determinant, add the products obtained by multiplying each element of any row or column by its corresponding cofactor. This method is called expansion by cofactors.

To apply this method it is helpful to follow these three steps:

1. Select a row or column. The work is simplified if the row or column contains a zero.
2. Multiply each element of the chosen row or column by its cofactor.
3. Evaluate the products and form their sum.

EXAMPLE 4 Evaluate this determinant by expansion of cofactors.

$$\begin{vmatrix} -2 & 10 & 2 \\ 3 & 5 & -4 \\ 2 & 0 & 8 \end{vmatrix}$$

Solution: Note that the second column and the third row contain a zero.

Step 1. We will expand about the second column.

Step 2. Multiply each element of the second column by its cofactor.

$$\begin{vmatrix} -2 & 10 & 2 \\ 3 & 5 & -4 \\ 2 & 0 & 8 \end{vmatrix} = -10 \begin{vmatrix} 3 & -4 \\ 2 & 8 \end{vmatrix} + 5 \begin{vmatrix} -2 & 2 \\ 2 & 8 \end{vmatrix} - 0 \begin{vmatrix} -2 & 2 \\ 3 & -4 \end{vmatrix}$$

Step 3. Evaluate the products and form their sum.

$$\begin{vmatrix} -2 & 10 & 2 \\ 3 & 5 & -4 \\ 2 & 0 & 8 \end{vmatrix} = -10 \begin{vmatrix} 3 & -4 \\ 2 & 8 \end{vmatrix} + 5 \begin{vmatrix} -2 & 2 \\ 2 & 8 \end{vmatrix} - 0 \begin{vmatrix} -2 & 2 \\ 3 & -4 \end{vmatrix}$$

$$= -10(24 + 8) + 5(-16 - 4) - 0$$
$$= -320 - 100 - 0$$
$$= -420$$

It is important to realize that a determinant can be evaluated by expanding about any row or column. The answer remains the same. To illustrate, we evaluate the determinant from Example 4 by expanding about its first row.

$$\begin{vmatrix} -2 & 10 & 2 \\ 3 & 5 & -4 \\ 2 & 0 & 8 \end{vmatrix} = -2 \begin{vmatrix} 5 & -4 \\ 0 & 8 \end{vmatrix} - 10 \begin{vmatrix} 3 & -4 \\ 2 & 8 \end{vmatrix} + 2 \begin{vmatrix} 3 & 5 \\ 2 & 0 \end{vmatrix}$$

$$= -2(40 + 0) - 10(24 + 8) + 2(0 - 10)$$
$$= -80 - 320 - 20$$
$$= -420$$

The answer remains the same, -420. You should verify this further by expanding this determinant about other rows and columns.

We now verify that expansion by cofactors produces the same rule as given by the definition of a third order determinant. That is,

$$\begin{vmatrix} a_1 & b_1 & c_1 \\ a_2 & b_2 & c_2 \\ a_3 & b_3 & c_3 \end{vmatrix} = a_1 b_2 c_3 - a_1 b_3 c_2 + a_3 b_1 c_2 - a_2 b_1 c_3 + a_2 b_3 c_1 - a_3 b_2 c_1$$

Expanding about the first row produces the following:

$$\begin{vmatrix} a_1 & b_1 & c_1 \\ a_2 & b_2 & c_2 \\ a_3 & b_3 & c_3 \end{vmatrix} = a_1 \begin{vmatrix} b_2 & c_2 \\ b_3 & c_3 \end{vmatrix} - b_1 \begin{vmatrix} a_2 & c_2 \\ a_3 & c_3 \end{vmatrix} + c_1 \begin{vmatrix} a_2 & b_2 \\ a_3 & b_3 \end{vmatrix}$$

$$= a_1(b_2 c_3 - c_2 b_3) - b_1(a_2 c_3 - c_2 a_3) + c_1(a_2 b_3 - b_2 a_3)$$
$$= a_1 b_2 c_3 - a_1 c_2 b_3 - b_1 a_2 c_3 + b_1 c_2 a_3 + c_1 a_2 b_3 - c_1 b_2 a_3$$
$$= a_1 b_2 c_3 - a_1 b_3 c_2 - a_2 b_1 c_3 + a_3 b_1 c_2 + a_2 b_3 c_1 - a_3 b_2 c_1$$
$$= a_1 b_2 c_3 - a_1 b_3 c_2 + a_3 b_1 c_2 - a_2 b_1 c_3 + a_2 b_3 c_1 - a_3 b_2 c_1$$

You should also verify that the definition is satisfied by expanding about other rows or columns.

Exercise 8.3

Evaluate the second order determinants.

1. $\begin{vmatrix} 3 & 0 \\ 2 & 3 \end{vmatrix}$
2. $\begin{vmatrix} -3 & 2 \\ 6 & -1 \end{vmatrix}$
3. $\begin{vmatrix} 2 & -3 \\ -4 & 1 \end{vmatrix}$
4. $\begin{vmatrix} -5 & -1 \\ 2 & 3 \end{vmatrix}$
5. $\begin{vmatrix} 1 & -2 \\ -1 & 2 \end{vmatrix}$
6. $\begin{vmatrix} -9 & -13 \\ 12 & 8 \end{vmatrix}$
7. $\begin{vmatrix} \frac{3}{4} & \frac{1}{2} \\ -\frac{1}{2} & \frac{1}{4} \end{vmatrix}$
8. $\begin{vmatrix} 1 & 3 \\ -4 & 7 \end{vmatrix} + \begin{vmatrix} 3 & 1 \\ 7 & -4 \end{vmatrix}$
9. $\begin{vmatrix} a & b \\ c & d \end{vmatrix} + \begin{vmatrix} b & a \\ d & c \end{vmatrix}$
10. $\begin{vmatrix} a & b \\ 0 & 1 \end{vmatrix} \cdot \begin{vmatrix} \frac{1}{a} & \frac{1}{b} \\ 0 & 1 \end{vmatrix}$ where $a \ne 0$, $b \ne 0$

Evaluate the cofactor of the element indicated.

11. $\begin{vmatrix} 2 & -1 & 3 \\ 4 & 2 & 1 \\ 0 & 1 & 2 \end{vmatrix}$ first row, third column

12. $\begin{vmatrix} 2 & 1 & 4 \\ 3 & 2 & 1 \\ 1 & 2 & 3 \end{vmatrix}$ second row, second column

13. $\begin{vmatrix} 4 & -3 & 2 \\ 1 & 2 & -1 \\ -2 & -3 & -4 \end{vmatrix}$ third row, first column

14. $\begin{vmatrix} 1 & 2 & 3 \\ -2 & -3 & -4 \\ 3 & 2 & 1 \end{vmatrix}$ second row, first column

15. $\begin{vmatrix} 6 & -8 & 7 \\ -3 & 2 & 4 \\ 3 & 5 & 2 \end{vmatrix}$ first row, second column

16. $\begin{vmatrix} 3 & -6 & 4 \\ 2 & -1 & 2 \\ 3 & 3 & 2 \end{vmatrix}$ third row, second column

Evaluate by expansion of cofactors.

17. $\begin{vmatrix} 0 & 2 & 1 \\ 2 & 0 & 1 \\ 2 & -1 & 4 \end{vmatrix}$
18. $\begin{vmatrix} 0 & 2 & 0 \\ 2 & 0 & 1 \\ 2 & -1 & 3 \end{vmatrix}$
19. $\begin{vmatrix} 2 & 6 & -5 \\ 0 & 0 & 4 \\ 4 & 7 & 8 \end{vmatrix}$
20. $\begin{vmatrix} 1 & 1 & 1 \\ 1 & 3 & 2 \\ 1 & -2 & 1 \end{vmatrix}$
21. $\begin{vmatrix} 2 & 4 & 8 \\ 5 & 1 & 5 \\ 2 & 3 & 1 \end{vmatrix}$
22. $\begin{vmatrix} 0 & 0 & 0 \\ 3 & -6 & 4 \\ 2 & -1 & -2 \end{vmatrix}$

23. $\begin{vmatrix} -2 & 2 & -1 \\ 1 & 1 & 3 \\ 2 & -1 & 1 \end{vmatrix}$

24. $\begin{vmatrix} 1 & 1 & -1 \\ 1 & 1 & 1 \\ 1 & -1 & 1 \end{vmatrix}$

25. $\begin{vmatrix} 1 & 2 & -1 \\ 2 & -1 & 3 \\ 3 & -2 & 3 \end{vmatrix}$

26. $\begin{vmatrix} 1 & 0 & 3 \\ 0 & 2 & 1 \\ 1 & 2 & -1 \end{vmatrix}$

27. $\begin{vmatrix} 1 & 6 & -1 \\ 2 & -13 & 3 \\ 3 & -16 & 3 \end{vmatrix}$

28. $\begin{vmatrix} x & 1 & 1 \\ y & 2 & 0 \\ z & 3 & 1 \end{vmatrix}$

29. $\begin{vmatrix} 1 & 0 & 1 \\ a & b & c \\ 2 & 2 & 2 \end{vmatrix}$

30. $\begin{vmatrix} x & 0 & 0 \\ 0 & x & 0 \\ 0 & 0 & x \end{vmatrix}$

31. $\begin{vmatrix} 1 & -1 & 1 \\ -1 & 3 & -1 \\ 2 & 4 & 2 \end{vmatrix}$

32. $\begin{vmatrix} 1 & 2 & 3 \\ 3 & 2 & 1 \\ 1 & 1 & 1 \end{vmatrix}$

33. $\begin{vmatrix} 3 & 2 & 3 \\ 1 & -4 & 2 \\ 0 & 0 & 1 \end{vmatrix}$

34. $\begin{vmatrix} 3 & 9 & 6 \\ 4 & -4 & 4 \\ 10 & 5 & 20 \end{vmatrix}$

Solve the following for x by first evaluating the determinant.

35. $\begin{vmatrix} x & 1 \\ 4 & 2 \end{vmatrix} = 5$

36. $\begin{vmatrix} -3 & x \\ 4 & x \end{vmatrix} = 20$

37. $\begin{vmatrix} -8 & 4x \\ 1 & -x \end{vmatrix} = 27$

38. $\begin{vmatrix} 2x & 1 \\ 3 & x \end{vmatrix} = -x$

39. $\begin{vmatrix} x & 0 & 0 \\ 2 & 1 & 3 \\ 0 & 1 & 4 \end{vmatrix} = 6$

40. $\begin{vmatrix} x^2 & 0 & 1 \\ 2 & -1 & 3 \\ 3 & 2 & 0 \end{vmatrix} = 4$

41. $\begin{vmatrix} 0 & -4 & 3 \\ x & 2 & 0 \\ -1 & 5 & x \end{vmatrix} = -8$

42. $\begin{vmatrix} 2x & 1 \\ 2 & 2x \end{vmatrix} = \begin{vmatrix} 3x & 1 \\ x & x \end{vmatrix}$

CRAMER'S RULE FOR SOLVING SYSTEMS OF LINEAR EQUATIONS

8.4

In this section you will learn how to solve linear systems in two and three variables using second and third order determinants. This method is called Cramer's Rule.

First we solve the following general system of two linear equations in two variables by elimination. This provides us with a general solution to be used for any such system.

$$a_1 x + b_1 y = c_1 \qquad (1)$$

$$a_2 x + b_2 y = c_2 \qquad (2)$$

To solve for x we eliminate y by multiplying b_2 times equation (1) and $-b_1$ times equation (2). Then, the equations are added and the result is solved for x.

$$a_1b_2x + b_1b_2y = c_1b_2$$
$$-a_2b_1x - b_1b_2y = -c_2b_1$$
$$\overline{a_1b_2x - a_2b_1x + \quad 0 = c_1b_2 - c_2b_1}$$
$$(a_1b_2 - a_2b_1)x = c_1b_2 - c_2b_1$$
$$x = \frac{c_1b_2 - c_2b_1}{a_1b_2 - a_2b_1}$$

Similarly, to solve for y we eliminate x by multiplying equation (1) by $-a_2$ and equation (2) by a_1. The equations are added, then solved for y.

$$-a_1a_2x - a_2b_1y = -a_2c_1$$
$$a_1a_2x + a_1b_2y = a_1c_2$$
$$\overline{0 - a_2b_1y + a_1b_2y = a_1c_2 - a_2c_1}$$
$$(a_1b_2 - a_2b_1)y = a_1c_2 - a_2c_1$$
$$y = \frac{a_1c_2 - a_2c_1}{a_1b_2 - a_2b_1}$$

Thus, the general solution is

$$x = \frac{c_1b_2 - c_2b_1}{a_1b_2 - a_2b_1} \text{ and } y = \frac{a_1c_2 - a_2c_1}{a_1b_2 - a_2b_1}$$

(Note that the denominators are identical.)

The common denominator and the numerators of the general solution can be written as second order determinants. The common denominator is symbolized D, where

$$D = \begin{vmatrix} a_1 & b_1 \\ a_2 & b_2 \end{vmatrix} = a_1b_2 - a_2b_1$$

The numerator of the x value is symbolized D_x, where

$$D_x = \begin{vmatrix} c_1 & b_1 \\ c_2 & b_2 \end{vmatrix} = c_1b_2 - c_2b_1$$

And, the numerator of the y value is symbolized D_y, where

$$D_y = \begin{vmatrix} a_1 & c_1 \\ a_2 & c_2 \end{vmatrix} = a_1c_2 - a_2c_1$$

Thus, using determinants, the general solution becomes

$$x = \frac{D_x}{D} = \frac{\begin{vmatrix} c_1 & b_1 \\ c_2 & b_2 \end{vmatrix}}{\begin{vmatrix} a_1 & b_1 \\ a_2 & b_2 \end{vmatrix}} \text{ and } y = \frac{D_y}{D} = \frac{\begin{vmatrix} a_1 & c_1 \\ a_2 & c_2 \end{vmatrix}}{\begin{vmatrix} a_1 & b_1 \\ a_2 & b_2 \end{vmatrix}}$$

CRAMER'S RULE FOR SOLVING SYSTEMS OF LINEAR EQUATIONS

where the elements of D are the coefficients of the variables in the given system (equations (1) and (2)). The elements of D_x are identical to those in D except the coefficients of x have been replaced by their corresponding constants, c_1 and c_2. Similarly, the elements of D_y are identical to those in D except the coefficients of y have been replaced by their respective constants c_1 and c_2. The use of these determinants to solve linear systems is known as Cramer's Rule.

Cramer's Rule (Two Variables)

Given the linear system

$a_1x + b_1y = c_1$

$a_2x + b_2y = c_2$

then

$$x = \frac{D_x}{D} = \frac{\begin{vmatrix} c_1 & b_1 \\ c_2 & b_2 \end{vmatrix}}{\begin{vmatrix} a_1 & b_1 \\ a_2 & b_2 \end{vmatrix}} \quad \text{and} \quad y = \frac{D_y}{D} = \frac{\begin{vmatrix} a_1 & c_1 \\ a_2 & c_2 \end{vmatrix}}{\begin{vmatrix} a_1 & b_1 \\ a_2 & b_2 \end{vmatrix}}$$

where $D \neq 0$

Use Cramer's Rule to solve this system:

$3x + 4y = -2$

$2x - 3y = -7$

EXAMPLE 1

Solution: Find the three determinants, D, D_x and D_y. Then form the appropriate quotients.

$$D = \begin{vmatrix} 3 & 4 \\ 2 & -3 \end{vmatrix} = -9 - 8 = -17$$

> To form D_x, the constants c_1 and c_2 are placed in the first column.

$$D_x = \begin{vmatrix} -2 & 4 \\ -7 & -3 \end{vmatrix} = 6 - (-28) = 34$$

> To form D_y, the constants c_1 and c_2 are placed in the second column.

$$D_y = \begin{vmatrix} 3 & -2 \\ 2 & -7 \end{vmatrix} = -21 - (-4) = -17$$

Therefore, by Cramer's Rule,

$$x = \frac{D_x}{D} = \frac{34}{-17} = -2 \text{ and } y = \frac{D_y}{D} = \frac{-17}{-17} = 1.$$

The solution is $(-2, 1)$. You should check it by substituting into the original system.

In similar fashion, Cramer's Rule can be used to solve a system of three linear equations in three variables.

Cramer's Rule (Three Variables)

Given the linear system

$a_1x + b_1y + c_1z = d_1$
$a_2x + b_2y + c_2z = d_2$
$a_3x + b_3y + c_3z = d_3$

then

$$x = \frac{D_x}{D}, \; y = \frac{D_y}{D}, \; z = \frac{D_z}{D}$$

where

$$D = \begin{vmatrix} a_1 & b_1 & c_1 \\ a_2 & b_2 & c_2 \\ a_3 & b_3 & c_3 \end{vmatrix}, \; D_x = \begin{vmatrix} d_1 & b_1 & c_1 \\ d_2 & b_2 & c_2 \\ d_3 & b_3 & c_3 \end{vmatrix}$$

$$D_y = \begin{vmatrix} a_1 & d_1 & c_1 \\ a_2 & d_2 & c_2 \\ a_3 & d_3 & c_3 \end{vmatrix}, \; D_z = \begin{vmatrix} a_1 & b_1 & d_1 \\ a_2 & b_2 & d_2 \\ a_3 & b_3 & d_3 \end{vmatrix}$$

and $D \neq 0$

Observe that the elements of the denominator determinant D, are the coefficients of the variables in the given system. And, the numerator determinants are formed from D by replacing the coefficients of the respective variable by the constants d_1, d_2, and d_3.

EXAMPLE 2

Use Cramer's Rule to solve this system:

$x + y + 2z = 2$
$2x + 3y - 2z = -1$
$x + 2y + 4z = 1$

Solution: Form the four determinants D, D_x, D_y, and D_z. Evaluate, then form the appropriate quotients. Each determinant will be expanded about row 1.

CRAMER'S RULE FOR SOLVING SYSTEMS OF LINEAR EQUATIONS

$$D = \begin{vmatrix} 1 & 1 & 2 \\ 2 & 3 & -2 \\ 1 & 2 & 4 \end{vmatrix} = 1\begin{vmatrix} 3 & -2 \\ 2 & 4 \end{vmatrix} - 1\begin{vmatrix} 2 & -2 \\ 1 & 4 \end{vmatrix} + 2\begin{vmatrix} 2 & 3 \\ 1 & 2 \end{vmatrix}$$

$$= 1(16) - 1(10) + 2(1)$$
$$= 8$$

> To form D_x, the constants d_1, d_2 and d_3 are placed in the first column.

$$D_x = \begin{vmatrix} 2 & 1 & 2 \\ -1 & 3 & -2 \\ 1 & 2 & 4 \end{vmatrix} = 2\begin{vmatrix} 3 & -2 \\ 2 & 4 \end{vmatrix} - 1\begin{vmatrix} -1 & -2 \\ 1 & 4 \end{vmatrix} + 2\begin{vmatrix} -1 & 3 \\ 1 & 2 \end{vmatrix}$$

$$= 2(16) - 1(-2) + 2(-5)$$
$$= 24$$

> To form D_y, the constants d_1, d_2, and d_3 are placed in the second column.

$$D_y = \begin{vmatrix} 1 & 2 & 2 \\ 2 & -1 & -2 \\ 1 & 1 & 4 \end{vmatrix} = 1\begin{vmatrix} -1 & -2 \\ 1 & 4 \end{vmatrix} - 2\begin{vmatrix} 2 & -2 \\ 1 & 4 \end{vmatrix} + 2\begin{vmatrix} 2 & -1 \\ 1 & 1 \end{vmatrix}$$

$$= 1(-2) - 2(10) + 2(3)$$
$$= -16$$

> To form D_z, the constants d_1, d_2, and d_3 are placed in the third column.

$$D_z = \begin{vmatrix} 1 & 1 & 2 \\ 2 & 3 & -1 \\ 1 & 2 & 1 \end{vmatrix} = 1\begin{vmatrix} 3 & -1 \\ 2 & 1 \end{vmatrix} - 1\begin{vmatrix} 2 & -1 \\ 1 & 1 \end{vmatrix} + 2\begin{vmatrix} 2 & 3 \\ 1 & 2 \end{vmatrix}$$

$$= 1(5) - 1(3) + 2(1)$$
$$= 4$$

Therefore, by Cramer's Rule,

$$x = \frac{D_x}{D} = \frac{24}{8} = 3, \; y = \frac{D_y}{D} = \frac{-16}{8} = -2, \; z = \frac{D_z}{D} = \frac{4}{8} = \frac{1}{2}$$

The solution is $\left(3, -2, \frac{1}{2}\right)$. You should check by substituting into the original system.

Cramer's Rule does not apply if $D = 0$. For a system of two variables, this means that the equations are either inconsistent or dependent; you can determine which by using elimination or substitution. For a system of three variables, we will state that there is no unique solution.

Exercise 8.4 Solve the following systems by Cramer's Rule.

1. $2x + y = 1$
   $3x - 2y = -9$

2. $2x + 3y = 1$
   $3x + 5y = -2$

3. $x - 3y = -5$
   $2x + 3y = 5$

4. $5x + 4y = 3$
   $2x + 3y = 4$

5. $3x + 2y = 4$
   $x - 2y = 8$

6. $2x - 3y + 4 = 0$
   $5x + 7y - 1 = 0$

7. $3x + 5y + 4 = 0$
   $4x + 3y - 2 = 0$

8. $8x + 3y = 7$
   $5x - 4y = 2$

9. $5x + 2y + 3 = 0$
   $4x - 7y - 3 = 0$

10. $x + y = a$
    $x - y = b$

11. $\frac{1}{4}x - \frac{1}{3}y = -1$
    $\frac{1}{2}x + \frac{1}{3}y = -2$

12. $\frac{1}{2}x + \frac{1}{4}y = 4$
    $\frac{1}{3}x - \frac{1}{2}y = 0$

13. $2x - y + z = -6$
    $x + 2y - z = 3$
    $3x + 2y + 2z = 2$

14. $2x - y + z = 8$
    $x + 2y + 3z = 9$
    $4x + y - 2z = 1$

15. $-x - y + z = 0$
    $x + y - z = 0$
    $x - y + z = \frac{1}{2}$

16. $x + y + z = 6$
    $2x - y - z = -3$
    $x - 3y + 2z = 1$

17. $x + y + z = 2$
    $x + y - z = 0$
    $x - y - z = -1$

18. $x + y + 2z = -3$
    $2x + y - 2z = -4$
    $x + y + z = 1$

19. $2x + y - z = -5$
    $-5x - 3y + 2z = 7$
    $x + 4y - 3z = 0$

20. $x + y - 3z = -2$
    $2x - y + z = 9$
    $3x + y - z = 6$

21. $x + 2y = 7$
    $4y - z = 3$
    $x - z = 4$

22. $y - 2z = -3$
    $x - y = 2$
    $3x + z = 11$

23. $y + 3z = 8$
    $x - 2y = 4$
    $2x - z = 1$

24. $2x - y + 3z = 20$
    $2x + y = 11$
    $y = 3$

USING SYSTEMS OF LINEAR EQUATIONS TO SOLVE APPLIED PROBLEMS

8.5 Applied problems often contain more than one unknown quantity and can be solved more easily using a system of equations rather than a single equation. To do this, we first determine the number of unknowns and then use the data to form a system having the same number of equations. The system is solved using methods studied in preceding sections.

USING SYSTEMS OF LINEAR EQUATIONS TO SOLVE APPLIED PROBLEMS 347

EXAMPLE 1

A rectangular steel plate has a perimeter of 80 centimeters. Find its dimensions if its length is 4 centimeters more than its width.

Solution: The problem has two unknowns, length and width. Assign the variables, letting

x = the length
y = the width

The perimeter is the distance around the plate, and is given by the formula $P = 2l + 2w$, where l = length and w = width. Since the perimeter is 80 centimeters, we write

80 = 2x + 2y

The length is 4 centimeters more than width. Thus,

x = 4 + y

We now have the following system of two equations in two unknowns, which we solve by substitution.

80 = 2x + 2y (1)
x = 4 + y (2)

We substitute (4 + y) for x in equation (1), then solve for y.

80 = 2(4 + y) + 2y
80 = 8 + 2y + 2y
80 = 8 + 4y
72 = 4y
18 = y

To find x, replace 18 for y in equation (2).

x = 4 + y
x = 4 + 18
x = 22

Since x = 22 and y = 18, the rectangular plate has a length of 22 centimeters and a width of 18 centimeters.

EXAMPLE 2

A total of $10,000 is invested in two bonds. One bond pays a dividend of 11% while a safer bond pays a dividend of 9%. How much is invested at each rate if the total annual return is $1,060?

Solution: The problem has two unknowns, the portion invested at 11% and the portion invested at 9%. Assign the variables, letting

x = portion at 11%
y = portion at 9%

The sum of both portions is $10,000. So,

x + y = 10,000

Annual dividend (interest) is the product of the principal and its corresponding rate. Thus, the dividend generated by x is .11x, while the dividend generated by y is .09y. The sum produces the total return of $1,060. Hence,

$$.11x + .09y = 1,060$$

We form the following system and solve by elimination:

$$x + y = 10,000 \quad (1)$$
$$.11x + .09y = 1,060 \quad (2)$$

To eliminate y, we multiply equation (1) by −9. Equation (2) is multiplied by 100 to eliminate decimals. The resulting equations are added.

$$-9x - 9y = -90,000$$
$$\underline{11x + 9y = 106,000}$$
$$2x = 16,000$$
$$x = 8,000$$

To find y, replace 8,000 for x in equation (1).

$$x + y = 10,000$$
$$8,000 + y = 10,000$$
$$y = 2,000$$

Since $x = 8,000$ and $y = 2,000$, we find that $8,000 is invested in the 11% bond and $2,000 is invested in the 9% bond.

EXAMPLE 3

A cyclist who travels with the wind can cover a distance of 35 miles in 2 hours. The return trip against the wind takes 3 hours. What is the speed of the cyclist in still air and what is the speed of the wind?

Solution: The problem has two variables, the cyclist's speed in still air and the wind speed. We assign the variables, letting

x = cyclist's speed in still air

y = speed of the wind

Traveling with the wind, the cyclist's rate is the sum of the speed in still air and the speed of the wind, written algebraically as $x + y$. The return trip is against the wind, so the wind speed subtracts from the cyclist's speed in still air to give a rate of $x - y$. Remembering that distance is the product of rate and time ($D = RT$), we form the following table:

	D	R	T	Equations D = RT
With the wind	35	x + y	2	35 = (x + y)2
Against the wind	35	x − y	3	35 = (x − y)3

USING SYSTEMS OF LINEAR EQUATIONS TO SOLVE APPLIED PROBLEMS

Simplifying the two equations produces this system, which we solve by Cramer's Rule:

$2x + 2y = 35$
$3x - 3y = 35$

$D = \begin{vmatrix} 2 & 2 \\ 3 & -3 \end{vmatrix} = -6 - 6 = -12$

$D_x = \begin{vmatrix} 35 & 2 \\ 35 & -3 \end{vmatrix} = -105 - 70 = -175$

$D_y = \begin{vmatrix} 2 & 35 \\ 3 & 35 \end{vmatrix} = 70 - 105 = -35$

$x = \dfrac{D_x}{D} = \dfrac{-175}{-12} = 14\dfrac{7}{12}$

$y = \dfrac{D_y}{D} = \dfrac{-35}{-12} = 2\dfrac{11}{12}$

The cyclist's speed in still air is $14\dfrac{7}{12}$ miles per hour. The speed of the wind is $2\dfrac{11}{12}$ miles per hour.

EXAMPLE 4

Find three numbers whose sum is 30, if the first number is 3 less than twice the second and also 1 less than the third.

Solution: The problem has three unknowns. We assign the variables, letting

x = the first number
y = the second number
z = the third number

The sum of the three numbers is 30, so we write,

$x + y + z = 30$

Since the first number is 3 less than twice the second we have,

$x = 2y - 3$ or $x - 2y = -3$

The first number is also 1 less than the third. Thus,

$x = z - 1$ or $x - z = -1$

These three equations form the following system in three unknowns, which we solve by elimination.

$x + y + z = 30$ (1)
$x - 2y = -3$ (2)
$x - z = -1$ (3)

We eliminate z by adding equation (1) and equation (3).

$$x + y + z = 30$$
$$\underline{x - z = -1}$$
$$2x + y = 29 \tag{4}$$

This produces equation (4), which we place with equation (2) to form a system of two equations involving x and y.

$$2x + y = 29 \tag{4}$$
$$x - 2y = -3 \tag{2}$$

We eliminate y by multiplying equation (4) by 2, then adding.

$$4x + 2y = 58$$
$$\underline{x - 2y = -3}$$
$$5x = 55$$
$$x = 11$$

To find y we replace x by 11 in equation (2).

$$x - 2y = -3$$
$$11 - 2y = -3$$
$$-2y = -14$$
$$y = 7$$

To find z we replace x by 11 in equation (3).

$$x - z = -1$$
$$11 - z = -1$$
$$12 = z$$

Since $x = 11$, $y = 7$, and $z = 12$, the three numbers are 11, 7, and 12.

EXAMPLE 5   Determine the values of a, b, and c in order that the ordered pairs $(-2, 14)$, $(1, 2)$, and $(3, 4)$ each satisfy the equation
$y = ax^2 + bx + c$

Solution: We form a system of three equations in three unknowns by substituting each ordered pair into the equation $y = ax^2 + bx + c$.
$14 = a(-2)^2 + b(-2) + c$
$2 = a(1)^2 + b(1) + c$
$4 = a(3)^2 + b(3) + c$

Simplifying and rearranging terms gives the following system, which we solve by Cramer's Rule. Each determinant is expanded about the first column.

USING SYSTEMS OF LINEAR EQUATIONS TO SOLVE APPLIED PROBLEMS   351

$$4a - 2b + c = 14$$
$$a + b + c = 2$$
$$9a + 3b + c = 4$$

$$D = \begin{vmatrix} 4 & -2 & 1 \\ 1 & 1 & 1 \\ 9 & 3 & 1 \end{vmatrix} = 4\begin{vmatrix} 1 & 1 \\ 3 & 1 \end{vmatrix} - 1\begin{vmatrix} -2 & 1 \\ 3 & 1 \end{vmatrix} + 9\begin{vmatrix} -2 & 1 \\ 1 & 1 \end{vmatrix}$$
$$= 4(1 - 3) - 1(-2 - 3) + 9(-2 - 1)$$
$$= -8 + 5 - 27$$
$$= -30$$

$$D_a = \begin{vmatrix} 14 & -2 & 1 \\ 2 & 1 & 1 \\ 4 & 3 & 1 \end{vmatrix} = 14\begin{vmatrix} 1 & 1 \\ 3 & 1 \end{vmatrix} - 2\begin{vmatrix} -2 & 1 \\ 3 & 1 \end{vmatrix} + 4\begin{vmatrix} -2 & 1 \\ 1 & 1 \end{vmatrix}$$
$$= 14(1 - 3) - 2(-2 - 3) + 4(-2 - 1)$$
$$= -28 + 10 - 12$$
$$= -30$$

$$D_b = \begin{vmatrix} 4 & 14 & 1 \\ 1 & 2 & 1 \\ 9 & 4 & 1 \end{vmatrix} = 4\begin{vmatrix} 2 & 1 \\ 4 & 1 \end{vmatrix} - 1\begin{vmatrix} 14 & 1 \\ 4 & 1 \end{vmatrix} + 9\begin{vmatrix} 14 & 1 \\ 2 & 1 \end{vmatrix}$$
$$= 4(2 - 4) - 1(14 - 4) + 9(14 - 2)$$
$$= -8 - 10 + 108$$
$$= 90$$

$$D_c = \begin{vmatrix} 4 & -2 & 14 \\ 1 & 1 & 2 \\ 9 & 3 & 4 \end{vmatrix} = 4\begin{vmatrix} 1 & 2 \\ 3 & 4 \end{vmatrix} - 1\begin{vmatrix} -2 & 14 \\ 3 & 4 \end{vmatrix} + 9\begin{vmatrix} -2 & 14 \\ 1 & 2 \end{vmatrix}$$
$$= 4(4 - 6) - 1(-8 - 42) + 9(-4 - 14)$$
$$= -8 + 50 - 162$$
$$= -120$$

To solve for a, b, and c, we form the appropriate quotients.

$$a = \frac{D_a}{D} = \frac{-30}{-30} = 1$$

$$b = \frac{D_b}{D} = \frac{90}{-30} = -3$$

$$c = \frac{D_c}{D} = \frac{-120}{-30} = 4$$

Since $a = 1$, $b = -3$, and $c = 4$, the equation $y = ax^2 + bx + c$ becomes $y = x^2 - 3x + 4$.

Solve each word problem by using a system of equations.   **Exercise 8.5**

1. The sum of two angles is 180°. Their difference is 50°. Find the angles.
2. Find two numbers that differ by 8 if their sum is 18.

3. The sum of two numbers is 19. One number is 2 less than twice the other. Find the numbers.
4. The difference of two numbers is 14 and one of the numbers is 1 more than 2 times the other. Find the numbers.
5. If $\frac{1}{3}$ of an integer is added to $\frac{1}{2}$ the next consecutive integer, the sum is 13. Find the integer.
6. Find a fraction that becomes $\frac{3}{5}$ when 1 is added to its numerator and denominator and becomes $\frac{1}{2}$ when 1 is subtracted from its numerator and denominator.
7. Ten dollars is to be divided between two people in the ratio of 9 to 11. How much will each receive?
8. 111 coins, made up of nickels and pennies, are worth $2.91. How many of each are there?
9. 55 coins, made up of dimes and nickels, are worth $3.75. How many dimes are there?
10. A collection of 43 coins worth $7.60 is made up of dimes and quarters. How many quarters are there?
11. A theater charges $3.00 for adults' tickest and half that amount for children's tickets. If 600 tickets are sold and $1,350 is taken in, how many children's tickets were sold?
12. A person has $5,000 invested, part at 6% and the rest at 8%. The annual income is $364. How much is invested at 6%?
13. A man invested $6,000 and $7,500 in two different accounts. He received $1,515 total interest for the year. The interest rate on the $7,500 was 4% higher than that on the $6,000. What was the rate on the $7,500 account?
14. A baseball team bought 4 dozen balls, some being practice balls at $5.25 each, and the others being game balls at $6.50 each. The total bill was $274.50. Find the number of practice balls.
15. The perimeter of a field is 192 meters. If the length is 8 meters more than the width, find the dimensions.
16. The width of a rectangle is 9 inches more than its width. Its perimeter is 66 inches. Find the dimensions.
17. The perimeter of a rectangle is 60 inches. If the length is increased by 4 inches and the width decreased by 4 inches, the new rectangle will be twice as long as it is wide. Find the dimensions of the original rectangle.
18. How many liters of a 10% acid solution should be mixed with a 5% acid solution to form 20 liters of an 8% acid solution?
19. A swimmer goes 2 kilometers in 15 minutes downstream, while upstream he needs 20 minutes. Find the rate of the swimmer and the current.

20. If it takes a plane $2\frac{1}{2}$ hours to fly 1,200 miles against the wind and 2 hours to return with the wind, find the speed of the plane in still air.
21. An airplane flies 3,500 kilometers with a tailwind in 5 hours, but requires 7 hours on the return trip against the wind. What is the speed of the plane in still air and the wind speed?
22. The speed of a plane is 143 miles per hour faster than that of a car. The car travels 50 miles in the same time the plane travels 180 miles. Find the speed of the car and plane.
23. The sum of three numbers is 18. The third number is 7 more than the second and the second is 1 more than the first. Find the numbers.
24. Three numbers have a sum of 30. The third number is 10 more than the sum of the first and second and the second is 3 less than $\frac{1}{2}$ the sum of the first and third. Find the numbers.
25. A collection of 17 coins, consisting of nickels, dimes, and quarters, is worth $1.50. There are twice as many nickels as dimes. How many quarters are there?
26. The sum of twice a first number, three times a second, and four times a third, is 8. The sum of the first and second is 8 more than the third. The sum of the second and third number is equal to the first. Find the numbers.
27. Tom, Dick, and Harry have $360 among them. Tom has five times as much as Harry. Dick and Harry together have $40 less than Tom. How much does each have?

Summary

KEY WORDS AND PHRASES

Consistent System of Linear Equations in Two Variables—a system of two lines that intersect at a single point.
Inconsistent System of Linear Equations in Two Variables—a system of two parallel lines with no common solution.
Dependent System of Linear Equations in Two Variables—a system of two equivalent equations with infinitely many solutions.
Matrix—a rectangular array of real numbers arranged in horizontal rows and vertical columns.
Square Matrix—a matrix of *n* rows and *n* columns.
Determinant of a Matrix—a real number associated with each square matrix by a particular rule.
Minor of an Element—the determinant that remains after deleting the row and column in which the element appears.
Cofactor of an Element—the associated minor prefixed by a corresponding sign.

Definition of Second Order Determinants—

$$\begin{vmatrix} a_1 & b_1 \\ a_2 & b_2 \end{vmatrix} = a_1 b_2 - a_2 b_1$$

Evaluating Determinants by Cofactors—add the products obtained by multiplying each element of any row or column by its corresponding cofactor.

Cramer's Rule for Two Variables—

Given $a_1 x + b_1 y = c_1$
$a_2 x + b_2 y = c_2$ then

$$x = \frac{D_x}{D} = \frac{\begin{vmatrix} c_1 & b_1 \\ c_2 & b_2 \end{vmatrix}}{\begin{vmatrix} a_1 & b_1 \\ a_2 & b_2 \end{vmatrix}} \text{ and } y = \frac{D_y}{D} = \frac{\begin{vmatrix} a_1 & c_1 \\ a_2 & c_2 \end{vmatrix}}{\begin{vmatrix} a_1 & b_1 \\ a_2 & b_2 \end{vmatrix}} \text{ where } D \neq 0$$

Self-Checking Exercise

Determine whether the following systems are consistent, inconsistent, or dependent.

1. $8x + 2y = 7$
   $4x + y = -7$

2. $3x + y = 4$
   $9x - 3y = 11$

3. $2x - 3y = 4$
   $x + 2y = 7$

4. $\frac{1}{2}x + y = 4$
   $2x + 4y = 16$

Solve each system by the addition or substitution method.

5. $2x + y = 9$
   $5x + 2y = 1$

6. $3x + 3y = 10$
   $6x - y = 6$

7. $\frac{1}{3}x - \frac{1}{2}y = \frac{2}{3}$
   $\frac{1}{2}x + \frac{1}{4}y = \frac{5}{2}$

8. $2x - 5y = -5$
   $3x + 4y = 27$

9. $x - 4y + 5z = 0$
   $3x + y - 2z = 9$
   $4x - 2y + z = 9$

10. $-4x + y = 5$
    $x = \frac{2}{3}y$

11. $3x + y - z = -1$
    $4x - 3y + z = -4$
    $x - 2y + z = 0$

12. $x + y - z = 7$
    $x - 2y + z = 3$
    $2x - 3y + z = -8$

Evaluate the cofactor of the given element.

13. $\begin{vmatrix} -3 & 2 & -2 \\ -1 & 1 & 4 \\ 4 & 5 & 6 \end{vmatrix}$ third row, second column

14. $\begin{vmatrix} 0 & 1 & -2 \\ 5 & 8 & -3 \\ 7 & 4 & 4 \end{vmatrix}$ second row, third column

Evaluate each determinant.

15. $\begin{vmatrix} -3 & 4 \\ 4 & -3 \end{vmatrix}$

16. $\begin{vmatrix} \frac{1}{2} & \frac{2}{3} \\ -\frac{2}{3} & 1 \end{vmatrix}$

17. $\begin{vmatrix} 2 & 0 & 1 \\ 3 & -2 & 4 \\ 1 & 1 & 2 \end{vmatrix}$

18. $\begin{vmatrix} 2 & 1 & -1 \\ -5 & -3 & 2 \\ 1 & 4 & -3 \end{vmatrix}$

Solve the following for x.

19. $\begin{vmatrix} x & 4 \\ -4x & 6 \end{vmatrix} = 8$

20. $\begin{vmatrix} 1 & -2 & 3 \\ -1 & 2 & x \\ x & 1 & 2 \end{vmatrix} = -12$

Use Cramer's Rule to solve each system.

21. $2x + 3y = 1$
    $3x + 5y = -2$

22. $3x - 2y = 11$
    $2x + 3y = 3$

23. $2x + y - z = -5$
    $-5x - 3y + 2z = 7$
    $x + 4y - 3z = 0$

24. $2x + 3y - 2z = 1$
    $3x - 2y - z = -6$
    $x - 4y + z = -3$

25. $x - z = -4$
    $x - 3y = 0$
    $2y + z = -1$

26. $\frac{1}{2}x = \frac{2}{3}y$
    $\frac{1}{4}z = \frac{1}{2}x$
    $y = z - 10$

Solve the following verbal problems.

27. A jar of 110 coins contains pennies and nickels worth $2.50. How many nickels are in the jar?
28. It takes a man 4 hours to paddle his canoe upstream 10 miles. The return trip downstream takes only 1 hour. What is the rate of the current?
29. How many gallons of 87 octane gasoline and 93 octane gasoline should be mixed to obtain 100 gallons of 91 octane gasoline? (Note: octane number represents a percentage of volume.)
30. The length of a rectangle is 2 centimeters less than 3 times its width, and its perimeter is 132 centimeters. Find the dimensions of the rectangle.

Solutions to Self-Checking Exercise

1. inconsistent
2. consistent
3. consistent
4. dependent
5. $(-17, 43)$
6. $\left(\frac{4}{3}, 2\right)$
7. $\left(\frac{17}{4}, \frac{3}{2}\right)$
8. $(5, 3)$
9. $(3, 2, 1)$
10. $(-2, -3)$
11. $(3, 13, 23)$
12. $(21, 32, 46)$
13. 14
14. 7
15. -7
16. $\frac{17}{18}$
17. -11
18. 6
19. $x = \frac{4}{11}$
20. $x = 1$ or $-\frac{9}{2}$
21. $(11, -7)$
22. $(3, -1)$

SOLUTIONS TO SELF-CHECKING EXERCISE

23. $(-2, 5, 6)$

25. $(-3, -1, 1)$

27. 35 nickels

29. $33\frac{1}{3}$ gallons of 87 octane

$66\frac{2}{3}$ gallons of 93 octane

24. $\left(\frac{15}{4}, 4, \frac{37}{4}\right)$

26. $(8, 6, 16)$

28. $\frac{15}{4}$ miles per hour

30. 17 centimeters width
49 centimeters length

THE CONIC SECTIONS

9

9.1   Graphing Quadratic Equations in Two Variables—Parabolas
9.2   The Quadratic Function
9.3   The Circle
9.4   The Ellipse
9.5   The Hyperbola
9.6   Conic Sections and Curve Recognition
9.7   Non-Linear Systems of Equations
9.8   Second Degree Inequalities

GRAPHING QUADRATIC EQUATIONS IN TWO VARIABLES—PARABOLAS

9.1

In this section we graph the following two types of quadratic equations.

(1) $y = ax^2 + bx + c$

(2) $x = ay^2 + by + c$

where $a, b, c \in R$ and $a \neq 0$.

As with other equations in two variables, we graph by obtaining ordered pairs that satisfy the given relation, then plot the corresponding points on the rectangular coordinate system. We are especially careful to plot numerous points, widely dispersed, so that the pattern is clearly evident before connecting them with a smooth curve. To illustrate, we will graph four quadratic equations, then summarize our observations.

EXAMPLE 1 Graph $y = x^2$

Solution: Form a table of values and plot the corresponding points to produce the graph in Figure 9.1.

Table of Values

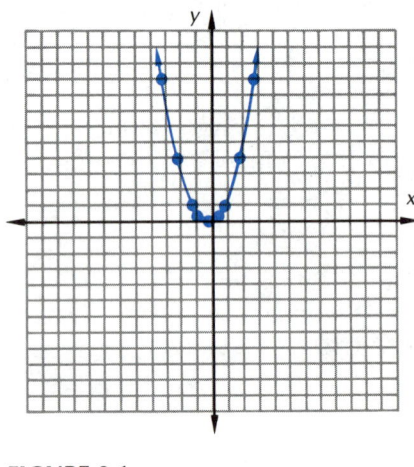

x	y
-3	9
-2	4
-1	1
$-\frac{1}{2}$	$\frac{1}{4}$
0	0
$\frac{1}{2}$	$\frac{1}{4}$
1	1
2	4
3	9

FIGURE 9.1

The graph of $y = x^2$ is a U-shaped curve continually becoming wider. It is an example of a parabola.

Parabolas are very useful and have many applications. For example, cross sections of spotlight reflectors and radar dishes are parabolically shaped. Disregarding air resistance, the non-vertical path of an object thrown upward is a parabola. Also, cables supporting a suspension bridge hang in the form of a parabola.

EXAMPLE 2 Graph $y = -x^2 + 2$

Solution: We again form a table of values and plot the corresponding points. This produces a parabola opening downward, shown in Figure 9.2.

Table of Values

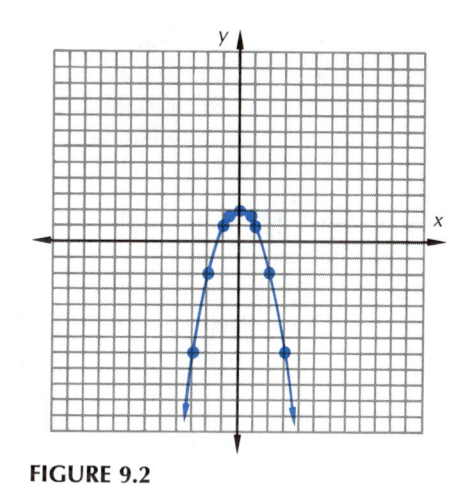

x	y
−3	−7
−2	−2
−1	1
−$\frac{1}{2}$	$\frac{7}{4}$
0	2
$\frac{1}{2}$	$\frac{7}{4}$
1	1
2	−2
3	−7

FIGURE 9.2

The preceding examples illustrate quadratic equations of the form

$$y = ax^2 + bx + c \quad (a \neq 0)$$

This form always defines functional relationships, since each value of x produces exactly one corresponding value of y. Thus, the graph of a quadratic function is a parabola that is either concave up (opening upward) or concave down (opening downward). A parabola is concave up when $a > 0$, and concave down when $a < 0$. Quadratic functions are discussed more comprehensively in the next section.

We continue our analysis of quadratic equations by graphing two examples that are not functions.

Graph $x = y^2 + 2y - 3$

EXAMPLE 3

Solution: Since the variable y is in the more complicated state, the table of values is formed by assigning arbitrary values to y and solving for the corresponding value of x. For example, if $y = 2$, then $x = 2^2 + 2(2) - 3$ or 5, producing the ordered pair (5, 2). The resulting parabola in Figure 9.3 opens to the right.

Table of Values

x	y
5	2
0	1
−3	0
−4	−1
−3	−2
0	−3
5	−4

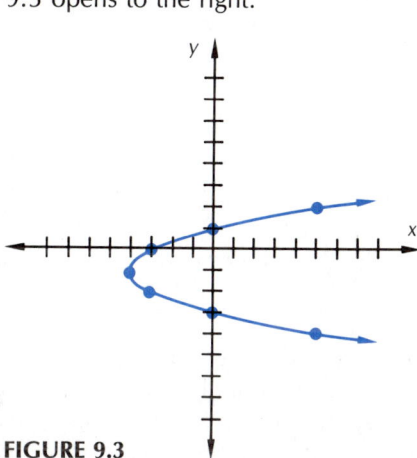

FIGURE 9.3

THE CONIC SECTIONS

When graphing an equation of two variables, it is helpful to determine where, if at all, the graph intersects the x and y-axes. These points are called the x and y-intercepts. Recall that x-intercepts are found by letting y = 0, and y-intercepts are found by letting x = 0.

EXAMPLE 4   Graph $x = -2y^2 + 2y + 4$

Solution: We form a table of values, being sure to include the x and y-intercepts as determined below:

x-intercept: Let y = 0 and solve for x, obtaining
$$x = -2(0)^2 + 2(0) + 4 \text{ or}$$
$$x = 4$$

y-intercepts: Let x = 0 and solve for y, obtaining
$$0 = -2y^2 + 2y + 4$$
$$0 = y^2 - y - 2$$
$$0 = (y - 2)(y + 1)$$
$$y - 2 = 0 \text{ or } y + 1 = 0$$
$$y = 2 \text{ and } y = -1$$

The parabola, opening to the left, is shown in Figure 9.4.

Table of Values

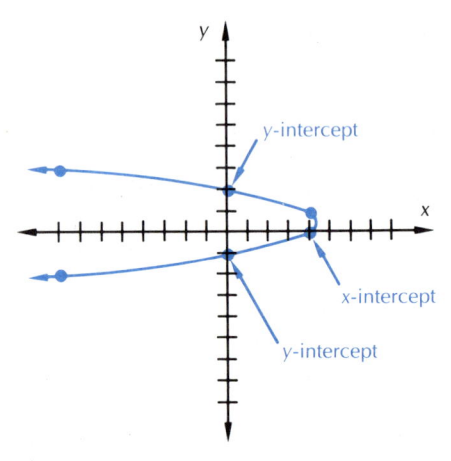

x	y
−8	−2
0	−1
4	0
4	1
0	2
−8	3

FIGURE 9.4

As illustrated by Examples 3 and 4, quadratic equations of the form
$$x = ay^2 + by + c \quad (a \neq 0)$$
produce parabolas opening either to the right or to the left. If $a > 0$, the parabola opens to the right; if $a < 0$, the parabola opens to the left. As shown in Figure 9.5, parabolas of this type do not represent functions because the

vertical line test indicates that for all but one x in the domain, there are two corresponding values of y.

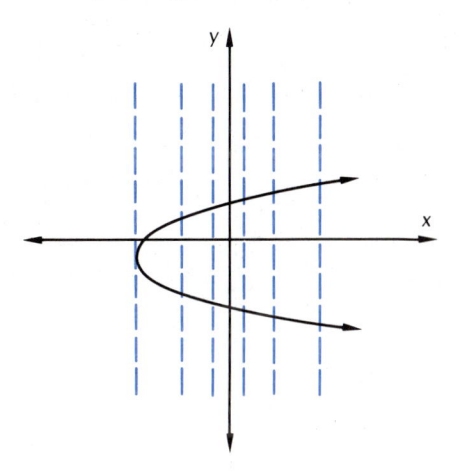

FIGURE 9.5

Our observations from this section are summarized in the Table 9.1.

TABLE 9.1 Summary of Quadratic Equations in Two Variables
(1) Quadratic equations of the form $$y = ax^2 + bx + c \quad (a \neq 0)$$ define functions. Their graphs are parabolas either concave up or concave down.   concave up ($a > 0$)   concave down ($a < 0$)
(2) Quadratic equations of the form $$x = ay^2 + by + c \quad (a \neq 0)$$

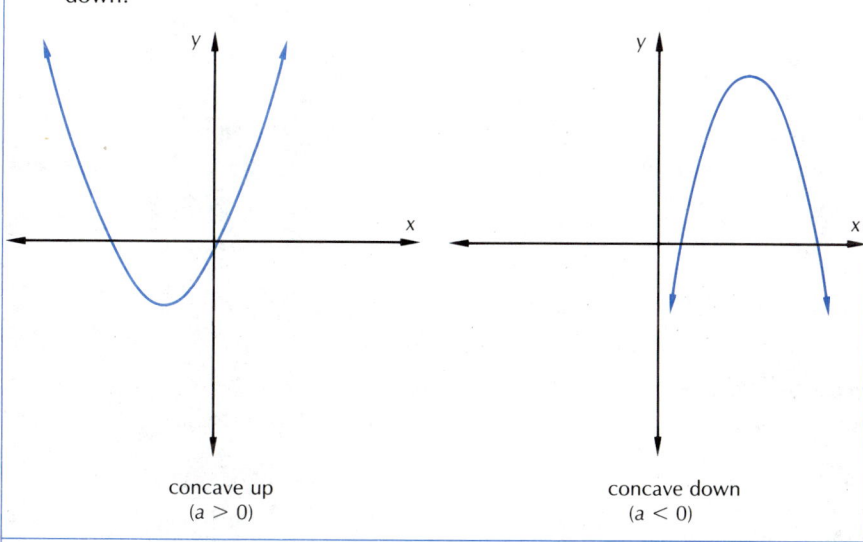

do not define functions. Their graphs are parabolas opening either to the right or to the left.

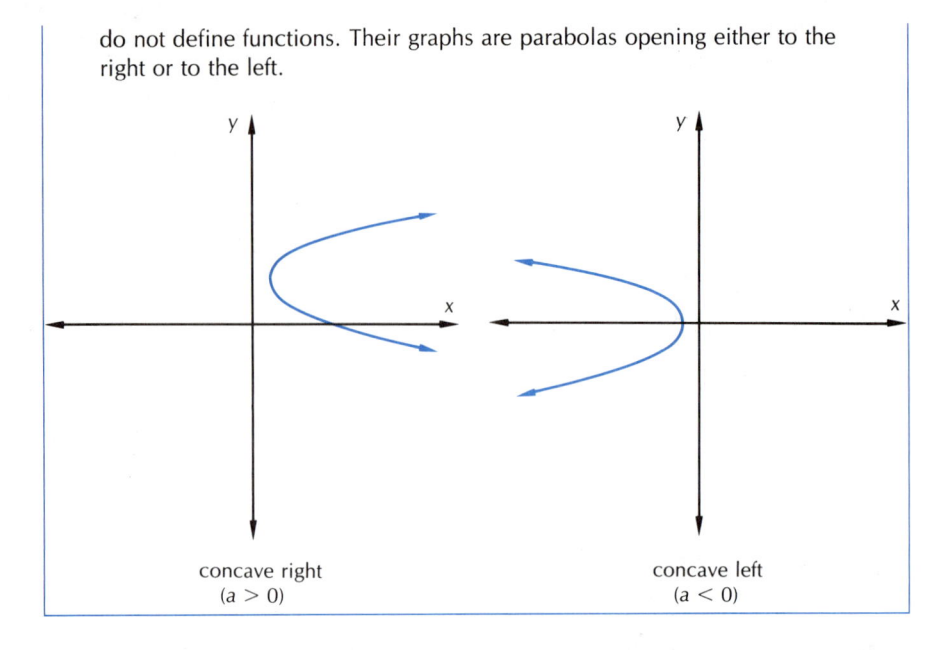

concave right
(a > 0)

concave left
(a < 0)

When graphing a quadratic equation containing two variables, it is helpful to follow these steps:

1. Determine whether the parabola opens up, down, right or left.
2. Form a table of values, being sure to obtain the x and y-intercepts, if any.
3. Plot the corresponding points and sketch in the parabola.

Exercise 9.1

Find the x and y-intercepts of each of the following.
1. $y = x^2 - 4$
2. $y = -x^2 - 8x$
3. $y = 2x^2 - 4x + 3$
4. $x = y^2 + 2y$
5. $y = -2x^2 + 12x - 10$
6. $y = -x^2 + 5x - 6$
7. $x = y^2 - 4y + 3$
8. $y = 3x^2 + 5x + 1$

Determine whether these parabolas open up, down, right, or left.
9. $y = x^2 - x - 6$
10. $x = y^2 + 1$
11. $x = y^2 - 10y$
12. $y = -3x^2 + 6x - 3$
13. $x = -y^2 + 5y - 6$
14. $y = x^2 + 4x + 4$
15. $y = 5 - x - x^2$
16. $y^2 + x = 0$

Graph the following.
17. $y = x^2 - 3$
18. $x = y^2$
19. $y = x^2 + 3$
20. $y = (x - 3)^2$

21. $x = -2y^2 + 4$
22. $y = x^2 + x - 6$
23. $y = -\frac{1}{2}x^2$
24. $y = -x^2 + 1$
25. $x = y^2 + 3y - 4$
26. $x = -2y^2 + 2y + 4$
27. $y = x^2 + 4x - 4$
28. $y = -x^2 + 5x$
29. $x = -y^2 - 2$
30. $y = -x^2 - 3$
31. $x = 3y^2 + 6y - 2$
32. $y = x^2 + 6x$
33. $y = \frac{1}{2}x^2 - 1$
34. $x = y^2 - 4y$

THE QUADRATIC FUNCTION

9.2

Any equation that can be written in the form

$$f(x) = ax^2 + bx + c$$

where $a, b, c \in R$ and $a \neq 0$, is said to define a *quadratic function*. As discussed in section 9.1, the graph of a quadratic function is a parabola that is either concave up or concave down. Concavity is determined by the sign of a (the coefficient of x^2). A parabola is concave up when $a > 0$, and concave down when $a < 0$.

A quadratic function always defines a parabola having a y-intercept, but it may or may not have x-intercepts. Parabolas having no x-intercepts, as shown in Figure 9.6, lie entirely above or entirely below the x-axis, and their discriminants are negative ($b^2 - 4ac < 0$).

concave up ($a > 0$)
$b^2 - 4ac < 0$

concave down ($a < 0$)
$b^2 - 4ac < 0$

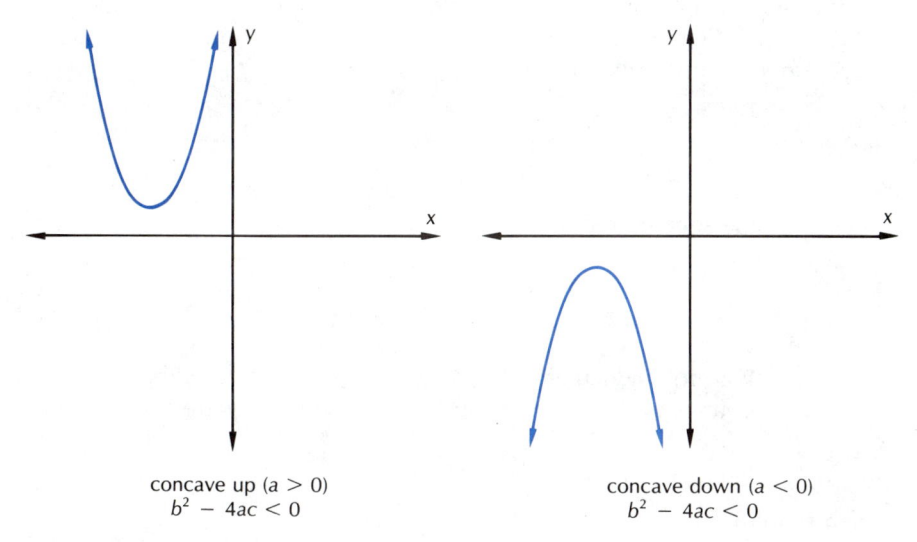

FIGURE 9.6

THE CONIC SECTIONS

The highest or lowest point of a parabola is called its vertex. The vertex is said to be the minimum point when the parabola is concave up (a > 0), and the maximum point when the parabola is concave down (a < 0).

The axis of a parabola defined by a quadratic function is a vertical line passing through the vertex. Every parabola is symmetric about its axis, meaning that its two halves are mirror images.

Table 9.2 provides a visual summary of the foregoing information.

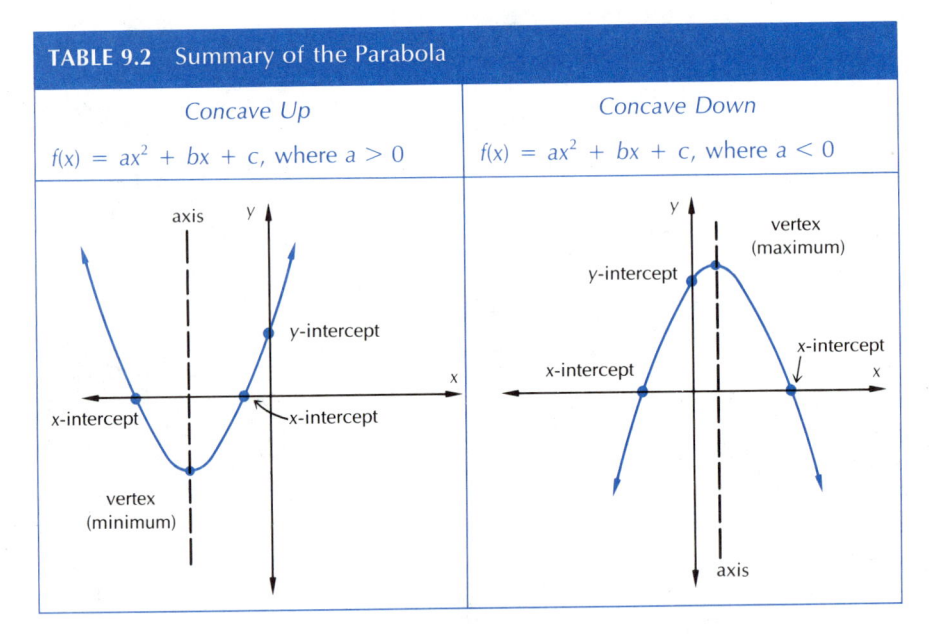

TABLE 9.2 Summary of the Parabola

Concave Up	Concave Down
$f(x) = ax^2 + bx + c$, where $a > 0$	$f(x) = ax^2 + bx + c$, where $a < 0$

A useful point on a parabola is the vertex, which can be found by the procedure of completing the square, first discussed in section 6.2. To illustrate, we consider the quadratic function

$$y = 3x^2 - 12x + 9 \qquad (1)$$

and rewrite it as

$$y = 3(x^2 - 4x \quad\quad) + 9$$

Completing the square produces

$$y = 3(x^2 - 4x + 4) + 9 - (3)(4)$$

$$\left(\frac{1}{2} \cdot -4\right)^2$$

where 3 · 4 was both added and subtracted on the right side of the equation, giving a net change of 0. The trinomial is now a perfect square and is rewritten as

$$y = 3(x - 2)^2 + 9 - (3)(4)$$

which simplifies to

$$y = 3(x - 2)^2 - 3$$

When x = 2, the binomial, x − 2, takes on its least value of 0; hence, as shown below, y has a minimum value of −3.

$$y = 3(x - 2)^2 - 3$$
$$y = 3(2 - 2)^2 - 3$$
$$= (3)(0) - 3$$
$$= -3$$

Thus, the vertex is (2, −3).

To graph the quadratic function defined by equation (1), we also determine its x and y-intercepts. Letting x = 0 and solving for y, we have

$$y = 3x^2 - 12x + 9$$
$$y = (3)(0^2) - (12)(0) + 9 = 9$$

Thus, the y-intercept is 9. Next, we let y = 0 and solve for x by factoring.

$$y = 3x^2 - 12x + 9$$
$$0 = 3x^2 - 12x + 9$$
$$0 = x^2 - 4x + 3 \quad \text{[dividing both sides by 3]}$$
$$0 = (x - 3)(x - 1)$$
$$x = 3 \text{ or } x = 1$$

This gives x-intercepts of 3 and 1. In Figure 9.7, we plot the vertex and intercepts, which are sufficient to sketch the graph.

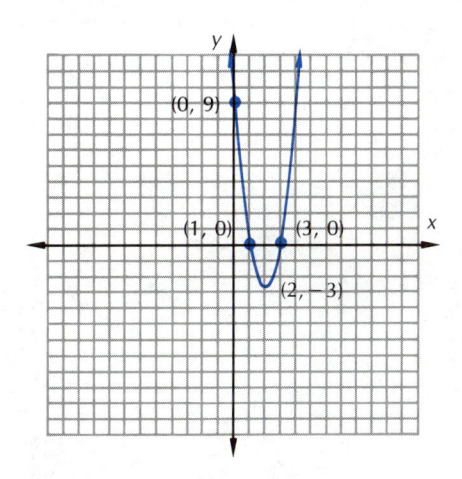

FIGURE 9.7

To simplify the procedure for finding the vertex, we derive a formula by completing the square on the general equation for a quadratic function, $y = ax^2 + bx + c$. Rewriting, we have

$$y = a\left(x^2 + \frac{b}{a}x \quad \right) + c$$

THE CONIC SECTIONS

Completing the square produces
$$y = a\left(x^2 + \frac{b}{a}x + \frac{b^2}{4a^2}\right) + c - (a)\left(\frac{b^2}{4a^2}\right)$$

where $a \cdot \frac{b^2}{4a^2}$ was both added and subtracted, giving a net change of 0. The trinomial is a perfect square and is rewritten as

$$y = a\left(x + \frac{b}{2a}\right)^2 + c - (a)\left(\frac{b^2}{4a^2}\right)$$

which simplifies to

$$y = a\left(x + \frac{b}{2a}\right)^2 + \frac{4ac - b^2}{4a}$$

If $a > 0$, the least value of y is $\frac{4ac - b^2}{4a}$. This occurs when $a\left(x + \frac{b}{2a}\right)^2 = 0$ or when $x = \frac{-b}{2a}$. However, if $a < 0$, the greatest value of y is $\frac{4ac - b^2}{4a}$, and this occurs when $x = \frac{-b}{2a}$. Thus, the graph of $y = ax^2 + bx + c$ has a maximum or minimum point (vertex) at $x = \frac{-b}{2a}$. The corresponding y value is $\frac{4ac - b^2}{4a}$. However, it is easier to simply calculate $f\left(\frac{-b}{2a}\right)$.

The Vertex of a Parabola

The vertex of a parabola defined by
$$f(x) = ax^2 + bx + c$$
is the point described by $\left(\frac{-b}{2a}, f\left(\frac{-b}{2a}\right)\right)$.

(1) If $a > 0$, the vertex is a minimum point. The minimum value is $f\left(\frac{-b}{2a}\right)$.

(2) If $a < 0$, the vertex is a maximum point. The maximum value is $f\left(\frac{-b}{2a}\right)$.

EXAMPLE 1 Find the vertex and minimum value of $f(x) = x^2 - 2x - 8$.

Solution: The x value of the vertex is given by the formula

$x = \dfrac{-b}{2a}$. Since $a = 1$ and $b = -2$, we have

$$x = \dfrac{-b}{2a}$$
$$= \dfrac{2}{2 \cdot 1}$$
$$= 1$$

The y value of the vertex is $f(1)$. Since $f(x) = x^2 - 2x - 8$, we have

$$f(1) = 1^2 - 2(1) - 8 = -9$$

Thus, the vertex is $(1, -9)$. Since $a = 1$, the parabola is concave up, producing a minimum value of -9.

The graph of a quadratic function (a parabola) can usually be sketched very quickly by finding its vertex, axis, and y-intercept. If the vertex and y-intercept happen to coincide, simply find another point.

Sketch the graph of $f(x) = 2x^2 + 8x + 5$

EXAMPLE 2

Solution: Find the vertex where $a = 2$ and $b = 8$.

$$x = \dfrac{-b}{2a} = \dfrac{-8}{2 \cdot 2} = -2$$
$$y = f(-2) = 2(-2)^2 + 8(-2) + 5 = -3$$

The vertex is $(-2, -3)$.

Find the y-intercept by letting $x = 0$.
$$f(0) = 2(0)^2 + 8(0) + 5 = 5$$
The y-intercept is 5.

Plot the vertex and y-intercept, locate the axis, and use symmetry to plot a companion point. Then, sketch the parabola as shown in Figure 9.8.

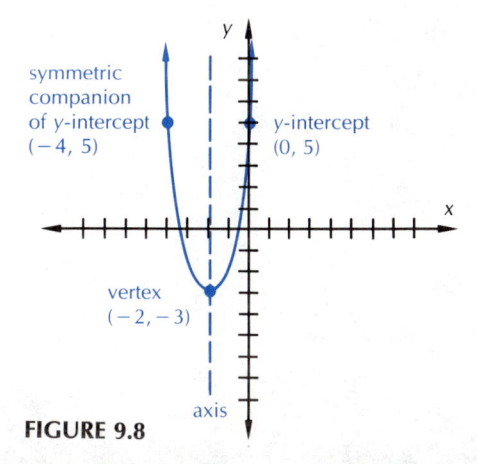

FIGURE 9.8

Many applied problems are concerned with finding a maximum or minimum value of some quantity. If that quantity can be represented by a quadratic function, the maximum or minimum value can be found using the vertex.

EXAMPLE 3 A farmer wants to fence a rectangular plot of land adjacent to a river. He decides to use 100 yards of fencing. Find the maximum area he can enclose and the dimensions of the plot.

Solution: We sketch a diagram in Figure 9.9 and note that he does not need to fence along the river.

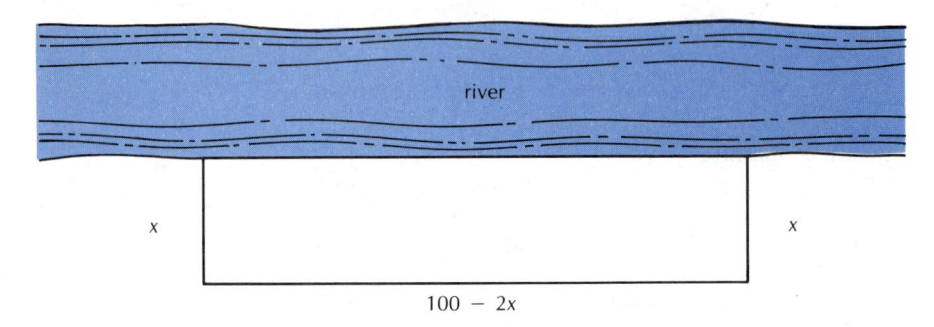

FIGURE 9.9

If x represents the width, then $100 - 2x$ gives the length. The area (A) is the product of the length and width. Thus,

$A = (100 - 2x)x$

Rewriting shows that we have a quadratic function.

$A = 100x - 2x^2$
$ = -2x^2 + 100x$

Since $a = -2$, the graph is a concave-down parabola having a maximum value. So, we find the vertex, using $a = -2$ and $b = 100$.

$x = \dfrac{-b}{2a} = \dfrac{-100}{2(-2)} = 25$

$A = -2x^2 + 100x$
$ = -2(25)^2 + 100(25)$
$ = -1{,}250 + 2{,}500$
$ = 1{,}250$

The vertex is (25, 1,250).

The maximum value is 1,250 square yards, and occurs when the width is 25 yards. The length, represented by $100 - 2x$, is 50 yards.

Domain of a Quadratic Function

Since a parabola becomes wide beyond all bounds, the domain of every quadratic function is the set of real numbers,

$$\{x \mid x \in R\}$$

However, as illustrated in Figure 9.10, the range is dependent on the maximum or minimum value. If $a < 0$, the function has a maximum and the range consists of all y values below or equal to the maximum,

$$\left\{y \mid y \leq f\left(\frac{-b}{2a}\right)\right\}$$

On the other hand, if $a > 0$, the function has a minimum and the range consists of all y values above or equal to the minimum,

$$\left\{y \mid y \geq f\left(\frac{-b}{2a}\right)\right\}$$

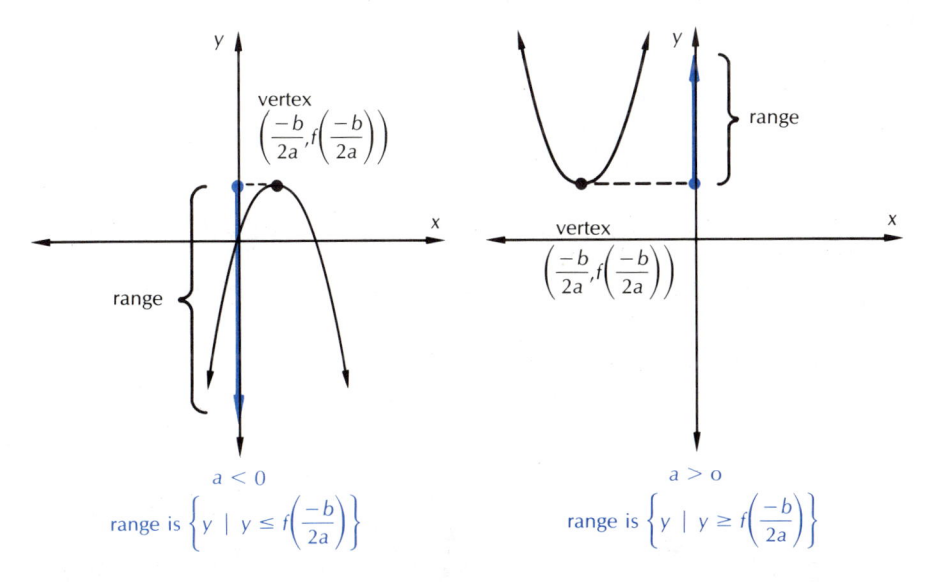

FIGURE 9.10

Find the domain and range of $f(x) = -x^2 + 4x - 5$.

EXAMPLE 4

Solution: The domain is $\{x \mid x \in R\}$.

To find the range we obtain the vertex:

$$x = \frac{-b}{2a} = \frac{-4}{2(-1)} = 2$$

$$y = f(2) = -(2)^2 + 4(2) - 5 = -1$$

The vertex is $(2, -1)$.

Since $a = -1$, the parabola is concave down and the maximum value is -1. Thus, the range is $\{y \mid y \leq -1\}$. A sketch is provided in Figure 9.11.

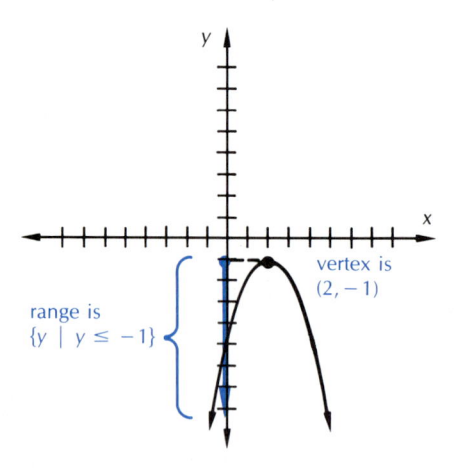

FIGURE 9.11

Exercise 9.2

Determine whether each parabola is concave up or down and whether it has a maximum or minimum point.

1. $f(x) = 3x^2 + x - 2$
2. $f(x) = -x^2 + 2$
3. $f(x) = -3 - x + x^2$
4. $f(x) = 5 + x - 3x^2$

Find the x and y-intercepts of each parabola.

5. $f(x) = x^2 - 6x + 9$
6. $f(x) = x^2 - 3x$
7. $f(x) = -x^2 - \dfrac{3}{2}x$
8. $f(x) = x^2 - 3x - 10$
9. $f(x) = -x^2 + 5x - 4$
10. $f(x) = -x^2 - 4$

Find the vertex and maximum or minimum value of the parabola.

11. $f(x) = x^2 - 4x + 2$
12. $f(x) = -x^2 + x - 6$
13. $f(x) = 2x^2 + 6x - 3$
14. $f(x) = x^2 + 6$
15. $f(x) = -x^2 + x$
16. $f(x) = 6 + x - x^2$
17. $f(x) = -2x^2 - 6x + 3$
18. $f(x) = -2x^2 - 3$
19. $f(x) = x^2 - 5x + 4$
20. $f(x) = x^2 + x - 12$
21. $f(x) = 5x^2 + 3x - 2$
22. $f(x) = 3x^2 - 3x + 2$

Graph each parabola. Make use of its vertex, axis and y-intercept.

23. $f(x) = 2x^2 - 4$
24. $f(x) = 1 - x^2$
25. $f(x) = x^2 + 4x - 2$
26. $f(x) = -x^2 + 5x$
27. $f(x) = 2x^2 - 4x - 1$
28. $f(x) = -x^2 + 8x - 10$

Find the domain and range of each quadratic function.

29. $f(x) = -4x^2$
30. $f(x) = 4x^2 + 4x$
31. $f(x) = 5 - 2x^2 + 4x$
32. $f(x) = -x^2 - x$
33. $f(x) = x^2 + 4x - 1$
34. $f(x) = 3x^2 - 4x - 3$

Represent the following applied problems with a quadratic function and solve.

35. Find two numbers whose sum is 16 and whose product is as large as possible.
36. Find two numbers whose sum is 30 and whose product is as large as possible.
37. A projectile is propelled upward so its height h in feet after x seconds is given by $h = 3200x - 16x^2$. Find the maximum height that it attains.
38. An arrow is propelled upward so its height h in feet after t seconds is given by $h = 128t - 16t^2$. Find the maximum height that it attains and the number of seconds it takes to attain that height.
39. A farmer has 80 feet of fencing. He wants to put a fence around a rectangular plot of land adjacent to a river. Find the maximum area he can enclose. (Note: the side next to the river will not be fenced.)
40. Find the largest possible product that can be formed by two numbers whose sum is 38.

THE CIRCLE

9.3

In previous sections we discussed two types of second degree equations whose graphs are parabolas. In this section we examine second degree equations whose graphs are circles. To do this, we must first use the Pythagorean Theorem to develop a formula for finding the distance between two points in the rectangular coordinate system.

The Pythagorean Theorem

In a right triangle, the square of the length of the longest side (hypotenuse) is equal to the sum of the squares of the lengths of the remaining two sides (legs).

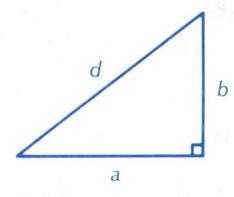

Using symbols,

$$d^2 = a^2 + b^2$$

EXAMPLE 1 Use the Pythagorean Theorem to find the distance between (5, 1) and $(-4, -6)$.

Solution: As shown in Figure 9.12, we graph the two points and use them to construct a right triangle. Since $a = |-4 - 5| = 9$, and $b = |-6 - 1| = 7$, we use the Pythagorean Theorem to find the distance (d) between (5, 1) and $(-4, -6)$.

$$d^2 = a^2 + b^2$$
$$d^2 = 9^2 + 7^2$$
$$d = \sqrt{81 + 49}$$
$$= \sqrt{130}$$

The distance between the given points is $\sqrt{130}$.

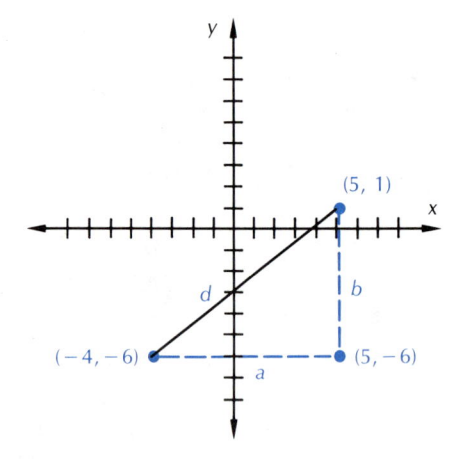

FIGURE 9.12

Generalizing from Example 1, we derive a formula for finding the distance between any two points, (x_1, y_1) and (x_2, y_2). These two points are shown in Figure 9.13. We construct a right triangle and note that $a = |x_2 - x_1|$ and $b = |y_2 - y_1|$. Using the Pythagorean Theorem, we have

$$d^2 = a^2 + b^2$$
$$= |x_2 - x_1|^2 + |y_2 - y_1|^2$$

Since $|x_2 - x_1|^2 = (x_2 - x_1)^2$ and $|y_2 - y_1|^2 = (y_2 - y_1)^2$, we write
$$d^2 = (x_2 - x_1)^2 + (y_2 - y_1)^2$$

Taking the square root of both sides produces the final result, called the distance formula.

$$d = \sqrt{(x_2 - x_1)^2 + (y_2 - y_1)^2}$$

THE CIRCLE

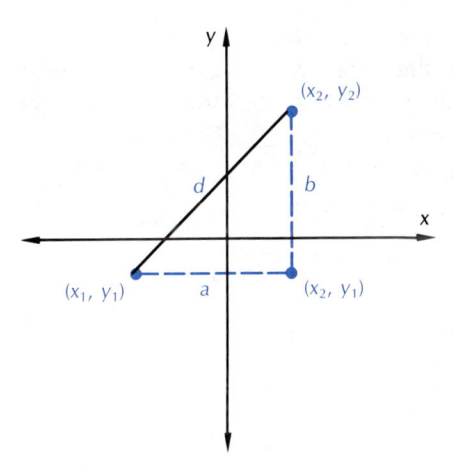

FIGURE 9.13

The Distance Formula

The distance (d) between any two points (x_1, y_1) and (x_2, y_2) is given by the equation

$$d = \sqrt{(x_2 - x_1)^2 + (y_2 - y_1)^2}$$

Since $(x_2 - x_1)^2 = (x_1 - x_2)^2$ and $(y_2 - y_1)^2 = (y_1 - y_2)^2$, order is not important in the distance formula. It merely states to square a difference of the abscissas, square a difference of the ordinates, add the squares and take the square root.

Find the distance between $(-2, -7)$ and $(-3, 1)$.

EXAMPLE 2

Solution: Use the distance formula:
$$d = \sqrt{(x_2 - x_1)^2 + (y_2 - y_1)^2}$$
$$= \sqrt{(-2 - (-3))^2 + (-7 - 1)^2}$$
$$= \sqrt{1^2 + (-8)^2}$$
$$= \sqrt{65}$$

The distance formula can be used to derive the equation of a circle.

Definition of a Circle

A circle is the set of all points in a plane that lie a given distance from a fixed point. The fixed point is called the center and the given distance is called the radius.

THE CONIC SECTIONS

Given a circle, as shown in Figure 9.14, having a center at (h, k) and a radius r, then for any point (x, y) on the circle, the distance from (x, y) to the center (h, k) must be equal to the radius r. Thus, by the distance formula,

$$r = \sqrt{(x - h)^2 + (y - k)^2}$$

Squaring both sides gives the equation of the circle.

$$r^2 = (x - h)^2 + (y - k)^2$$

The Equation of a Circle

The standard form of the equation of a circle with center (h, k) and radius r is

$$(x - h)^2 + (y - k)^2 = r^2$$

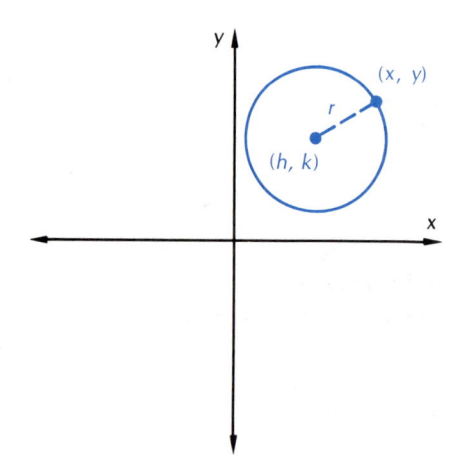

FIGURE 9.14

EXAMPLE 3   Write the equation of a circle with radius 4 and center at the origin, $(0, 0)$.

Solution: Since $r = 4$ and $(h, k) = (0, 0)$, we use the standard form of the circle and replace 4 for r, 0 for h, and 0 for k.

$$(x - h)^2 + (y - k)^2 = r^2$$
$$(x - 0)^2 + (y - 0)^2 = 4^2$$
$$x^2 + y^2 = 16$$

The graph is shown in Figure 9.15.

THE CIRCLE 377

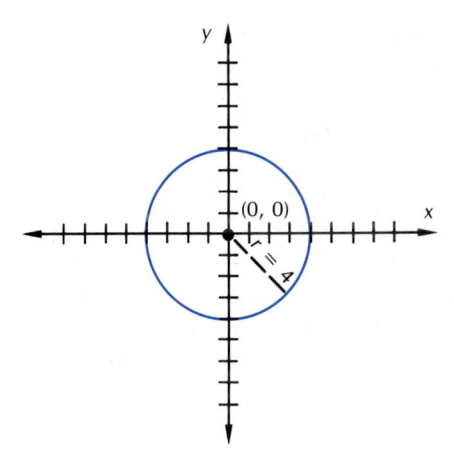

FIGURE 9.15

This example suggests that the equation of a circle with center at the origin and radius r is

$$x^2 + y^2 = r^2$$

Write the equation of a circle with center at $(-3, 2)$ and radius 3.   **EXAMPLE 4**

Solution: We use the standard form for the circle, letting $h = -3$, $k = 2$, and $r = 3$.

$(x - h)^2 + (y - k)^2 = r^2$
$(x + 3)^2 + (y - 2)^2 = 9$

Multiplying and collecting like terms gives

$x^2 + 6x + 9 + y^2 - 4y + 4 = 9$
$x^2 + y^2 + 6x - 4y + 4 = 0$

The graph is shown in Figure 9.16.

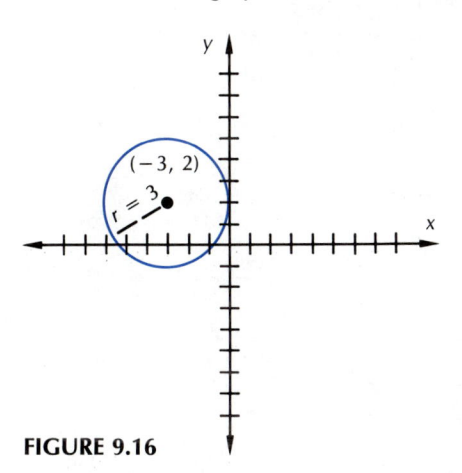

FIGURE 9.16

THE CONIC SECTIONS

As illustrated by the final equation in Example 4, the equation of a circle may be expanded so that it is not in standard form. When this occurs, the numerical coefficients of x^2 and y^2 are equal. Given such an equation, the center and radius are found by completing the square on x and on y.

EXAMPLE 5 Graph $x^2 + y^2 + 6x - 2y - 15 = 0$

Solution: Since the numerical coefficients of x^2 and y^2 are equal, the graph will likely be a circle. To find its center and radius, rewrite the equation so that the variable terms are on one side and the constant is on the other. Then, complete the square on x and on y.

$$x^2 + y^2 + 6x - 2y - 15 = 0$$
$$x^2 + y^2 + 6x - 2y = 15$$

Group the x terms together and the y terms together. Then, complete the squares, being sure to add the numbers to both sides of the equation.

$$(x^2 + 6x \quad) + (y^2 - 2y \quad) = 15$$
$$(x^2 + 6x + 9) + (y^2 - 2y + 1) = 15 + 1 + 9$$

$$\left(\frac{1}{2} \cdot 6\right)^2 \quad \left(\frac{1}{2} \cdot -2\right)^2$$

Write each group as the square of a binomial.

$$(x + 3)^2 + (y - 1)^2 = 25$$

This equation shows that the graph is a circle with center at $(-3, 1)$ and radius 5. The graph is shown in Figure 9.17.

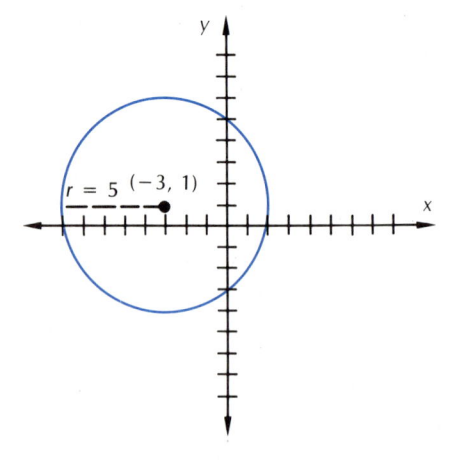

FIGURE 9.17

THE CIRCLE 379

Find the distance between the given points. **Exercise 9.3**
1. (0, 0) (6, −2)
2. (−2, 3) (2, −6)
3. (6, −3) (2, −9)
4. (3, −1) (3, −5)
5. (3, 2) (−2, −8)
6. (3, 6) (−3, −3)
7. (x, x − y) (y, x − y)
8. (a + b, b) (a + b, a)

Find the perimeter of the triangle formed by joining the three given points.
9. (−2, 1), (2, 5), and (1, −2)
10. (1, 3), (1, −2), and (2, −4)
11. (−2, −1), (1, 6), and (8, 1)
12. (4, −1), (−2, 3), and (1, −2)

Give the center and radius of each circle and draw its graph.
13. $x^2 + y^2 = 9$
14. $(x − 1)^2 + y^2 = 16$
15. $(x − 2)^2 + (y + 2)^2 = 9$
16. $(x − 3)^2 + (y − 1)^2 = 16$
17. $(x − 1)^2 + (y − 5)^2 − 10 = 0$
18. $(x + 3)^2 + y^2 − 5 = 0$

Write an equation in standard form for the circle with the given center and radius.
19. center (2, −4), radius 16
20. center (−3, 0), radius $\sqrt{6}$
21. center (−1, 4), radius 2
22. center (4, 4), radius 4

Find the center and radius of each circle by completing the square.
23. $x^2 + y^2 + 2x − 8y + 1 = 0$
24. $x^2 + y^2 + 6x − 8y + 21 = 0$
25. $x^2 + y^2 − 10x + 10y + 22 = 0$
26. $x^2 + y^2 + 8x + 4y − 40 = 0$
27. $x^2 − 5x + y^2 + 2y = 12$
28. $x^2 − x + y^2 − 8y = 5$

Miscellaneous.
29. Determine if the points (2, −4), (6, −8), and (10, 4) form a right triangle.
30. The point (8, 3) is on a circle with center at (5, −1). Find the diameter.
31. Find x if the distance from (x, 5) to (3, 4) is $\sqrt{2}$.
32. Find the equation of a circle with center at (0, 1) and y-intercepts (0, 5) and (0, −3).
33. Find the equation of a circle with center at (−1, 1) if the point (3, 3) is on the circle.

THE ELLIPSE

9.4

An ellipse is the set of all points in a plane the sum of whose distances from two fixed points is a constant. The two fixed points are called focal points or foci. Figure 9.18 shows an ellipse with foci F_1 and F_2. If P and Q name any two points on the ellipse and if c is the constant, then

$$F_1P + F_2P = c$$

and similarly,

$$F_1Q + F_2Q = c$$

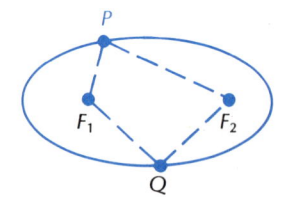

FIGURE 9.18

Ellipses have many applications in astrophysics. The planets of our solar system have elliptical orbits where the sun is one of the foci. Satellites also travel in elliptical orbits. As illustrated in Figure 9.19, you can easily draw an ellipse by attaching a piece of string to two tacks stuck in a piece of cardboard. The tacks act as the foci and the length of the string represents the constant. Take a pencil, pull the string taut and swing it around to form the ellipse.

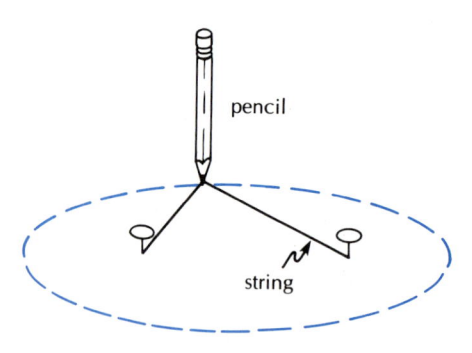

FIGURE 9.19

An ellipse is said to be in standard position when its foci are on either the x or y-axis and are equidistant from the origin. Figure 9.20 shows an ellipse in standard position whose foci are at $(-c, 0)$ and $(c, 0)$, having x-intercepts of $-a$ and a, and y-intercepts of b and $-b$.

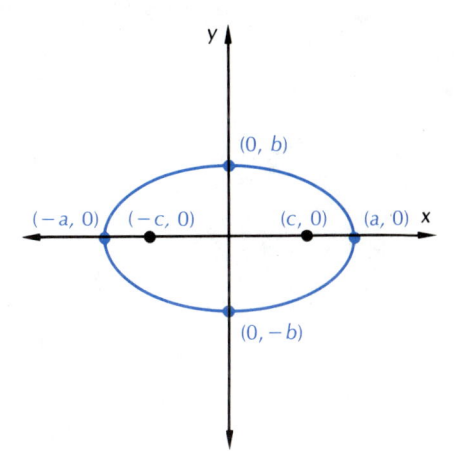

FIGURE 9.20

The distance formula can be used to derive the following equation of an ellipse in standard position.

Equation of an Ellipse in Standard Position

The equation of an ellipse in standard position with x-intercepts of a and $-a$ and with y-intercepts of b and $-b$ is

$$\frac{x^2}{a^2} + \frac{y^2}{b^2} = 1, \text{ where } a \neq b$$

(Note: if $a = b$, the equation would describe a circle.)

This equation is said to be the standard form of the equation of an ellipse.

Find the intercepts and sketch the graph of the ellipse whose equation is

$$\frac{x^2}{25} + \frac{y^2}{9} = 1$$

EXAMPLE 1

Solution: Since $a^2 = 25$, the x-intercepts are 5 and -5. And, because $b^2 = 9$, the y-intercepts are 3 and -3. As shown in Figure 9.21, we plot the intercepts and sketch the graph.

THE CONIC SECTIONS

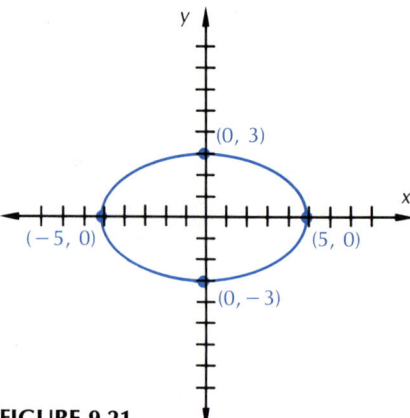

FIGURE 9.21

EXAMPLE 2 Graph $\dfrac{x^2}{4} + \dfrac{y^2}{36} = 1$

Solution: The equation defines an ellipse where $a^2 = 4$ and $b^2 = 36$. Thus, the x-intercepts are 2 and -2 while the y-intercepts are 6 and -6. In Figure 9.22, we plot the intercepts and sketch the ellipse.

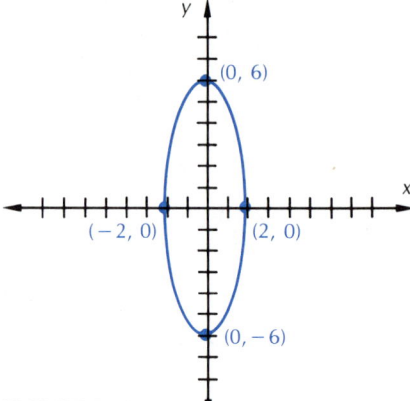

FIGURE 9.22

A variation of the equation for an ellipse in standard position can be obtained, as follows, by multiplying both sides of the equation by a common denominator, such as a^2b^2.

$$\frac{x^2}{a^2} + \frac{y^2}{b^2} = 1$$

$$a^2b^2 \cdot \left[\frac{x^2}{a^2} + \frac{y^2}{b^2}\right] = 1 \cdot a^2b^2$$

$$b^2x^2 + a^2y^2 = a^2b^2$$

THE ELLIPSE

Or, letting $A = b^2$, $B = a^2$, and $C = a^2b^2$, we write the equation more simply as:

$$Ax^2 + By^2 = C$$

where A, B, and C have the same sign and $A \neq B$.

Each of these equations is a variation of the standard form describing an ellipse. Note in each case that A, B, and C have the same sign and $A \neq B$.

EXAMPLE 3

(a) $4x^2 + y^2 = 4$; $A = 4$, $B = 1$, $C = 4$
(b) $9x^2 + 16y^2 = 144$; $A = 9$, $B = 16$, $C = 144$
(c) $-16x^2 - 4y^2 = -16$; $A = -16$, $B = -4$, $C = -16$

To graph a variation of the standard form describing an ellipse, we convert the equation back to standard form by dividing both sides by the constant, C.

Graph $x^2 + 4y^2 = 64$

EXAMPLE 4

Solution: Divide both sides of the equation by 64.

$$x^2 + 4y^2 = 64$$

$$\frac{x^2}{64} + \frac{4y^2}{64} = \frac{64}{64}$$

$$\frac{x^2}{64} + \frac{y^2}{16} = 1$$

Since $a^2 = 64$ and $b^2 = 16$, the x-intercepts are ± 8 and the y-intercepts are ± 4. The ellipse is sketched in Figure 9.23.

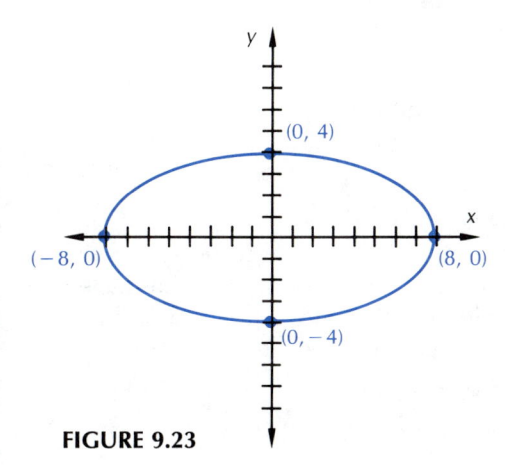

FIGURE 9.23

Find the x and y-intercepts of each ellipse.

Exercise 9.4

1. $x^2 + 4y^2 = 4$
2. $9x^2 + y^2 = 36$
3. $4x^2 + y^2 = 100$
4. $16x^2 + y^2 = 16$

5. $\dfrac{x^2}{9} + \dfrac{y^2}{25} = 1$ 6. $\dfrac{x^2}{4} + y^2 = 1$

7. $\dfrac{x^2}{20} + \dfrac{y^2}{4} = 1$ 8. $3x^2 + y^2 = 5$

Graph the following ellipses.

9. $\dfrac{x^2}{9} + \dfrac{y^2}{4} = 1$ 10. $\dfrac{x^2}{4} + \dfrac{y^2}{25} = 1$

11. $\dfrac{x^2}{16} + y^2 = 1$ 12. $\dfrac{x^2}{49} + \dfrac{y^2}{81} = 1$

13. $8x^2 + 4y^2 = 32$ 14. $4x^2 + y^2 - 4 = 0$
15. $2x^2 + 10y^2 - 10 = 0$ 16. $x^2 + 2y^2 - 8 = 0$
17. $5x^2 + 9y^2 = 45$ 18. $6x^2 + y^2 = 6$

19. $10x^2 + 2y^2 = 20$ 20. $\dfrac{x^2}{-16} + \dfrac{y^2}{-9} = -1$

21. $\dfrac{x^2}{\tfrac{25}{16}} + \dfrac{y^2}{\tfrac{9}{4}} = 1$ 22. $\dfrac{x^2}{\tfrac{49}{4}} + \dfrac{y^2}{25} = 1$

Graph the following on the same set of axes.

23. $x^2 + y^2 = 4$; $x^2 + 4y^2 = 4$
24. $x^2 + y^2 = 4$; $4x^2 + y^2 = 4$
25. $(x - 1)^2 + y^2 = 16$; $4x^2 + y^2 = 16$
26. $x^2 + (y - 2)^2 = 9$; $4x^2 + 9y^2 = 36$
27. $(x + 2)^2 + y^2 = 16$; $\dfrac{x^2}{9} + \dfrac{y^2}{4} = 1$
28. $9x^2 + 16y^2 = 144$; $16x^2 + 9y^2 = 144$

THE HYPERBOLA

9.5

A *hyperbola* is the set of all points in a plane the difference of whose distances from two fixed points is a constant. The two fixed points are called *foci*. Figure 9.24 shows a hyperbola with foci F_1 and F_2. If P and Q name any two points on the hyperbola and if c represents the constant, then

$$|F_1P - F_2P| = |F_1Q - F_2Q| = c$$

Observe that a hyperbola has two branches opening in opposite directions.

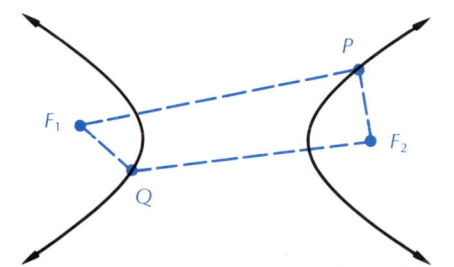

FIGURE 9.24

THE HYPERBOLA

A hyperbola is said to be in standard position when its foci are on either the x-axis or y-axis and are equidistant from the origin. Figure 9.25 shows two hyperbolas in standard position, one having its foci on the x-axis with x-intercepts of a and −a; and the other having its foci on the y-axis with y-intercepts of a and −a. Note that a hyperbola in standard position has only one pair of intercepts, either x-intercepts or y-intercepts, but not both.

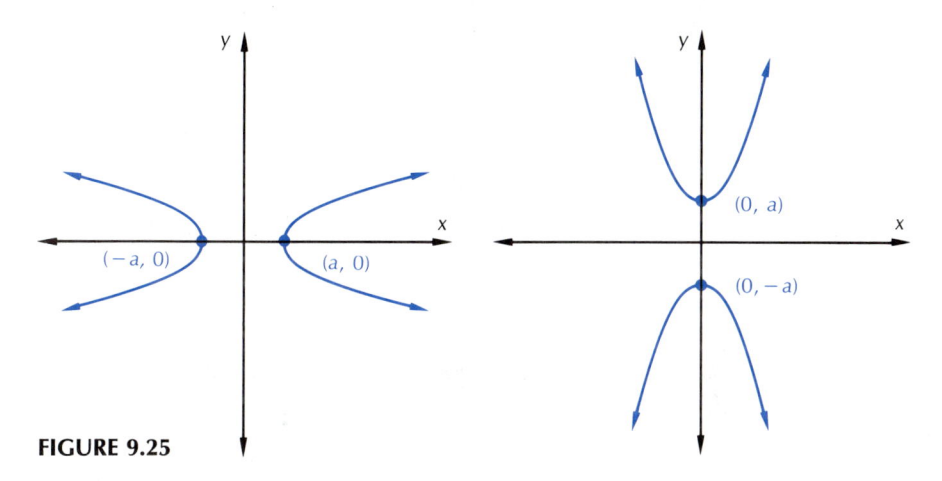

FIGURE 9.25

Using the definition together with the distance formula, it can be shown that hyperbolas in standard position are described by the following types of equations.

Equations of Hyperbolas in Standard Position

(1) The equation of a hyperbola in standard position with foci on the x-axis and x-intercepts of a and −a is

$$\frac{x^2}{a^2} - \frac{y^2}{b^2} = 1$$

(b) The equation of a hyperbola in standard position with foci on the y-axis and y-intercepts of a and −a is

$$\frac{y^2}{a^2} - \frac{x^2}{b^2} = 1$$

These equations are called the standard forms of the equations of a hyperbola.

Graph the hyperbola whose equation is $\frac{x^2}{25} - \frac{y^2}{16} = 1$

EXAMPLE 1

Solution: The equation fits the standard form for a hyperbola having its foci on the x-axis. Since $a^2 = 25$, the x-intercepts are 5 and −5. With

THE CONIC SECTIONS

the aid of a few more points from a table of values, we sketch the graph in Figure 9.26.

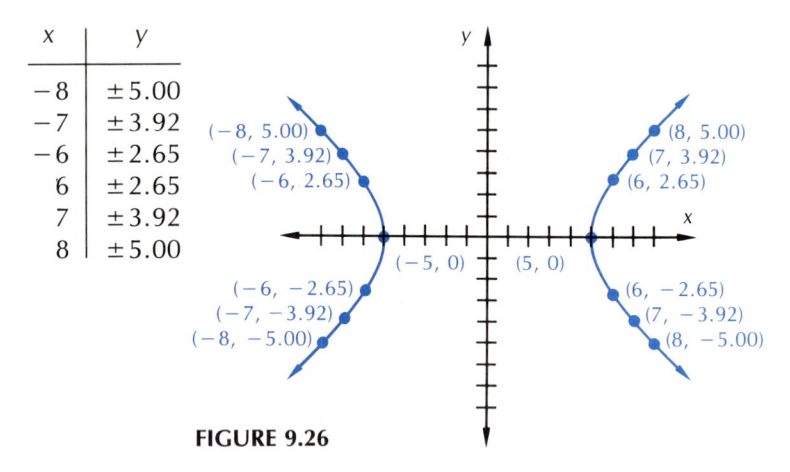

x	y
−8	±5.00
−7	±3.92
−6	±2.65
6	±2.65
7	±3.92
8	±5.00

FIGURE 9.26

A hyperbola in standard position can be sketched quickly without forming a table of values. Given the standard form

$$\frac{x^2}{a^2} - \frac{y^2}{b^2} = 1$$

we simply plot the four points defined by $(a, \pm b)$ and $(-a, \pm b)$ and form a rectangle as shown in Figure 9.27. The hyperbola opens horizontally from its x-intercepts and approaches the lines formed by the diagonals of the rectangle. These lines are called the *asymptotes* of the hyperbola.

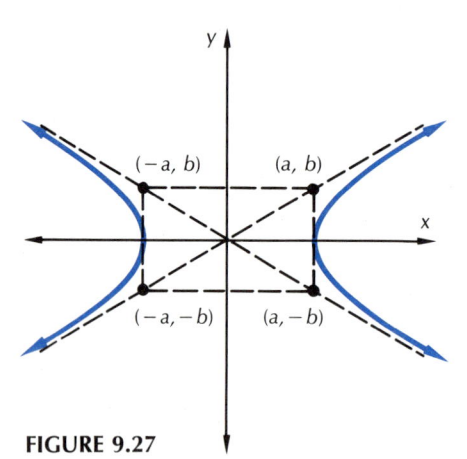

FIGURE 9.27

EXAMPLE 2   Graph $\dfrac{x^2}{9} - \dfrac{y^2}{4} = 1$

Solution: This equation defines a hyperbola having its foci on the x-axis. Since $a^2 = 9$ and $b^2 = 4$, the x-intercepts are ±3 and the vertices of the rectangle are given by $(3, \pm 2)$ and $(-3, \pm 2)$. The diagonals are drawn and the hyperbola is sketched in Figure 9.28.

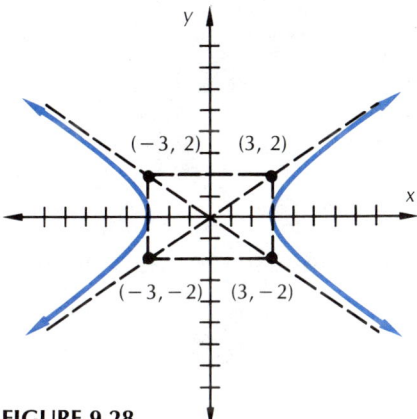

FIGURE 9.28

Similarly, as illustrated by the following example, a hyperbola of the form

$$\frac{y^2}{a^2} - \frac{x^2}{b^2} = 1$$

can be sketched by drawing the diagonals of a rectangle whose vertices are given by $(b, \pm a)$ and $(-b, \pm a)$. A hyperbola of this form has its foci on the y-axis and will open vertically from its y-intercepts of $\pm a$.

Graph $\dfrac{y^2}{9} - \dfrac{x^2}{16} = 1$

EXAMPLE 3

Solution: This equation defines a hyperbola having its foci on the y-axis. Since $a^2 = 9$ and $b^2 = 16$, the y-intercepts are ± 3 and the vertices of the rectangle are given by $(4, \pm 3)$ and $(-4, \pm 3)$. The asymptotes are drawn and the hyperbola is sketched in Figure 9.29.

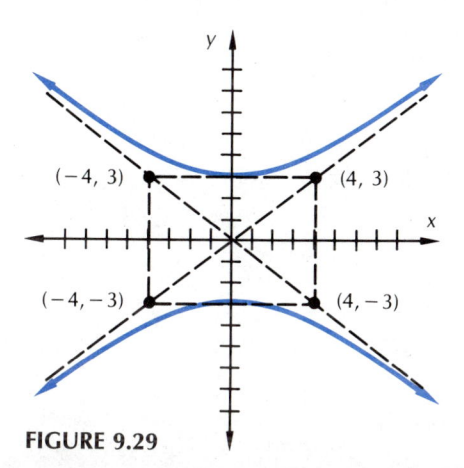

FIGURE 9.29

Variations of the standard forms describing hyperbolas can be formed, as follows, by multiplying both sides of the equations by a common denominator, such as a^2b^2.

$$\frac{x^2}{a^2} - \frac{y^2}{b^2} = 1$$

$$a^2b^2 \cdot \left[\frac{x^2}{a^2} - \frac{y^2}{b^2}\right] = 1 \cdot a^2b^2$$

$$b^2x^2 - a^2y^2 = a^2b^2$$

Or, letting $A = b^2$, $B = a^2$, and $C = a^2b^2$, we write the equation more simply as:

$$Ax^2 - By^2 = C$$

where $A, B, C > 0$.

Similarly, the standard form

$$\frac{y^2}{a^2} - \frac{x^2}{b^2} = 1$$

can be transformed to

$$By^2 - Ax^2 = C$$

where $A, B, C > 0$.

EXAMPLE 4   Each of these equations is a variation of the standard form of a hyperbola.

(a) $25x^2 - 4y^2 = 100$; $A = 25$, $B = 4$, $C = 100$
(b) $9y^2 - 16x^2 = 144$; $B = 9$, $A = 16$, $C = 144$
(c) $x^2 - y^2 = 9$; $A = 1$, $B = 1$, $C = 9$

To graph an equation that is a variation, we convert it to standard form by dividing both sides by the constant, C.

EXAMPLE 5   Graph $25x^2 - 4y^2 = 100$

Solution: Convert to standard form by dividing both sides of the equation by 100.

$$25x^2 - 4y^2 = 100$$

$$\frac{25x^2}{100} - \frac{4y^2}{100} = \frac{100}{100}$$

$$\frac{x^2}{4} - \frac{y^2}{25} = 1$$

THE HYPERBOLA

As seen in Figure 9.30, this equation describes a hyperbola having its foci on the x-axis with x-intercepts of ±2. The vertices of the rectangle are defined by (2, ±5) and (−2, ±5).

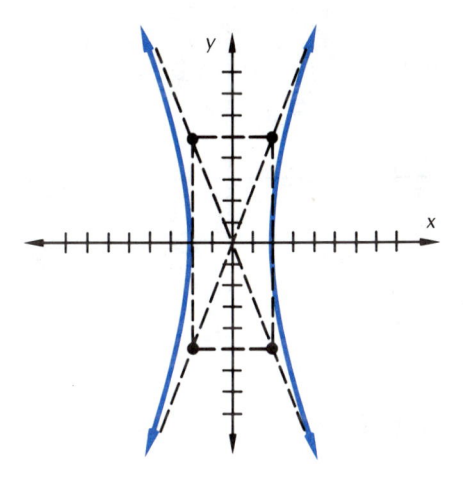

FIGURE 9.30

Next we examine another form of a hyperbola. This form is described by the equation

$$xy = c, \text{ where } c \neq 0$$

To illustrate, we plot points to graph $xy = 4$ in Figure 9.31 and $xy = -6$ in Figure 9.32.

$xy = 4$ or $y = \dfrac{4}{x}$

x	y
0.5	8
1	4
2	2
4	1
8	0.5
−0.5	−8
−1	−4
−2	−2
−4	−1
−8	−0.5

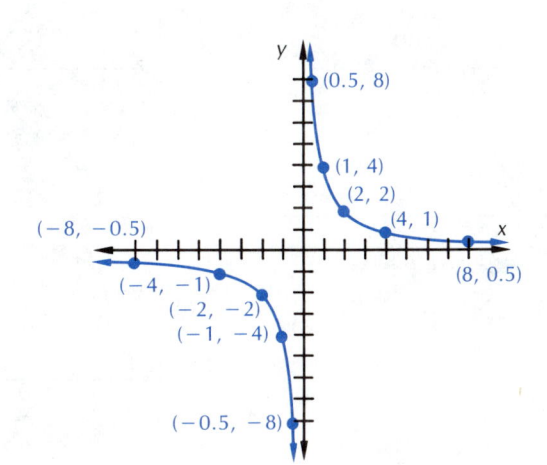

FIGURE 9.31

$xy = -6$ or $y = \dfrac{-6}{x}$

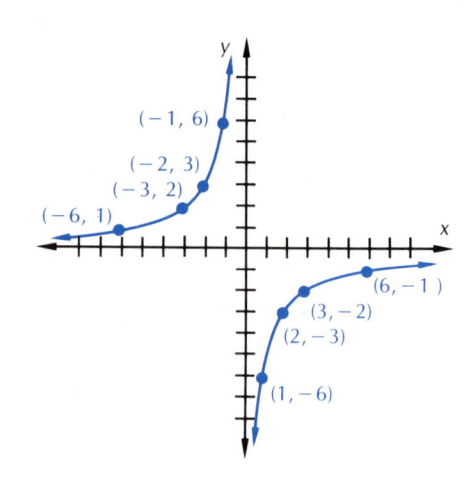

x	y
−0.5	12
−1	6
−2	3
−3	2
−6	1
−12	0.5
0.5	−12
1	−6
2	−3
3	−2
6	−1
12	−0.5

FIGURE 9.32

Observe that hyperbolas having equations of the form $xy = c$ approach the x and y-axes as their asymptotes. Also, if $c > 0$, the hyperbola is contained in the first and third quadrants; and if $c < 0$, the hyperbola is in the second and fourth quadrants.

Summary of Hyperbolas

$\dfrac{x^2}{a^2} - \dfrac{y^2}{b^2} = 1$ or

$Ax^2 - By^2 = C,$
where $A, B, C > 0$

$\dfrac{y^2}{a^2} - \dfrac{x^2}{b^2} = 1$ or

$By^2 - Ax^2 = C,$
where $A, B, C > 0$

$xy = c$ or $y = \dfrac{c}{x}$,

where $c > 0$

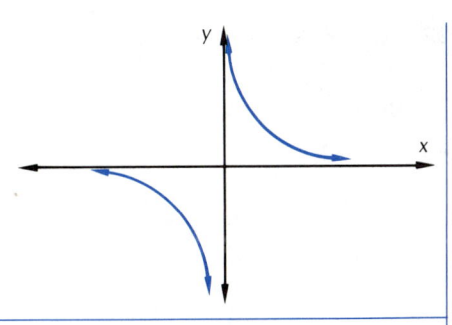

$xy = c$ or $y = \dfrac{c}{x}$,

where $c < 0$

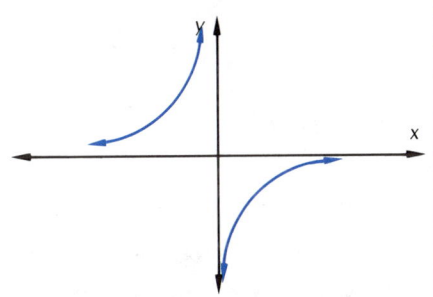

Find any x or y-intercepts that these hyperbolas may have.

Exercise 9.5

1. $x^2 - y^2 = 16$
2. $9x^2 - y^2 = 36$
3. $xy = 10$
4. $16x^2 - y^2 = 16$
5. $y^2 - 4x^2 = 12$
6. $\dfrac{y^2}{4} - \dfrac{x^2}{9} = 1$
7. $3y^2 - x^2 = 6$
8. $x = \dfrac{-6}{y}$

Find the coordinates of the special rectangle associated with each hyperbola.

9. $\dfrac{x^2}{9} - \dfrac{y^2}{4} = 1$
10. $\dfrac{y^2}{16} - \dfrac{x^2}{9} = 1$
11. $4x^2 - 25y^2 = 100$
12. $x^2 - y^2 = 1$
13. $y^2 - 16x^2 = 16$
14. $x^2 - y^2 = 3$

Graph the following hyperbolas.

15. $\dfrac{x^2}{16} - \dfrac{y^2}{9} = 1$
16. $\dfrac{y^2}{4} - \dfrac{x^2}{9} = 1$
17. $\dfrac{1}{4}x^2 - y^2 = 1$
18. $\dfrac{x^2}{25} - \dfrac{y^2}{9} = 1$
19. $4y^2 - 5x^2 = 20$
20. $x^2 - y^2 = 16$
21. $y^2 - \dfrac{x^2}{4} = 1$
22. $y^2 - x^2 = 16$

23. $x^2 - 2y^2 - 8 = 0$

24. $25x^2 - 4y^2 = 100$

25. $\frac{1}{9}x^2 - y^2 = 1$

26. $\frac{y^2}{5} - x^2 = 1$

27. $xy = 8$

28. $xy = -6$

29. $y = \frac{9}{x}$

30. $y = \frac{-4}{x}$

31. $y^2 = 9 + x^2$

32. $4y^2 - 9x^2 - 36 = 0$

Graph both equations on the same set of axes.

33. $\frac{x^2}{4} - \frac{y^2}{9} = 1$; $\quad \frac{x^2}{4} + \frac{y^2}{9} = 1$

34. $xy = 4$; $\quad x^2 + y^2 = 4$

35. $y^2 - 4x^2 = 16$; $\quad y^2 + 4x^2 = 16$

36. $4x^2 - y^2 = 16$; $\quad 4x^2 + y^2 = 16$

37. $y^2 - x^2 = 4$; $\quad x^2 - y^2 = 4$

CONIC SECTIONS AND CURVE RECOGNITION

9.6

If a plane intersects a surface, the resulting curve of intersection is called a section. If the surface is a right circular cone, the curve of intersection is said to be a conic section. The conic sections include circles, ellipses, parabolas, hyperbolas, and straight lines. As illustrated in Figure 9.33, a circle results when a plane intersects the cone perpendicular to its axis. An ellipse is formed when the plane is slightly titled. A parabola is obtained when the plane is tilted so that it is parallel to a surface element. A hyperbola is formed when the plane intersects the cone parallel to its axis. And, a straight line results when the plane coincides with a surface element.

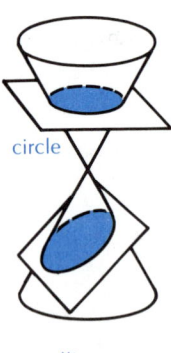

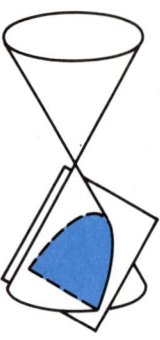

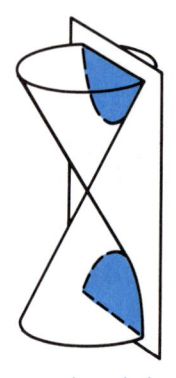

FIGURE 9.33

Each of these conic sections has been studied analytically in some detail. Thus, we provide a summary which will help you identify a graph by simply looking at its equation.

Summary of Equations Describing Conic Sections

Straight Line
$Ax + By + C = 0$

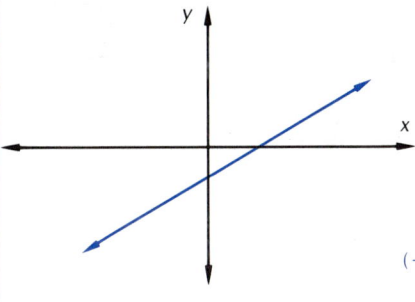

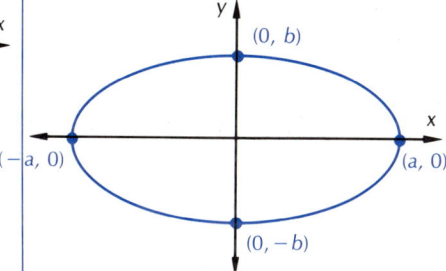

Parabola (concave up or down)
$f(x) = ax^2 + bx + c$, where $a \neq 0$

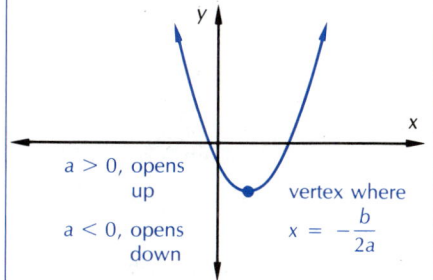

$a > 0$, opens up
$a < 0$, opens down
vertex where $x = -\dfrac{b}{2a}$

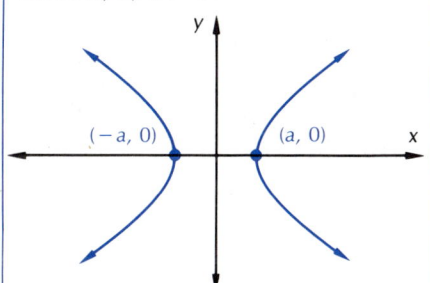

Parabola (opening left or right)
$x = ay^2 + by + c$, where $a \neq 0$

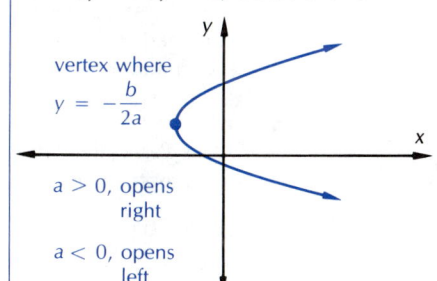

vertex where $y = -\dfrac{b}{2a}$
$a > 0$, opens right
$a < 0$, opens left

Ellipse
$\dfrac{x^2}{a^2} + \dfrac{y^2}{b^2} = 1$ or $Ax^2 + By^2 = C$,

where A, B, C have the same sign and $A \neq B$.

$(0, b)$, $(-a, 0)$, $(a, 0)$, $(0, -b)$

Hyperbola (opening horizontally)
$\dfrac{x^2}{a^2} - \dfrac{y^2}{b^2} = 1$ or $Ax^2 - By^2 = C$,

where $A, B, C > 0$

$(-a, 0)$, $(a, 0)$

Hyperbola (opening vertically)
$\dfrac{y^2}{a^2} - \dfrac{x^2}{b^2} = 1$ or $By^2 - Ax^2 = C$,

where $A, B, C > 0$

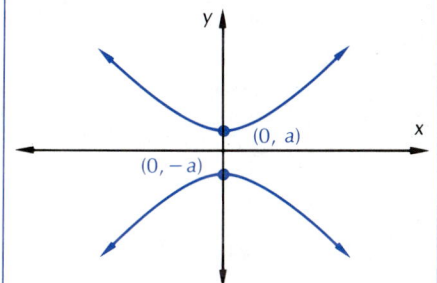

$(0, a)$, $(0, -a)$

THE CONIC SECTIONS

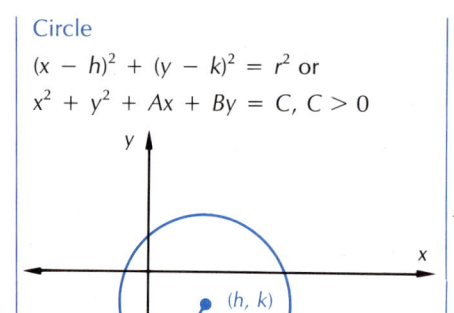

Circle

$(x - h)^2 + (y - k)^2 = r^2$ or
$x^2 + y^2 + Ax + By = C, C > 0$

Hyperbola (contained in quadrants)

$xy = c$ or $y = \dfrac{c}{x}$, where $c \neq 0$

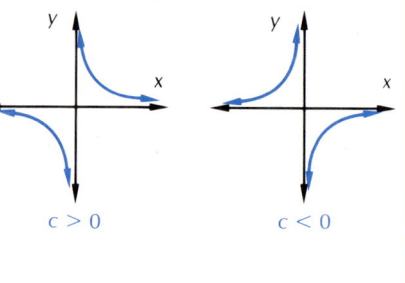

$c > 0$ $\qquad$ $c < 0$

Using this summary, you should be able to examine an equation of a conic section, identify its graph, write the standard form, and list its major characteristics. Practice by studying Table 9.3.

TABLE 9.3

Equation	Graph	Standard Form	Characteristics
$4x^2 - 9y^2 = 36$	Hyperbola	$\dfrac{x^2}{9} - \dfrac{y^2}{4} = 1$	Opens horizontally, x-intercepts = ± 3
$4x - 25y = 100$	Line	$4x - 25y - 100 = 0$	x-intercept = 25, y-intercept = -4
$x + 2y^2 = 6$	Parabola	$x = -2y^2 + 6$	Opens left, vertex at (6, 0)
$x^2 + 9y^2 = 81$	Ellipse	$\dfrac{x^2}{81} + \dfrac{y^2}{9} = 1$	x-intercepts = ± 9, y-intercepts = ± 3
$x^2 + y^2 = 36$	Circle	$(x - 0)^2 + (y - 0)^2 = 6^2$	Center at (0, 0), radius = 6
$y = -\dfrac{10}{x}$	Hyperbola	$xy = -10$	Contained in the second and fourth quadrants
$y^2 - x^2 = 1$	Hyperbola	$\dfrac{y^2}{1} - \dfrac{x^2}{1} = 1$	Opens vertically, y-intercepts = ± 1
$x^2 + y^2 - 2x + 4y + 1 = 0$	Circle	$(x - 1)^2 + (y + 2)^2 = 4$	Center at (1, 2), radius = 2
$2x^2 + 8x - y + 1 = 0$	Parabola	$y = 2x^2 + 8x + 1$	Concave up, vertex at $(-2, -7)$
$49x^2 + 4y^2 = 196$	Ellipse	$\dfrac{x^2}{4} + \dfrac{y^2}{49} = 1$	x-intercepts = ± 2, y-intercepts = ± 7

Sometimes it is necessary to find the domain and range of a relation defining a conic section. This can be accomplished by projecting its graph onto the x and y-axes.

Find the domain and range of $3x^2 + 25y^2 = 75$

EXAMPLE 1

Solution: This equation defines an ellipse. In standard form the equation becomes

$$\frac{x^2}{25} + \frac{y^2}{3} = 1$$

The x-intercepts are ± 5 and the y-intercepts are $\pm\sqrt{3}$ (approximately ± 1.7). The ellipse is sketched in Figure 9.34, and the domain and range are read from the indicated projections on the axes.

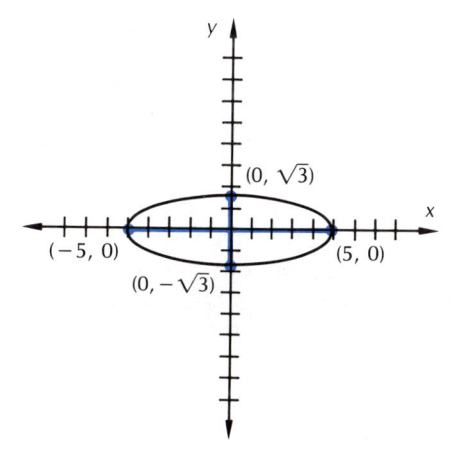

FIGURE 9.34

The domain is $\{x \mid -5 \leq x \leq 5\}$.
The range is $\{y \mid -\sqrt{3} \leq y \leq \sqrt{3}\}$.

Identify the graph of the conic section, write its standard form, and give its characteristics.

Exercise 9.6

1. $9x^2 - y^2 = 9$
2. $5x^2 + 5y^2 = 20$
3. $y^2 + 3x^2 = 9$
4. $(x - 3)^2 = 64 - y^2$
5. $9x^2 - y^2 = 81$
6. $x = \dfrac{-4}{y}$
7. $x - y = 1$
8. $x^2 + 4y = 4$
9. $y^2 - 4x^2 = 4$
10. $x^2 + 36y^2 = 36$
11. $4x^2 + 4y^2 = 1$
12. $-5x^2 - 2y^2 = -10$
13. $x^2 - y = 9$
14. $2xy = 16$

15. $x^2 - 2x + y^2 + 4y + 1 = 0$
16. $y = 2 - 6x - x^2$
17. $y = x - 4$
18. $x - 1 = y^2 - 2y$
19. $y = -3x^2$
20. $\dfrac{x}{4} = 1 + y$

Find the domain and range of each of these conic sections.
21. $y = x^2 - 4$
22. $x^2 + y^2 = 36$
23. $\dfrac{x^2}{9} + \dfrac{y^2}{16} = 1$
24. $4x^2 - 9y^2 = 36$
25. $x^2 + y^2 = 12$
26. $x^2 - y^2 = 16$
27. $\dfrac{y^2}{4} - \dfrac{x^2}{5} = 1$
28. $\dfrac{x^2}{8} + \dfrac{y^2}{12} = 1$
29. $x^2 + 25y^2 = 25$
30. $xy = -2$
31. $y^2 - x^2 = 9$
32. $x + y^2 = 4$

NON-LINEAR SYSTEMS OF EQUATIONS

9.7

A non-linear system of equations contains at least one equation that is not linear. These systems are solved using the substitution method or the elimination method first discussed in Section 8.1. As shown in Example 1, the substitution method is particularly effective when one of the equations is linear.

EXAMPLE 1

Solve the system:
$x^2 + y^2 = 16$ (1)
$x - 2y = 4$ (2)

Solution: Solve the linear equation for x and substitute into the non-linear equation to produce an equation in one variable. We solve equation (2) for x.

$$x - 2y = 4$$
$$x = 2y + 4 \qquad (3)$$

Substitute $2y + 4$ for x in equation (1).
$$x^2 + y^2 = 16$$
$$(2y + 4)^2 + y^2 = 16$$
$$4y^2 + 16y + 16 + y^2 = 16$$
$$5y^2 + 16y = 0$$

The resulting equation is quadratic and is solved by factoring.
$$y(5y + 16) = 0$$
$$y = 0 \quad \text{or} \quad y = -\dfrac{16}{5}$$

NON-LINEAR SYSTEMS OF EQUATIONS

To find the corresponding values of x, we substitute the y values into equation (3), which is linear.

$x = 2y + 4$ $x = 2y + 4$
$x = 2(0) + 4$ $x = 2\left(-\dfrac{16}{5}\right) + 4$
$x = 4$
         $x = -\dfrac{12}{5}$

The solutions of the system are $(4, 0)$ and $\left(-\dfrac{12}{5}, -\dfrac{16}{5}\right)$. The graph of this system is the intersection of a straight line with a circle, as illustrated in Figure 9.35.

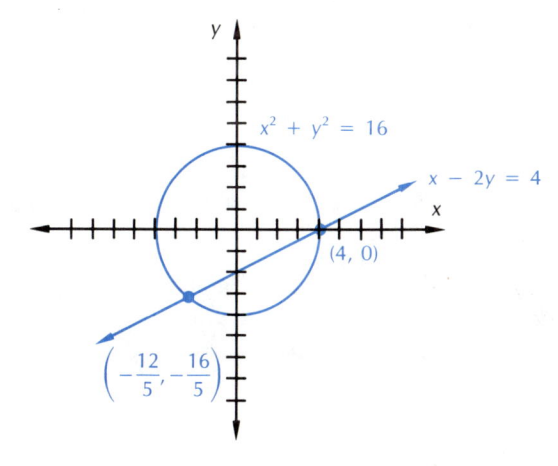

FIGURE 9.35

As shown by the following example, the substitution method is also effective when one of the equation is of the form $xy = c$. The equation is solved for either x or y and the result is substituted into the other equation.

Solve the system: **EXAMPLE 2**
$x^2 - 4y^2 = 5$ (1)
$xy = 3$ (2)

Solution: Solve equation (2) for y.
$$xy = 3$$
$$y = \dfrac{3}{x}$$ (3)

Substitute $\dfrac{3}{x}$ for y in equation (1) to produce an equation in one variable.

$$x^2 - 4y^2 = 5$$
$$x^2 - 4\left(\frac{3}{x}\right)^2 = 5$$
$$x^2 - \frac{36}{x^2} = 5$$
$$x^4 - 36 = 5x^2$$
$$x^4 - 5x^2 - 36 = 0$$

Since this equation is quadratic in form, we solve by factoring:
$$(x^2 - 9)(x^2 + 4) = 0$$
$$x^2 = 9 \text{ or } x^2 = -4$$

Since $x^2 = -4$ has no real solution, the only solutions for x are ± 3. The corresponding values of y are found by substituting 3 and -3 into equation (3).

$$y = \frac{3}{x} \qquad\qquad y = \frac{3}{x}$$
$$y = \frac{3}{3} \qquad\qquad y = \frac{3}{-3}$$
$$y = 1 \qquad\qquad y = -1$$

The solutions of the system are $(3, 1)$ and $(-3, -1)$. The graph is the intersection of two hyperbolas, as illustrated in Figure 9.36.

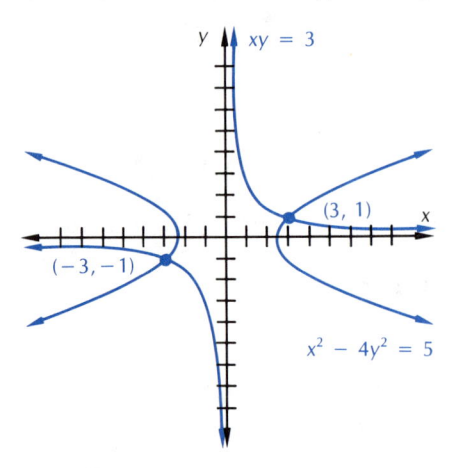

FIGURE 9.36

The elimination method, as illustrated by the next example, is most useful when the system contains two second degree equations.

EXAMPLE 3   Solve the system:
$$4x^2 - 5y^2 = 16 \qquad (1)$$
$$3x^2 + 2y^2 = 35 \qquad (2)$$

Solution: To eliminate the y terms, we multiply equation (1) by 2 and equation (2) by 5. The two equations are then added.

$$4x^2 - 5y^2 = 16 \xrightarrow{\text{multiply by 2}} 8x^2 - 10y^2 = 32$$
$$3x^2 + 2y^2 = 35 \xrightarrow{\text{multiply by 5}} \underline{15x^2 + 10y^2 = 175}$$
$$23x^2 \qquad\qquad = 207$$

The resulting equation has only one variable and is solved for x.

$$23x^2 = 207$$
$$x^2 = \frac{207}{23} = 9$$
$$x = \pm 3$$

The corresponding values of y are found by substituting 3 and -3 into equation (2).

$$3x^2 + 2y^2 = 35 \qquad\qquad 3x^2 + 2y^2 = 35$$
$$3(3)^2 + 2y^2 = 35 \qquad\qquad 3(-3)^2 + 2y^2 = 35$$
$$27 + 2y^2 = 35 \qquad\qquad 27 + 2y^2 = 35$$
$$y^2 = 4 \qquad\qquad y^2 = 4$$
$$y = \pm 2 \qquad\qquad y = \pm 2$$

The system has four solutions: $(3, 2)$, $(3, -2)$, $(-3, 2)$, and $(-3, -2)$. The graph is the intersection of a hyperbola with an ellipse, as shown in Figure 9.37.

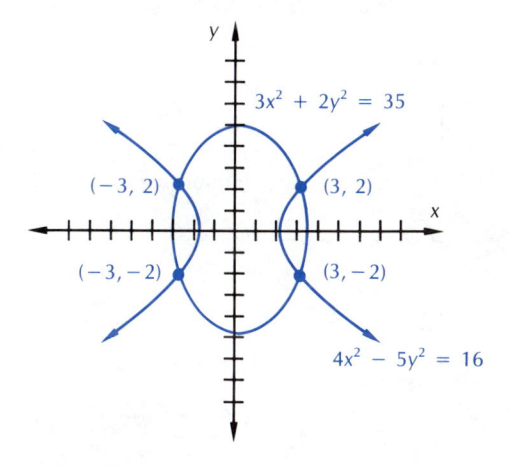

FIGURE 9.37

As indicated in the next example, the elimination method can often be abbreviated by keeping the plus or minus sign ($\pm$) when substituting the initial values back into one of the original equations.

EXAMPLE 4

Solve the system:

$$x^2 + y^2 = 10 \quad (1)$$
$$x^2 - y^2 = 4 \quad (2)$$

Solution: Using the elimination method, we add the two equations to eliminate the terms containing y^2.

$$\begin{aligned} x^2 + y^2 &= 10 \\ x^2 - y^2 &= 4 \\ \hline 2x^2 &= 14 \\ x^2 &= 7 \\ x &= \pm\sqrt{7} \end{aligned}$$

To find the corresponding y values, we substitute into equation (1).

$$x^2 + y^2 = 10$$
$$(\pm\sqrt{7})^2 + y^2 = 10$$
$$7 + y^2 = 10$$
$$y^2 = 3$$
$$y = \pm\sqrt{3}$$

The solutions are $(\sqrt{7}, \sqrt{3})$, $(-\sqrt{7}, \sqrt{3})$, $(\sqrt{7}, -\sqrt{3})$, and $(-\sqrt{7}, -\sqrt{3})$.

Summary of Solving Non-Linear Systems	
Type of System	Method
$Ax^2 + By^2 = C$ $Dx + Ey = F$	Substitution
$Ax^2 + By^2 = C$ $xy = K$	Substitution
$Ax + By = C$ $xy = K$	Substitution
$Ax^2 + By^2 = C$ $Dx^2 + Ey^2 = F$	Elimination

Exercise 9.7

Solve the following systems.

1. $x^2 + y^2 = 13$
   $x - y = 1$

2. $x^2 - y^2 = 25$
   $x - y = 5$

3. $x^2 + y^2 = 37$
   $x - y = 5$
4. $x^2 + y^2 = 35$
   $y - x = 7$
5. $x^2 - y^2 = 4$
   $x + y = 2$
6. $x^2 - y^2 = 21$
   $x + y = 7$
7. $x^2 + y^2 = 25$
   $y + 3x = -5$
8. $x^2 + y^2 = 25$
   $x + y = 1$
9. $y = x^2 - 4x + 4$
   $x + y = 2$
10. $2x + 3y = 2$
    $xy = -8$
11. $xy = 2$
    $x - 3y = 5$
12. $x^2 - 4y^2 = 5$
    $xy = 3$
13. $xy = 12$
    $y + 2x = 10$
14. $xy = -12$
    $x - 2y = 10$
15. $2x - y = 9$
    $xy = -10$
16. $x^2 + y^2 = 25$
    $xy = -12$
17. $x^2 - y^2 = 4$
    $x = 2 - y$
18. $x^2 = 4y$
    $x - y = 1$
19. $4x + 2y = 10$
    $y = x^2 - 10$
20. $x^2 + 4y + 2x = 11$
    $y - x = 5$
21. $y = x^2 + 1$
    $y = 2x$
22. $4x^2 + y^2 = 25$
    $x^2 - y^2 = -5$
23. $x^2 - y^2 = 1$
    $x^2 + y^2 = 3$
24. $5x^2 - y^2 = 3$
    $x^2 + 2y^2 = 5$
25. $x^2 - y^2 = 5$
    $x^2 + y^2 = 11$
26. $x^2 - 2y^2 = 8$
    $x^2 + y^2 = 20$
27. $x^2 + y^2 = 25$
    $x^2 - y^2 = 25$
28. $2x^2 + 3y^2 = 6$
    $x^2 + 3y^2 = 3$
29. $x^2 + y^2 = 9$
    $x + 2y = 3$
30. $x^2 + y^2 = 5$
    $y = x^2 - 3$
31. $y = x^2 - 3$
    $x^2 + y^2 = 9$
32. $x^2 - y^2 = 16$
    $2x^2 - 3y^2 = 28$
33. $y^2 - x^2 = 4$
    $4x^2 = 64 - y^2$
34. $25x^2 + 4y^2 = 100$
    $x^2 = 4 - y^2$

SECOND DEGREE INEQUALITIES

9.8

Each of the conic sections shown in Figure 9.38 is defined by a second degree equation. Observe that each curve separates the plane into three regions: the set of points on the curve called the boundary, the region inside the curve, and the region outside the curve.

Since each boundary is defined by a second degree equation, the inner and outer regions are described by second degree inequalities.

THE CONIC SECTIONS

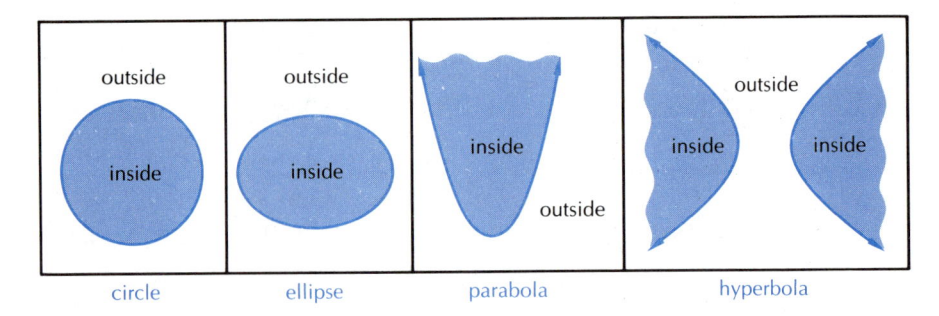

FIGURE 9.38

Second degree inequalities are graphed in the same manner as linear inequalities studied in Section 7.6. The boundary equation is graphed first. A solid curve is used when the inequality contains an equal sign ($\leq$ or $\geq$); otherwise it is dashed. Next, we select a test point not on the boundary and substitute its coordinates into the inequality. If the result is true, we shade the region containing the test point; otherwise we shade the opposing region.

EXAMPLE 1 Graph $x^2 + y^2 \leq 16$

Solution: The boundary is described by the equation $x^2 + y^2 = 16$. As seen in Figure 9.39, it is a circle of radius 4 having its center at the origin. The boundary is drawn solid since it is included in the inequality.

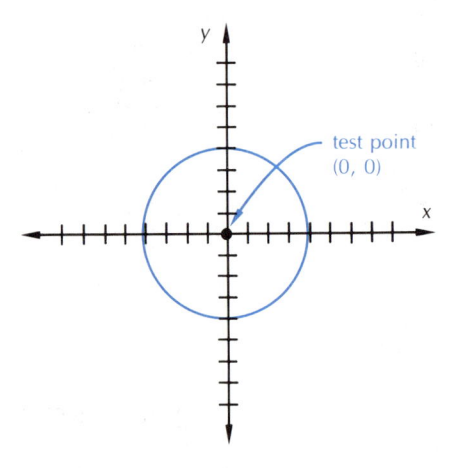

FIGURE 9.39

We select a test point of (0, 0) and substitute into the original inequality.

$x^2 + y^2 \leq 16$
$0^2 + 0^2 \leq 16$
$0 \leq 16$ (true)

SECOND DEGREE INEQUALITIES

Since the result is true, we shade the inner region containing the test point, as illustrated in Figure 9.40.

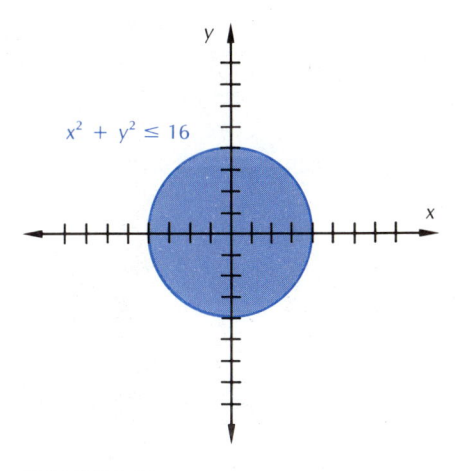

$x^2 + y^2 \leq 16$

FIGURE 9.40

Graph $x^2 - 9y^2 > 9$

EXAMPLE 2

Solution: The boundary is described by the equation $x^2 - 9y^2 = 9$ or $\dfrac{x^2}{9} - \dfrac{y^2}{1} = 1$.

As seen in Figure 9.41, it is a hyperbola with x-intercepts of ± 3. The asymptotes are the diagonals of the rectangle defined by $(3, \pm 1)$ and $(-3, \pm 1)$. The boundary is dashed since it is not included in the inequality.

test point (0, 0)

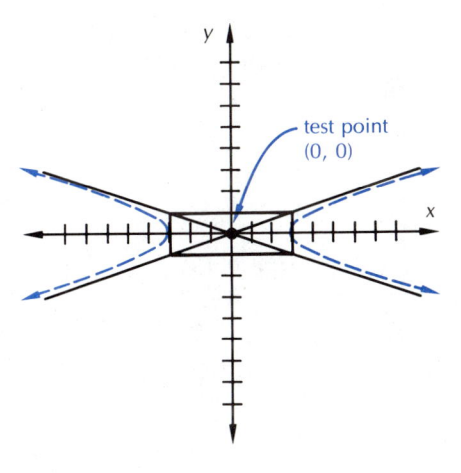

FIGURE 9.41

We select the test point (0, 0) and substitute into the original inequality.

$$x^2 - 9y^2 > 9$$
$$0^2 - 9(0)^2 > 9$$
$$0 > 9 \quad \text{(false)}$$

Since the result is false, we shade the opposing region as illustrated in Figure 9.42.

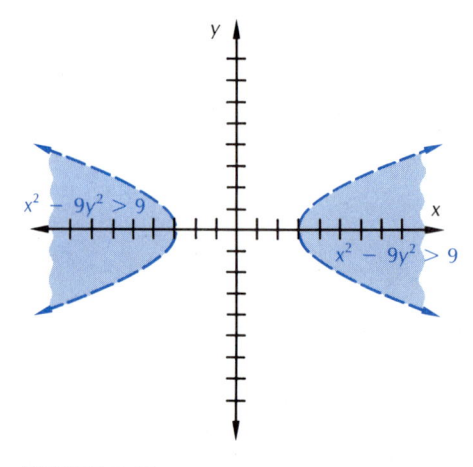

FIGURE 9.42

A system of second degree inequalities consists of two or more second degree inequalities and is solved by finding the intersection of their corresponding regions.

EXAMPLE 3 Solve the system:

$$x^2 + y^2 > 9$$
$$\frac{x^2}{25} + \frac{y^2}{4} < 1$$

Solution: The graph of $x^2 + y^2 > 9$, shown in Figure 9.43, is the region outside the circle $x^2 + y^2 = 9$. The graph of $\frac{x^2}{25} + \frac{y^2}{4} < 1$, shown in Figure 9.44, is the region inside the ellipse $\frac{x^2}{25} + \frac{y^2}{4} = 1$.

SECOND DEGREE INEQUALITIES

FIGURE 9.43 — $x^2 + y^2 > 9$

FIGURE 9.44 — $\frac{x^2}{25} + \frac{y^2}{4} < 1$

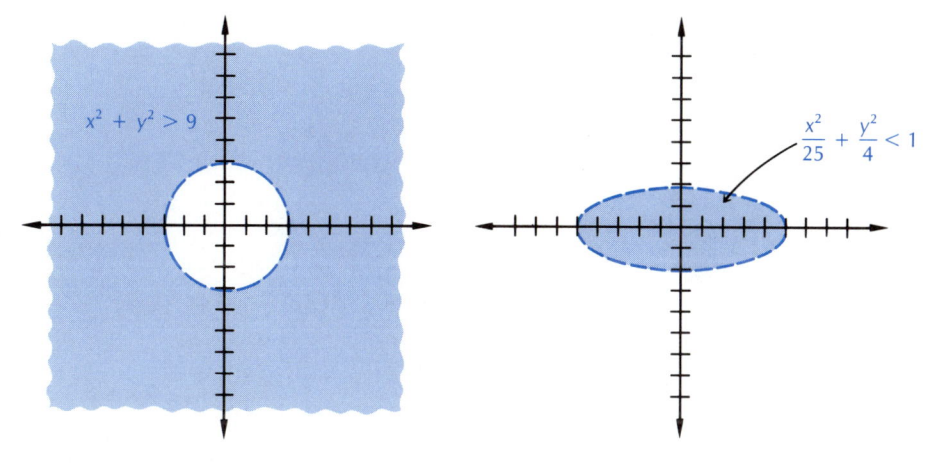

Both graphs are placed on the same coordinate system, and the solution is the shaded intersection in Figure 9.45. Note that the boundaries are not included.

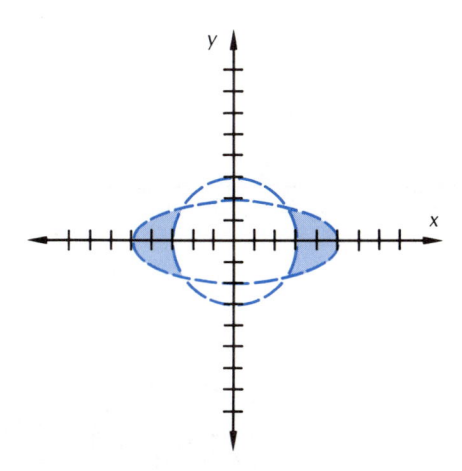

FIGURE 9.45

Graph the following inequalities.

Exercise 9.8

1. $y \geq x^2$
2. $xy \geq 4$
3. $y < x^2 - 4$
4. $y > -x^2 + 2$
5. $x^2 + y^2 \leq 16$
6. $x^2 + y^2 > 9$
7. $\frac{x^2}{4} - \frac{y^2}{9} < 1$
8. $y \leq -x^2 + 4$

9. $x^2 + 4y^2 > 36$

10. $x^2 + y^2 > 36$

11. $\dfrac{x^2}{9} + \dfrac{y^2}{25} \leq 1$

12. $\dfrac{y^2}{9} - \dfrac{x^2}{9} \leq 1$

13. $\dfrac{x^2}{9} - \dfrac{y^2}{25} > 1$

14. $\dfrac{x^2}{9} + \dfrac{y^2}{25} < 1$

Graph the systems.

15. $x^2 + y^2 > 9$
    $\dfrac{x^2}{4} + \dfrac{y^2}{25} < 1$

16. $x^2 + y^2 < 16$
    $4x^2 + 36y^2 > 144$

17. $x^2 - y^2 \leq 9$
    $y \geq x^2 - 4$

18. $y^2 - 4x^2 \leq 36$
    $y^2 + 4x^2 \leq 36$

19. $x^2 + y^2 \geq 9$
    $x + 2y \leq 4$

20. $xy < 4$
    $\dfrac{x^2}{16} + \dfrac{y^2}{9} > 1$

21. $y^2 - x^2 \leq 9$
    $x^2 + y^2 \leq 9$

22. $x^2 + y^2 > 9$
    $16x^2 + 36y^2 < 576$

23. $x^2 + y^2 \leq 25$
    $x^2 + y^2 \geq 4$

24. $9x^2 + 4y^2 > 36$
    $9x^2 - 4y^2 > 36$

25. $y \geq x^2 + 6x + 4$
    $y \leq 4 - 2x - x^2$

26. $xy > -6$
    $\dfrac{x^2}{25} + \dfrac{y^2}{36} < 1$

Summary

CHARACTERISTICS OF QUADRATIC EQUATIONS IN TWO VARIABLES

(1) *Form:* $y = ax^2 + bx + c,\ a \neq 0$
    Graph: parabola; concave up if $a > 0$, concave down if $a < 0$
(2) *Form:* $x = ay^2 + by + c,\ a \neq 0$
    Graph: parabola; opens to right if $a > 0$, opens to left if $a < 0$

CHARACTERISTICS OF THE QUADRATIC FUNCTION

(1) *Form:* $f(x) = ax^2 + bx + c$, where $a > 0$
    Intercepts: has y-intercept at c; may or may not have x-intercepts
    Graph: parabola, concave up
    Vertex: the lowest point or minimum point; the point $\left(\dfrac{-b}{2a},\ f\left(\dfrac{-b}{2a}\right)\right)$
    Axis: a vertical line passing through the vertex
    Domain: $\{x \mid x \in R\}$
    Range: $\left\{y \mid y \geq f\left(\dfrac{-b}{2a}\right)\right\}$

(2) *Form:* $f(x) = ax^2 + bx + c$, where $a < 0$
Intercepts: has y-intercept at c; may or may not have x-intercepts
Graph: parabola, concave down
Vertex: the highest or maximum point; the point $\left(\dfrac{-b}{2a}, f\left(\dfrac{-b}{2a}\right)\right)$
Axis: a vertical line passing through the vertex
Domain: $\{x \mid x \in R\}$
Range: $\left\{y \mid y \leq f\left(\dfrac{-b}{2a}\right)\right\}$

CHARACTERISTICS OF THE CIRCLE

Definition: the set of all points in a plane that lie a given distance from a fixed point. The given distance is called the radius and the fixed point is called the center.
Standard Form: $(x - h)^2 + (y - k)^2 = r^2$, where the center is (h, k) and the radius is r.

CHARACTERISTICS OF THE ELLIPSE

Definition: the set of all points in a plane the sum of whose distances from two fixed points is a constant. The two fixed points are called focal points or foci.
Standard Form: $\dfrac{x^2}{a^2} + \dfrac{y^2}{b^2} = 1$, where $a \neq b$; the x-intercepts are a and $-a$, the y-intercepts are b and $-b$.
Variation of the Standard Form: $Ax^2 + By^2 = C$, where A, B, and C have the same sign and $A \neq B$.

CHARACTERISTICS OF THE HYPERBOLA

Definition: the set of all points in a plane the difference of whose distances from two fixed points is a constant. The two fixed points are called foci.
Standard Position: a hyperbola whose foci are on either the x-axis or the y-axis.
Equations of Standard Form:

$\dfrac{x^2}{a^2} - \dfrac{y^2}{b^2} = 1$, foci on x-axis and x-intercepts of a and $-a$.

$\dfrac{y^2}{a^2} - \dfrac{x^2}{b^2} = 1$, foci on y-axis and y-intercepts of a and $-a$.

Variation of the Standard Form: $Ax^2 - By^2 = C$, where A, B, $C > 0$ or $By^2 - Ax^2 = C$, where A, B, $C > 0$.

Another Form of the Hyperbola: $xy = c$ or $y = \dfrac{c}{x}$; if $c > 0$, the hyperbola is contained in the first and third quadrants; if $c < 0$, the hyperbola is contained in the second and fourth quadrants.

THE CONIC SECTIONS

Straight Line	$Ax + By + C = 0$
Parabola	$f(x) = ax^2 + bx + c, a \neq 0$ $x = ay^2 + by + c, a \neq 0$
Circle	$(x - h)^2 + (y - k)^2 = r^2$ or $x^2 + y^2 + Ax + By = C, C > 0$
Ellipse	$\dfrac{x^2}{a^2} + \dfrac{y^2}{b^2} = 1$ or $Ax^2 + By^2 = C$, where A, B, C have the same sign and $A \neq B$
Hyperbola	$\dfrac{x^2}{a^2} - \dfrac{y^2}{b^2} = 1$ or $Ax^2 - By^2 = C$, where $A, B, C > 0$ $\dfrac{y^2}{a^2} - \dfrac{x^2}{b^2} = 1$ or $By^2 - Ax^2 = C$, where $A, B, C > 0$ $xy = c$ or $y = \dfrac{c}{x}, c \neq 0$

FORMULAS

Pythagorean Theorem: $d^2 = a^2 + b^2$, where d is the length of the hypotenuse of a right triangle and a and b are the lengths of the legs.

Distance Formula: $d = \sqrt{(x_2 - x_1)^2 + (y_2 - y_1)^2}$

DEFINITIONS

Conic Section: the resulting curve when a plane intersects a right circular cone.

Asymptotes of a Hyperbola: lines that a hyperbola approaches, formed by the diagonals of the rectangle with vertices at $(a, \pm b)$ and $(-a, \pm b)$ for the hyperbola $\dfrac{x^2}{a^2} - \dfrac{y^2}{b^2} = 1$, and vertices at $(b, \pm a)$ and $(-b, \pm a)$ for the hyperbola $\dfrac{y^2}{a^2} - \dfrac{x^2}{b^2} = 1$.

Self-Checking Exercise

Find the x and y-intercepts of each parabola.
1. $y = x^2 - 6x + 5$
2. $y = -x^2 + 2x$
3. $x = y^2 - 2y - 3$
4. $x = -2y^2 + y + 3$

Determine if the parabola is concave up or concave down and find its vertex.
5. $y = -4x^2$
6. $y = 4x^2 - 6x - 1$
7. $y = x^2 + 4x - 2$
8. $y = -x^2 + 3x + 2$

Find the maximum or minimum value of each parabola.
9. $y = -x^2 + 5x - 1$
10. $y = 1 + x + x^2$
11. $y = 2x^2 + 4x$
12. $y = -2x^2 - 6x + 11$

Graph each parabola.
13. $y = x^2 - 6x + 5$
14. $y = -x^2 + 2x$
15. $x = y^2 - 2y - 3$
16. $x = -2y^2 + y + 3$

Find the distance between the given points.
17. (6, 2) (5, 8)
18. (−2, 5) (3, 12)
19. (−2, −5) (6, 8)
20. (0, 9) (6, 0)

Write the equation of each circle in standard form.
21. center at (−4, 3), radius of 6
22. center at (3, −1), radius of $2\sqrt{2}$

Complete the square and determine the center and radius of each circle.
23. $x^2 + y^2 + 6x - 8y + 21 = 0$
24. $x^2 + y^2 - 10x - 2y - 5 = 0$

Identify the following conic sections.
25. $x^2 - y^2 = 1$
26. $x = 9 - y^2$
27. $6x^2 + 4y^2 = 12$
28. $x = 36 + y^2$
29. $xy = -2$
30. $\dfrac{x^2}{9} + \dfrac{y^2}{16} = 1$
31. $\dfrac{y^2}{9} + \dfrac{x^2}{16} = 1$
32. $xy = 5$
33. $3y^2 + 3x^2 = 8$
34. $6y = 4 - x$

Graph and give the domain and range.
35. $\dfrac{x^2}{9} + \dfrac{y^2}{16} = 1$
36. $xy = -4$
37. $25x^2 + 4y^2 = 100$
38. $y^2 - 4 = x^2$
39. $x^2 + y^2 = 16$
40. $y = x^2 - 4x + 4$

THE CONIC SECTIONS

Graph the inequalities.

41. $x^2 + y^2 \leq 9$

42. $y \geq 4 - x^2$

43. $4x^2 - 9y^2 > 36$

44. $\dfrac{x^2}{25} + \dfrac{y^2}{9} > 1$

Graph the systems.

45. $3x^2 + y^2 \leq 12$
    $x^2 + y^2 \geq 4$

46. $x^2 + 4y^2 < 16$
    $\dfrac{x^2}{9} - \dfrac{y^2}{16} < 1$

Solve the systems.

47. $x + 4y = -10$
    $xy = 4$

48. $x^2 + 2y^2 = 6$
    $2x^2 + y^2 = 5$

49. $y = x^2 - 1$
    $x^2 + y^2 = 13$

50. $x^2 + y^2 = 4$
    $x = 2 - y$

Applied Problems

51. Find two numbers whose sum is -2 and whose product is 1.
52. Find the dimensions of a rectangle whose area is 15 square meters and whose perimeter is 16 meters.
53. A rancher has 8 miles of fencing with which to enclose a rectangular pasture. One side is bounded by a neighbor's fence, so he needs to enclose only three sides. Find the dimensions that would give the greatest area.
54. Find a number such that the sum of the number and its square is a minimum.

Solutions to Self-Checking Exercise

1. x-intercepts: 1 and 5
   y-intercept: 5
2. x-intercepts: 0 and 2
   y-intercept: 0
3. x-intercept: -3
   y-intercepts: -1 and 3
4. x-intercept: 3
   y-intercepts: -1 and $\dfrac{3}{2}$
5. concave down; $(0, 0)$
6. concave up; $\left(\dfrac{3}{4}, -\dfrac{13}{4}\right)$
7. concave up; $(-2, -6)$
8. concave down; $\left(\dfrac{3}{2}, \dfrac{17}{4}\right)$
9. maximum of $\dfrac{21}{4}$
10. minimum of $\dfrac{3}{4}$

SOLUTIONS TO SELF-CHECKING EXERCISE 411

11. minimum of -2

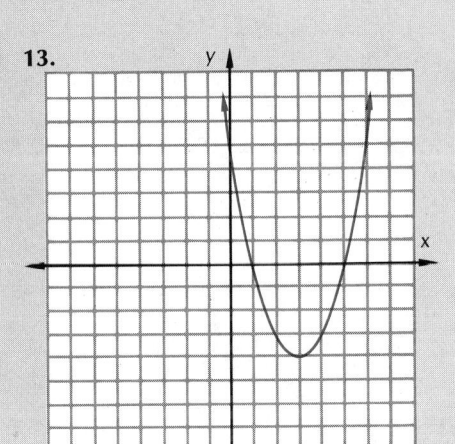

12. maximum of $\dfrac{31}{2}$

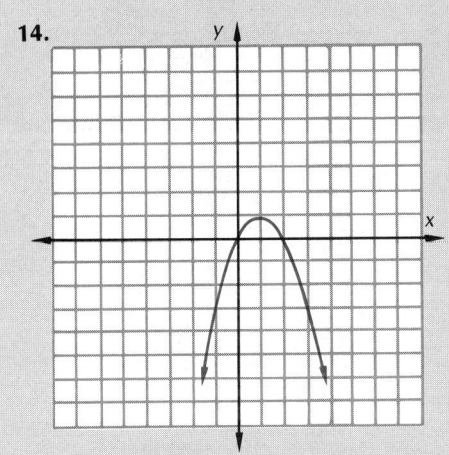

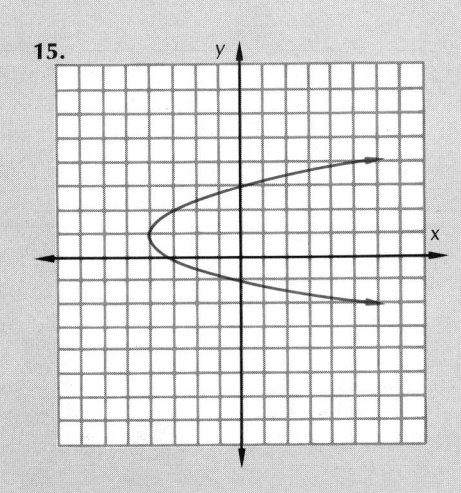

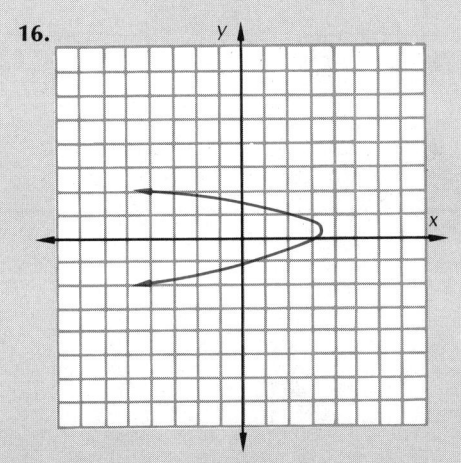

13.

14.

15.

16.

17. $\sqrt{37}$

18. $\sqrt{74}$

19. $\sqrt{233}$

20. $3\sqrt{13}$

21. $(x + 4)^2 + (y - 3)^2 = 36$

22. $(x - 3)^2 + (y + 1)^2 = 8$

23. center at $(-3, 4)$
radius is 2

24. center at $(5, 1)$
radius is $\sqrt{31}$

25. hyperbola

26. parabola

27. ellipse

28. parabola

29. hyperbola

30. ellipse

31. ellipse

32. hyperbola

33. circle

34. straight line

THE CONIC SECTIONS

35.

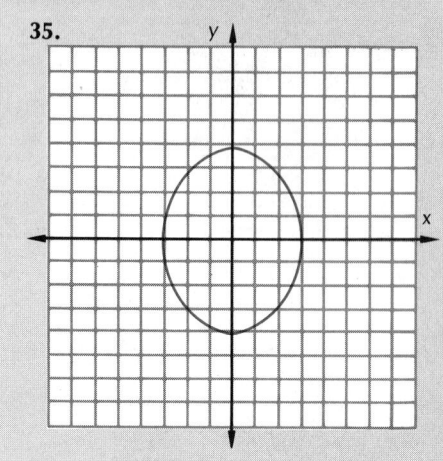

domain: $\{x \mid -3 \leq x \leq 3\}$
range: $\{y \mid -4 \leq y \leq 4\}$

36.

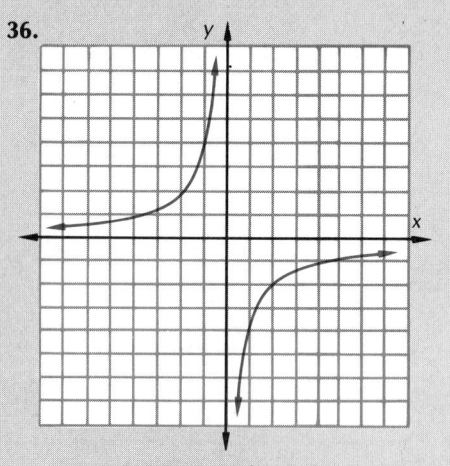

domain: $\{x \mid x \neq 0\}$
range: $\{y \mid y \neq 0\}$

37.

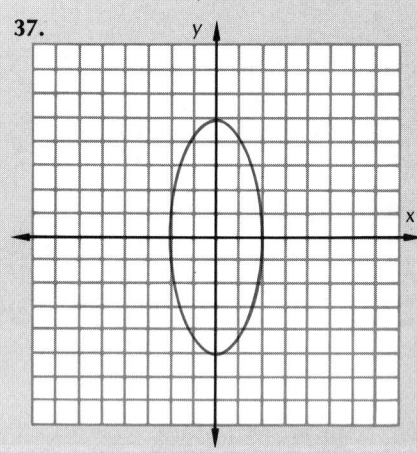

domain: $\{x \mid -2 \leq x \leq 2\}$
range: $\{y \mid -5 \leq y \leq 5\}$

38.

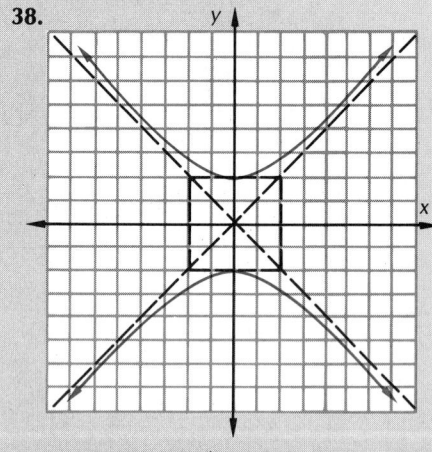

domain: $\{x \mid x \in R\}$
range: $\{y \mid y \geq 2 \text{ or } y \leq -2\}$

39.

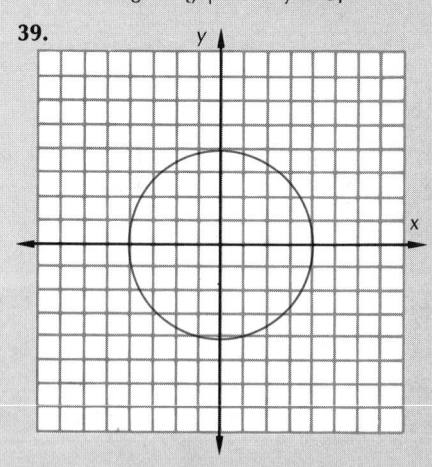

domain: $\{x \mid -4 \leq x \leq 4\}$
range: $\{y \mid -4 \leq y \leq 4\}$

40.

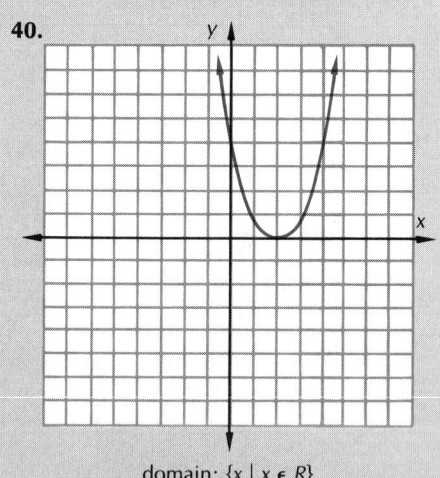

domain: $\{x \mid x \in R\}$
range: $\{y \mid y \geq 0\}$

SOLUTIONS TO SELF-CHECKING EXERCISE

41.

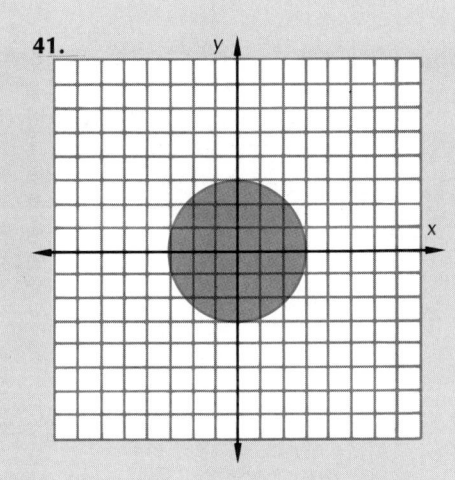

42.

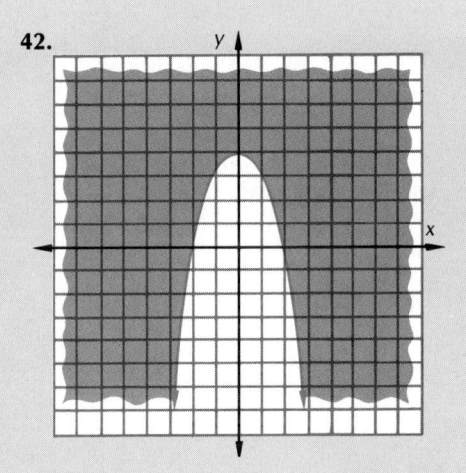

43.

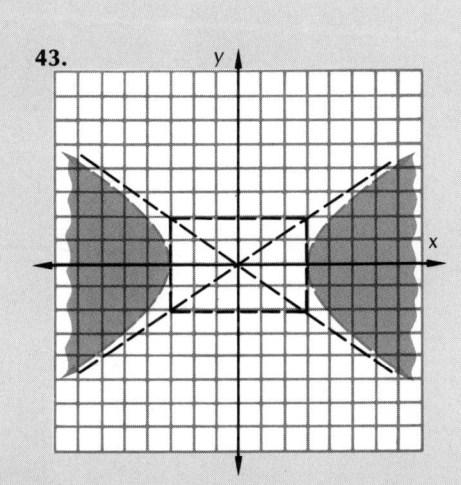

44.

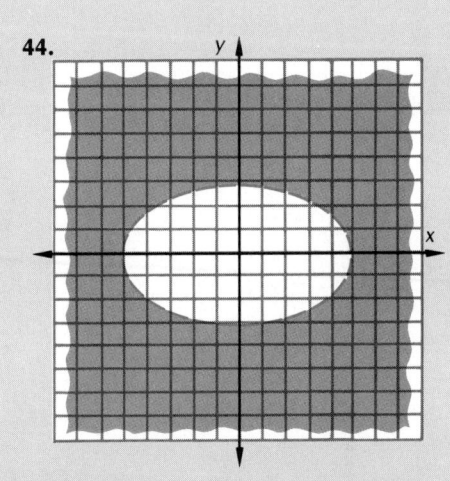

45.

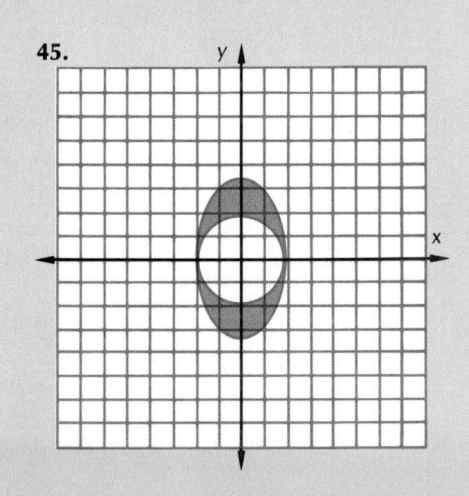

46.

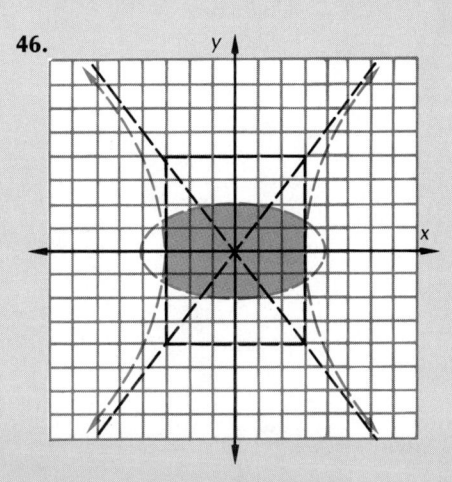

47. $(-2, -2)$, $\left(-8, -\dfrac{1}{2}\right)$
48. $\left(\dfrac{2\sqrt{3}}{3}, \dfrac{\sqrt{21}}{3}\right)$, $\left(\dfrac{2\sqrt{3}}{3}, -\dfrac{\sqrt{21}}{3}\right)$, $\left(-\dfrac{2\sqrt{3}}{3}, \dfrac{\sqrt{21}}{3}\right)$, $\left(-\dfrac{2\sqrt{3}}{3}, -\dfrac{\sqrt{21}}{3}\right)$

49. $(2, 3)$, $(-2, 3)$
50. $(2, 0)$, $(0, 2)$
51. -1 and -1
52. length is 5 meters, width is 3 meters.   or   length 3 meters, width 5 meters.
53. 2 miles wide, 4 miles long
54. $-\dfrac{1}{2}$

EXPONENTIAL AND LOGARITHMIC FUNCTIONS

10

10.1 Exponential Functions
10.2 Applications of Exponential Functions and the Base e
10.3 Logarithmic Functions
10.4 Properties of Logarithms
10.5 Computing with Common Logarithms
10.6 Logarithms, Exponentials, and the Calculator
10.7 Exponential and Logarithmic Equations

EXPONENTIAL FUNCTIONS

10.1

In Section 5.3 we defined rational exponents such that
$$a^{m/n} = \sqrt[n]{a^m} = (\sqrt[n]{a})^m$$
where $a \geq 0$ if n is an even integer. We now assume that irrational exponents also exist and follow the same properties as rational exponents. Thus, for positive bases a and b, we assume that the following properties hold for all real exponents, m and n.

1. $a^m \cdot a^n = a^{m+n}$
2. $\dfrac{a^m}{a^n} = a^{m-n}$, where $a \neq 0$
3. $(a^m)^n = a^{m \cdot n}$
4. $(ab)^n = a^n b^n$ and $\left(\dfrac{a}{b}\right)^n = \dfrac{a^n}{b^n}$, where $b \neq 0$

We are now able to define a new class of functions called *exponential functions*. Exponential functions differ greatly from any other type we have studied. Previously, we were concerned with algebraic functions which are defined using only the basic operations of addition, subtraction, multiplication, and division. Exponential functions are not algebraic since, as seen in the following definition, they contain a variable in the exponent.

Definition of an Exponential Function

An exponential function is of the form
$$f(x) = a^x$$
where $a > 0$ and $a \neq 1$. The constant a is called the base and the variable x may assume any real value.

The base of an exponential function is not allowed to be 1, because for all real x,
$$1^x = 1$$
This is simply a constant function defined by $f(x) = 1$.

To become acquainted with exponential functions, we graph several examples by plotting points and connecting them with a smooth curve.

EXAMPLE 1   Graph $y = 2^x$

Solution: Form a table of values and plot the corresponding points to obtain the graph in Figure 10.1.

EXPONENTIAL FUNCTIONS

x	y = 2^x
−3	$2^{-3} = \frac{1}{8}$
−2	$2^{-2} = \frac{1}{4}$
−1	$2^{-1} = \frac{1}{2}$
0	$2^0 = 1$
$\frac{1}{2}$	$2^{1/2} = \sqrt{2} \approx 1.4$
1	$2^1 = 2$
$\frac{3}{2}$	$2^{3/2} = \sqrt{8} \approx 2.8$
2	$2^2 = 4$
3	$2^3 = 8$
4	$2^4 = 16$

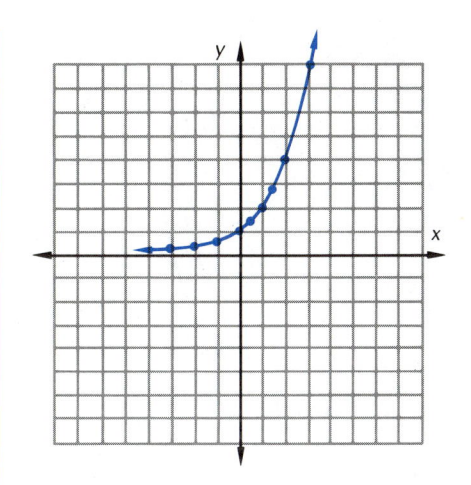

FIGURE 10.1

The graph in Figure 10.1 typifies the graphs of exponential functions of the form $f(x) = a^x$, where the base is greater than 1 ($a > 1$). As x gets larger, y becomes larger at an increasing rate. Thus, exponential functions of this type are said to be increasing functions.

This graph also shows the domain and range. Since the horizontal sweep covers the entire x-axis, the domain is the set of real numbers,

$$\{x \mid x \in R\}$$

The vertical sweep covers the positive y-axis; thus the range consists of the positive real numbers,

$$\{y \mid y > 0\}$$

In Figure 10.2, we have graphed several exponential functions of the form $y = a^x$, where $a > 1$. (You should check each graph by forming a table of values.) Note that as the base becomes larger, the graphs rise even faster. Also, you should see that each graph passes through the point (0, 1).

Next, we graph an exponential function that will typify the form $f(x) = a^x$, where the base is between 0 and 1, ($0 < a < 1$).

EXPONENTIAL AND LOGARITHMIC FUNCTIONS

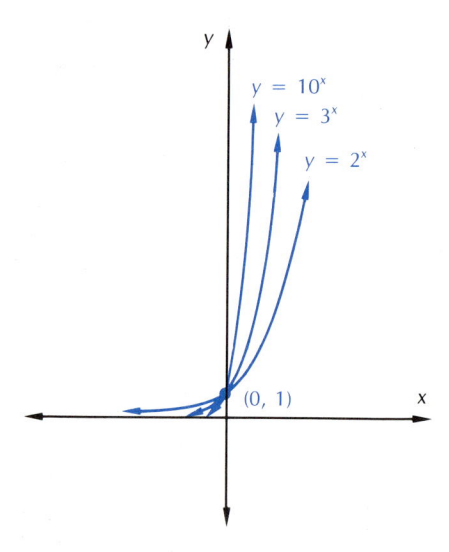

FIGURE 10.2

EXAMPLE 2 Graph $y = \left(\dfrac{1}{2}\right)^x$

Solution: Again we form a table of values and plot the corresponding points. As seen in Figure 10.3, the graph is similar to $y = 2^x$, except it is reversed. As x gets larger, y becomes smaller. Thus, we have a decreasing function.

x	$y = \left(\dfrac{1}{2}\right)^x$
-4	$\left(\dfrac{1}{2}\right)^{-4} = 16$
-3	$\left(\dfrac{1}{2}\right)^{-3} = 8$
-2	$\left(\dfrac{1}{2}\right)^{-2} = 4$
$-\dfrac{3}{2}$	$\left(\dfrac{1}{2}\right)^{-3/2} = \sqrt{8} \approx 2.8$
-1	$\left(\dfrac{1}{2}\right)^{-1} = 2$

x	$y = \left(\dfrac{1}{2}\right)^x$
$-\dfrac{1}{2}$	$\left(\dfrac{1}{2}\right)^{-1/2} = \sqrt{2} \approx 1.4$
0	$\left(\dfrac{1}{2}\right)^{0} = 1$
1	$\left(\dfrac{1}{2}\right)^{1} = \dfrac{1}{2}$
2	$\left(\dfrac{1}{2}\right)^{2} = \dfrac{1}{4}$
3	$\left(\dfrac{1}{2}\right)^{3} = \dfrac{1}{8}$

EXPONENTIAL FUNCTIONS

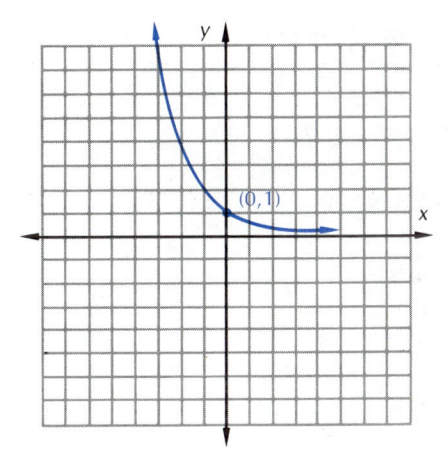

FIGURE 10.3

As illustrated by the graph in Figure 10.3, exponential functions of the form $f(x) = a^x$, where $0 < a < 1$, are decreasing functions. Their domain is the set of real numbers,

$$\{x \mid x \in R\}$$

and their range is the set of positive real numbers,

$$\{y \mid y > 0\}$$

In Figure 10.4 we have graphed several decreasing exponential functions. (Again, you should check these graphs by forming tables of values.) Note that as the base becomes smaller, the graphs fall even faster and, as before, they all share the point (0, 1).

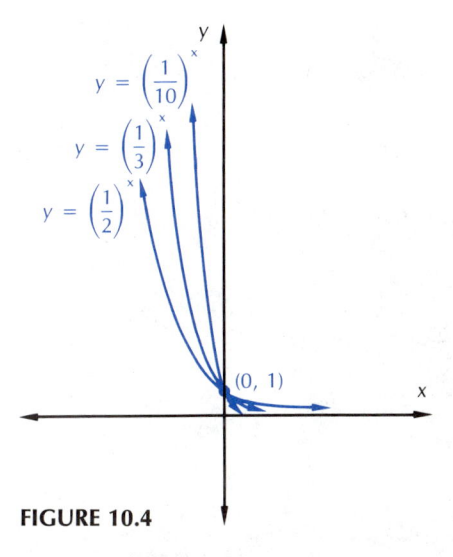

FIGURE 10.4

Sometimes decreasing exponential functions are written in an alternate form involving a minus sign in the exponent. For example,

$$y = \left(\frac{1}{2}\right)^x$$
$$= (2^{-1})^x$$
$$= 2^{-x}$$

EXAMPLE 3 The following are equivalent forms.

(a) $y = 3^{-x}$ and $y = \left(\frac{1}{3}\right)^x$

(b) $f(x) = \left(\frac{1}{10}\right)^x$ and $f(x) = 10^{-x}$

(c) $g(x) = \left(\frac{3}{2}\right)^{-x}$ and $g(x) = \left(\frac{2}{3}\right)^x$

It should be clear, by thinking of a graph, that a given exponential function will always be one to one. That is, it will never produce the same value of y for different values of x. This suggests the following property, which is used to solve many equations where the variable is contained in an exponent. Such equations are called exponential equations.

Equality Property of Exponentials
For $a > 0$ and $a \neq 1$, If $a^m = a^n$, then $m = n$

EXAMPLE 4 Solve $2^x = 8$

Solution: Write 8 as a power of 2, then use the equality property of exponentials.
$$2^x = 8$$
$$2^x = 2^3$$
$$x = 3$$

EXAMPLE 5 Solve $25^x = 125$

Solution: Since each base is a power of 5, we write 25 as 5^2 and 125 as 5^3.
$$25^x = 125$$
$$(5^2)^x = 5^3$$
$$5^{2x} = 5^3$$

Use the equality property of exponentials to set the exponents equal.
$$2x = 3$$
$$x = \frac{3}{2}$$

EXPONENTIAL FUNCTIONS   **421**

Summary of Exponential Functions	
$f(x) = a^x$, where $a > 1$, is an increasing function	$f(x) = a^x$, where $0 < a < 1$, is a decreasing function
(graph of increasing exponential through (0, 1))	*(graph of decreasing exponential through (0, 1))*
1. The greater the value of a, the faster the function increases. 2. The domain is $\{x \mid x \in R\}$. 3. The range is $\{y \mid y > 0\}$.	1. The smaller the value of a, the faster the function decreases. 2. The domain is $\{x \mid x \in R\}$. 3. The range is $\{y \mid y > 0\}$. 4. The function may be written with a negative exponent and a base greater than 1. For example, $y = \left(\dfrac{1}{2}\right)^x$ is $y = 2^{-x}$.

Exercise 10.1

Graph the following by forming a table of values and plotting the corresponding points.

1. $y = 4^x$
2. $y = 4^{-x}$
3. $y = \left(\dfrac{1}{2}\right)^x$
4. $y = \left(\dfrac{1}{3}\right)^x$
5. $y = -2^x$
6. $y = \left(\dfrac{1}{2}\right)^{-x}$

Give the domain and range and state whether the function is increasing or decreasing.

7. $f(x) = 4^x$
8. $f(x) = \left(\dfrac{1}{2}\right)^x$

9. $f(x) = 3^{-x}$

10. $f(x) = \left(\dfrac{1}{2}\right)^{-x}$

11. $f(x) = a^x$, $a > 1$

12. $f(x) = a^x$, $0 < a < 1$

Write an equivalent form with no minus sign in the exponent.

13. $y = 2^{-x}$

14. $y = \left(\dfrac{3}{4}\right)^{-x}$

15. $y = \left(\dfrac{1}{5}\right)^{-x}$

16. $y = 4^{-x}$

Solve for x.

17. $5^x = 25$

18. $2^x = \dfrac{1}{8}$

19. $4^x = 32$

20. $4^x = 64$

21. $10^x = .01$

22. $3^x = \dfrac{1}{27}$

23. $5^{-x} = \dfrac{1}{5}$

24. $2^x = 1$

25. $3^{10} = 3^{5x}$

26. $\left(\dfrac{1}{2}\right)^x = 8$

27. $\left(\dfrac{3}{4}\right)^x = \dfrac{16}{9}$

28. $2^x = 8^{x+2}$

29. $3^x = 9^{x-2}$

30. $2^x = 4^{x+1}$

31. $\left(\dfrac{2}{3}\right)^x = \dfrac{27}{8}$

32. $3^{-x} = 243$

33. A certain strain of bacteria takes 1 day to reproduce by dividing in half. If there are 50 bacteria present to begin with, the total number present after x days is given by $f(x) = 50(2)^x$. Find
 a) the number present after 2 days
 b) the number present after 3 days
 c) the number present after 4 days
 d) how many days must elapse before 51,200 bacteria are present

34. If the pulse rate of a normal healthy adult is approximated by $P = 94h^{-(1/2)}$, where P is the pulse rate in beats per minute and h is the height in meters, find to the nearest whole number the pulse rate of a person:
 a) 5 feet 8 inches tall (1.7272 meters)
 b) 5 feet 0 inches tall (1.524 meters)
 c) 6 feet 2 inches tall (1.8796 meters)

APPLICATIONS OF EXPONENTIAL FUNCTIONS AND THE BASE e

10.2

Exponential functions occur in many formulas used to describe a wide variety of practical problems involving either growth or decline. In this section we discuss three such applications: compound interest, population growth, and radioactive decay.

In finance, many transactions involve compound interest. This happens when interest also earns interest by being credited to principal at regular time intervals. The amount of money accumulated in an account earning compound interest is given by the following formula.

Compound Interest

If P represents the principal invested at an annual interest rate r, compounded t times yearly, then the amount A accumulated in n years is given by

$$A = P\left(1 + \frac{r}{t}\right)^{tn}$$

EXAMPLE 1

What is the accumulated amount in a savings account where $5,000 is invested for two years at 10% compounded annually?

Solution: The principal P is 5,000 and the annual interest rate r is 0.1. Since interest is compounded once a year, t is 1. And, n is 2 because the accumulation is for two years. We substitute into the compound interest formula and solve for A as follows:

$$A = P\left(1 + \frac{r}{t}\right)^{tn}$$
$$= 5{,}000\left(1 + \frac{0.1}{1}\right)^{1 \cdot 2}$$
$$= 5{,}000(1.1)^2$$
$$= 5{,}000(1.21)$$
$$= 6{,}050$$

The accumulated amount is $6,050.

Problems complicated by larger exponents can be worked on a calculator, as demonstrated by the next example.

EXAMPLE 2

Find the accumulated amount of a savings account after 5 years 6 months, where $10,000 has been invested at 13% compounded quarterly.

Solution: $P = 10{,}000$, $r = 0.13$, $t = 4$, and $n = 5.5$.

We substitute into the compound interest formula and use a calculator as follows:

$$A = P\left(1 + \frac{r}{t}\right)^{tn}$$

$$= 10{,}000\left(1 + \frac{0.13}{4}\right)^{4(5.5)}$$

$$= 10{,}000\left(1 + \frac{0.13}{4}\right)^{22}$$

Calculator Sequence: (10,000) (×) (((((+) (.13) (÷) (4) ( ))
(y^x) (22) (=) 20,210.699

The accumulated amount is $20,210.70.

Many equations describing growth or decline involve an expression similar to

$$\left(1 + \frac{1}{m}\right)^m$$

In Table 10.1, we evaluate this expression for progressively larger and larger values of m to show that it does not exceed a number which is approximately 2.7182818, symbolized by e. The number e is irrational and is called the natural base.

TABLE 10.1

m	1	5	10	100	1,000	10,000	100,000	1,000,000
$\left(1 + \frac{1}{m}\right)^m$	2.0	2.48832	2.59374	2.70481	2.71692	2.7181	2.71828	2.7182818

As illustrated in succeeding examples, the number e has many applications since it describes continuous growth (or decay) at a rate proportional to the amount present. In fact, it is used as a base so often that it is convenient to have a table giving positive and negative powers of e. These are found in Table II in the appendix beginning on page A-3.

Many calculators give powers of e by merely entering the exponent, then pressing the key labeled (e^x). Otherwise, the exponential key (y^x) can be used with a base of 2.7182818.

EXAMPLE 3 The following powers of e have been found by using either Table II or a calculator.

(a) $e^{1.27} = 3.561$ [from Table II]

(b) $e^{1.27} = 3.5608526$ [by Calculator]

(c) $e^{-0.39} = 0.677$ [from Table II]
(d) $e^{-0.39} = 0.67705687$ [by Calculator]

We realize that these powers of e have been rounded off and are approximations. However, it is traditional to still use the equal sign.

Many types of continuous growth or decline (decay) are described by exponential equations of the form

$$y = y_0 e^{kt}$$

where t is time, y_0 is the initial amount present when $t = 0$, and k is a constant representing the rate of increase or decrease.

EXAMPLE 4

A city that had a population of 25,000 in 1987 projects an increase of 8% per year. What population does the city expect in the year 2002?

Solution: t represents 15 years, y_0 is the initial population of 25,000, and k is 0.08, representing the 8% yearly increase. We substitute these values into the formula and solve for y.

$$y = y_0 e^{kt}$$
$$= 25{,}000 e^{(.08)(15)}$$
$$= 25{,}000 e^{1.2}$$
$$= 25{,}000(3.320) \quad \text{[from Table II]}$$
$$= 83{,}000$$

The population in the year 2002 is expected to be 83,000.

Next, by looking at radioactive decay, we illustrate how the same type of exponential equation describes a decline rather than a growth. To better understand this example, it is helpful to know that in radioactive material some of the atoms continually cease to be radioactive. This is called radioactive decay and causes a decrease in the amount of radioactive material.

EXAMPLE 5

The decay of strontium 90 is described by

$$y = y_0 e^{-0.028t},$$

where y_0 is the initial amount of radioactive material and y is the amount left after t years. If we begin with 5 grams, how much will be radioactive after 10 years?

Solution: Substitute 5 for y_0 and 10 for t, then solve for y.

$$y = y_0 e^{-0.028t}$$
$$= 5 e^{(-0.028)(10)}$$
$$= 5 e^{-0.28}$$
$$= 5(0.756) \quad \text{[from Table II]}$$
$$= 3.78$$

After 10 years, 3.78 grams is still radioactive.

Exercise 10.2

Find the following powers of e, first giving the answers using Table II; and then giving the answers using a calculator.

1. $e^{1.2}$
2. $e^{.68}$
3. $e^{2.68}$
4. $e^{-.29}$
5. $e^{-1.62}$
6. $e^{-2.67}$
7. $e^{1.24}$
8. $e^{4.5}$

Using the compound interest formula $A = P\left(1 + \dfrac{r}{t}\right)^{tn}$, where P is the amount invested, r the annual interest rate, t the number of times the compounding occurs in a year, n the number of years, and A the accumulated amount, solve the following problems.

9. Find the amount accumulated if $200 is invested for 4 years at 6% compounded annually.
10. Find the amount accumulated if $500 is invested for 3 years at 8% compounded quarterly.
11. Find the amount accumulated if $4,000 is invested for 2 years at 12% compounded semi-annually.
12. After 4 years 6 months, find the accumulated amount when $5,500 is invested in a savings account at 11.5% interest compounded quarterly.

The scrap value of an item is the value of the item at the end of its useful life. If S is the scrap value, then $S = P(1 - r)^n$, where P is the purchase price, r is the annual rate of depreciation, and n is the number of years. Use the scrap value formula to solve the following.

13. A truck costs $15,000 and is used 6 years. It is depreciated 15% each year. What is its scrap value?
14. A salesman purchased a $10,000 car and depreciated it 10% for 4 years. What was its scrap value at the end of 4 years?
15. A business man purchased some equipment for his office for $4,000 and depreciated it for 5 years at 15% a year. What is the scrap value at the end of 5 years?
16. A business firm bought a small computer for $20,000 and depreciated it at 20% a year for 4 years. What is the scrap value at the end of 4 years?

The atmospheric pressure P in inches of mercury is given by $P = 30(10)^{-.09a}$, where a is the altitude in miles above sea level. Solve the following to the nearest tenth.

17. Find the pressure in inches of mercury at sea level.
18. Find the pressure in inches of mercury 3 miles above sea level.
19. Find the pressure in inches of mercury $\dfrac{1}{2}$ mile above sea level.
20. Find the pressure in inches of mercury in the mile-high city of Denver.

APPLICATIONS OF EXPONENTIAL FUNCTIONS AND THE BASE e 427

The atmospheric pressure P in pounds per square inch is given by $P = 14.7e^{-.21a}$, where a is the altitude in miles above sea level. Solve the following to the nearest tenth.

21. Find the pressure in pounds per square inch at sea level.
22. Find the pressure in pounds per square inch 3 miles above sea level.
23. Find the pressure in pounds per square inch $\frac{1}{2}$ mile above sea level.
24. Find the pressure in pounds per square inch in the mile-high city of Denver.

Population is described by $y = y_0 e^{kt}$, where t is time, y_0 is the initial amount present when $t = 0$, and k is a constant representing the rate of increase or decrease.

25. A city that had a population of 40,000 in 1987 projects an increase of 5% per year. What population does the city expect in 1997?
26. California in 1970 had a population of 15,717,000. Assuming a rate of growth of 2.4%, estimate the population in 1990, rounding the answer to the nearest thousand.
27. If the population of the United States in 1980 was 250 million, estimate to the nearest million the population in the year 2000, assuming a growth of 1.8% per year.
28. If Australia had a 1973 population of 13,268,000 and assumes a growth of 1.84% a year, estimate the population in 1988, rounding the answer to the nearest thousand.

The monthly payment M is given by $M = \dfrac{P \cdot i}{1 - \dfrac{1}{(1 + i)^n}}$, where P is the dollars borrowed, n is the number of months, and i is the monthly interest rate.

29. Find the monthly payment if $50,000 is borrowed for 20 years at 12% per year.
30. Find the monthly payment for buying a used car for $4,000 if it is financed for 3 years at 14% per year.
31. Find the monthly payment for borrowing $80,000 to buy a house at 15% for 30 years.
32. A new car costs $12,850. The down payment was $2,000. The balance was financed at 14% for 5 years. What is the monthly payment?

The amount A of carbon 14 that a non-living substance contains is given by the formula $A = A_0 \left(\dfrac{1}{2}\right)^{t/5600}$, where A_0 is the amount of carbon 14 the substance contains at a given time and t is the number of years later. Solve the following to the nearest hundredth.

428   EXPONENTIAL AND LOGARITHMIC FUNCTIONS

33. A substance contains 4 micrograms of carbon 14. How much will be left after 300 years?
34. A substance contains 3 micrograms of carbon 14. How much will be left after 2,000 years?
35. A substance contains 5 micrograms of carbon 14. How much will it contain after 10,000 years?
36. In how many years would a substance that contains 1 microgram of carbon 14 contain $\frac{1}{2}$ microgram of carbon 14?

The decay of strontium 90 is given by $y = y_0 e^{-0.028t}$, where y_0 is the initial amount of radioactive material and y is the amount left after t years. Solve the following to the nearest hundredth.

37. If 4 grams of strontium 90 is radioactive, how much will be radioactive after 50 years?
38. If 2 grams of strontium 90 is radioactive, how much will be radioactive after 100 years?

LOGARITHMIC FUNCTIONS

10.3

The inverse of an exponential function of the form $y = a^x$, where $a > 0$ and $a \neq 1$, is called a logarithmic function.

In Section 7.3 you learned that an inverse function is found by interchanging x with y and then solving for y. The domain and range are also interchanged. Recall that the graph of an inverse function is the reflection across the line $y = x$.

To obtain inverses of exponential functions we proceed as follows:

1. The form $y = a^x$, where $a > 0$ and $a \neq 1$, describes all exponential functions. Each of these functions is one to one, so the inverse function will exist. The domain of each exponential function is

$$\{x \mid x \in R\}$$

The range is

$$\{y \mid y > 0\}$$

2. To form the inverse, called a logarithmic function, we interchange x and y, producing the equation

$$x = a^y$$

where $a > 0$ and $a \neq 1$. The domain is
$$\{x \mid x > 0\}$$
The range is
$$\{y \mid y \in R\}$$

3. The equation $x = a^y$ should now be solved for y. However, at present we have no direct way of doing this. So, we introduce the new notation, $y = \log_a x$. This is read, "y equals the logarithm, base a, of x." It means that y is the exponent on the base a that produces x.

Definition of Logarithmic Functions

For $a > 0$ and $a \neq 1$, $y = \log_a x$ is equivalent to $x = a^y$. The domain is $\{x \mid x > 0\}$ and the range is $\{y \mid y \in R\}$.

The statements $y = \log_a x$ and $x = a^y$ are two different forms defining the same function. Thus, logarithmic statements can be written in exponential form and exponential statements can be written in logarithmic form. The following chart shows several equivalent statements. Study them carefully.

Logarithmic Form	*Exponential Form*
$\log_a x = y$	$a^y = x$
$\log_2 8 = 3$	$2^3 = 8$
$\log_{10} 1000 = 3$	$10^3 = 1000$
$\log_{10} 1 = 0$	$10^0 = 1$
$\log_5 \dfrac{1}{5} = -1$	$5^{-1} = \dfrac{1}{5}$
$\log_{1/3} 9 = -2$	$\left(\dfrac{1}{3}\right)^{-2} = 9$
$\log_4 8 = \dfrac{3}{2}$	$4^{3/2} = 8$

As illustrated by the graphs in Figure 10.5, logarithmic functions, like exponential functions, are one to one and either increasing or decreasing, depending on their base. Note that graphs of logarithmic functions pass through the point (1, 0).

430 EXPONENTIAL AND LOGARITHMIC FUNCTIONS

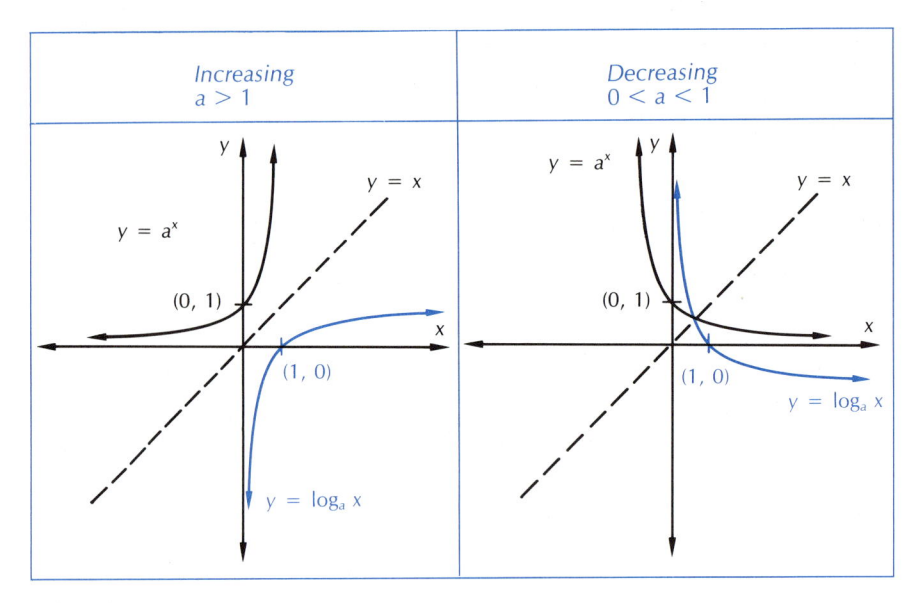

FIGURE 10.5

EXAMPLE 1 Graph $y = \log_2 x$

Solution: Write the equation in exponential form and make a table of values. Since $y = \log_2 x$ is equivalent to $x = 2^y$, it will be easier to choose y values and then find the corresponding values of x. The graph is shown in Figure 10.6. It is increasing because $a > 1$.

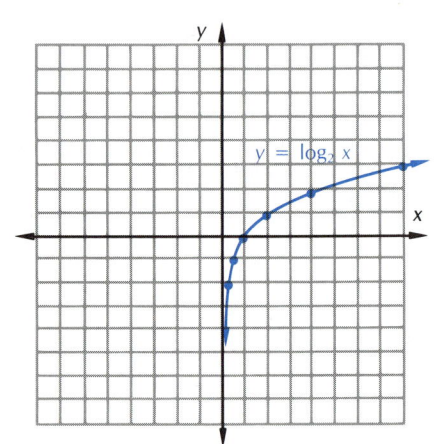

$x = 2^y$	y
$\frac{1}{4}$	-2
$\frac{1}{2}$	-1
1	0
2	1
4	2
8	3

FIGURE 10.6

EXAMPLE 2 Graph $y = \log_{1/2} x$

Solution: We again write the equation in exponential form, $x = \left(\dfrac{1}{2}\right)^y$, and make a table of values. The graph, as shown in Figure 10.7, is decreasing because $0 < a < 1$.

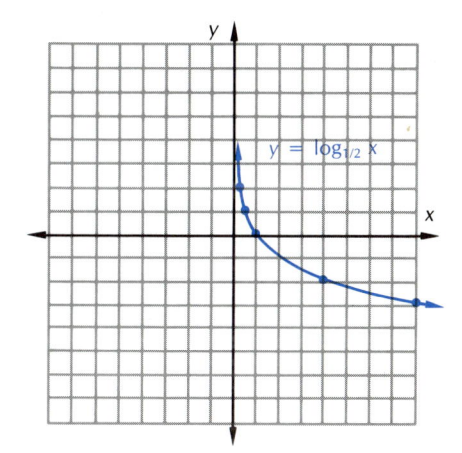

$x = \left(\dfrac{1}{2}\right)^y$	y
8	-3
4	-2
2	-1
1	0
$\dfrac{1}{2}$	1
$\dfrac{1}{4}$	2

FIGURE 10.7

Equations containing logarithms are called *logarithmic equations*. Often they can be solved easily by writing them in exponential form.

Solve $\log_6 x = -2$ **EXAMPLE 3**

Solution: $\log_6 x = -2$ is equivalent to $6^{-2} = x$. Thus,

$$x = 6^{-2} = \frac{1}{6^2} = \frac{1}{36}$$

Solve $\log_x 25 = 2$ **EXAMPLE 4**

Solution: $\log_x 25 = 2$ is equivalent to $x^2 = 25$. Thus,

$x^2 = 25$

$x = \pm 5$

However, the base of a logarithm can never be negative. Hence, the only solution is 5.

Graph the following (1–6). **Exercise 10.3**

1. $y = \log_3 x$
2. $y = \log_4 x$
3. $y = \log_{1/4} x$
4. $y = \log_{1/3} x$
5. $y = \log_2(x - 1)$
6. $y = \log_3(x - 2)$
7. Give the domain and range of $y = \log_3 x$. Does it increase or decrease?
8. Give the domain and range of $y = \log_{1/3} x$. Does it increase or decrease?

Change to log form.

9. $2^6 = 64$
10. $10^0 = 1$
11. $4^3 = 64$
12. $3^{-2} = \frac{1}{9}$
13. $4^{-2} = \frac{1}{16}$
14. $3^3 = 27$
15. $.01 = 10^{-2}$
16. $27 = 81^{3/4}$

Change to exponential form.

17. $\log_5 25 = 2$
18. $\log_7 49 = 2$
19. $\log_2 \frac{1}{2} = -1$
20. $\log_2 32 = 5$
21. $\log_3 27 = 3$
22. $\log_6 1 = 0$
23. $\log_8 2 = \frac{1}{3}$
24. $\log_{.1} .01 = 2$

Find the value.

25. $\log_6 36$
26. $\log_{10} 100$
27. $\log_4 2$
28. $\log_9 3$
29. $\log_\pi \pi$
30. $\log_2 16$
31. $\log_8 2$
32. $\log_8 8$
33. $\log_9 \frac{1}{9}$
34. $\log_7 \sqrt{7}$
35. $\log_2 8 - \log_4 2$
36. $\log_4 \frac{1}{2} - \log_6 \frac{1}{6}$
37. $\log_3 9 + \log_9 3$
38. $\log_3 \frac{1}{9} + \log_9 \frac{1}{3} + \log_3 1$

Solve the following equations.

39. $\log_2 x = -1$
40. $\log_{10} x = -4$
41. $\log_b 25 = -2$
42. $\log_8 x = \frac{2}{3}$
43. $\log_3 27 = x$
44. $\log_8 2 = x$
45. $\log_4 x = \frac{3}{2}$
46. $\log_6 1 = x$
47. $\log_{25} 125 = x$
48. $\log_b 81 = 2$

PROPERTIES OF LOGARITHMS

10.4

Historically, logarithms have been an important aid in simplifying numerical calculations. In this section we develop several important properties which allow us to apply logarithms. Since logarithms, by definition, are exponents, these properties will parallel the rules of exponents studied in earlier chapters.

Addition Property of Logarithms

For identical bases, the logarithm of a product is equal to the sum of the logarithms of its factors. Thus,

$$\log_a(m \cdot n) = \log_a m + \log_a n$$

where $m, n > 0$ and a is any logarithm base.

Example: $\log_{10}(10 \cdot 100) = \underbrace{\log_{10} 10}_{1} + \underbrace{\log_{10} 100}_{2}$

$\underbrace{\log_{10} 1{,}000}_{3}$

To prove this property, we let $\log_a m = x$ and $\log_a n = y$. Then, changing to exponential form gives

$$a^x = m \text{ and } a^y = n$$

By the addition property of exponents,

$$m \cdot n = a^x \cdot a^y = a^{x+y}$$

In logarithmic form, the equation $m \cdot n = a^{x+y}$ becomes

$$\log_a(m \cdot n) = x + y$$

Substituting for x and y gives

$$\log_a(m \cdot n) = \log_a m + \log_a n$$

which completes the proof.

EXAMPLE 1

We use the addition property of logarithms to rewrite the following (assume $x > 0$).

(a) $\log_2(5 \cdot 11) = \log_2 5 + \log_2 11$
(b) $\log_7 9 + \log_7 5 = \log_7(9 \cdot 5) = \log_7 45$
(c) $\log_8(5x) = \log_8 5 + \log_8 x$
(d) $\log_{10} x^2 = \log_{10}(x \cdot x)$

$\qquad = \log_{10} x + \log_{10} x$

$\qquad = 2 \log_{10} x$

(e) $\log_{10}(100x) = \log_{10} 100 + \log_{10} x$

$\qquad = 2 + \log_{10} x$, since $\log_{10} 100 = 2$

434 EXPONENTIAL AND LOGARITHMIC FUNCTIONS

> **Subtraction Property of Logarithms**
>
> For identical bases, the logarithm of a quotient is equal to the logarithm of the numerator minus the logarithm of the denominator. Thus,
>
> $$\log_a \frac{m}{n} = \log_a m - \log_a n$$
>
> where $m, n > 0$ and a is any logarithm base.
>
> Example: $\log_{10} \dfrac{1{,}000}{100} = \underbrace{\log_{10} 1{,}000}_{3} - \underbrace{\log_{10} 100}_{2}$
>
> $\underbrace{\phantom{\log_{10} 10}}_{1}$ $\log_{10} 10$

To prove this property, we again let $\log_a m = x$ and $\log_a n = y$. Changing to exponential form gives

$$a^x = m \text{ and } a^y = n$$

By the subtraction property of exponents,

$$\frac{m}{n} = \frac{a^x}{a^y} = a^{x-y}$$

In logarithmic form, the equation $\dfrac{m}{n} = a^{x-y}$ becomes

$$\log_a \frac{m}{n} = x - y$$

Substituting for x and y produces

$$\log_a \frac{m}{n} = \log_a m - \log_a n$$

which completes the proof.

EXAMPLE 2 The following have been rewritten using the subtraction property of logarithms.

(a) $\log_3 \dfrac{15}{2} = \log_3 15 - \log_3 2$

(b) $\log_8 13 - \log_8 17 = \log_8 \dfrac{13}{17}$

The addition and subtraction properties can be combined to rewrite certain logarithmic expressions.

EXAMPLE 3 The following expressions have been rewritten using the previous properties (assume $x, y > 0$).

(a) $\log_{10} \dfrac{xy}{3} = \log_{10} xy - \log_{10} 3$

$= \log_{10} x + \log_{10} y - \log_{10} 3$

(b) $\log_7 5 - (\log_7 2 - \log_7 x) = \log_7 5 - \log_7 2 + \log_7 x$

$$= \log_7 \frac{5}{2} + \log_7 x$$

$$= \log_7 \left[\frac{5}{2} \cdot x\right]$$

$$= \log_7 \frac{5x}{2}$$

Power Property of Logarithms

For identical bases, the logarithm of a number to a power is equal to the power (or exponent) times the logarithm of the number. Thus,

$$\log_a m^n = n \cdot \log_a m$$

where $m > 0$, $n \in R$, and a is any logarithm base.

Example: $\underbrace{\log_{10} 10^3}_{\log_{10} 1{,}000 \;=\; 3} = \underbrace{3 \cdot \log_{10} 10}_{3 \cdot 1}$

To prove this property, let $\log_a m = x$. Changing to exponential form gives

$$a^x = m$$

Next we raise both sides to the power n and apply the power to a power property of exponents. This gives

$$(a^x)^n = m^n \text{ or } a^{n \cdot x} = m^n$$

In logarithmic form, the equation $a^{n \cdot x} = m^n$ becomes

$$\log_a m^n = n \cdot x$$

Substituting for x produces

$$\log_a m^n = n \cdot \log_a m$$

which completes the proof.

EXAMPLE 4

These expressions have been rewritten using the power property of logarithms (assume $x > 0$).

(a) $\log_3 25^6 = 6 \cdot \log_3 25$

(b) $3 \log_{10} x = \log_{10} x^3$

As a special case, the power property of logarithms also holds for roots. The expression $\log_a \sqrt[r]{m}$ means $\log_a m^{1/r}$.

Letting $n = \dfrac{1}{r}$ in the power property gives

$$\log_a \sqrt[r]{m} = \frac{1}{r} \cdot \log_a m$$

436   EXPONENTIAL AND LOGARITHMIC FUNCTIONS

EXAMPLE 5   The power property of logarithms has been applied to these expressions involving roots (assume $x > 0$).

(a) $\log_5 \sqrt{x} = \log_5 x^{1/2} = \dfrac{1}{2} \log_5 x$

(b) $\log_{10} \sqrt[4]{x^3} = \log_{10} x^{3/4} = \dfrac{3}{4} \log_{10} x$

(c) $\dfrac{5}{3} \log_2 7 = \log_2 \sqrt[3]{7^5}$

The properties of logarithms can be used in conjunction with each other to expand a logarithm to sums or differences, or to simplify expressions to a single logarithm. (For the remainder of our work with logarithms, we shall assume that variables obey the necessary restrictions.)

EXAMPLE 6   Expand $\log_a \dfrac{2\sqrt[3]{x}}{y^2}$

Solution: The expression is a basic quotient, so we begin by applying the subtraction property of logarithms.

$\log_a \dfrac{2\sqrt[3]{x}}{y^2} = \log_a(2\sqrt[3]{x}) - \log y^2$ [subtraction property]

$\phantom{\log_a \dfrac{2\sqrt[3]{x}}{y^2}} = \log_a 2 + \log_a \sqrt[3]{x} - \log y^2$ [addition property]

$\phantom{\log_a \dfrac{2\sqrt[3]{x}}{y^2}} = \log_a 2 + \dfrac{1}{3} \log_a x - 2 \log y$ [power property]

EXAMPLE 7   Simplify to a single logarithm:

$\log_a 3 + 4 \log_a x - \dfrac{1}{2} \log_a y$

Solution: First apply the power property so that the coefficient of each logarithm is 1.

$\log_a 3 + 4 \log_a x - \dfrac{1}{2} \log_a y = \log_a 3 + \log_a x^4 - \log_a \sqrt{y}$ [power property]

$\phantom{\log_a 3 + 4 \log_a x - \dfrac{1}{2} \log_a y} = \log_a 3x^4 - \log_a \sqrt{y}$ [addition property]

$\phantom{\log_a 3 + 4 \log_a x - \dfrac{1}{2} \log_a y} = \log_a \dfrac{3x^4}{\sqrt{y}}$ [subtraction property]

EXAMPLE 8   Given that $\log_5 2 = 0.4307$ and $\log_5 7 = 1.2091$, the properties of logarithms have been applied to evaluate the following.

(a) $\log_5 14$; since $14 = 2 \cdot 7$, we use the addition property:
$$\log_5 14 = \log_5(2 \cdot 7)$$
$$= \log_5 2 + \log_5 7$$
$$= 0.4307 + 1.2091$$
$$= 1.6398$$

(b) $\log_5 3.5$; since $3.5 = \frac{7}{2}$, we use the subtraction property:
$$\log_5 3.5 = \log_5 \frac{7}{2}$$
$$= \log_5 7 - \log_5 2$$
$$= 1.2091 - 0.4307$$
$$= 0.7784$$

(c) $\log_5 8$; since $8 = 2^3$, we use the power property:
$$\log_5 8 = \log_5 2^3$$
$$= 3 \log_5 2$$
$$= 3(0.4307)$$
$$= 1.2921$$

Since logarithmic functions are one to one, we can state a final property that will be particularly useful in later units.

Equality Property for Logarithms

For $m, n > 0$ and where a is any logarithm base,
$$\text{if } m = n, \text{ then } \log_a m = \log_a n$$

This property allows us to take logarithms of both sides of an equation, providing they represent positive numbers.

Exercise 10.4

Rewrite the following using the addition property of logarithms.
1. $\log_4 7x$
2. $\log_{10} 5xy$
3. $\log_3 2\pi r$
4. $\log_{10} 100xyz$

Expand the following using the addition and/or subtraction property of logarithms.
5. $\log_{10} \frac{R}{V}$
6. $\log_3 \frac{xy}{z}$

7. $\log_2 \dfrac{x}{yz}$

8. $\log_{10} \dfrac{4\pi}{3r}$

Expand the following using the power, addition, and/or subtraction properties of logarithms.

9. $\log_{10} x^5$

10. $\log_2 5xy^4$

11. $\log_{10} \sqrt{7}$

12. $\log_3 \sqrt[3]{x^2}$

13. $\log_4 \dfrac{4x}{y}$

14. $\log_{10} \dfrac{x^2}{y^3}$

Expand using the properties of logarithms.

15. $\log_{10} \dfrac{5}{(4x)^3}$

16. $\log_3 \dfrac{2\sqrt{3}}{5}$

17. $\log_5 \sqrt{\dfrac{xz}{y}}$

18. $\log_b \sqrt[3]{\dfrac{x^5}{y^3 z^2}}$

19. $\log_3 \dfrac{\sqrt[3]{x}\sqrt[4]{y^3}}{\sqrt[5]{z}}$

20. $\log_{10} \dfrac{(x+1)^2}{(x-1)}$

21. $\log_{10} \sqrt{s(s-a)(s-b)(s-c)}$

22. $\log_{10} 2\pi \sqrt{\dfrac{l}{g}}$

Write as a single logarithm and simplify where possible.

23. $\dfrac{1}{2}\log_2 5 + \log_2 7 - 2\log_2 15$

24. $\log_b 2 - \dfrac{1}{2}\log_b 3 + 4\log_b 5$

25. $3\log_6 x - 2\log_6 t - \log_6 y$

26. $-\log_{10} k - \log_{10} m$

27. $2\log_b 3xy - \log_b 3x^3y^2 + \log_b 2y$

28. $3\log_b 2xy - \log_b 2xy^2 + 2\log_b 3x$

29. $\dfrac{1}{2}(\log_b x + \log_b y - 4\log_b z)$

30. $-\dfrac{1}{2}\log_{10} x - \log_{10} y$

Give that $\log_{10} 3 = .4771$, find the value of each logarithm by using the properties of logarithms.

31. $\log_{10} 30$

32. $\log_{10} .3$

33. $\log_{10} 27$

34. $\log_{10} \sqrt{300}$

Given that $\log_{10} 2 = .3010$ and $\log_{10} 3 = .4771$, find the value of each logarithm by using the properties of logarithms.

35. $\log_{10} 6$

36. $\log_{10} 81$

37. $\log_{10} 36$

38. $\log_{10} 72$

Given that $\log_{10} 2 = .3010$, $\log_{10} 3 = .4771$, and $\log_{10} 5 = .6990$, find the value of each logarithm by using the properties of logarithms.

39. $\log_{10} 200$

40. $\log_{10} \dfrac{15}{2}$

41. $\log_{10} \sqrt{15}$

42. $\log_{10} \dfrac{5}{6}$

Proofs.

43. Prove: $\log_a a = 1$, where $a > 0$, $a \neq 1$.

44. Prove: $\log_a \dfrac{1}{x} = -\log x$, where $a > 0$, $a \neq 1$.

COMPUTING WITH COMMON LOGARITHMS

10.5

Base ten logarithms are known as common logarithms. For convenience, $\log_{10} x$ is written $\log x$.

Common logarithms were developed before the advent of computers and calculators in order to ease the complexity of certain types of calculations. Today, computing with logarithms is primarily of historical interest, although it is still helpful for finding powers and roots on calculators not having such keys. Also, computing with logarithms is an aid to understanding their properties which are still of importance.

In order for you to compute with logarithms, we have provided Table IV, Common Logarithms, beginning on page A-8 in the Appendix. This table can be used to find a common logarithm (to the nearest ten-thousandth) of any positive real number rounded to three significant digits. To use this table, the number must be written in scientific notation. To take an example, to find log 16,700, write the number in scientific notation and apply the properties of logarithms. Thus,

$$\begin{aligned}\log 16{,}700 &= \log(1.67 \times 10^4) \\ &= \log 1.67 + \log 10^4 \\ &= \log 1.67 + 4 \\ &= 4 + \log 1.67\end{aligned}$$

Next, to find log 1.67 refer to Table IV, which is partially reproduced here. Read down on the left to the row coinciding with 1.6, then across to the column headed by 7. You should find the decimal .2227. Thus,

$$\begin{aligned}\log 16{,}700 &= 4 + .2227 \\ &= 4.2227\end{aligned}$$

Common Logarithms (from Table IV, Appendix)

N	0	1	2	3	4	5	6	7	8	9
1.0	.0000	.0043	.0086	.0128	.0170	.0212	.0253	.0294	.0334	.0374
1.1	.0414	.0453	.0492	.0531	.0569	.0607	.0645	.0682	.0719	.0755
1.2	.0792	.0828	.0864	.0899	.0934	.0969	.1004	.1038	.1072	.1106
1.3	.1139	.1173	.1206	.1239	.1271	.1303	.1335	.1367	.1399	.1430
1.4	.1461	.1492	.1523	.1553	.1584	.1614	.1644	.1673	.1703	.1732
1.5	.1761	.1790	.1818	.1847	.1875	.1903	.1931	.1959	.1987	.2014
1.6	.2041	.2068	.2095	.2122	.2148	.2175	.2201	.2227	.2253	.2279
1.7	.2304	.2330	.2355	.2380	.2405	.2430	.2455	.2480	.2504	.2529
1.8	.2553	.2577	.2601	.2625	.2648	.2672	.2695	.2718	.2742	.2765
1.9	.2788	.2810	.2833	.2856	.2878	.2900	.2923	.2945	.2967	.2989

As seen below, the integer part of this logarithm, 4, is called the characteristic, while the decimal part, .2227, is called the mantissa.

log 16,700 = 4.2227

        ↗   ↖

characteristic   mantissa

Most logarithms are irrational numbers, so the numbers we obtain are approximations. However, it is traditional to use the equal sign.

All mantissas in Table IV are positive. This is particularly important to remember when a characteristic is negative. For example, to find the log 0.00595, we again write the number in scientific notation and use Table IV. Thus,

$$\log 0.00595 = \log(5.95 \times 10^{-3})$$
$$= \log 5.95 + \log 10^{-3}$$
$$= \log 5.95 + (-3)$$
$$= -3 + \log 5.95$$

The characteristic is -3 and the mantissa, as shown below, is .7745.

N	0	1	2	3	4	5	6	7	8	9
5.5	.7404	.7412	.7419	.7427	.7435	.7443	.7451	.7459	.7466	.7474
5.6	.7482	.7490	.7497	.7505	.7513	.7520	.7528	.7536	.7543	.7551
5.7	.7559	.7566	.7574	.7582	.7589	.7597	.7604	.7612	.7619	.7627
5.8	.7634	.7642	.7649	.7657	.7664	.7672	.7679	.7686	.7694	.7701
5.9	.7709	.7716	.7723	.7731	.7738	.7745	.7752	.7760	.7767	.7774

Thus,

log 0.00595 = -3 + .7745

It is incorrect to write -3.7745 because the minus sign would imply that both parts were negative:

$$-3.7745 = -3 - .7745 \neq -3 + .7745$$

COMPUTING WITH COMMON LOGARITHMS

When a characteristic is negative, we avoid confusion by placing it to the right of the mantissa. So,
$$\log 0.00595 = 0.7745 - 3$$
Alternate forms also exist by adding and subtracting the same number to keep the net sum of the first and last digit equal to the characteristic. Thus,

$$\log 0.00595 = \underbrace{0.7745 - 3}_{\text{net sum is } -3}$$

$$= \underbrace{2.7745 - 5}_{\text{net sum is } -3} \quad \text{[add and subtract 2]}$$

$$= \underbrace{7.7745 - 10}_{\text{net sum is } -3} \quad \text{[add and subtract 7]}$$

EXAMPLE 1 Use scientific notation together with Table IV in the Appendix to find the following logarithms. (Round to three significant digits before using Table IV.)

(a) $\log 869 = \log(8.69 \times 10^2)$
$= 2.9390$

(b) $\log 7.926 \approx \log 7.93$
$= \log(7.93 \times 10^0)$
$= 0.8993$

(c) $\log 0.0702 = \log(7.02 \times 10^{-2})$
$= 0.8463 - 2$

(d) $\log 0.6 = \log(6.00 \times 10^{-1})$
$= 0.7782 - 1$

Often when computing with logarithms, you will need to find a number corresponding to a given logarithm. This is called finding an *antilogarithm*. Thus,
$$N = \text{antilog } x \quad \text{means} \quad \log N = x$$
or more simply,
$$\text{antilog } x = 10^x$$

EXAMPLE 2 Study the following relationships.

(a) If $\log 847 = 2.9279$, then antilog $2.9279 = 847$.
(b) If $\log 0.0261 = 0.4166 - 2$, then antilog $(0.4166 - 2) = 0.0261$.
(c) antilog $2 = 100$ because $10^2 = 100$.

Table IV in the Appendix is also used to find antilogarithms. For example, to find antilog 3.1367, we locate the mantissa, .1367, within the body of the table, as illustrated below. This corresponds with the number 1.37. Since the characteristic is 3, the antilogarithm is 1.37×10^3. So,
$$\text{antilog } 3.1367 = 1.37 \times 10^3 \text{ or } 1,370.$$

EXPONENTIAL AND LOGARITHMIC FUNCTIONS

N	0	1	2	3	4	5	6	7	8	9
1.0	.0000	.0043	.0086	.0128	.0170	.0212	0.253	.0294	.0334	.0374
1.1	.0414	.0453	.0492	.0531	.0569	.0607	.0645	.0682	.0719	.0755
1.2	.0792	.0828	.0864	.0899	.0934	.0969	.1004	.1038	.1072	.1106
1.3	.1139	.1173	.1206	.1239	.1271	.1303	.1335	.1367	.1399	.1430
1.4	.1461	.1492	.1523	.1553	.1584	.1614	.1644	.1673	.1703	.1732

If the exact mantissa is not in the table, simply locate the closest. For example, to find antilog (0.7908 − 2), we look for the mantissa, .7908. As shown below, the closest value is .7910, corresponding to the number 6.18. Since the characteristic is −2, the antilogarithm is approximately 6.18×10^{-2}. So,

$$\text{antilog } (0.7908 - 2) \approx \text{antilog } (0.7910 - 2)$$
$$= 6.18 \times 10^{-2} \text{ or } 0.0618$$

N	0	1	2	3	4	5	6	7	8	9
6.0	.7782	.7789	.7796	.7803	.7810	.7818	.7825	.7832	.7839	.7846
6.1	.7853	.7860	.7868	.7875	.7882	.7889	.7896	.7903	.7910	.7917
6.2	.7924	.7931	.7938	.7945	.7952	.7959	.7966	.7973	.7980	.7987
6.3	.7993	.8000	.8007	.8014	.8021	.8028	.8035	.8041	.8048	.8055
6.4	.8062	.8069	.8075	.8082	.8089	.8096	.8102	.8109	.8116	.8122

EXAMPLE 3 Find the following antilogarithms using Table IV in the Appendix. (Round mantissas to four significant digits before using the table.)

(a) antilog $4.6191 = 4.16 \times 10^4$ or 41,600

(b) antilog $1.73261 \approx$ antilog 1.7324
$$= 5.40 \times 10^1 \text{ or } 54.0$$

(c) antilog $(0.9042 - 3) = 8.02 \times 10^{-3}$ or 0.00802

(d) antilog $(8.9792 - 10) \approx$ antilog $(0.9791 - 2)$
$$= 9.53 \times 10^{-2} \text{ or } 0.0953$$

(e) Find N if log $N = 2.1614$; this is equivalent to
$$N = \text{antilog } 2.1614 = 1.45 \times 10^2 \text{ or } 145$$

The following examples illustrate how the properties of logarithms are used with Table IV to perform numerical calculations.

EXAMPLE 4 Evaluate $(3,960)(18.2)$.

Solution: Let N equal the expression to be evaluated. Then, take the common logarithm of each side and expand.

$$N = (3,960(18.2)$$
$$\log N = \log(3,960)(18.2)$$
$$= \log 3,960 + \log 18.2$$

COMPUTING WITH COMMON LOGARITHMS

The logarithms are found in Table IV and added as follows:

$$\begin{aligned}\log 3{,}960 &= 3.5977 \\ \log 18.2\ &=\ 1.2601 \end{aligned}\Bigg\} \text{add}$$

$\log N\ \ \ = 4.8578$

The answer is the antilogarithm. Thus,

$N = \text{antilog } 4.8578$

From Table IV, 7.21 is the number whose mantissa is closest to .8578. Since the characteristic is 4, the answer is

$N = 7.21 \times 10^4$ or 72,100

EXAMPLE 5

Evaluate $\dfrac{(5.62)^3}{0.638}$

Solution: Let $N = \dfrac{(5.62)^3}{0.638}$, then

$$\begin{aligned}\log N &= \log \frac{(5.62)^3}{0.638} \\ &= \log (5.62)^3 - \log 0.638 \\ &= 3 \log 5.62 - \log 0.638\end{aligned}$$

The logarithms are found in Table IV, and the calculations are performed as follows:

$$\begin{aligned}3 \log 5.62 &= 3(0.7497) = 2.2491 \\ \log 0.638 &= .8048 - 1\end{aligned}\Bigg\}\text{subtract}$$

$$\begin{aligned}\log N &= 1.4443 + 1 \\ &= 2.4443\end{aligned}$$

We use the body of Table IV to find the antilogarithm, using the closest mantissa. Thus, the answer is

$$\begin{aligned}N &= \text{antilog } 2.4443 \\ &\approx \text{antilog } 2.4440 \\ &= 2.78 \times 10^2 \text{ or } 278\end{aligned}$$

EXAMPLE 6

Evaluate $\sqrt[3]{0.0694}$

Solution: Let $N = \sqrt[3]{0.0694}$, then

$$\begin{aligned}\log N &= \log \sqrt[3]{0.0694} \\ &= \frac{1}{3} \log 0.0694\end{aligned}$$

Using Table IV, $\log 0.0694$ is found to be $0.8414 - 2$. However, multiplying by $\dfrac{1}{3}$ (or dividing by 3) would produce a fractional characteristic. To keep the characteristic an integer, we add and subtract 1 to write $0.8414 - 2$ as $1.8414 - 3$. Thus,

$$\log N = \frac{1}{3}(1.8414 - 3) = 0.6138 - 1$$

Using the body of Table IV, we find the antilogarithm, giving the answer

$$N = \text{antilog}\,(0.6138 - 1)$$
$$= 4.11 \times 10^{-1} \text{ or } 0.411$$

Exercise 10.5

Give the characteristic and mantissa of each logarithm. Use scientific notation along with Table IV in the Appendix.

1. log 55.2
2. log 4250
3. log .34
4. log .00623

Use scientific notation along with Table IV to find the following logarithms. Round numbers to three significant digits before using the table.

5. log 1.34
6. log 623
7. log .0476
8. log .00381
9. log 12500
10. log .878
11. log .17982
12. log 50636
13. log .91128
14. log 48.264

Find antilogs using Table IV. Round mantissas to four significant digits before using the table.

15. $\log N = 2.5211$
16. $\log N = .7348 - 2$
17. $\log N = 3.7959$
18. $\text{antilog}\,(0.66943) = N$
19. $\text{antilog}\,(.39091 - 1) = N$
20. $\log N = 8.6770 - 10$
21. $\text{antilog}\,(1.76436) = N$
22. $\text{antilog}\,(.6146 - 2) = N$
23. $\log N = 9.3592 - 10$
24. $\log N = 8.94486 - 10$

Evaluate, using logarithms and Table IV.

25. $(47.2)(.897)$
26. $(50.4)(.0036)(1.68)$
27. $(2.56)^4$
28. $\sqrt{47.6}$
29. $\sqrt[3]{.802}$
30. $33.7\sqrt{7.58}$
31. $\dfrac{50.6}{3.14}$
32. $\dfrac{(2.83)^2}{.814}$
33. $\dfrac{50.6}{\sqrt{84.3}}$
34. $\dfrac{(88.3)^2}{(9.64)(.516)}$
35. $\dfrac{\sqrt[4]{8.26}}{(3.47)^3}$
36. $\dfrac{(8.65)(13.1)^2}{(.551)(92.6)}$

37. $\dfrac{\sqrt[3]{9200}}{45.5\sqrt{21.6}}$

38. $\sqrt{\dfrac{(4.17)(61.3)^2}{.0352}}$

39. $\dfrac{\sqrt{(2.86)^3(.87)}}{.035}$

40. $\dfrac{\sqrt{83.6}\sqrt[3]{392}}{\sqrt[4]{9.16}}$

LOGARITHMS, EXPONENTIALS, AND THE CALCULATOR

10.6

Many calculators are particularly useful because they will give values for logarithmic and exponential functions. Obviously, this is more efficient and accurate than using tables.

As you recall from earlier sections, most logarithmic and exponential values are irrational. Calculators, like tables, give approximations. However, as before, we continue to use the equal sign. Also, we will round all final answers to five significant digits.

Common Logarithms

To find the common (base 10) logarithm, your calculator will need a key labeled (LOG). To use this key simply enter the number first, then push the (LOG) key.

Find log 391.63 **EXAMPLE 1**

Calculator Sequence: (391.63) (LOG) 2.592876

So, log 391.63 = 2.5929. Remember, this means $10^{2.5929} = 391.63$.

Find log 0.00682 **EXAMPLE 2**

Calculator Sequence: (.00682)(LOG) − 2.1662156

Thus, log 0.00682 = −2.1662.

When finding the common logarithm of a number less than 1, the calculator gives a negative answer. On the other hand, a table produces an answer that is part positive (mantissa) and part negative (characteristic). To illustrate, using the calculator in Example 2, we obtained

$$\log 0.00682 = -2.1662$$

But, using Table IV in the Appendix, we obtain

$$\log 0.00682 = 0.8338 - 3$$

These results look different, but as shown below, they are actually equivalent:

0.8338 − 3 means − 3 + .8338.

Adding, we have   −3.0000

+ .8338

−2.1662

446 EXPONENTIAL AND LOGARITHMIC FUNCTIONS

This suggests the following procedure for changing table logarithms to calculator logarithms.

> To change a logarithm having a positive mantissa and a negative characteristic to a completely negative logarithm, simply add the two components.

EXAMPLE 3 Change $0.1036 - 4$ to a negative logarithm.

Solution: Add the two components:

$$-4.0000$$
$$+\ \ 0.1036$$
$$-3.8964$$

Thus, $0.1036 - 4 = -3.8964$.

Given a negative logarithm, it is sometimes necessary to obtain its characteristic and mantissa. For example, by calculator,

$$\log 0.0203 = -1.6925$$

To find its characteristic and mantissa, -1.6925 must be written in table form. This is accomplished by adding and subtracting the same positive integer. This positive integer should be at least as large as the absolute value of the negative logarithm. Since

$$|-1.6925| = 1.6925$$

we add and subtract the positive integer, 2. So,

$$-1.6925 = 2 - 1.6925 - 2$$

Adding, we have

$$2.0000$$
$$-1.6925\ -\ 2$$
$$0.3075\ -\ 2$$

Thus, $\log 0.0203 = -1.6925 = 0.3075 - 2$. Its characteristic is -2 and its mantissa is .3075.

> To find the characteristic and mantissa of a negative logarithm, add and subtract the same positive integer. This positive integer should be at least as large as the absolute value of the negative logarithm.

EXAMPLE 4 Find $\log 0.0008731$ by calculator. Then, determine its characteristic and mantissa.

Calculator Sequence: (.0008731) (LOG) -3.058936

Thus, $\log 0.0008731 = -3.0589$
To find its characteristic and mantissa, add and subtract 4:

LOGARITHMS, EXPONENTIALS, AND THE CALCULATOR

$$-3.0589 = 4 - 3.0589 - 4$$

Adding, $\quad$ 4.0000
$$\underline{-3.0589 - 4}$$
$$0.9411 - 4$$

So, log 0.0008731 = -3.0589 = 0.9411 $-$ 4.
The characteristic is -4 and the mantissa is .9411.

Antilogarithms

Recall that an antilogarithm is the number whose logarithm is given. Thus,

$$\text{antilog } x = 10^x$$

For example, antilog 2 = 100 because log 100 = 2, or more simply because $10^2 = 100$.

To find an antilogarithm (base 10), you need a calculator with either an inverse or a second function key, (INV), together with a (LOG) key. Or, it may have a key labeled (10^x). If not, you can use the exponential key (y^x) by inserting a base of 10.

EXAMPLE 5

Find antilog 3.8041

Calculator Sequences: (a) (3.8041) (INV) (LOG) 6369.4217
(b) (3.8041) (10^x) 6369.4217
(c) (10) $(y^x)(3.8041)$ = 6369.4217

Thus, antilog 3.8041 = 6,369.4. Remember, this means $10^{3.8041}$ = 6,369.4.

With a calculator, the antilogarithm of a negative number can be found directly. Otherwise, it must first be evaluated and entered as a completely negative logarithm.

EXAMPLE 6

Find N if log $N = -3.6039$

Solution: Since the logarithm is negative, enter it in the calculator and find the antilogarithm.

Calculator Sequences: (a) (3.6039) $(+/-)$ (INV) (LOG) .00024894
(b) (3.6039) $(+/-)$ (10^x) .00024894
(c) (10) (y^x) (3.6039) $(+/-)$ $(=)$ .00024894

So, N = 0.00024894.

EXAMPLE 7

Find N if log $N = 0.4267 - 3$

Solution: Evaluate, then find its antilogarithm.

Calculator Sequences: (a) $(.4267)$ $(-)(3)(=)$ (INV) (LOG) .00267116
(b) $(.4267)$ $(-)(3)(=)$ (10^x) .00267116
(c) (10) (y^x) $(()$ $(.4267)$ $(-)(3)())$ $(=)$.00267116

Hence, N = 0.0026712.

448 EXPONENTIAL AND LOGARITHMIC FUNCTIONS

Natural Logarithms

Logarithms to the base e (2.7182818 . . .) are called natural logarithms. For convenience, $\log_e x$ is written ln x. To find a natural logarithm, you need a calculator with a key labeled (ln x).

EXAMPLE 8

Find ln 15.806

Calculator Sequence: (15.806) (ln x) 2.7603896

So, ln 15.806 = 2.76039. This means
$$e^{2.76039} = 15.806.$$

An antilogarithm (base e) can be written antiln x, but it simply means e^x. If your calculator does not have a key labeled (e^x), you may use the sequence (INV) (ln x). Or, the exponential key, (y^x), can be used with a base of 2.7182818.

EXAMPLE 9

Find N, if ln N = −1.2365.

Calculator Sequences: (a) (1.2365) (+/−) (e^x) .29039884
(b) (1.2365) (+/−) (INV) (ln x) .29039884
(c) (2.7182818) (y^x) (1.2365) (+/−) .29039884

Thus, N = 0.29040, which means that
$e^{-1.2365} = 0.29040.$

Logarithms to Any Base

We now derive a simple formula that enables us to find logarithms to any base by using the common logarithm key, (LOG).

Consider the equation
$$\log_a N = x$$

We convert to common logarithms by writing it in exponential form and taking a common logarithm of both sides, as follows:

$$\log_a N = x$$
$$a^x = N$$
$$\log a^x = \log N$$

We solve for x by using the power property of logarithms.

$$x \cdot \log a = \log N$$
$$x = \frac{\log N}{\log a}$$

Thus,
$$\log_a N = \frac{\log N}{\log a}$$

LOGARITHMS, EXPONENTIALS, AND THE CALCULATOR

> **Change of Base**
> For any logarithm base a,
> $$\log_a N = \frac{\log N}{\log a}$$
> Calculator Sequence: Ⓝ ⒧⒪⒢ ⊘ ⓐ ⒧⒪⒢ ⊜

Find $\log_2 13.641$ **EXAMPLE 10**

Solution: Since $a = 2$ and $N = 13.641$,
$$\log_2 13.641 = \frac{\log 13.641}{\log 2}$$

Calculator Sequence: ⒀⒔⒍⒋⒈ ⒧⒪⒢ ⊘ ② ⒧⒪⒢ ⊜ 3.7698775

So, $\log_2 13.641 = 3.7699$. This means $2^{3.7699} = 13.641$.

As illustrated by the next example, exponential equations can be solved quickly and easily by converting to logarithmic form and using your calculator.

Solve $31.605^x = 9.2183$ **EXAMPLE 11**

Solution: By the definition of logarithmic functions, this equation is equivalent to

$$\log_{31.605} 9.1283 = x$$

Since $a = 31.605$ and $N = 9.2183$, we convert to common logarithm form, obtaining

$$x = \log_{31.605} 9.2183 = \frac{\log 9.2183}{\log 31.605}$$

Calculator Sequence: ⒐⒉⒈⒏⒊ ⒧⒪⒢ ⊘ ⒈⒈⒍⒇⒌ ⒧⒪⒢ ⊜ .64320527

Thus $x = 0.64321$. To check, the calculator may be used to verify the equation
$$31.605^{0.64321} = 9.2183$$

Checking Sequence: ⒈⒈⒍⒇⒌ ⓨˣ ⒍⒋⒊⒉⒈ ⊜ 9.2184504

The small error is due to rounding.

Change the following logarithms to negative logarithms. **Exercise 10.6**

1. $.2614 - 2$
2. $.4382 - 1$
3. $.8824 - 3$
4. $.6445 - 2$
5. $.4288 - 4$
6. $.7649 - 3$

Find the characteristic and mantissa of the following negative logarithms.

7. -2.6183
8. $-.4822$
9. -1.5827
10. -3.6004
11. $-.2632$
12. -1.0686

Use a calculator to find the following logarithms.

13. log 58.32
14. log .4318
15. log .07124
16. log 86317
17. ln 12.701
18. ln 8.624
19. ln .8
20. ln 2.258
21. log .0006414
22. ln .623
23. log 623.07
24. log 10.4267

Use a calculator to find the following antilogs.

25. Find N if log $N = -1.6034$
26. Find N if log $N = .4133 - 1$
27. antilog 0.8342
28. antilog (2.6813)
29. Find N if ln $N = 4.2163$
30. antiln (.6176)
31. Find N if log $N = -2.6814$
32. Find N if log $N = .6157 - 2$
33. antiln (12.0835)
34. Find N if ln $N = .3384 - 1$
35. Find N if log $N = .4248 - 3$
36. antilog 1.6789
37. antilog (1.2116)
38. Find N if log $N = 8.6114 - 10$

Use the change of base formula and a calculator to find:

39. $\log_3 14.64$
40. $\log_2 20.1$
41. $\log_4 6.832$
42. $\log_{12} 15$
43. $\log_5 31.642$
44. $\log_{3.14} 6.214$
45. $\log_5 1000$
46. $\log_{4.812} .9218$
47. $\log_{1/4} 23$
48. $\log_{10}(\ln 5)$
49. $\log_5(\ln 5)$
50. $\log_2(e^{2.13})$

EXPONENTIAL AND LOGARITHMIC EQUATIONS

10.7

Exponential equations of the form
$$a^x = b, \text{ where } a > 0, a \neq 1$$
may be solved by taking logarithms of both sides of the equation. The power property of logarithms is then used to help isolate the variable.

EXAMPLE 1 Solve $2^x = 13$

Solution: Take common logarithms of both sides.
$$2^x = 13$$
$$\log 2^x = \log 13$$

Apply the power property of logarithms.

$x \cdot \log 2 = \log 13$

$$x = \frac{\log 13}{\log 2}$$

This is the exact solution. A decimal approximation is obtained by using a logarithm table or a calculator.

$$x \approx \frac{1.1139}{0.3010} = 3.7007$$

Note: Be sure to divide 1.1139 by 0.3010. A common mistake is to subtract, but

$$\frac{\log 13}{\log 2} \neq \log 13 - \log 2.$$

EXAMPLE 2

Solve $5^{3n-1} = 112$

Solution: Take common logarithms of both sides, then use the power property of logarithms.

$$5^{3n-1} = 112$$
$$\log 5^{3n-1} = \log 112$$
$$(3n - 1) \log 5 = \log 112$$
$$3n \log 5 - \log 5 = \log 112$$
$$3n \log 5 = \log 112 + \log 5$$
$$n = \frac{\log 112 + \log 5}{3 \log 5}$$

To approximate, we use a logarithm table or a calculator.

$$n \approx \frac{2.0492 + 0.6990}{3(0.699)} = 1.3105$$

Equations containing a logarithm of the variable are called logarithmic equations. They are generally solved by simplifying to a single logarithm and converting to exponential form. You must always check to be sure that the solution is in the domain.

EXAMPLE 3

Solve $\log_2(x - 2) = 3 - \log_2 x$

Solution: Rewrite the equation, isolating the logarithms on one side.

$$\log_2(x - 2) = 3 - \log_2 x$$

$\log_2(x - 2) + \log_2 x = 3$

Use the addition property of logarithms to write as a single logarithm.

$$\log_2[(x - 2)x] = 3$$

Convert to exponential form and solve for x.
$$(x - 2)x = 2^3$$
$$x^2 - 2x = 8$$
$$x^2 - 2x - 8 = 0$$
$$(x - 4)(x + 2) = 0$$
$$x = 4 \quad \text{or} \quad x = -2$$

From the original equation, x and $x - 2$ must both be positive. Thus, -2 is rejected, leaving 4 as the only solution.

EXAMPLE 4 Solve $\log_5(x + 1)^2 = 3$

Solution: Use the power property of logarithms.
$$\log_5(x + 1)^2 = 3$$
$$2 \log_5(x + 1) = 3$$
$$\log_5(x + 1) = \frac{3}{2}$$

Convert to exponential form and solve for x.
$$x + 1 = 5^{3/2}$$
$$x = -1 + 5^{3/2}$$
$$= -1 + \sqrt{5^3}$$
$$= -1 + 5\sqrt{5}$$

From the original equation, $(x + 1)^2$ is positive for all $x \neq -1$. Thus, the solution is $-1 + 5\sqrt{5}$. A decimal approximation is obtained by either using Table I in the Appendix or a calculator to evaluate $\sqrt{5}$.
$$x = -1 + 5\sqrt{5} \approx -1 + 5(2.236)$$
$$= 10.18$$

The following examples show that exponential and logarithmic equations are used to solve a wide variety of applied problems.

For Example 5, you will need the compound interest formula from section 10.2:

Compound Interest

If P represents the principal invested at an annual interest rate r, compounded t times yearly, then the amount accumulated in n years is given by

$$A = P\left(1 + \frac{r}{t}\right)^{tn}$$

EXPONENTIAL AND LOGARITHMIC EQUATIONS

EXAMPLE 5

Suppose that over the next several years the average annual inflation rate is projected to be 7%. How long would it take for the average price level to double?

Solution: Use the compound interest formula to determine when $1 will double to $2. $P = 1$, $A = 2$, $r = 0.07$, and $t = 1$. Substitute and solve for n.

$$A = P\left(1 + \frac{r}{t}\right)^{tn}$$

$$2 = 1\left(1 + \frac{.07}{1}\right)^{1 \cdot n}$$

$$2 = (1.07)^n$$

$$\log 2 = n \log 1.07$$

$$n = \frac{\log 2}{\log 1.07}$$

$$n \approx \frac{0.3010}{0.0294} = 10.24$$

An annual inflation rate of 7% will double the average price level approximately every 10 years. This also means that average prices would quadruple in 20 years and be eight times higher in 30 years.

For Example 6, we apply a continuous growth formula from section 10.2. We will also refer to Table III in the Appendix, which gives values of natural logarithms (base e).

Continuous Growth or Decay

Exponential equations of the form

$$y = y_0 e^{kt}$$

where t is time, y_0 is the initial amount present at $t = 0$, and k is a constant representing the rate of increase or decrease, describe continuous growth or decay.

EXAMPLE 6

A city increased in population from 28,000 in 1972 to 70,000 in 1987. Find the rate of increase represented by the constant k.

Solution: The initial population, y_0, is 28,000. The final population, y, is 70,000. And the time, t, is 15 years. Substitute into the growth formula and solve for k:

$$y = y_0 e^{kt}$$

$$70{,}000 = 28{,}000 e^{k \cdot 15}$$

$$\frac{70{,}000}{28{,}000} = e^{15k}$$

$$2.5 = e^{15k}$$

Since the base of the exponential is e, take a natural logarithm of each side:

$\ln 2.5 = \ln e^{15k}$

$\ln 2.5 = 15k \ln e$ [$\ln e = 1$, since $e^1 = e$]

$\ln 2.5 = 15k$

$k = \dfrac{\ln 2.5}{15}$

Using either Table III in the Appendix or a calculator, we see that

$k = \dfrac{\ln 2.5}{15} \approx \dfrac{0.9163}{15} = 0.061$

Thus, the yearly growth constant is 6.1%.

This final example shows how logarithms are used in chemistry to measure the acidity or alkalinity of a solution in terms of its pH.

---

pH of a Solution

The pH (hydronium ion potential) of solution is defined as

$$\text{pH} = \log \dfrac{1}{[H_3O^+]} = -\log(H_3O^+)$$

where $[H_3O^+]$ represents the hydronium ion concentration in aqueous solution given in moles per liter.

Water has a pH of 7. Acids have pH values less than 7, while alkaline solutions have pH values greater than 7.

EXAMPLE 7   Find the pH (to the nearest tenth) of vinegar that has a hydronium ion concentration of 1.25×10^{-3} moles per liter.

Solution: Substitute 1.25×10^{-3} for $[H_3O^+]$ in the pH formula:

$\text{pH} = -\log[H_3O^+]$

$\quad = -\log(1.25 \times 10^{-3})$

Apply the addition property of logarithms:

$\quad = -(\log 1.25 + \log 10^{-3})$

$\quad = -(0.0969 - 3)$

$\quad = -0.0969 + 3$

$\quad = 2.9031$

To the nearest tenth, pH = 2.9.

EXPONENTIAL AND LOGARITHMIC EQUATIONS

Solve, using a calculator or tables. Round answer to four decimal places.

Exercise 10.7

1. $3^x = 18$
2. $2^x = 3$
3. $5^x = 10$
4. $4^x = 75$
5. $6^x = 66$
6. $4.6^x = 100$
7. $8.71^x = 8.57$
8. $3^{2x} = 4$
9. $4^{x+2} = 7.6$
10. $2^{x-1} = 12$
11. $3^{x+1} = 8$
12. $3^{-x} = 15$
13. $10^{2x-1} = 20.2$
14. $2^{x-4} = 16$
15. $(2.44)^{x+1} = 8.44$
16. $(2.14)^{2x-1} = 23.1$
17. $(8.23)^{.624x} = 11.8$
18. $(11.8)^{.624x} = 8.23$

Solve.

19. $\log x + \log(7 - x) = \log 10$
20. $\log(x + 3) + \log(x - 2) = \log 6$
21. $\log x + \log(x - 3) = \log 4$
22. $\log(x + 4) - \log 3 = \log (x - 2)$
23. $\log x + \log 3 = 2$
24. $\log(x - 3) + \log x = 1$
25. $2 \log x + \log 2 - \log(2x - 5) = 1$
26. $\log(2x + 3) - 1 = \log x$
27. $\log_3(3x + 2) - \log_3 x = \log_3 4$
28. $\log_4(x + 3) - \log_4 x = 2$

For the following problems use the compound interest formula,
$$A = P\left(1 + \frac{r}{t}\right)^{tn}.$$

29. How long will it take $1,000 to double at 8% compounded semi-annually? (Give the answer to the nearest tenth of a year.)
30. How much money must have been invested 40 years ago to give $1,000,000 today? Assume 5% interest compounded quarterly.
31. If over the next several years the average annual inflation rate is projected to be 5%, how long would it take for the average price level to double? (Give the answer to the nearest tenth of a year.)
32. If over the next several years the average annual inflation rate is projected to be 9%, how long would it take for the average price level to double? (Give the answer to the nearest tenth of a year.)

For the following problems, use the population formula, $y = y_0 e^{kt}$.

33. San Francisco had a 1980 population of 647,063. Between 1960 and 1980 the city had been declining in population at an annual rate of 0.47%. Estimate the population of San Francisco in 1990.
34. If a city's population is growing at a rate of 3% a year, how long will it take for the population to double? (Give the answer to the nearest tenth.)

35. A city increased in population from 60,000 in 1970 to 100,000 in 1985. Find the annual rate of increase to the nearest tenth of a percent.
36. If Florida had a 1970 population of 4,952,000 and a 1980 population of 6,789,000, predict its population in 1990. (Round your answer to the nearest thousand.)

Use the pH formula, $pH = -\log[H_3O^+]$, where $[H_3O^+]$ represents the hydronium ion concentration in aqueous solution given in moles per liter, to find either the pH value to the nearest tenth or the hydronium ion concentration as indicated.

37. Seawater; 3.16×10^{-9} is the hydronium ion concentration
38. Grapefruit; 7.9×10^{-4} is the hydronium ion concentration
39. Lemons; .01 is the hydronium ion concentration
40. Rainwater; 6.31×10^{-7} is the hydronium ion concentration
41. If wine has a pH of 3.4, find the hydronium ion concentration.
42. If beer has a pH of 4.8, find the hydronium ion concentration.

For the following problems, use the atmospheric pressure formula, $P = 30(10)^{-.09a}$, where a is the altitude in miles and P is the pressure in inches of mercury.

43. What is the altitude if the pressure is 4.8 inches of mercury? Give the answer to the nearest hundredth.
44. Find the altitude to the nearest hundredth if the pressure is 10.1 inches of mercury.

Use the atmospheric pressure formula, $P = 14.7e^{-.21a}$, where a is the altitude in miles and P is the pressure in pounds per square inch.

45. The pressure at the top of a mountain is 6.6 pounds per square inch. Find the altitude of the mountain to the nearest hundredth.
46. The pressure at the top of Mt. Ararat in Turkey is 7.5 pounds per square inch. Find the height of Mt. Ararat to the nearest hundred feet (1 mile = 5,280 feet).

The Richter Scale number is the base 10 logarithm of the amount of energy released by an earthquake. The San Francisco quake of 1906 measured 8.25 on the Richter Scale, while the Seattle quake of 1965 measured 7.0. How many more times severe was the San Francisco quake? To answer this question, let x = energy released by a quake of magnitude y. Then $y = \log_{10} x$ or $10^y = x$.

	San Francisco quake	Seattle quake
Energy	$10^{8.25}$	10^7

The San Francisco quake was $\dfrac{10^{8.25}}{10^7}$ or 17.78 times as severe.

47. The India quake of 1897 measured 8.7 on the Richter Scale, while the Alaska quake of 1964 measured 8.4. How many more times severe was the India quake?

48. The Chile quake of 1960 measured 8.5 on the Richter Scale, while the Alaska quake of 1964 measured 8.4. How much more powerful was the Chilean quake?

49. How much more powerful is a recording of 7 on the Richter Scale over a recording of 6?

Summary

DEFINITIONS

Exponential Function: $f(x) = a^x$, with $a > 0$, $a \neq 1$

domain: $\{x \mid x \in R\}$

range: $\{f(x) \mid f(x) > 0\}$

increasing function if $a > 1$
decreasing function if $0 < a < 1$

Logarithmic Function: $y = \log_a x$ is equivalent to $x = a^y$, with $a > 0$, $a \neq 1$

domain: $\{x \mid x > 0\}$

range: $\{y \mid y \in R\}$

Common Logarithms: base 10 logarithms, written $\log_{10} x$ or $\log x$

Natural Logarithms: base e logarithms, written $\log_e x$ or $\ln x$

e: a number approximately 2.7182818; the base for natural logarithms.

Mantissa: the decimal part of a logarithm when the logarithm is written as the sum of an integer and a non-negative decimal.

Characteristic: the integral part of a logarithm where the mantissa is non-negative.

Antilog: the number corresponding to a given logarithm.

PROPERTIES

Equality Property For Exponentials:
 If $a > 0$, $a \neq 1$ and $a^m = a^n$, then $m = n$.

Addition Property of Logarithms:
 $\log_a(m \cdot n) = \log_a m + \log_a n$, where $m, n > 0$ and a is any log base.

Subtraction Property of Logarithms:
 $\log_a \dfrac{m}{n} = \log_a m - \log_a n$, where $m, n > 0$ and a is any log base.

Power Property of Logarithms:
 $\log_a m^n = n \cdot \log_a m$, where $m > 0$, $n \in R$, and a is any log base.

Equality Property for Logarithms:
 If $m, n > 0$ and a is any log base and $m = n$, then $\log_a m = \log_a n$.

FORMULAS

Compound Interest:

$$A = P\left(1 + \frac{r}{t}\right)^{tn},$$ where P is the principal, r the annual interest rate, t the number of times compounded per year, and n the number of years.

Growth or Decay: $y = y_0 e^{kt}$, where t is the time in years, y_0 the initial amount present, and k the constant representing rate of increase or decrease.

pH of a Solution

$$pH = \log \frac{1}{[H_3O+]} = -\log[H_3O+],$$ where $[H_3O+]$ represents the hydronium ion concentration in aqueous solution given in moles per liter.

Change of Base

$$\log_a N = \frac{\log N}{\log a},$$ where $N > 0$ and a is any log base.

Self-Checking Exercise

Graph and give the domain and range; determine if the function is increasing or decreasing.

1. $y = 3^x$
2. $y = \left(\dfrac{1}{2}\right)^x$
3. $y = \log_4 x$
4. $y = \log_{1/2} x$

Solve each of the following for x.

5. $6^x = 36$
6. $3^x = \dfrac{1}{81}$
7. $4^x = \dfrac{1}{32}$
8. $10^x = 0.001$
9. $3^x = 9^{x+1}$
10. $4^{x+1} = 16$

Change to log form.

11. $2^3 = 8$
12. $5^{-2} = \dfrac{1}{25}$
13. $16^{1/2} = 4$
14. $4^{5/2} = 32$

Change to exponential form.

15. $\log_{10} 0.01 = -2$
16. $\log_{10} 1 = 0$
17. $\log_2 \left(\dfrac{1}{8}\right) = -3$
18. $\log_{4/9}\left(\dfrac{2}{3}\right) = \dfrac{1}{2}$

Rewrite using the properties of logarithms.

19. $\log_b c\sqrt{d}$
20. $\log_3 \dfrac{5x}{3}$
21. $\log_4 \dfrac{a^2}{bc}$
22. $\log_b \dfrac{y\sqrt{x}}{z^3}$

Write as a single logarithm.

23. $4 \log_b a - 2 \log_b d$
24. $\dfrac{1}{2} \log_3 6 + \dfrac{3}{2} \log_3 7 - 4 \log_3 8$
25. $3 \log_b x - 2 \log_b y + 4 \log_b z$
26. $\log\left(\dfrac{2}{5}\right) + \log\left(\dfrac{5}{2}\right)$

Given $\log 2 = 0.3010$, $\log 3 = 0.4771$, and $\log 5 = 0.6990$, find the following logs without tables or calculator.

27. $\log 60$
28. $\log 75$
29. $\log \dfrac{3}{2}$
30. $\log (0.02)^2$

EXPONENTIAL AND LOGARITHMIC FUNCTIONS

Find the characteristic and mantissa of each logarithm. Use scientific notation along with Table IV.

31. log 16.3
32. log 6280
33. log 0.0542
34. log 0.000267

Use Table IV to find the following logarithms or antilogarithms. Before using the table, round numbers to three significant digits and mantissas to four significant digits.

35. log 31.46
36. log 0.00246
37. log 60336
38. log $N = .4216 - 1$
39. antilog (0.49437)
40. antilog (2.3594)

Compute using logarithms and Table IV.

41. $(.324)^3$
42. $\sqrt[3]{48.6}$
43. $\dfrac{69.2}{\sqrt[5]{85.6}}$
44. $\sqrt{\dfrac{(27.4)(113.1)}{(2.6)^3}}$

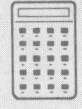

Give both the table and calculator result for each of the following. Use Tables II and III.

45. $e^{1.438}$
46. $e^{-.38}$
47. ln 0.6
48. ln(1.4)
49. antilog (-2.6835)
50. antiln (0.7893)

Use the change of base formula and a calculator to find the following. Round answers to three decimal places.

51. $\log_8 6$
52. $\log_e \pi$
53. $\log_{20} 432$
54. $\log_5 240$

Solve for the unknown. Round answers to 3 decimal places.

55. $\log_3 x - \log_3 2 = 1$
56. $5^x = 75$
57. $7^{2x-1} = 14.2$
58. $\log(x + 3) + \log x = \log 18$

Solve the following by using the appropriate formula.

59. How much will $5,000 amount to at the end of 4 years if it is invested at 8% compounded quarterly? $\left(A = P\left(1 + \dfrac{r}{t}\right)^{tn} \right)$

60. Find the scrap value of equipment costing $20,000, with a useful life of 8 years and an annual depreciation of 10%. $(S = P(1 - r)^n)$

61. Bacteria grows according to the formula $y = 6000 \, e^{.5t}$, where y is the number present at time t in hours.
 a) How many bacteria will be present in 4 hours?
 b) When will there be 12,000 bacteria?

62. Find the pH value of ammonia if the hydronium ion concentration is 1.8×10^{-5} (pH = $-\log[H_3O^+]$).

63. Austria had a 1968 population of 7,349,000 and a 1973 population of 7,520,000. If it continues to grow at the same rate, estimate its population in 1990 to the nearest thousand. ($y = y_0 e^{kt}$)

64. Find the monthly payment in borrowing $4,000 for 4 years at 15% interest. $\left(M = \dfrac{P \cdot (i)}{1 - \dfrac{1}{(1+i)^n}}, \text{ where } \begin{array}{l} P \text{ is the principal,} \\ i \text{ the monthly interest rate,} \\ n \text{ the number of months.} \end{array} \right)$

65. What is the pressure in pounds per square inch at Lake Arrowhead, California, where the altitude is .97 mile?
$$(P = 14.7 e^{-.21a})$$

66. If the average inflation rate over the next several years is projected to be 5%, how long will it take for the average price level to double? $\left(A = P\left(1 + \dfrac{r}{t}\right)^{tn} \right)$

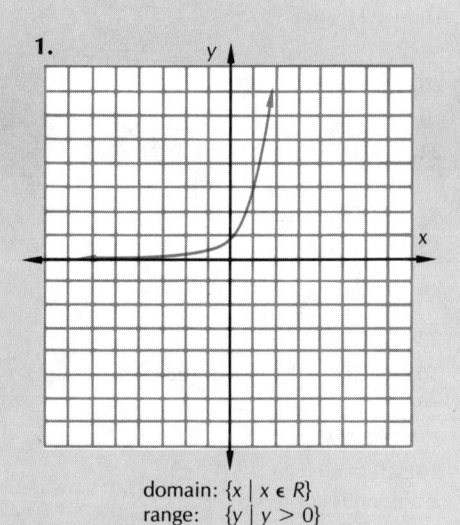

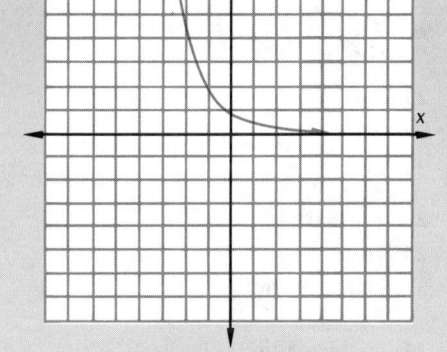

Solutions to Self-Checking Exercise

1.

domain: $\{x \mid x \in R\}$
range: $\{y \mid y > 0\}$
increasing

2.

domain: $\{x \mid x \in R\}$
range: $\{y \mid y > 0\}$
decreasing

3.

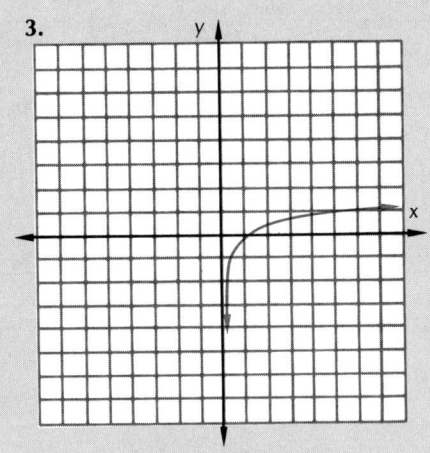

domain: $\{x \mid x > 0\}$
range: $\{y \mid y \in R\}$
increasing

4.

domain: $\{x \mid x > 0\}$
range: $\{y \mid y \in R\}$
decreasing

5. $x = 2$

6. $x = -4$

7. $x = -\dfrac{5}{2}$

8. $x = -3$

9. $x = -2$

10. $x = 1$

11. $\log_2 8 = 3$

12. $\log_5 \left(\dfrac{1}{25}\right) = -2$

13. $\log_{16} 4 = \dfrac{1}{2}$

14. $\log_4 32 = \dfrac{5}{2}$

15. $10^{-2} = 0.01$

16. $10^0 = 1$

17. $2^{-3} = \dfrac{1}{8}$

18. $\left(\dfrac{4}{9}\right)^{1/2} = \dfrac{2}{3}$

19. $\log_b c + \dfrac{1}{2} \log_b d$

20. $\log_3 5 + \log_3 x - 1$

21. $2 \log_4 a - \log_4 b - \log_4 c$

22. $\log_b y + \dfrac{1}{2} \log_b x - 3 \log_b z$

23. $\log_b \dfrac{a^4}{d^2}$

24. $\log_3 \dfrac{\sqrt{6} \cdot 7^{3/2}}{8^4}$

25. $\log_b \dfrac{x^3 \cdot z^4}{y^2}$

26. $\log_{10} 1$ (equals zero)

27. 1.7781

28. 1.8751

29. .1761

30. .6020 − 4 (or −3.398)

31. characteristic: 1
mantissa: .2122

32. characteristic: 3
mantissa: .7980

33. characteristic: −2
mantissa: .7340

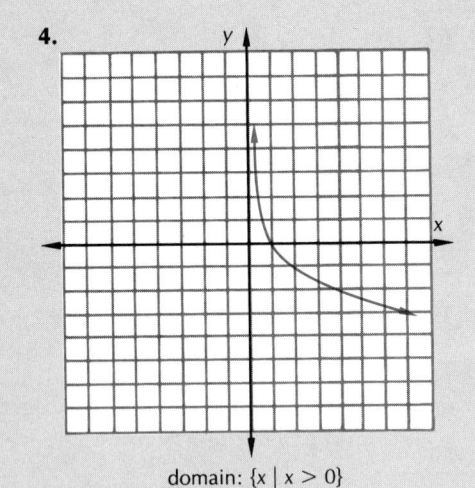

34. characteristic: −4
mantissa: .4265

35. 1.4983

36. .3909 − 3

37. 4.7803

38. 2.64×10^{-1} or .264

39. 3.12
41. .0340
43. 28.4
45. table: 4.221
    calculator: 4.2122629
47. table: −0.5108
    calculator: −0.5108256
49. table: 2.07×10^{-3} or 0.00207
    calculator: 0.00207252
51. 0.862
53. 2.026
55. $x = 6$
57. 1.182
59. $6,863.93
61. a) 44,334
    b) in about 1.4 hours
63. 8,132,000
65. 11.99 pounds per square inch

40. 2.29×10^2 or 229
42. 3.65
44. 13.0
46. table: 0.684
    calculator: 0.6838614
48. table: 0.3365
    calculator: 0.3364722
50. table: 2.2
    calculator: 2.2018546
52. 1.145
54. 3.405
56. $x = 2.683$
58. $x = 3$
60. $8,609.34
62. 4.74
64. $111.32
66. about 14.2 years

NATURAL NUMBER FUNCTIONS

11

11.1 Sequences and Series
11.2 Arithmetic Progressions
11.3 Geometric Progressions
11.4 The Binomial Expansion

SEQUENCES AND SERIES

11.1

In this chapter we will study patterns that describe successions of numbers such as

$$1, 4, 9, 16, 25, \ldots$$

Such a set is called a *sequence* and each number is called a *term*. Rewriting these terms as

$$1^2, 2^2, 3^2, 4^2, 5^2, \ldots$$

shows that this sequence is composed of the squares of successive natural numbers. Thus, any particular term, called the nth *term*, is described by the function a, where

$$a(n) = n^2$$

and n represents any natural number. Such a function is called a *sequence function*. Note, below, how this sequence function generates the first five terms:

$$a(1) = 1^2 = 1$$
$$a(2) = 2^2 = 4$$
$$a(3) = 3^2 = 9$$
$$a(4) = 4^2 = 16$$
$$a(5) = 5^2 = 25$$

It is convenient to write a sequence function with subscripts. So, $a(n)$ is usually written a_n. For example, the first four terms generated by the sequence function $a_n = 2n + 1$ are

$$a_1 = a(1) = 2(1) + 1 = 3$$
$$a_2 = a(2) = 2(2) + 1 = 5$$
$$a_3 = a(3) = 2(3) + 1 = 7$$
$$a_4 = a(4) = 2(4) + 1 = 9$$

The foregoing discussion suggests the following definition:

Definition of a Sequence Function

A sequence function is a function whose domain is the set of natural numbers, $\{1, 2, 3, \ldots\}$.

If a represents a sequence function, then

$a_1 = a(1)$ = first term of the sequence

$a_2 = a(2)$ = second term of the sequence

$a_3 = a(3)$ = third term of the sequence

.
.
.

SEQUENCES AND SERIES   467

$a_n = a(n) = n^{th}$ term of the sequence

. .
. .
. .

The nth term is also called the general term of the sequence. Since it defines the sequence function, it is used to determine the other terms of the sequence.

EXAMPLE 1 Write the first four terms of the sequence whose general term is given by $a_n = 3n - 1$.

Solution: To find the first four terms, we simply substitute 1, 2, 3, and 4 for n.

$$\text{First term} = a_1 = 3(1) - 1 = 2$$
$$\text{Second term} = a_2 = 3(2) - 1 = 5$$
$$\text{Third term} = a_3 = 3(3) - 1 = 8$$
$$\text{Fourth term} = a_4 = 3(4) - 1 = 11$$

The entire sequence is written as

$$2, 5, 8, 11, \ldots, 3n - 1, \ldots$$

EXAMPLE 2 Write the first four terms of the sequence defined by $a_n = \dfrac{2}{n+1}$.

Solution: We replace n with 1, 2, 3, and 4.

$$\text{First term} = a_1 = \frac{2}{1+1} = 1$$
$$\text{Second term} = a_2 = \frac{2}{2+1} = \frac{2}{3}$$
$$\text{Third term} = a_3 = \frac{2}{3+1} = \frac{1}{2}$$
$$\text{Fourth term} = a_4 = \frac{2}{4+1} = \frac{2}{5}$$

The entire sequence is written as

$$1, \frac{2}{3}, \frac{1}{2}, \frac{2}{5}, \ldots, \frac{2}{n+1}, \ldots$$

EXAMPLE 3 Find the seventh and eighth term of the sequence defined by $a_n = 2^{n-5}$.

Solution: Substitute 7 and 8 for n.

$$\text{Seventh term} = a_7 = 2^{7-5} = 4$$
$$\text{Eighth term} = a_8 = 2^{8-5} = 8$$

468 NATURAL NUMBER FUNCTIONS

A sequence is either infinite or finite, depending on whether or not it terminates. For example,

$$2, 4, 6, 8, 10$$

is a finite sequence of five terms. On the other hand,

$$2, 4, 6, 8, 10, \ldots, 2n, \ldots$$

is an infinite sequence of even numbers.

The terms of the sequence may be added. For example,

$$2 + 4 + 6 + 8 + 10 = 30.$$

When this is done, the indicated sum is called a series.

> **Definition of a Series**
>
> A series is the indicated sum of the terms of a sequence.

A series can be finite or infinite. For example,

$$2 + 4 + 6 + 8 + 10$$

is a finite series while

$$2 + 4 + 6 + 8 + 10 + \ldots$$

is an infinite series. In this section we will consider only finite series. Infinite series will be discussed in section 11.3.

Given the general term, a compact notation is used to express a finite series. It is called summation notation, or sigma notation since it involves the Greek letter sigma, Σ. For example, the series which is the sum of the first four terms of the sequence, $a_n = 2n + 3$, is written

$$\sum_{n=1}^{4} (2n + 3)$$

This notation indicates that we are to form the sum of all numbers $2n + 3$, where n is replaced by all consecutive natural numbers from 1 through 4 as follows:

$$\sum_{n=1}^{4} (2n + 3) = (2 \cdot 1 + 3) + (2 \cdot 2 + 3) + (2 \cdot 3 + 3) + (2 \cdot 4 + 3)$$
$$= 5 + 7 + 9 + 11$$
$$= 32$$

The variable n is called the index of summation, or simply the index. It is a "dummy" variable; any other letter can be used, such as i or k.

EXAMPLE 4 Expand and simplify $\sum_{i=1}^{5} i^2$

Solution: In the expression i^2, replace the index i with all consecutive natural numbers from 1 through 5.

$$\sum_{i=1}^{5} = 1^2 + 2^2 + 3^2 + 4^2 + 5^2$$
$$= 1 + 4 + 9 + 16 + 25$$
$$= 55$$

EXAMPLE 5 Expand and simplify $\sum_{k=3}^{6} (-2)^{k-4}$

Solution: In the expression $(-2)^{k-4}$, replace the index with all consecutive natural numbers from 3 through 6.

$$\sum_{k=3}^{6} (-2)^{k-4} = (-2)^{3-4} + (-2)^{4-4} + (-2)^{5-4} + (-2)^{6-4}$$
$$= (-2)^{-1} + (-2)^0 + (-2)^1 + (-2)^2$$
$$= -\frac{1}{2} + 1 - 2 + 4$$
$$= \frac{5}{2} \text{ or } 2\frac{1}{2}$$

Exercise 11.1

Write the first four terms of the sequence defined by the given general term.

1. $a_n = 2n - 1$
2. $a_n = (-2)^n$
3. $a_n = 2^{n+2}$
4. $a_n = 3^{n+1}$
5. $a_n = \dfrac{n+1}{n+2}$
6. $a_n = \dfrac{3}{n+1}$
7. $a_n = 1 + \dfrac{1}{n}$
8. $a_n = \dfrac{n+2}{n^2 - 2}$
9. $a_n = (-1)^n(2n + 1)$
10. $a_n = \log(n + 1)$

Expand and simplify.

11. $\sum_{n=1}^{4} (3n + 2)$
12. $\sum_{i=1}^{5} (2i - 1)$
13. $\sum_{n=3}^{7} (-2)^{n-1}$
14. $\sum_{n=1}^{4} \dfrac{n}{1+n}$
15. $\sum_{n=2}^{5} (-1)^n(n - 1)$
16. $\sum_{k=1}^{5} k^3$
17. $\sum_{i=1}^{4} \dfrac{(-1)^i}{2i}$
18. $\sum_{n=1}^{5} \sqrt{2n - 1}$
19. $\sum_{k=1}^{4} \left(\dfrac{1}{3}\right)^k$
20. $\sum_{i=1}^{3} \dfrac{2}{3i + 1}$

NATURAL NUMBER FUNCTIONS

Find a sigma notation for the following.

21. $1 + 4 + 9 + 16 + 25 + 36$

22. $\frac{1}{2} + \frac{2}{4} + \frac{3}{8} + \frac{4}{16} + \frac{5}{32}$

23. $\frac{1}{2} + \frac{1}{4} + \frac{1}{6} + \frac{1}{8}$

24. $(-2) + 4 + (-8) + 16$

25. $\frac{1}{2} + \frac{2}{3} + \frac{3}{4}$

26. $-1 + 1 + 3 + 5 + 7$

27. $\frac{1}{2} + \frac{\sqrt{2}}{3} + \frac{\sqrt{3}}{4} + \frac{2}{5} + \frac{\sqrt{5}}{6}$

28. $\frac{1}{2} + \frac{4}{3} + \frac{9}{4} + \frac{16}{5}$

ARITHMETIC PROGRESSIONS

11.2

A sequence in which each term after the first is obtained by adding the same constant to the preceding term is called an *arithmetic sequence* or *arithmetic progression* (abbreviated A.P.). The constant is called the *common difference*. For example, the sequence

$$-1, 2, 5, 8, \ldots$$

is an arithmetic progression since each term, except the first, is formed by adding 3 to the preceding term. The common difference is 3, and can be found by subtracting any term from the succeeding term. Thus,

$$2 - (-1) = 3$$
$$5 - 2 = 3$$
$$8 - 5 = 3$$

EXAMPLE 1 Find the common difference for the arithmetic progression
$4, -1, -6, -11, \ldots$

Solution: Subtract any term from its successor. Subtracting the first term, 4, from the second term, -1, we have

$$-1 - 4 = -5$$

The common difference is -5. When -5 is added to any term, we obtain the next term of the A.P.

In general, for an arithmetic progression in which a is the first term and d is the common difference, the sequence is

$$a, a + d, (a + d) + d, [(a + d) + d] + d, \ldots$$

or more simply

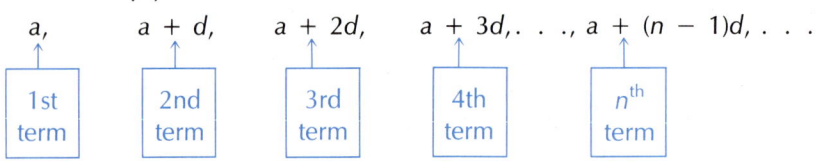

a, $a + d$, $a + 2d$, $a + 3d, \ldots, a + (n - 1)d, \ldots$

| 1st term | 2nd term | 3rd term | 4th term | n^{th} term |

ARITHMETIC PROGRESSIONS 471

This suggests the following definition:

> **Definition of an Arithmetic Progression**
> An arithmetic progression (A.P.) is a sequence of the form
> $$a, a + d, a + 2d, a + 3d, \ldots, a + (n - 1)d, \ldots$$
> where a is the first term, d is the common difference, and $a + (n - 1)d$ is the nth term.

Recall that the nth term is also known as the general term and defines the sequence function. Thus, the general term or sequence function of an arithmetic progression is

$$a_n = a + (n - 1)d$$

where a is the first term, d is the common difference, and n is a natural number specifying a particular term.

EXAMPLE 2 Find the general term for the arithmetic progression $-5, -1, 3, 7, \ldots$

Solution: The first term, a, is -5, and the common difference, d, is 4. Substituting into the formula for the general term, we have

$$\begin{aligned} a_n &= a + (n - 1)d \\ &= -5 + (n - 1) \cdot 4 \\ &= -5 + 4n - 4 \\ &= 4n - 9 \end{aligned}$$

EXAMPLE 3 Find the twelfth term of the arithmetic progression $-7, -2, 3, 8, \ldots$

Solution: The first term, a, is -7, and the common difference, d, is 5. We must find the 12th term, so n is 12. Substituting into the sequence function, we have

$$\begin{aligned} a_n &= a + (n - 1)d \\ a_{12} &= -7 + (12 - 1) \cdot 5 \\ &= -7 + 11 \cdot 5 \\ &= 48 \end{aligned}$$

EXAMPLE 4 Find the number of terms in the arithmetic progression

$$2, \frac{1}{2}, -1, -\frac{5}{2}, \ldots, -19$$

Solution: The nth term, a_n, is -19. The first term, a, is 2, and the common difference, d, is $-\frac{3}{2}$. We substitute into the formula for the general term and solve for n.

$$a_n = a + (n - 1)d$$
$$-19 = 2 + (n - 1)\left(-\frac{3}{2}\right)$$
$$-38 = 4 - (n - 1) \cdot 3$$
$$-38 = 4 - 3n + 3$$
$$-45 = -3n$$
$$15 = n$$

There are 15 terms in the given arithmetic progression.

EXAMPLE 5 Find the sixteenth term, a_{16}, of the arithmetic progression whose tenth term, a_{10}, is 13 and whose nineteenth term, a_{19}, is 40.

Solution: According to the formula for the general term, the tenth term is $a_{10} = a + 9d$, and the nineteenth term is $a_{19} = a + 18d$. Since these terms equal 13 and 40 respectively, we set up and solve the system

$$a + 9d = 13 \qquad (1)$$
$$a + 18d = 40 \qquad (2)$$

To solve for d we multiply equation (1) by -1 and add it to equation (2).

$$-a - 9d = -13$$
$$\underline{a + 18d = 40}$$
$$9d = 27$$
$$d = 3$$

We solve for a by replacing 3 for d in equation (1).

$$a + 9d = 13$$
$$a + 9 \cdot 3 = 13$$
$$a + 27 = 13$$
$$a = -14$$

To find the sixteenth term we use the general formula $a_n = a + (n - 1)d$, where $a = -14$ and $d = 3$. Thus,

$$a_{16} = a + 15d$$
$$= -14 + 15 \cdot 3$$
$$= -14 + 45$$
$$= 31$$

The sum of the first n terms of an arithmetic progression, denoted by S_n, is written

$$S_n = a + [a + d] + [a + 2d] + \ldots + a_n$$

This series can also be written in reverse as

$$S_n = a_n + [a_n - d] + [a_n - 2d] + \ldots + a$$

To derive a formula for S_n, we add the two equations term by term, obtaining

$$2S_n = (a + a_n) + (a + a_n) + (a + a_n) + \ldots + (a + a_n)$$

Since the right member of the equation contains n terms, it can be simplified to

$$2S_n = n(a + a_n)$$

or

$$S_n = \frac{n}{2}(a + a_n)$$

Sum of First n Terms of an A.P.

The sum of the first n terms of an A.P. is given by

$$S_n = \frac{n}{2}(a + a_n)$$

where a is the first term and a_n is the nth term.

EXAMPLE 6

Find the sum of the first twelve terms of the arithmetic progression 3, 7, 11, 15, . . .

Solution: The first term is 3 and the common difference is 4. To find the twelfth term, we use the formula for the general term,

$$a_n = a + (n - 1)d$$
$$a_{12} = 3 + (12 - 1) \cdot 4 = 47$$

Substituting 12 for n, 3 for a, and 47 for a_{12} into the formula

$$S_n = \frac{n}{2}(a + a_n)$$

produces

$$S_{12} = \frac{12}{2}(3 + 47)$$
$$= 6(50)$$
$$= 300$$

The sum of the first twelve terms is 300.

474   NATURAL NUMBER FUNCTIONS

EXAMPLE 7   Find $\sum_{i=1}^{9} (3i - 5)$

Solution: We are asked to find the sum of the first nine terms of the arithmetic progression having a general term of $3i - 5$, $n = 9$, $a = 3 \cdot 1 - 5 = -2$, and $a_9 = 3 \cdot 9 - 5 = 22$. Substituting these values into the formula

$$S_n = \frac{n}{2}(a + a_n)$$

produces

$$S_9 = \frac{9}{2}(-2 + 22)$$

$$= \frac{9}{2} \cdot 20$$

$$= 90$$

Thus, $\sum_{i=1}^{9} (3i - 5) = 90$.

The following reviews two formulas that apply to arithmetic progressions.

Formulas for Arithmetic Progressions

Given an arithmetic progression where a is the first term, d is the common difference, and n is a natural number specifying a particular term:

(1) The general term is

$$a_n = a + (n - 1)d$$

(2) The sum of the first n terms is

$$S_n = \frac{n}{2}(a + a_n)$$

Exercise 11.2   Find the common difference for the following arithmetic progressions and continue for two more terms.

     1. 5, 8, 11, . . .   **2.** $-6, -1, 4, \ldots$

     3. $a, a + 5, a + 10, \ldots$ **4.** $4, \frac{11}{2}, 7, \ldots$

Find the general term for the following arithmetic progressions.

     5. 2, 6, 10, . . .   **6.** 7, 9, 11, . . .

     7. $-1, 2, 5, \ldots$ **8.** 12, 8, 4, . . .

Find the indicated term of each arithmetic progression.

9. a_{10} for 6, 10, 14, . . .
10. a_{12} for .5, 1.5, 2.5, . . .
11. a_{12} for $-\frac{5}{4}, -\frac{1}{4}, \frac{3}{4}, \ldots$
12. a_{15} for 12, 9, 6, . . .
13. a_{41} for $-3, -\frac{11}{4}, -\frac{5}{2}, \ldots$
14. a_{47} for $-4, -1, 2, \ldots$

Find the indicated sum of each arithmetic progression.

15. S_{10} for 1, $1\frac{1}{2}$, 2, . . .
16. S_{20}, where $a_n = 9n$
17. S_{18}, where $a_n = 3n - 1$
18. S_{16}, where $a = 6$, $d = 4$

Miscellaneous.

19. If $a_6 = -2$ and $a_{40} = 100$, find a_{20}.
20. If $a = -12$, $n = 15$, and $S_n = 135$, find d.
21. Find the sum of the first 120 positive integers.
22. Find the sum of the first 120 positive even integers.
23. Find the sum of all even integers between 11 and 619.
24. Find the sum of all odd integers between 28 and 232.
25. If $S_n = 13$, $a = 1$, and $n = 3$, find a_n.
26. If 1¢ is saved the first day, 2¢ the second day, 3¢ the third day, etc., find the sum that will accumulate at the end of 365 days.
27. How many terms of 4, 6, 8, . . . add up to 928?
28. How many terms of an arithmetic progression add up to 54, if the first term is 20 and the last term is -14?
29. If $a = 2$ and $a_{30} = 60$, find a_{20}.
30. Find the sum of all positive integers between 13 and 312 inclusive.
31. Find k so that 8, k, 15 will form an arithmetic progression. (Hint: use the definition of arithmetic progression.)
32. Find k so that $2k + 2$, $5k - 11$, $7k - 13$ will form an arithmetic progression.
33. In an A.P., $a_3 = 9$ and $a_6 = 18$. Find S_{10}.
34. In an A.P., $a_4 = \frac{1}{2}$ and $a_8 = \frac{17}{2}$. Find S_{10}.
35. Insert two numbers between 5 and 25 so that all four numbers will form an arithmetic progression.
36. Insert two numbers between 3 and 21 so that all four numbers will form an arithmetic progression.

GEOMETRIC PROGRESSIONS

11.3

A sequence in which each term after the first is obtained by multiplying the same constant times the preceding term is called a geometric sequence or a geometric progression (abbreviated G.P.). The constant multiplier is called the common ratio. For example, the sequence

$$1, 3, 9, 27, \ldots$$

is a geometric progression in which each term, except the first, is formed by multiplying 3 times the preceding term. The common ratio is 3, and can be found by dividing any term by the preceding term. Thus,

$$\frac{3}{1} = 3$$

$$\frac{9}{3} = 3$$

$$\frac{27}{9} = 3$$

EXAMPLE 1   Find the common ratio for the geometric progression

$$\frac{1}{3}, \frac{1}{6}, \frac{1}{12}, \frac{1}{24}, \ldots$$

Solution: Divide any term by its predecessor. Dividing the second term by the first term, we have

$$\frac{1}{6} \div \frac{1}{3} = \frac{1}{6} \cdot \frac{3}{1} = \frac{1}{2}$$

The common ratio is 1/2. When it is multiplied times any term, we obtain the next term of the G.P.

In general, for a geometric progression in which a is the first term and r is the common ratio, the sequence is

$$a, ar, (ar)r, [(ar)r]r, \ldots$$

or more simply

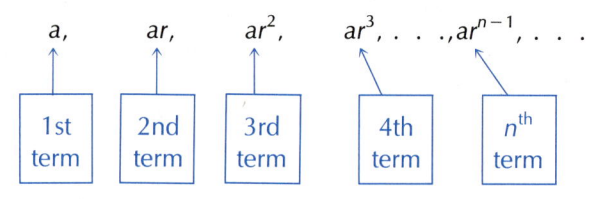

$$a, \quad ar, \quad ar^2, \quad ar^3, \ldots, ar^{n-1}, \ldots$$

| 1st term | 2nd term | 3rd term | 4th term | n^{th} term |

This suggests the following definition:

GEOMETRIC PROGRESSIONS

Definition of a Geometric Progression
A geometric progression (G.P.) is a sequence of the form
$$a, ar, ar^2, ar^3, \ldots, ar^{n-1}, \ldots$$
where a is the first term, r is the common ratio, and ar^{n-1} is the nth or general term.

Find the general term for the geometric progression 3, 6, 12, 24, . . . **EXAMPLE 2**

Solution: The first term is 3 and the common ratio is 2. Substituting into the formula for the general term, we have
$$a_n = ar^{n-1}$$
$$= 3 \cdot 2^{n-1}$$

Find the seventh term of the geometric progression $5, \dfrac{5}{2}, \dfrac{5}{4}, \dfrac{5}{8}, \ldots$ **EXAMPLE 3**

Solution: The first term is 5 and the common ratio is 1/2. We are asked to determine the seventh term, so n is 7. Substituting into the formula for the general term, we have
$$a_n = ar^{n-1}$$
$$a_7 = 5 \cdot \left(\frac{1}{2}\right)^{7-1}$$
$$= 5 \cdot \left(\frac{1}{2}\right)^6$$
$$= 5 \cdot \frac{1}{64}$$
$$= \frac{5}{64}$$

As with arithmetic progressions, the sum of the first n terms of a geometric progression is denoted by S_n. This sum is written
$$S_n = a + ar + ar^2 + \ldots + ar^{n-1} \qquad (1)$$
To derive a formula for S_n, we begin by multiplying both sides of equation (1) by r, producing
$$rS_n = ar + ar^2 + ar^3 + \ldots + ar^{n-1} + ar^n \qquad (2)$$
Next, we subtract equation (2) from equation (1).
$$S_n = a + ar + ar^2 + \ldots + ar^{n-1}$$
$$rS_n = \phantom{a + {}}ar + ar^2 + \ldots + ar^{n-1} + ar^n$$
$$\overline{S_n - rS_n = a + 0 + 0 + \ldots + 0 - ar^n} \text{ or}$$
$$S_n - rS_n = a - ar^n$$

NATURAL NUMBER FUNCTIONS

Factoring S_n from the left side and a from the right side, we have

$$S_n(1 - r) = a(1 - r^n)$$

Dividing both sides by $1 - r$ gives the formula

$$S_n = \frac{a(1 - r^n)}{1 - r}$$

> **Sum of *n* Terms of G.P.**
>
> The sum of the first *n* terms of a G.P. is given by
> $$S_n = \frac{a(1 - r^n)}{1 - r}$$
> where a is the first term and r is the common ratio.

EXAMPLE 4 Find the sum of the first six terms of the geometric progression 4, 8, 16, . . .

Solution: The first term is 4, the common ratio is 2, and n is 6. Substituting into the formula $S_n = \frac{a(1 - r^n)}{1 - r}$, we have

$$S_6 = \frac{4(1 - 2^6)}{1 - 2}$$
$$= \frac{4(1 - 64)}{-1}$$
$$= \frac{4(-63)}{-1}$$
$$= 252$$

EXAMPLE 5 Find $\sum_{i=1}^{4} 2(-3)^i$

Solution: Since $\sum_{i=1}^{4} 2(-3)^i = -6 + 18 - 54 + 162$, we are asked to find the sum of the first four terms of a geometric progression whose first term is -6 and whose common ratio is -3.

Substituting into the formula $S_n = \frac{a(1 - r^n)}{1 - r}$ produces

GEOMETRIC PROGRESSIONS

$$S_4 = \frac{-6[1-(-3)^4]}{1-(-3)}$$

$$= \frac{-6[1-81]}{4}$$

$$= \frac{-6(-80)}{4}$$

$$= 120$$

Thus, $\sum_{i=1}^{4} 2(-3)^i = 120$

We now consider the infinite geometric progressions, such as

$$1, \frac{1}{2}, \frac{1}{4}, \frac{1}{8}, \frac{1}{16}, \frac{1}{32}, \ldots$$

To find the sum, we note that each term is smaller than its predecessor and consequently contributes a smaller and smaller amount to the sum. As the number of terms increases without bound, the corresponding values become closer to zero. The following diagram gives the sum after each term of the series:

	$n=1$	$n=2$	$n=3$	$n=4$	$n=5$	$n=6$
	1	$+\;\frac{1}{2}$	$+\;\frac{1}{4}$	$+\;\frac{1}{8}$	$+\;\frac{1}{16}$	$+\;\frac{1}{32}$ $+\ldots$
Sum:	1	$1\frac{1}{2}$	$1\frac{3}{4}$	$1\frac{7}{8}$	$1\frac{15}{16}$	$1\frac{31}{32}$

As the number of terms increases without bound, the sum approaches 2. We say that this geometric series **converges** and that its sum is 2. In fact, an infinite geometric series will converge whenever its common ratio is between -1 and 1. That is, whenever $|r| < 1$. Thus, looking at the formula

$$S_n = \frac{a(1-r^n)}{1-r}$$

where $|r| < 1$, r^n will approach zero when n increases without bound. This is written as

$$S_\infty = \frac{a(1-0)}{1-r} = \frac{a}{1-r}$$

Sum of an Infinite G.P.

The sum of an infinite geometric progression with $|r| < 1$ is

$$S_\infty = \frac{a}{1-r}$$

EXAMPLE 6 Use the preceding formula to show that the sum of our original infinite G.P. is 2.

Solution: Our original G.P. is $1, \frac{1}{2}, \frac{1}{4}, \frac{1}{8}, \frac{1}{16}, \frac{1}{32}, \ldots$

The common ratio is $\frac{1}{2}$. Since $|r| < 1$, we use the formula

$$S_\infty = \frac{a}{1-r} \text{ to obtain}$$

$$S_\infty = \frac{1}{1 - \frac{1}{2}}$$

$$= \frac{1}{\frac{1}{2}}$$

$$= 2$$

An infinite geometric progression with $|r| \geq 1$ does not have a sum. For example, consider the infinite G.P.

$$3, 6, 12, 24, \ldots$$

where $r = 2$. As seen in the following diagram, each term contributes a larger and larger amount to the sum, forcing it to increase without bound. Such a series is said to **diverge**.

	n = 1	n = 2	n = 3	n = 4	
	3 +	6 +	12 +	24 +	...
Sum:	3	9	21	45	

EXAMPLE 7 Find $\sum_{n=1}^{\infty} 2\left(-\frac{1}{3}\right)^n$

Solution: Since $\sum_{n=1}^{\infty} 2\left(-\frac{1}{3}\right)^n = -\frac{2}{3} + \frac{2}{9} - \frac{2}{27} + \ldots$, we are asked to find the sum of an infinite geometric progression where $a = -\frac{2}{3}$ and $r = -\frac{1}{3}$. Note: $\left|-\frac{1}{3}\right| < 1$; therefore, we use the formula

$S_\infty = \dfrac{a}{1-r}$ to obtain

$S_\infty = \dfrac{-\dfrac{2}{3}}{1 - \left(-\dfrac{1}{3}\right)}$

$= \dfrac{-\dfrac{2}{3}}{\dfrac{4}{3}}$

$= -\dfrac{1}{2}$

Thus, $\displaystyle\sum_{n=1}^{\infty} 2\left(-\dfrac{1}{3}\right)^n = -\dfrac{1}{2}$

Find the fraction equivalent to the repeating decimal 0.535353 . . . **EXAMPLE 8**

Solution: The repeating decimal can be written as

.53 + .0053 + .000053 + . . . or

$\dfrac{53}{100} + \dfrac{53}{10{,}000} + \dfrac{53}{1{,}000{,}000} + \cdots$

This is an infinite geometric progression with $a = \dfrac{53}{100}$ and $r = \dfrac{1}{100}$. Since $|r| < 1$, we use the formula $S_\infty = \dfrac{a}{1-r}$ to obtain

$S_\infty = \dfrac{\dfrac{53}{100}}{1 - \dfrac{1}{100}}$

$= \dfrac{\dfrac{53}{100}}{\dfrac{99}{100}}$

$= \dfrac{53}{99}$

Thus, 0.535353 . . . $= \dfrac{53}{99}$

The following summarizes the formulas that apply to geometric progressions.

NATURAL NUMBER FUNCTIONS

Formulas for Geometric Progressions

Given a geometric progression in which a is the first term, r is the common ratio, and n is a natural number specifying a particular term:

(1) The general term is
$$a_n = ar^{n-1}$$

(2) The sum of the first n terms is
$$S_n = \frac{a(1-r^n)}{1-r}$$

(3) The sum of an infinite geometric progression with $|r| < 1$ is
$$S_\infty = \frac{a}{1-r}$$

Exercise 11.3

Find the common ratio for the following geometric progressions.

1. $3, -6, 12, \ldots$
2. $\frac{9}{2}, 6, 8, \ldots$
3. $-45, 15, -5, \ldots$
4. $\frac{18}{25}, \frac{6}{5}, 2, \ldots$

Find the general term for the following geometric progressions.

5. $-6, -12, -24, \ldots$
6. $2, 6, 18, \ldots$
7. $\frac{1}{9}, \frac{1}{3}, 1, \ldots$
8. $-3, \frac{3}{2}, -\frac{3}{4}, \ldots$

Find the indicated term of each geometric progression.

9. a_7 for $8, -4, 2, \ldots$
10. a_8 for $-81, -27, -9, \ldots$
11. a_6 for $3, -6, 12, \ldots$
12. a_5 for $-3, \frac{3}{2}, -\frac{3}{4}, \ldots$
13. a_5 for $\frac{1}{2}, \frac{1}{6}, \frac{1}{18}, \ldots$
14. a_6 for $-1, 3, -9, \ldots$
15. a_5 where $a_n = \left(\frac{1}{3}\right)^n$
16. a_5 for $\frac{1}{4}, \frac{1}{6}, \frac{1}{9}, \ldots$

Find the indicated sum of each geometric progression.

17. S_6 for $3, 3^2, 3^3, \ldots$
18. S_8 for $1, -2, 4, \ldots$
19. S_7 for $4, 12, 36, \ldots$
20. $\sum_{i=1}^{5} 3^i$

21. $\sum_{n=1}^{4} (-2)^n$
22. $\sum_{i=3}^{7} \left(\frac{1}{2}\right)^{i-2}$
23. S_5 for $-\frac{4}{3}, -\frac{4}{9}, -\frac{4}{27}, \ldots$
24. $\sum_{k=1}^{5} \left(\frac{1}{3}\right)^k$

Find the sum of the following infinite geometric progressions.

25. $6 + 3 + \frac{3}{2} + \ldots$
26. $3 + \frac{2}{3} + \frac{4}{27} + \ldots$
27. $8 + 6 + \frac{9}{2} + \ldots$
28. $\sum_{n=1}^{\infty} \left(-\frac{1}{2}\right)^n$
29. $\sum_{i=1}^{\infty} \left(\frac{1}{4}\right)^i$
30. $\sum_{k=1}^{\infty} 2\left(-\frac{1}{3}\right)^k$
31. $\frac{4}{3} + \frac{2}{3} + \frac{1}{3} + \ldots$
32. $10 + 6 + \frac{18}{5} + \ldots$

Find the fraction in lowest terms equivalent to each repeating decimal.
33. .777 . . .
34. .5252 . . .
35. 1.1212 . . .
36. 1.0202 . . .
37. .3131 . . .
38. .13636 . . .
39. 1.222 . . .
40. .191191 . . .

Miscellaneous.
41. In a geometric progression, $a = 2$ and $r = 3$. Find a_5.
42. In a geometric progression, $n = 8$, $r = -2$, and $a_n = -640$. Find a.
43. In a G.P., the third term is 9 and the first term is 1. Find S_4.
44. A certain ball when dropped from a height of 10 feet rebounds $\frac{2}{5}$ of the original height. How high will it rebound after the fourth bounce?

THE BINOMIAL EXPANSION

11.4

In this section we develop and apply a formula for expanding (multiplying out) binomials raised to any natural number power. To develop this formula, called the binomial expansion, we examine the following binomials that were expanded by regular multiplication.

$(x + y)^1 = x + y$
$(x + y)^2 = x^2 + 2xy + y^2$
$(x + y)^3 = x^3 + 3x^2y + 3xy^2 + y^3$
$(x + y)^4 = x^4 + 4x^3y + 6x^2y^2 + 4xy^3 + y^4$
$(x + y)^5 = x^5 + 5x^4y + 10x^3y^2 + 10x^2y^3 + 5xy^4 + y^5$

NATURAL NUMBER FUNCTIONS

To predict the form of the expansion of $(x + y)^n$, where $n \in N$, the preceding examples suggest the following conclusions:

1. The expansion has $n + 1$ terms.
2. The first term is x^n and the last term is y^n.
3. In each term, the sum of the exponents on x and y is n.
4. In each succeeding term after the first, the power of x decreases by 1 while the power of y increases by 1. Thus, the variables in the expansion of $(x + y)^n$ have this pattern:
$$x^n,\ x^{n-1}y,\ x^{n-2}y^2,\ x^{n-3}y^3,\ \ldots\ xy^{n-1},\ y^n$$
5. The numerical coefficients form the following array, called Pascal's triangle. The numbers at each end of a row are always 1. And, as shown by the shading, each of the other numbers is the sum of the two nearest numbers in the row directly above.

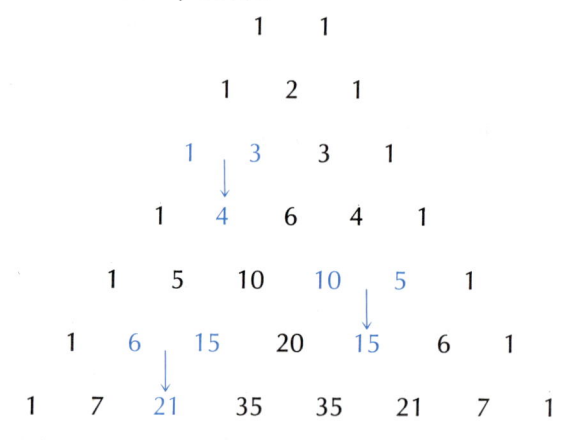

$$
\begin{array}{c}
1\quad 1\\
1\quad 2\quad 1\\
1\quad 3\quad 3\quad 1\\
1\quad 4\quad 6\quad 4\quad 1\\
1\quad 5\quad 10\quad 10\quad 5\quad 1\\
1\quad 6\quad 15\quad 20\quad 15\quad 6\quad 1\\
1\quad 7\quad 21\quad 35\quad 35\quad 21\quad 7\quad 1
\end{array}
$$

EXAMPLE 1 Use the preceding conclusions to expand $(x + 2)^4$.

Solution: Since $n = 4$, we conclude the following:
(1) The expansion will contain 5 terms.
(2) The first term is x^4 and the last term is 2^4.
(3) In each term, the sum of the exponents on x and 2 is 4.
(4) In each succeeding term after the first, the power of x increases by 1 while the power of 2 decreases by 1, giving this pattern:
$$x^4,\ x^3 \cdot 2,\ x^2 \cdot 2^2,\ x \cdot 2^3,\ 2^4$$
(5) The numerical coefficient of each term is found in the 4th row of Pascal's triangle:

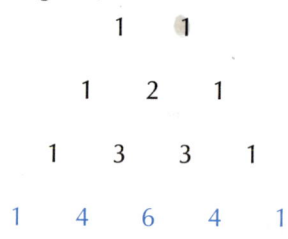

$$
\begin{array}{c}
1\quad 1\\
1\quad 2\quad 1\\
1\quad 3\quad 3\quad 1\\
1\quad 4\quad 6\quad 4\quad 1
\end{array}
$$

THE BINOMIAL EXPANSION

Thus,
$$(x + 2)^4 = 1 \cdot x^4 + 4 \cdot x^3 \cdot 2 + 6 \cdot x^2 \cdot 2^2 + 4 \cdot x \cdot 2^3 + 1 \cdot 2^4$$
$$= x^4 + 8x^3 + 24x^4 + 32x + 16$$

Expand $(2x - y)^5$ **EXAMPLE 2**

Solution: We think of $(2x - y)^5$ as $[(2x) + (-y)]^5$. Since $n = 5$, we conclude the following:

(1) The expansion will have 6 terms.
(2) The first term is $(2x)^5$ and the last term is $(-y)^5$.
(3) In each term the sum of the exponents on $(2x)$ and $(-y)$ is 5.
(4) The pattern of terms is
$$(2x)^5, (2x)^4(-y), (2x)^3(-y)^2, (2x)^2(-y)^3, (2x)(-y)^4, (-y)^5$$
(5) The numerical coefficient of each term is found in the 5th row of Pascal's triangle:

```
              1   1
            1   2   1
          1   3   3   1
        1   4   6   4   1
      1   5  10  10   5   1
```

Thus,
$$(2x - y)^5 = (2x)^5 + 5(2x)^4(-y) + 10(2x)^3(-y)^2 + 10(2x)^2(-y)^3 + 5(2x)(-y)^4 + (-y)^5$$
$$= 32x^5 + 5(16x^4)(-y) + 10(8x^3)(y^2) + 10(4x^2)(-y)^3 + 5(2x)(y^4) + (-y)^5$$
$$= 32x^5 - 80x^4y + 80x^3y^2 - 40x^2y^3 + 10xy^4 - y^5$$

(Note: when the middle sign of a binomial is minus, the terms of the expansion alternate in sign, beginning with a plus sign.)

When n is large, it is not convenient to find coefficients using Pascal's triangle. Instead, we use the following rule:

Numerical Coefficients of Binomial Expansions

To find the numerical coefficient of any term (except the first) in the expansion of $(x + y)^n$, go to the preceding term and multiply its numerical coefficient by the exponent on x and then divide by one more than the exponent on y.

The first three terms of the expansion of $(x + y)^{10}$ are **EXAMPLE 3**
$x^{10} + 10x^9y + 45x^8y^2$

Use the preceding rule to write the next (fourth) term.

486 NATURAL NUMBER FUNCTIONS

Solution: Multiply 45 (the numerical coefficient of the third term) by 8 (the exponent on x), then divide by 3 (one more than the exponent on y). Thus, the numerical coefficient of the fourth term is

$$\frac{(45)(8)}{3} = 120$$

and the entire fourth term is
$120x^7y^3$

EXAMPLE 4 Expand $(a + b)^6$

Solution: Since $n = 6$, we make the following observations.
(1) The expansion will contain 7 terms.
(2) The first term is a^6 and the last term is b^6.
(3) In each term the sum of the exponents on a and b is 6.
(4) The pattern of terms is
$$a^6, a^5b, a^4b^2, a^3b^3, a^2b^4, ab^5, b^6$$
(5) The numerical coefficients will be found using the preceding rule.

So, $(a + b)^6 = 1a^6 + 6a^5b^1 + 15a^4b^2 + 20a^3b^3 + 15a^2b^4 + 6a^1b^5 + 1b^6$

$$\frac{1\cdot 6}{1} \quad \frac{6\cdot 5}{2} \quad \frac{15\cdot 4}{3} \quad \frac{20\cdot 3}{4} \quad \frac{15\cdot 2}{5} \quad \frac{6\cdot 1}{6}$$

$$= a^6 + 6a^5b + 15a^4b^2 + 20a^3b^3 + 15a^2b^4 + 6ab^5 + b^6$$

Using the rule for finding numerical coefficients, we are now able to summarize the procedure for expanding binomials by writing a formula called the binomial theorem or binomial expansion.

The Binomial Theorem

For $n \in N$,

$$(x + y)^n = x^n + \frac{n}{1}x^{n-1}y + \frac{n(n-1)}{2\cdot 1}x^{n-2}y^2 + \frac{n(n-1)(n-2)}{3\cdot 2\cdot 1}x^{n-3}y^3 + \ldots + y^n$$

To write a simpler version of the binomial theorem, we introduce new symbolism called factorial notation.

Definition of Factorials

The symbol $n!$ is read n factorial and means to find the product of all consecutive natural numbers from 1 to n. Thus, for $n \in N$,

$$n! = n \cdot (n-1) \cdot (n-2) \ldots (2) \cdot (1)$$

THE BINOMIAL EXPANSION   **487**

EXAMPLE 5

The following factorials have been evaluated:
(a) $1! = 1$
(b) $2! = 2 \cdot 1 = 2$
(c) $3! = 3 \cdot 2 \cdot 1 = 6$
(d) $4! = 4 \cdot 3 \cdot 2 \cdot 1 = 24$
(e) $5! = 5 \cdot 4 \cdot 3 \cdot 2 \cdot 1 = 120$

From the definition of $n!$, it is evident that

$$n! = n(n - 1)!$$

That is, $7! = 7 \cdot 6!$ or $7 \cdot (6 \cdot 5 \cdot 4 \cdot 3 \cdot 2 \cdot 1)$. We find it convenient to define $0!$ in such a manner that the recursive relationship $n! = n(n - 1)!$ holds true. Letting $n = 1$, we have

$$1! = 1 \cdot 0! = 1$$

Thus, $0!$ must be defined to be 1.

Definition of Zero Factorial

Since $n! = n(n - 1)!$ for $n \in N$, then

$$0! = 1$$

EXAMPLE 6

Evaluate $\dfrac{8! \cdot 0!}{7!}$

Solution: First we note that $0! = 1$. Then we use the recursive relationship $n! = n(n - 1)!$ to write $8!$ as $8 \cdot 7!$. This enables us to reduce the fraction quickly and efficiently.

$$\frac{8! \cdot 0!}{7!} = \frac{8!}{7!}$$

$$= \frac{8 \cdot \cancel{7!}}{\cancel{7!}}$$

$$= 8$$

The binomial theorem can now be written using factorial notation. Recall that

$$(x + y)^n = x^n + \frac{n}{1}x^{n-1}y + \frac{n(n-1)}{2 \cdot 1}x^{n-2}y^2 + \frac{n(n-1)(n-2)}{3 \cdot 2 \cdot 1}x^{n-3}y^3 + \ldots + y^n$$

Using factorial notation to describe the denominators, we write the binomial theorem as follows:

The Binomial Theorem with Factorial Notation

$$(x + y)^n = x^n + \frac{n}{1!} x^{n-1}y + \frac{n(n-1)}{2!} x^{n-2}y^2 + \frac{n(n-1)(n-2)}{3!} x^{n-3}y^3 + \ldots + y^n,$$

where $n \in N$

EXAMPLE 7 Expand $(m + n)^8$ using factorial notation.

Solution: $(m + n)^8 = m^8 + \dfrac{8}{1!}m^7n + \dfrac{8 \cdot 7}{2!}m^6n^2 + \dfrac{8 \cdot 7 \cdot 6}{3!}m^5n^3 +$

$\dfrac{8 \cdot 7 \cdot 6 \cdot 5}{4!}m^4n^4 + \dfrac{8 \cdot 7 \cdot 6 \cdot 5 \cdot 4}{5!}m^3n^5 +$

$\dfrac{8 \cdot 7 \cdot 6 \cdot 5 \cdot 4 \cdot 3}{6!}m^2n^6 +$

$\dfrac{8 \cdot 7 \cdot 6 \cdot 5 \cdot 4 \cdot 3 \cdot 2}{7!}mn^7 + n^8$

Simplifying the numerical coefficients produces

$(m + n)^8 = m^8 + 8m^7n + 28m^6n^2 + 56m^5n^3 + 70m^4n^4$
$+ 56m^3n^5 + 28m^2n^6 + 8mn^7 + n^8$

We can also develop a formula which gives any particular term of a binomial expansion. To do this we examine the sixth term of the expansion $(m + n)^8$ from Example 7,

$$\frac{8 \cdot 7 \cdot 6 \cdot 5 \cdot 4}{5!} m^3n^5$$

and note the following relationships:

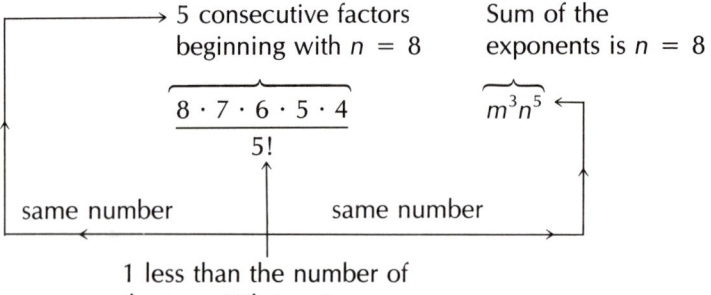

→ 5 consecutive factors beginning with $n = 8$ Sum of the exponents is $n = 8$

$$\underbrace{\frac{8 \cdot 7 \cdot 6 \cdot 5 \cdot 4}{5!}} \qquad m^3n^5$$

same number   same number

1 less than the number of the term (6th term).

Next, we generalize for the rth term of the expansion $(x + y)^n$.

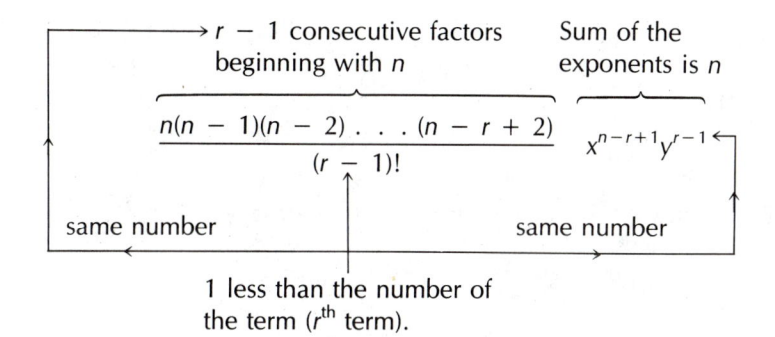

THE BINOMIAL EXPANSION

$r - 1$ consecutive factors beginning with n — Sum of the exponents is n

$$\underbrace{\frac{n(n-1)(n-2)\ldots(n-r+2)}{(r-1)!}}_{\text{same number}} \underbrace{x^{n-r+1}y^{r-1}}$$

same number ← → same number

1 less than the number of the term (r^{th} term).

---

r^{th} Term of a Binomial Expansion

Given $(x + y)^n$, where $n \in N$, the rth term is given by

$$\frac{n(n-1)(n-2)\ldots(n-r+2)}{(r-1)!} x^{n-r+1} y^{r-1}$$

As illustrated by the following example, this formula can be applied in successive steps.

Find the fourth term of $(a + b)^7$

EXAMPLE 8

Solution: (1) Write a fraction bar and two sets of parentheses:

$$\text{———} (\ )(\ )$$

(2) Place the first term of the binomial in the first set of parentheses and the second term in the second set of parentheses:

$$\text{———} (a)(b)$$

(3) In the denominator write a factorial which is 1 less than the number of the term, r.

$$\frac{}{3!} (a)(b)$$

(4) In the numerator write consecutive factors beginning with $n = 7$, decreasing each time by 1, until the number of factors is equal to $r - 1$ (in this case 3).

$$\frac{7 \cdot 6 \cdot 5}{3!} (a)(b)$$

(5) The number of the factorial in the denominator is the same as the exponent of the factor in the second set of parentheses.

$$\frac{7 \cdot 6 \cdot 5}{3!} (a)(b)^3$$

490   NATURAL NUMBER FUNCTIONS

(6) The exponent for the first set of parenthesis is found by subtracting the second exponent from n. Since $n = 7$, the first exponent is $7 - 3$, or 4.

$$\frac{7 \cdot 6 \cdot 5}{3!} (a)^4 (b)^3$$

(7) Simplifying, we obtain the fourth term of $(a + b)^7$.

$$\frac{7 \cdot 6 \cdot 5}{3!} (a)^4 (b)^3 = \frac{7 \cdot \cancel{6} \cdot 5}{\cancel{3} \cdot \cancel{2} \cdot 1} a^4 b^3$$
$$= 35 a^4 b^3$$

EXAMPLE 9 Find the middle term of $(x^2 - 1)^8$.

Solution: The expansion will have 9 terms. The middle term is the fifth term, because four terms come before it and four terms come after it. Thus, $r = 5$. Also, we think of $(x^2 - 1)^8$ as being $[(x^2) + (-1)]^8$.

(1)	——— ( )( )
(2)	——— $(x^2)(-1)$
(3)	$\dfrac{\quad}{4!}$ $(x^2)(-1)$
(4)	$\dfrac{8 \cdot 7 \cdot 6 \cdot 5}{4!}$ $(x^2)(-1)$
(5)	$\dfrac{8 \cdot 7 \cdot 6 \cdot 5}{4!}$ $(x^2)(-1)^4$
(6)	$\dfrac{8 \cdot 7 \cdot 6 \cdot 5}{4!}$ $(x^2)^4 (-1)^4$
(7)	Simplify to obtain the middle term. $\dfrac{8 \cdot 7 \cdot 6 \cdot 5}{4!} (x^2)^4 (-1)^4 = \dfrac{8 \cdot 7 \cdot 6 \cdot 5}{4 \cdot 3 \cdot 2 \cdot 1} x^8 \cdot 1$ $= 70 x^8$

With experience, you will be able to condense these steps and write the answer directly.

Exercise 11.4   Fill in the blanks with the correct response.
1. The expansion of $(x + y)^n$, where $n \in N$, contains _____ terms.
2. In the third term of the expansion of $(x + y)^8$, the sum of the exponents of x and y is _____.

SUMMARY

3. In the fourth term of the expansion of $(x + y)^7$, the exponent on x is _____ and the exponent on y is _____.
4. The power of x in the middle term of the expansion of $(x + y)^8$ is _____.

Use the binomial theorem to expand and simplify.

5. $(x + y)^4$
6. $(x - 1)^3$
7. $(x - 1)^4$
8. $(x + y)^5$
9. $(1 - x)^5$
10. $(x + y^2)^3$
11. $(x - y^3)^4$
12. $(2x + 1)^4$
13. $(2x - 1)^5$
14. $(x^2 + 2)^5$
15. $(2x^2 - y)^3$
16. $(x + h)^6$
17. $\left(x - \dfrac{1}{2}\right)^5$
18. $\left(\dfrac{x}{2} + 2\right)^5$
19. $(x^2 + y^2)^3$
20. $(x^3 - y^3)^3$
21. $\left(x - \dfrac{1}{x}\right)^4$
22. $(x + 1)^n$, where $n \in N$

Write and simplify the first four terms only, of the following expansions.

23. $(x + 1)^{13}$
24. $(2x - 1)^{10}$
25. $(1 + 3x^3)^8$
26. $(2a + 1)^9$

Evaluate.

27. $\dfrac{3!}{0!}$
28. $\dfrac{9!}{6!}$
29. $\dfrac{8!4!}{6!10!}$
30. $\dfrac{5!7!}{8!}$
31. $\dfrac{6!}{7! - 6!}$
32. $5! + 3! - 2!$

Find the indicated term of the binomial expansion.

33. 4th term of $(x + y)^{10}$
34. 6th term of $(a^2 + b)^8$
35. 5th term of $(x^3 - 2)^6$
36. middle term of $(x^2 - 1)^8$
37. middle term of $(2x^2 - 3)^6$
38. 4th term of $(2x^2 - y)^9$
39. middle term of $\left(y^2 - \dfrac{1}{2}\right)^8$
40. 8th term of $(x - 1)^{10}$

Summary

KEY WORDS AND PHRASES

Sequence—a function whose domain is the set of natural numbers.
Term—each number in a sequence.
Finite Sequence—a sequence that terminates.
Series—the indicated sum of the terms of a sequence.
Infinite Sequence—a sequence that does not terminate.
Arithmetic Progression (A.P.)—a sequence where each term after the first is obtained by adding the same constant to the preceding term.

NATURAL NUMBER FUNCTIONS

Geometric Progression (G.P.)—a sequence where each term after the first is obtained by multiplying the same constant times the preceding term.

Factorial n—the symbol $n!$, read *n factorial*, means to find the product of all consecutive natural numbers from 1 to n.

FORMULAS

A.P.

general term: $a_n = a + (n - 1)d$

sum: $S_n = \dfrac{n}{2}(a + a_n)$

G.P.

general term: $a_n = ar^{n-1}$

sum: $S_n = \dfrac{a(1 - r^n)}{1 - r}$

sum of infinite G.P.: $S_\infty = \dfrac{a}{1 - r}$ with $|r| < 1$

Factorial:

$n! = n \cdot (n - 1) \cdot (n - 2) \ldots (2) \cdot (1)$, where $n \in N$

$0! = 1$

Binomial Theorem:

For $n \in N$,

$$(x + y)^n = x^n + \frac{n}{1}x^{n-1}y + \frac{n(n - 1)}{2 \cdot 1}x^{n-2}y^2 + \frac{n(n - 1)(n - 2)}{3 \cdot 2 \cdot 1}x^{n-3}y^3 + \ldots + y^n$$

Binomial Theorem with Factorial Notation:

For $n \in N$,

$$(x + y)^n = x^n + \frac{n}{1!}x^{n-1}y + \frac{n(n - 1)}{2!}x^{n-2}y^2 + \frac{n(n - 1)(n - 2)}{3!}x^{n-3}y^3 + \ldots + y^n$$

r^{th} **Term of a Binomial Expansion:**

$$\frac{n(n - 1)(n - 2) \ldots (n - r + 2)}{(r - 1)!} x^{n-r+1}y^{r-1}$$

Self-Checking Exercise

Write the first four terms of the sequence defined by the given general term.

1. $a_n = \dfrac{n^2 - 1}{n^2 + 1}$
2. $a_n = \dfrac{2n + 1}{n^2}$
3. $a_n = n - \dfrac{1}{n}$
4. $a_n = 2n + \dfrac{(-1)^n}{n}$

Expand and simplify.

5. $\displaystyle\sum_{n=1}^{4} \dfrac{1}{2n}$
6. $\displaystyle\sum_{i=1}^{4} \dfrac{1}{2i - 1}$
7. $\displaystyle\sum_{i=1}^{5} (-1)^i (-3)^{i-1}$
8. $\displaystyle\sum_{k=4}^{6} \log k$

Find a sigma notation for the following.

9. $\dfrac{3}{4} + \dfrac{4}{5} + \dfrac{5}{6} + \dfrac{6}{7}$
10. $\dfrac{1}{\sqrt{2}} + \dfrac{1}{\sqrt{3}} + \dfrac{1}{2} + \dfrac{1}{\sqrt{5}}$

Determine which of the following are arithmetic, which are geometric, and which are neither.

11. 4, 8, 12, . . .
12. 4, −8, 16, . . .
13. 25, 19, 13, . . .
14. $\dfrac{4}{3}, \dfrac{2}{3}, \dfrac{1}{3}, \ldots$
15. 2, 0, 1, 4, . . .
16. $-4, 3, -\dfrac{9}{4}, \ldots$

Find the general term for the following arithmetic or geometric progressions.

17. 5, 10, 20, . . .
18. 5, 10, 15, . . .
19. 20, 17, 14, . . .
20. $1, -\dfrac{1}{3}, \dfrac{1}{9}, \ldots$

Find the indicated term or sum of each A.P.

21. a_{12} for −1, 2, 5, 8, . . .
22. a_{30} for −2, −1, 0, . . .
23. S_{22} for 2, 5, 8, . . .
24. S_{16} for $6, \dfrac{9}{2}, 3, \ldots$

Find the indicated term or sum of each G.P.

25. a_6 for 4, 2, 1, . . .
26. S_6 for 12, 6, 3, . . .
27. S_5 for $2, -\dfrac{4}{3}, \dfrac{8}{9}, \ldots$
28. a_7 for −4, −2, −1, . . .

Find the sum of the following infinite G.P.'s

29. $1 + .01 + .0001 + \ldots$
30. $1 - \dfrac{1}{3} + \dfrac{1}{9} - \dfrac{1}{27} + \ldots$
31. $\displaystyle\sum_{i=2}^{\infty} \left(\dfrac{1}{2}\right)^i$
32. $\dfrac{4}{5} + \dfrac{1}{5} + \dfrac{1}{20} + \ldots$

Find the fraction in lowest terms equivalent to the given repeating decimal.
33. $.26\overline{26}$
34. $.43\overline{3}$

Miscellaneous problems dealing with A.P.'s or G.P.'s.
35. In an A.P., $a = 5$, $d = -3$, $a_n = -76$. Find n.
36. In an A.P., $a = 8$, $a_{11} = 26$. Find a_6.
37. Find the sum of all even numbers between 10 and 100.
38. How many multiples of 5 are between 50 and 500?
39. The sum of a geometric progression with three terms is 39. If the first term is 3, find the third term. (Hint: let the second term be $3r$ and the third term be $3r^2$.)
40. Insert two numbers between 3 and 375 so that all four numbers will be in a G.P.
41. Find the value of $\sum_{i=1}^{7} \left(\frac{1}{2}\right)^{i-1}$.
42. Find the fraction in lowest terms equivalent to $2.31\overline{2312}$.

Evaluate.
43. $\dfrac{6!}{4!0!}$
44. $\dfrac{8!}{6!(4-2)!}$
45. $\dfrac{9!6!}{10!5!}$
46. $\dfrac{1}{0!} + \dfrac{1}{1!} + \dfrac{1}{2!} + \dfrac{1}{3!}$

Expand and simplify by the binomial theorem.
47. $(x - 1)^5$
48. $(x^2 + 1)^4$
49. $(3x + y)^3$
50. $\left(2 - \dfrac{x}{2}\right)^5$

Find the indicated term.
51. 5th term of $(x + y)^{10}$
52. 4th term of $(x - 1)^{12}$
53. middle term of $(x - 2y)^6$
54. the next to last term of $(a + 2b)^5$

Solutions to Self-Checking Exercise

1. $0, \dfrac{3}{5}, \dfrac{4}{5}, \dfrac{15}{17}$
2. $3, \dfrac{5}{4}, \dfrac{7}{9}, \dfrac{9}{16}$
3. $0, \dfrac{3}{2}, \dfrac{8}{3}, \dfrac{15}{4}$
4. $1, \dfrac{9}{2}, \dfrac{17}{3}, \dfrac{33}{4}$
5. $\dfrac{25}{24}$
6. $\dfrac{176}{105}$
7. -121
8. $\log 4 + \log 5 + \log 6 \approx 2.0792$

9. $\sum_{i=1}^{4} \dfrac{2+i}{3+i}$ 10. $\sum_{n=1}^{4} \dfrac{1}{\sqrt{n+1}}$

11. A.P. 12. G.P.
13. A.P. 14. G.P.
15. neither 16. G.P.
17. $S_n = 5(2)^{n-1}$ 18. $a_n = 5n$
19. $a_n = 23 - 3n$ 20. $a_n = \left(-\dfrac{1}{3}\right)^{n-1}$

21. 32 22. 27
23. 737 24. -84
25. $\dfrac{1}{8}$ 26. $\dfrac{189}{8}$
27. $\dfrac{110}{81}$ 28. $-\dfrac{1}{16}$
29. $\dfrac{100}{99}$ 30. $\dfrac{3}{4}$
31. $\dfrac{1}{2}$ 32. $\dfrac{16}{15}$
33. $\dfrac{26}{99}$ 34. $\dfrac{13}{30}$
35. 28 36. 17
37. 2,420 38. 89
39. 27 or 48 40. 15 and 75
41. $\dfrac{127}{64}$ 42. $\dfrac{770}{333}$
43. 30 44. 28
45. $\dfrac{3}{5}$ 46. $\dfrac{8}{3}$
47. $x^5 - 5x^4 + 10x^3 - 10x^2 + 5x - 1$
48. $x^8 + 4x^6 + 6x^4 + 4x^2 + 1$
49. $27x^3 + 27x^2y + 9xy^2 + y^3$
50. $32 - 40x + 20x^2 - 5x^3 + \dfrac{5x^4}{8} - \dfrac{x^5}{32}$
51. $210x^6y^4$
52. $-220x^9$
53. $-160x^3y^3$
54. $80ab^4$

Cumulative Review Exercise

Part I. True or False.

1. The graph of $3x - y = 6$ passes through $(4, 6)$.
2. $(x + y)^2 = x^2 + y^2$
3. $2^m \cdot 2^n = 4^{m+n}$
4. If $-m = 8$, then m is a negative number.
5. $(x^{-1} + y^{-1})^{-1} = x + y$, where $x \neq 0$, $y \neq 0$.
6. If $x = 5 - 3 \cdot 2$, then $x = 4$.
7. Zero is an integer.
8. $\sqrt{9} = \pm 3$
9. $\dfrac{x}{x} = 1$ for all x.
10. $-1 < -100$
11. $-x$ is a negative number.
12. A function is a relation.
13. Every real number has a reciprocal.
14. The line through $(3, 2)$ with slope zero is the line $x = 2$.
15. $(-1)^{-1} = -1$
16. The statement $|x| < 3$ is equivalent to $-3 < x < 3$.
17. The graph of $x^2 + 2y^2 = 4$ is a hyperbola.
18. $\log_{10} 1 = 0$
19. The parabola $y = x^2 + 2x - 1$ has a minimum value.
20. $-10, -4, 2, 8, \ldots$ is an arithmetic progression.

Part II. Perform the indicated operations and simplify.

21. $(2x^2 - 2x + 5) + (-2x^2 - x + 5) + (3x^2 - 5x + 2)$
22. $(-4x^2 + 2x + 5) - (-2x^2 + x + 7)$
23. $(5x + 3)(2x - 5)$
24. $(x^2 - 9x + 20) \div (x - 5)$
25. $(x^3 y^2)^4$
26. $\left(\dfrac{2^2 b^{-3}}{ab} \right)^{-1}$
27. $\dfrac{x^2 + 4x - 5}{x^2 + 5x + 6} \cdot \dfrac{x^2 + 3x - 4}{(x - 1)^2}$
28. $\dfrac{\dfrac{x}{y} + y}{x - \dfrac{y}{x}}$
29. $\dfrac{1}{x} + \dfrac{2}{y} + \dfrac{3}{z}$
30. Find the value of $\dfrac{3^0 x + 4x^{-1}}{x^{-(2/3)}}$ if $x = 8$.
31. $\dfrac{-2}{2 - \sqrt{3}}$
32. $\sqrt{27} + \sqrt{48} - \sqrt{12}$

Part III. Solve each equation.

33. $2x - 7 = 11$
34. $5(x - 2) + 4 = -2(x + 1)$
35. $x - \dfrac{3x - 6}{2} = 5 - \dfrac{x - 3}{3}$
36. $|2x - 3| = 9$
37. $2x^2 = 3 - x$
38. $3x^2 + 5x - 1 = 0$
39. $x^4 - 13x^2 + 36 = 0$
40. $2\sqrt{x + 3} = -x - 4$
41. $x = \log_8 2$
42. $(x - 2)(x + 3) = 6$

Part IV. Solve each inequality.

43. $-2(x - 4) \geq x - 10$
44. $-6 < -2x + 1 < 10$
45. $|3x - 2| > 4$
46. $|x - 3| < 2$

Part V. Solve each system.

47. $2x - 3y = -1$
    $3x + y = 4$

48. $x + 2y - z = 6$
    $2x - y + 3z = -13$
    $3x - 2y + 3z = -16$

49. $x^2 - 2y^2 = 1$
    $3x^2 + 4y^2 = 43$

50. $x^2 + y^2 = 16$
    $xy = 8$

Part VI. Miscellaneous.

51. Write in $a + bi$ form: $5 - 2i - (3 + i)^2$
52. Find the distance between $(-3, 5)$ and $(6, -8)$.
53. Find the slope of the line passing through $(-2, 3)$ and $(1, -4)$.
54. Evaluate: $\begin{vmatrix} 4 & 2 & -1 \\ 0 & 3 & 2 \\ -1 & 1 & 2 \end{vmatrix}$
55. Find the 4th term of $(2x + 1)^8$.
56. Expand $(2x - 1)^4$ by the binomial theorem.
57. Find the fraction in lowest terms equivalent to $.22\overline{2}\ldots$
58. If $f(x) = 2x^3 - x^2 + x - 1$, find $f(-2)$.
59. Identify the graphs of
    a) $x^2 - 2y = 4$ b) $2x^2 - y^2 = 4$ c) $x^2 + y^2 = 4$
60. If y varies directly as the cube of x and inversely as the square of t, and $x = 4$ when $y = 8$ and $t = 2$, find y when $x = 2$ and $t = 10$.

Part VII. Solve the following applied problems.

61. The width of a rectangle is 5 feet less than its length. Find the dimensions if its area is 50 square feet.
62. Bob has as many quarters as dimes. He also has some nickels. If the value of his 68 coins is $8.15, how many dimes does he have?

498 CUMULATIVE REVIEW EXERCISE

63. Gus walked east at 3 miles per hour and returned at 4 miles per hour. If the round trip took $3\frac{1}{2}$ hours, how many miles did he walk one way?

64. How many ounces of a 10% salt solution and how many ounces of a 20% salt solution should be mixed together to form 50 ounces of a 15% salt solution?

65. The sum of two numbers is 51 and their difference is 7. Find the two numbers.

66. The sum of three consecutive even integers is 222. What are the integers?

Part VIII. Graph the following on the real number line.
67. $|x| < 3$
68. $|x| = 4$
69. $x < -1$ or $x \geq 3$

Part IX. Graph the following on the rectangular coordinate system.
70. $-2x + y = 4$
71. $x - y < -2$
72. $f(x) = x^2 - 4x$
73. $x^2 + y^2 = 16$
74. $25x^2 + 9y^2 = 225$
75. $16y^2 - 9x^2 = 144$
76. $y = 2^x$
77. $y = |x - 3|$
78. $y = \log_3 x$

Solutions to Cumulative Review Exercise

1. True, since $3(4) - 6 = 6$
2. False; equals $x^2 + 2xy + y^2$
3. False; equals 2^{m+n}
4. True, since $m = -8$
5. False; equals $\dfrac{xy}{y + x}$
6. False; equals -1
7. True
8. False; equals 3
9. False; undefined for $x = 0$
10. False; since -1 is to the right of -100 on the number line
11. False; can't tell until x is known
12. True
13. False; zero has no reciprocal
14. False; it is the line $y = 2$
15. True; equals $\dfrac{1}{-1}$
16. True
17. False; graph is an ellipse
18. True, since $10^0 = 1$
19. True, since coefficient of x^2 is positive
20. True, since common difference is 6
21. $3x^2 - 8x + 12$
22. $-2x^2 + x - 2$
23. $10x^2 - 19x - 15$
24. $\dfrac{(x-5)(x-4)}{(x-5)} = x - 4$
25. $x^{12}y^8$
26. $\dfrac{ab^4}{4}$

SOLUTIONS TO CUMULATIVE REVIEW EXERCISE

27. $\dfrac{(x+5)(x+4)}{(x+3)(x+2)}$

28. $\dfrac{x^2 + xy^2}{x^2y - y^2}$ or $\dfrac{x(x+y^2)}{y(x^2-y)}$

29. $\dfrac{yz + 2xz + 3xy}{xyz}$

30. $\dfrac{1 \cdot 8 + 4 \cdot \frac{1}{8}}{\frac{1}{8^{2/3}}} = 34$

31. $\dfrac{-2(2+\sqrt{3})}{(2-\sqrt{3})(2+\sqrt{3})} = -2(2+\sqrt{3})$

32. $5\sqrt{3}$

33. $x = 9$

34. $x = \dfrac{4}{7}$

35. $x = -18$

36. $x = -3$ or $x = 6$

37. $x = 1$ or $x = -\dfrac{3}{2}$

38. $x = \dfrac{-5 \pm \sqrt{37}}{6}$

39. $(x^2 - 9)(x^2 - 4) = 0 \rightarrow$ $x = \pm 3$ or $x = \pm 2$

40. $x = -2$ does not check; there is no solution

41. $8^x = 2 \rightarrow x = \dfrac{1}{3}$

42. $x^2 + x - 6 = 6 \rightarrow x = -4$ or $x = 3$

43. $x \leq 6$

44. $-\dfrac{9}{2} < x < \dfrac{7}{2}$

45. $x > 2$ or $x < -\dfrac{2}{3}$

46. $1 < x < 5$

47. $(1, 1)$

48. $(-1, 2, -3)$

49. $(3, 2)(3, -2)(-3, 2)(-3, -2)$

50. $(2\sqrt{2}, 2\sqrt{2}), (-2\sqrt{2}, -2\sqrt{2})$

51. $-3 - 8i$ or $-3 + (-8)i$

52. $d = \sqrt{[6-(-3)]^2 + [-8-5]^2}$
    $d = 5\sqrt{10}$

53. $m = \dfrac{-4-3}{1-(-2)} = \dfrac{-7}{3}$

54. 9

55. $\dfrac{8 \cdot 7 \cdot 6}{3!}(2x)^5(1)^3 = 1792x^5$

56. $(2x)^4 + 4(2x)^3(-1) + 6(2x)^2(-1)^2 + 4(2x)(-1)^3 + (-1)^4 =$
    $16x^4 - 32x^3 + 24x^2 - 8x + 1$

57. $\dfrac{\frac{2}{10}}{1 - \frac{1}{10}} = \dfrac{2}{9}$

58. -23

59. a) parabola
    b) hyperbola
    c) circle

60. $\dfrac{1}{25}$

61.

$x(x - 5) = 50$
length: 10 feet
width: 5 feet

(rectangle with width x and height $x - 5$)

SOLUTIONS TO CUMULATIVE REVIEW EXERCISE

62. x: number of quarters
x: number of dimes
68 − 2x: number of nickels

$25x + 10x + 5(68 - 2x) = 815$
Bob has 19 dimes

63.

	D	r	t
going	x	3	$\frac{x}{3}$
returning	x	4	$\frac{x}{4}$

$\frac{x}{3} + \frac{x}{4} = \frac{7}{2}$

Gus walked 6 miles

64. [x ounces / 10% salt] + [(50 − x) ounces / 20% salt] = [50 ounces / 15% salt]

$.10x + .20(50 - x) = .15(50)$

25 ounces of 10% salt
25 ounces of 20% salt

65. $x + y = 51$
$x - y = 7$. The two numbers are 22 and 29.

66. $x + (x + 2) + (x + 4) = 222$. The integers are 72, 74, and 76.

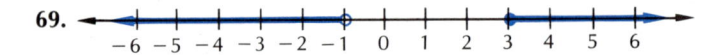

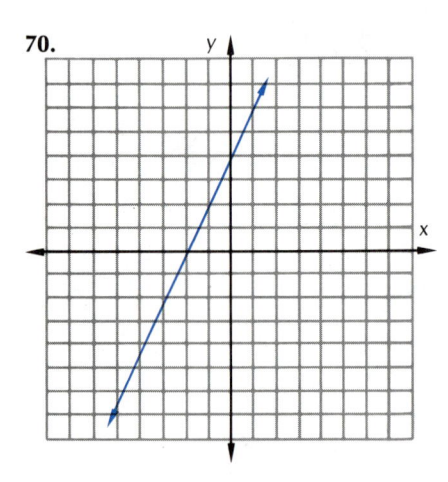

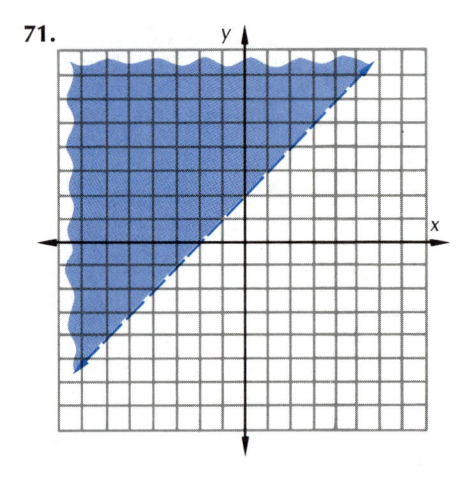

67. number line from −6 to 6, open circle at −2, solid line to open circle at 3

68. number line from −6 to 6, closed dots at −4 and 4

69. number line from −6 to 6, ray left through open circle at −1, ray right from closed dot at 3

70. graph of a line with positive slope through the origin region

71. graph showing shaded triangular region in the first quadrant

SOLUTIONS TO CUMULATIVE REVIEW EXERCISE

72.

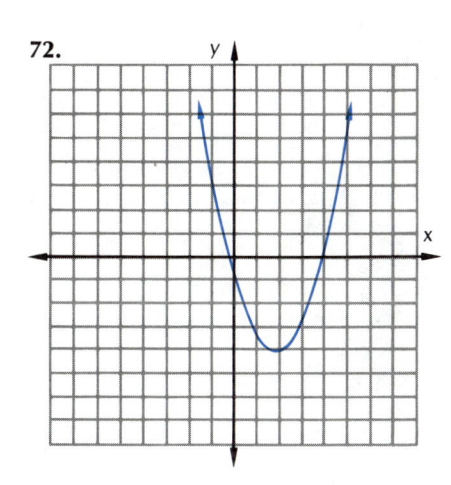

73.

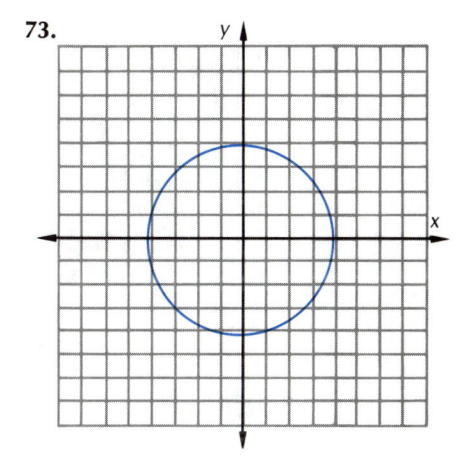

74.

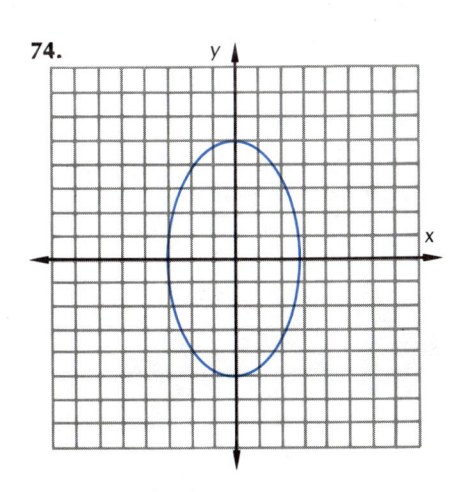

75.

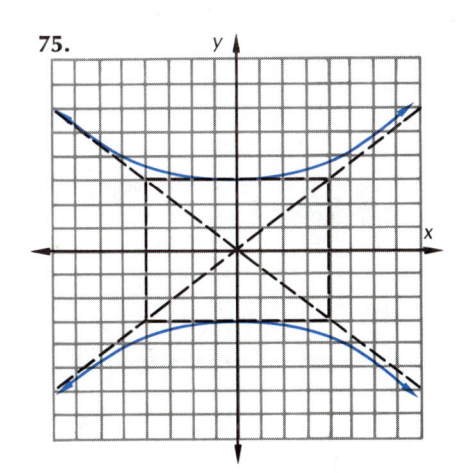

76.

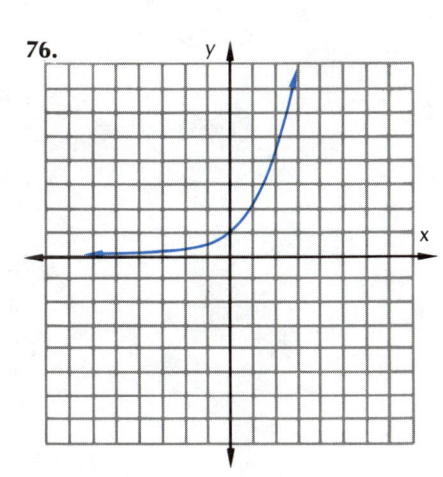

77.

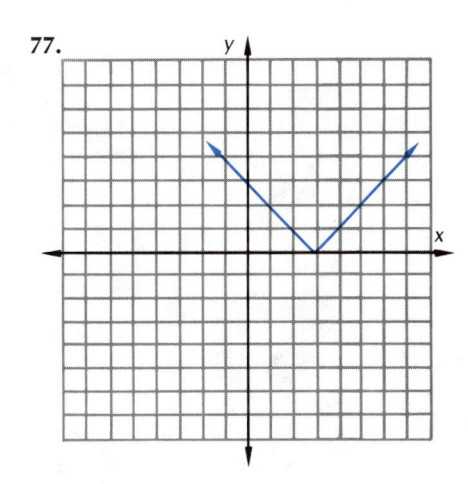

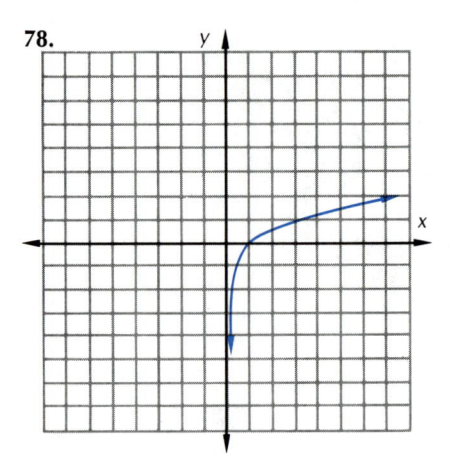

78.

APPENDIX/TABLES

TABLE I Powers and Roots
TABLE II Powers of e
TABLE III Natural Logarithms
TABLE IV Common Logarithms
Answers to Odd-Numbered Exercises

TABLE I Powers and Roots

n	n^2	n^3	$\sqrt{n}$	$\sqrt[3]{n}$	n	n^2	n^3	$\sqrt{n}$	$\sqrt[3]{n}$
0	0	0	0.000	0.000	50	2 500	125 000	7.071	3.684
1	1	1	1.000	1.000	51	2 601	132 651	7.141	3.708
2	4	8	1.414	1.260	52	2 704	140 608	7.211	3.733
3	9	27	1.732	1.442	53	2 809	148 877	7.280	3.756
4	16	64	2.000	1.587	54	2 916	157 464	7.348	3.780
5	25	125	2.236	1.710	55	3 025	166 375	7.416	3.803
6	36	216	2.449	1.817	56	3 136	175 616	7.483	3.826
7	49	343	2.646	1.913	57	3 249	185 193	7.550	3.849
8	64	512	2.828	2.000	58	3 364	195 112	7.616	3.871
9	81	729	3.000	2.080	59	3 481	205 379	7.681	3.893
10	100	1 000	3.162	2.154	60	3 600	216 000	7.746	3.915
11	121	1 331	3.317	2.224	61	3 721	226 981	7.810	3.936
12	144	1 728	3.464	2.289	62	3 844	238 328	7.874	3.958
13	169	2 197	3.606	2.351	63	3 969	250 047	7.937	3.979
14	196	2 744	3.742	2.410	64	4 096	262 144	8.000	4.000
15	225	3 375	3.873	2.466	65	4 225	274 625	8.062	4.021
16	256	4 096	4.000	2.520	66	4 356	287 496	8.124	4.041
17	289	4 913	4.123	2.571	67	4 489	300 763	8.185	4.062
18	324	5 832	4.243	2.621	68	4 624	314 432	8.246	4.082
19	361	6 859	4.359	2.668	69	4 761	328 509	8.307	4.102
20	400	8 000	4.472	2.714	70	4 900	343 000	8.367	4.121
21	441	9 261	4.583	2.759	71	5 041	357 911	8.426	4.141
22	484	10 648	4.690	2.802	72	5 184	373 248	8.485	4.160
23	529	12 167	4.796	2.844	73	5 329	389 017	8.544	4.179
24	576	13 824	4.899	2.884	74	5 476	405 224	8.602	4.198
25	625	15 625	5.000	2.924	75	5 625	421 875	8.660	4.217
26	676	17 576	5.099	2.962	76	5 776	438 976	8.718	4.236
27	729	19 683	5.196	3.000	77	5 929	456 533	8.775	4.254
28	784	21 952	5.292	3.037	78	6 084	474 552	8.832	4.273
29	841	24 389	5.385	3.072	79	6 241	493 039	8.888	4.291
30	900	27 000	5.477	3.107	80	6 400	512 000	8.944	4.309
31	961	29 791	5.568	3.141	81	6 561	531 441	9.000	4.327
32	1 024	32 768	5.657	3.175	82	6 724	551 368	9.055	4.344
33	1 089	35 937	5.745	3.208	83	6 889	571 787	9.110	4.362
34	1 156	39 304	5.831	3.240	84	7 056	592 704	9.165	4.380
35	1 225	42 875	5.916	3.271	85	7 225	614 125	9.220	4.397
36	1 296	46 656	6.000	3.302	86	7 396	636 056	9.274	4.414
37	1 369	50 653	6.083	3.332	87	7 569	658 503	9.327	4.431
38	1 444	54 872	6.164	3.362	88	7 744	681 472	9.381	4.448
39	1 521	59 319	6.245	3.391	89	7 921	704 969	9.434	4.465
40	1 600	64 000	6.325	3.420	90	8 100	729 000	9.487	4.481
41	1 681	68 921	6.403	3.448	91	8 281	753 571	9.539	4.498
42	1 764	74 088	6.481	3.476	92	8 464	778 688	9.592	4.514
43	1 849	79 507	6.557	3.503	93	8 649	804 357	9.644	4.531
44	1 936	85 184	6.633	3.530	94	8 836	830 584	9.695	4.547
45	2 025	91 125	6.708	3.557	95	9 025	857 375	9.747	4.563
46	2 116	97 336	6.782	3.583	96	9 216	884 736	9.798	4.579
47	2 209	103 823	6.856	3.609	97	9 409	912 673	9.849	4.595
48	2 304	110 592	6.928	3.634	98	9 604	941 192	9.899	4.610
49	2 401	117 649	7.000	3.659	99	9 801	970 299	9.950	4.626
					100	10 000	1 000 000	10.000	4.642

TABLE II Powers of e

x	e^x	e^{-x}	x	e^x	e^{-x}	x	e^x	e^{-x}
0.00	1.000	1.000	0.50	1.649	0.607	1.00	2.718	0.368
0.01	1.010	0.990	0.51	1.665	0.600	1.01	2.746	0.364
0.02	1.020	0.980	0.52	1.682	0.595	1.02	2.773	0.361
0.03	1.031	0.970	0.53	1.699	0.589	1.03	2.801	0.357
0.04	1.041	0.961	0.54	1.716	0.583	1.04	2.829	0.353
0.05	1.051	0.951	0.55	1.733	0.577	1.05	2.858	0.350
0.06	1.062	0.942	0.56	1.751	0.571	1.06	2.886	0.346
0.07	1.073	0.932	0.57	1.768	0.566	1.07	2.915	0.343
0.08	1.083	0.923	0.58	1.786	0.560	1.08	2.945	0.340
0.09	1.094	0.914	0.59	1.804	0.554	1.09	2.974	0.336
0.10	1.105	0.905	0.60	1.822	0.549	1.10	3.004	0.333
0.11	1.116	0.896	0.61	1.840	0.543	1.11	3.034	0.330
0.12	1.127	0.887	0.62	1.859	0.538	1.12	3.065	0.326
0.13	1.139	0.878	0.63	1.878	0.533	1.13	3.096	0.323
0.14	1.150	0.869	0.64	1.896	0.527	1.14	3.127	0.320
0.15	1.162	0.861	0.65	1.916	0.522	1.15	3.158	0.317
0.16	1.174	0.852	0.66	1.935	0.517	1.16	3.190	0.313
0.17	1.185	0.844	0.67	1.954	0.512	1.17	3.222	0.310
0.18	1.197	0.835	0.68	1.974	0.507	1.18	3.254	0.307
0.19	1.209	0.827	0.69	1.994	0.502	1.19	3.287	0.304
0.20	1.221	0.819	0.70	2.014	0.497	1.20	3.320	0.301
0.21	1.234	0.811	0.71	2.034	0.492	1.21	3.353	0.298
0.22	1.246	0.803	0.72	2.054	0.487	1.22	3.387	0.295
0.23	1.259	0.795	0.73	2.075	0.482	1.23	3.421	0.292
0.24	1.271	0.787	0.74	2.096	0.477	1.24	3.456	0.289
0.25	1.284	0.779	0.75	2.117	0.472	1.25	3.496	0.287
0.26	1.297	0.771	0.76	2.138	0.468	1.26	3.525	0.284
0.27	1.310	0.763	0.77	2.160	0.463	1.27	3.561	0.281
0.28	1.323	0.756	0.78	2.182	0.458	1.28	3.597	0.278
0.29	1.336	0.748	0.79	2.203	0.454	1.29	3.633	0.275
0.30	1.350	0.741	0.80	2.226	0.449	1.30	3.669	0.273
0.31	1.363	0.733	0.81	2.248	0.445	1.31	3.706	0.270
0.32	1.377	0.726	0.82	2.270	0.440	1.32	3.743	0.267
0.33	1.391	0.719	0.83	2.293	0.436	1.33	3.781	0.264
0.34	1.405	0.712	0.84	2.316	0.432	1.34	3.819	0.262
0.35	1.419	0.705	0.85	2.340	0.427	1.35	3.857	0.259
0.36	1.433	0.698	0.86	2.363	0.423	1.36	3.896	0.257
0.37	1.448	0.691	0.87	2.387	0.419	1.37	3.935	0.254
0.38	1.462	0.684	0.88	2.411	0.415	1.38	3.975	0.252
0.39	1.477	0.677	0.89	2.435	0.411	1.39	4.015	0.249
0.40	1.492	0.670	0.90	2.460	0.407	1.40	4.055	0.247
0.41	1.507	0.664	0.91	2.484	0.403	1.41	4.096	0.244
0.42	1.522	0.657	0.92	2.509	0.399	1.42	4.137	0.242
0.43	1.537	0.651	0.93	2.535	0.395	1.43	4.179	0.239
0.44	1.553	0.644	0.94	2.560	0.391	1.44	4.221	0.237
0.45	1.568	0.638	0.95	2.586	0.387	1.45	4.263	0.235
0.46	1.584	0.631	0.96	2.612	0.383	1.46	4.306	0.232
0.47	1.600	0.625	0.97	2.638	0.379	1.47	4.349	0.230
0.48	1.616	0.619	0.98	2.664	0.375	1.48	4.393	0.228
0.49	1.632	0.613	0.99	2.691	0.372	1.49	4.437	0.225

TABLE II Powers of e *(continued)*

x	e^x	e^{-x}	x	e^x	e^{-x}	x	e^x	e^{-x}
1.50	4.482	0.223	2.00	7.389	0.135	2.50	12.182	0.082
1.51	4.527	0.221	2.01	7.463	0.134	2.51	12.305	0.081
1.52	4.572	0.219	2.02	7.538	0.133	2.52	12.429	0.080
1.53	4.618	0.217	2.03	7.614	0.131	2.53	12.554	0.080
1.54	4.665	0.214	2.04	7.691	0.130	2.54	12.680	0.079
1.55	4.712	0.212	2.05	7.768	0.129	2.55	12.807	0.078
1.56	4.759	0.210	2.06	7.846	0.127	2.56	12.936	0.077
1.57	4.807	0.208	2.07	7.925	0.126	2.57	13.066	0.077
1.58	4.855	0.206	2.08	8.004	0.125	2.58	13.197	0.076
1.59	4.904	0.204	2.09	8.085	0.124	2.59	13.330	0.075
1.60	4.953	0.202	2.10	8.166	0.122	2.60	13.464	0.074
1.61	5.003	0.200	2.11	8.248	0.121	2.61	13.599	0.074
1.62	5.053	0.198	2.12	8.331	0.120	2.62	13.736	0.073
1.63	5.104	0.196	2.13	8.415	0.119	2.63	13.874	0.072
1.64	5.155	0.194	2.14	8.499	0.118	2.64	14.013	0.071
1.65	5.207	0.192	2.15	8.585	0.116	2.65	14.154	0.071
1.66	5.239	0.190	2.16	8.671	0.115	2.66	14.296	0.070
1.67	5.312	0.188	2.17	8.758	0.114	2.67	14.440	0.069
1.68	5.366	0.186	2.18	8.848	0.113	2.68	14.585	0.069
1.69	5.420	0.185	2.19	8.935	0.112	2.69	14.732	0.068
1.70	5.474	0.183	2.20	9.025	0.111	2.70	14.880	0.067
1.71	5.529	0.181	2.21	9.116	0.110	2.71	15.029	0.067
1.72	5.585	0.179	2.22	9.207	0.109	2.72	15.180	0.066
1.73	5.641	0.177	2.23	9.300	0.108	2.73	15.333	0.065
1.74	5.697	0.176	2.24	9.303	0.106	2.74	15.487	0.065
1.75	5.755	0.174	2.25	9.488	0.105	2.75	15.643	0.064
1.76	5.812	0.172	2.26	9.583	0.104	2.76	15.800	0.063
1.77	5.871	0.170	2.27	9.679	0.103	2.77	15.959	0.063
1.78	5.930	0.169	2.28	9.777	0.102	2.78	16.119	0.062
1.79	5.989	0.167	2.29	9.873	0.101	2.79	16.281	0.061
1.80	6.050	0.165	2.30	9.974	0.100	2.80	16.445	0.061
1.81	6.110	0.164	2.31	10.074	0.099	2.81	16.610	0.060
1.82	6.172	0.162	2.32	10.176	0.098	2.82	16.777	0.060
1.83	6.234	0.160	2.33	10.278	0.097	2.83	16.945	0.059
1.84	6.297	0.159	2.34	10.381	0.096	2.84	17.116	0.058
1.85	6.360	0.157	2.35	10.486	0.095	2.85	17.288	0.058
1.86	6.424	0.156	2.36	10.591	0.094	2.86	17.462	0.057
1.87	6.488	0.154	2.37	10.697	0.093	2.87	17.637	0.057
1.88	6.553	0.153	2.38	10.805	0.093	2.88	17.814	0.056
1.89	6.619	0.151	2.39	10.913	0.092	2.89	17.993	0.056
1.90	6.686	0.150	2.40	11.023	0.091	2.90	18.174	0.055
1.91	6.753	0.148	2.41	11.134	0.090	2.91	18.357	0.054
1.92	6.821	0.147	2.42	11.246	0.089	2.92	18.541	0.054
1.93	6.890	0.145	2.43	11.359	0.088	2.93	18.728	0.053
1.94	6.959	0.144	2.44	11.473	0.087	2.94	18.916	0.053
1.95	7.029	0.142	2.45	11.588	0.086	2.95	19.106	0.052
1.96	7.099	0.141	2.46	11.705	0.085	2.96	19.298	0.052
1.97	7.171	0.139	2.47	11.822	0.085	2.97	19.492	0.051
1.98	7.243	0.138	2.48	11.941	0.084	2.98	19.688	0.051
1.99	7.316	0.137	2.49	12.081	0.083	2.99	19.886	0.050

TABLE II Powers of e *(continued)*

x	e^x	e^{-x}	x	e^x	e^{-x}	x	e^x	e^{-x}
3.00	20.086	0.050	3.50	33.115	0.030	4.00	54.598	0.018
3.01	20.287	0.049	3.51	33.448	0.030	4.01	55.147	0.018
3.02	20.491	0.049	3.52	33.784	0.030	4.02	55.701	0.018
3.03	20.697	0.048	3.53	34.124	0.029	4.03	56.261	0.018
3.04	20.905	0.048	3.54	34.467	0.029	4.04	56.826	0.018
3.05	21.115	0.047	3.55	34.813	0.029	4.05	57.397	0.017
3.06	21.328	0.047	3.56	35.163	0.028	4.06	57.974	0.017
3.07	21.542	0.046	3.57	35.517	0.028	4.07	58.557	0.017
3.08	21.758	0.046	3.58	35.874	0.028	4.08	59.145	0.017
3.09	21.977	0.046	3.59	36.234	0.028	4.09	59.740	0.017
3.10	22.198	0.045	3.60	36.596	0.027	4.10	60.340	0.017
3.11	22.421	0.045	3.61	36.966	0.027	4.11	60.947	0.016
3.12	22.646	0.044	3.62	37.338	0.027	4.12	61.559	0.016
3.13	22.874	0.044	3.63	37.713	0.027	4.13	62.178	0.016
3.14	23.104	0.043	3.64	38.092	0.026	4.14	62.803	0.016
3.15	23.336	0.043	3.65	38.475	0.026	4.15	63.434	0.016
3.16	23.571	0.042	3.66	38.861	0.026	4.16	64.072	0.016
3.17	23.807	0.042	3.67	39.252	0.025	4.17	64.715	0.015
3.18	24.047	0.042	3.68	39.646	0.025	4.18	65.366	0.015
3.19	24.288	0.041	3.69	40.045	0.025	4.19	66.023	0.015
3.20	24.533	0.041	3.70	40.447	0.025	4.20	66.686	0.015
3.21	24.779	0.040	3.71	40.854	0.024	4.21	67.357	0.015
3.22	24.028	0.040	3.72	41.264	0.024	4.22	68.033	0.015
3.23	25.280	0.040	3.73	41.679	0.024	4.23	68.717	0.015
3.24	25.534	0.039	3.74	42.098	0.024	4.24	69.406	0.014
3.25	25.790	0.039	3.75	42.521	0.024	4.25	70.105	0.014
3.26	26.050	0.038	3.76	42.948	0.023	4.26	70.810	0.014
3.27	26.311	0.038	3.77	43.380	0.023	4.27	71.522	0.014
3.28	26.576	0.038	3.78	43.816	0.023	4.28	72.240	0.014
3.29	26.843	0.037	3.79	44.256	0.023	4.29	72.966	0.014
3.30	27.113	0.037	3.80	44.701	0.022	4.30	73.700	0.014
3.31	27.385	0.037	3.81	44.150	0.022	4.31	74.440	0.013
3.32	27.660	0.036	3.82	45.604	0.022	4.32	75.189	0.013
3.33	27.038	0.036	3.83	46.063	0.022	4.33	75.944	0.013
3.34	27.219	0.035	3.84	46.525	0.021	4.34	76.706	0.013
3.35	28.503	0.035	3.85	46.993	0.021	4.35	77.478	0.013
3.36	28.789	0.035	3.86	47.465	0.021	4.36	78.257	0.013
3.37	29.079	0.034	3.87	47.942	0.021	4.37	79.044	0.013
3.38	29.371	0.034	3.88	48.424	0.021	4.38	79.838	0.013
3.39	29.666	0.034	3.89	48.911	0.020	4.39	80.640	0.012
3.40	29.964	0.033	3.90	49.402	0.020	4.40	81.451	0.012
3.41	30.265	0.033	3.91	49.899	0.020	4.41	82.269	0.012
3.42	30.569	0.033	3.92	50.400	0.020	4.42	83.096	0.012
3.43	30.877	0.032	3.93	50.907	0.020	4.43	83.931	0.012
3.44	31.187	0.032	3.94	51.419	0.019	4.44	84.775	0.012
3.45	31.500	0.032	3.95	51.935	0.019	4.45	85.627	0.012
3.46	31.817	0.031	3.96	52.457	0.019	4.46	86.488	0.012
3.47	32.137	0.031	3.97	52.985	0.019	4.47	86.357	0.011
3.48	32.460	0.031	3.98	53.517	0.019	4.48	86.235	0.011
3.49	32.786	0.031	3.99	54.055	0.019	4.49	89.121	0.011

TABLE II Powers of e *(continued)*

x	e^x	e^{-x}	x	e^x	e^{-x}	x	e^x	e^{-x}
4.5	90.017	0.011	4.9	134.29	0.0074	8	2981.0	0.0003
4.6	99.484	0.0101	5	148.41	0.0067	9	8103.1	0.0001
4.7	109.95	0.0091	6	403.43	0.0025	10	22026	0.00005
4.8	121.51	0.0082	7	1096.6	0.0009			

TABLE III Natural Logarithms

x	ln x	x	ln x	x	ln x
0.1	−2.3026	4.0	1.3863	8.0	2.0794
0.2	−1.6094	4.1	1.4110	8.1	2.0919
0.3	−1.2040	4.2	1.4351	8.2	2.1041
0.4	−0.9163	4.3	1.4586	8.3	2.1163
0.5	−0.6931	4.4	1.4816	8.4	2.1282
0.6	−0.5108	4.5	1.5041	8.5	2.1401
0.7	−0.3567	4.6	1.5261	8.6	2.1518
0.8	−0.2231	4.7	1.5476	8.7	2.1633
0.9	−0.1054	4.8	1.5686	8.8	2.1748
1.0	0.0000	4.9	1.5892	8.9	2.1861
1.1	0.0953	5.0	1.6094	9.0	2.1972
1.2	0.1823	5.1	1.6292	9.1	2.2083
1.3	0.2624	5.2	1.6487	9.2	2.2192
1.4	0.3365	5.3	1.6677	9.3	2.2300
1.5	0.4055	5.4	1.6864	9.4	2.2407
1.6	0.4700	5.5	1.7047	9.5	2.2513
1.7	0.5306	5.6	1.7228	9.6	2.2618
1.8	0.5878	5.7	1.7405	9.7	2.2721
1.9	0.6419	5.8	1.7579	9.8	2.2824
2.0	0.6931	5.9	1.7750	9.9	2.2925
2.1	0.7419	6.0	1.7918	10	2.3026
2.2	0.7885	6.1	1.8083	11	2.3979
2.3	0.8329	6.2	1.8245	12	2.4849
2.4	0.8755	6.3	1.8405	13	2.5649
2.5	0.9163	6.4	1.8563	14	2.6391
2.6	0.9555	6.5	1.8718	15	2.7081
2.7	0.9933	6.6	1.8871	16	2.7726
2.8	1.0296	6.7	1.9021	17	2.8332
2.9	1.0647	6.8	1.9169	18	2.8904
3.0	1.0986	6.9	1.9315	19	2.9444
3.1	1.1314	7.0	1.9459	20	2.9957
3.2	1.1632	7.1	1.9601	25	3.2189
3.3	1.1939	7.2	1.9741	30	3.4012
3.4	1.2238	7.3	1.9879	35	3.5553
3.5	1.2528	7.4	2.0015	40	3.6889
3.6	1.2809	7.5	2.0149	45	3.8067
3.7	1.3083	7.6	2.0281	50	3.9120
3.8	1.3350	7.7	2.0412	55	4.0073
3.9	1.3610	7.8	2.0541	60	4.0943
		7.9	2.0669	65	4.1744

TABLE IV Common Logarithms

N	0	1	2	3	4	5	6	7	8	9
1.0	.0000	.0043	.0086	.0128	.0170	.0212	.0253	.0294	.0334	.0374
1.1	.0414	.0453	.0492	.0531	.0569	.0607	.0645	.0682	.0719	.0755
1.2	.0792	.0828	.0864	.0899	.0934	.0969	.1004	.1038	.1072	.1106
1.3	.1139	.1173	.1206	.1239	.1271	.1303	.1335	.1367	.1399	.1430
1.4	.1461	.1492	.1523	.1553	.1584	.1614	.1644	.1673	.1703	.1732
1.5	.1761	.1790	.1818	.1847	.1875	.1903	.1931	.1959	.1987	.2014
1.6	.2041	.2068	.2095	.2122	.2148	.2175	.2201	.2227	.2253	.2279
1.7	.2304	.2330	.2355	.2380	.2405	.2430	.2455	.2480	.2504	.2529
1.8	.2553	.2577	.2601	.2625	.2648	.2672	.2695	.2718	.2742	.2765
1.9	.2788	.2810	.2833	.2856	.2878	.2900	.2923	.2945	.2967	.2989
2.0	.3010	.3032	.3054	.3075	.3096	.3118	.3139	.3160	.3181	.3201
2.1	.3222	.3243	.3263	.3284	.3304	.3324	.3345	.3365	.3385	.3404
2.2	.3424	.3444	.3464	.3483	.3502	.3522	.3541	.3560	.3579	.3598
2.3	.3617	.3636	.3655	.3674	.3692	.3711	.3729	.3747	.3766	.3784
2.4	.3802	.3820	.3838	.3856	.3874	.3892	.3909	.3927	.3945	.3962
2.5	.3979	.3997	.4014	.4031	.4048	.4065	.4082	.4099	.4116	.4133
2.6	.4150	.4166	.4183	.4200	.4216	.4232	.4249	.4265	.4281	.4298
2.7	.4314	.4330	.4346	.4362	.4378	.4393	.4409	.4425	.4440	.4456
2.8	.4472	.4487	.4502	.4518	.4533	.4548	.4564	.4579	.4594	.4609
2.9	.4624	.4639	.4654	.4669	.4683	.4698	.4713	.4728	.4742	.4757
3.0	.4771	.4786	.4800	.4814	.4829	.4843	.4857	.4871	.4886	.4900
3.1	.4914	.4928	.4942	.4955	.4969	.4983	.4997	.5011	.5024	.5038
3.2	.5051	.5065	.5079	.5092	.5105	.5119	.5132	.5145	.5159	.5172
3.3	.5185	.5198	.5211	.5224	.5237	.5250	.5263	.5276	.5289	.5302
3.4	.5315	.5328	.5340	.5353	.5366	.5378	.5391	.5403	.5416	.5428
3.5	.5441	.5453	.5465	.5478	.5490	.5502	.5514	.5527	.5539	.5551
3.6	.5563	.5575	.5587	.5599	.5611	.5623	.5635	.5647	.5658	.5670
3.7	.5682	.5694	.5705	.5717	.5729	.5740	.5752	.5763	.5775	.5786
3.8	.5798	.5809	.5821	.5832	.5843	.5855	.5866	.5877	.5888	.5899
3.9	.5911	.5922	.5933	.5944	.5955	.5966	.5977	.5988	.5999	.6010
4.0	.6021	.6031	.6042	.6053	.6064	.6075	.6085	.6096	.6107	.6117
4.1	.6128	.6138	.6149	.6160	.6170	.6180	.6191	.6201	.6212	.6222
4.2	.6232	.6243	.6253	.6263	.6274	.6284	.6294	.6304	.6314	.6325
4.3	.6335	.6345	.6355	.6365	.6375	.6385	.6395	.6405	.6415	.6425
4.4	.6435	.6444	.6454	.6464	.6474	.6484	.6493	.6503	.6513	.6522
4.5	.6532	.6542	.6551	.6561	.6571	.6580	.6590	.6599	.6609	.6618
4.6	.6628	.6637	.6646	.6656	.6665	.6675	.6684	.6693	.6702	.6712
4.7	.6721	.6730	.6739	.6749	.6758	.6767	.6776	.6785	.6794	.6803
4.8	.6812	.6821	.6830	.6839	.6848	.6857	.6866	.6875	.6884	.6893
4.9	.6902	.6911	.6920	.6928	.6937	.6946	.6955	.6964	.6972	.6981
5.0	.6990	.6998	.7007	.7016	.7024	.7033	.7042	.7050	.7059	.7067
5.1	.7076	.7084	.7093	.7101	.7110	.7118	.7126	.7135	.7143	.7152
5.2	.7160	.7168	.7177	.7185	.7193	.7202	.7210	.7218	.7226	.7235
5.3	.7243	.7251	.7259	.7267	.7275	.7284	.7292	.7300	.7308	.7316
5.4	.7324	.7332	.7340	.7348	.7356	.7364	.7372	.7380	.7388	.7396
5.5	.7404	.7412	.7419	.7427	.7435	.7443	.7451	.7459	.7466	.7474
5.6	.7482	.7490	.7497	.7505	.7513	.7520	.7528	.7536	.7543	.7551
5.7	.7559	.7566	.7574	.7582	.7589	.7597	.7604	.7612	.7619	.7627
5.8	.7634	.7642	.7649	.7657	.7664	.7672	.7679	.7686	.7694	.7701
5.9	.7709	.7716	.7723	.7731	.7738	.7745	.7752	.7760	.7767	.7774

TABLE IV Common Logarithms (continued)

N	0	1	2	3	4	5	6	7	8	9
6.0	.7782	.7789	.7796	.7803	.7810	.7818	.7825	.7832	.7839	.7846
6.1	.7853	.7860	.7868	.7875	.7882	.7889	.7896	.7903	.7910	.7917
6.2	.7924	.7931	.7938	.7945	.7952	.7959	.7966	.7973	.7980	.7987
6.3	.7993	.8000	.8007	.8014	.8021	.8028	.8035	.8041	.8048	.8055
6.4	.8062	.8069	.8075	.8082	.8089	.8096	.8102	.8109	.8116	.8122
6.5	.8129	.8136	.8142	.8149	.8156	.8162	.8169	.8176	.8182	.8189
6.6	.8195	.8202	.8209	.8215	.8222	.8228	.8235	.8241	.8248	.8254
6.7	.8261	.8267	.8274	.8280	.8287	.8293	.8299	.8306	.8312	.8319
6.8	.8325	.8331	.8338	.8344	.8351	.8357	.8363	.8370	.8376	.8382
6.9	.8388	.8395	.8401	.8407	.8414	.8420	.8426	.8432	.8439	.8445
7.0	.8451	.8457	.8463	.8470	.8476	.8482	.8488	.8494	.8500	.8506
7.1	.8513	.8519	.8525	.8531	.8537	.8543	.8549	.8555	.8561	.8567
7.2	.8573	.8579	.8585	.8591	.8597	.8603	.8609	.8615	.8621	.8627
7.3	.8633	.8639	.8645	.8651	.8657	.8663	.8669	.8675	.8681	.8686
7.4	.8692	.8698	.8704	.8710	.8716	.8722	.8727	.8733	.8739	.8745
7.5	.8751	.8756	.8762	.8768	.8774	.8779	.8785	.8791	.8797	.8802
7.6	.8808	.8814	.8820	.8825	.8831	.8837	.8842	.8848	.8854	.8859
7.7	.8865	.8871	.8876	.8882	.8887	.8893	.8899	.8904	.8910	.8915
7.8	.8921	.8927	.8932	.8938	.8943	.8949	.8954	.8960	.8965	.8971
7.9	.8976	.8982	.8987	.8993	.8998	.9004	.9009	.9015	.9020	.9025
8.0	.9031	.9036	.9042	.9047	.9053	.9058	.9063	.9069	.9074	.9079
8.1	.9085	.9090	.9096	.9101	.9106	.9112	.9117	.9122	.9128	.9133
8.2	.9138	.9143	.9149	.9154	.9159	.9165	.9170	.9175	.9180	.9186
8.3	.9191	.9196	.9201	.9206	.9212	.9217	.9222	.9227	.9232	.9238
8.4	.9243	.9248	.9253	.9258	.9263	.9269	.9274	.9279	.9284	.9289
8.5	.9294	.9299	.9304	.9309	.9315	.9320	.9325	.9330	.9335	.9340
8.6	.9345	.9350	.9355	.9360	.9365	.9370	.9375	.9380	.9385	.9390
8.7	.9395	.9400	.9405	.9410	.9415	.9420	.9425	.9430	.9435	.9440
8.8	.9445	.9450	.9455	.9460	.9465	.9469	.9474	.9479	.9484	.9489
8.9	.9494	.9499	.9504	.9509	.9513	.9518	.9523	.9528	.9533	.9538
9.0	.9542	.9547	.9552	.9557	.9562	.9566	.9571	.9576	.9581	.9586
9.1	.9590	.9595	.9600	.9605	.9609	.9614	.9619	.9624	.9628	.9633
9.2	.9638	.9643	.9647	.9652	.9657	.9661	.9666	.9671	.9675	.9680
9.3	.9685	.9689	.9694	.9699	.9703	.9708	.9713	.9717	.9722	.9727
9.4	.9731	.9736	.9741	.9745	.9750	.9754	.9759	.9763	.9768	.9773
9.5	.9777	.9782	.9786	.9791	.9795	.9800	.9805	.9809	.9814	.9818
9.6	.9823	.9827	.9832	.9836	.9841	.9845	.9850	.9854	.9859	.9863
9.7	.9868	.9872	.9877	.9881	.9886	.9890	.9894	.9899	.9903	.9908
9.8	.9912	.9917	.9921	.9926	.9930	.9934	.9939	.9943	.9948	.9952
9.9	.9956	.9961	.9965	.9969	.9974	.9978	.9983	.9987	.9991	.9996

ANSWERS TO ODD-NUMBERED EXERCISES

Exercise 1.1

1. false
3. false
5. false
7. the set consisting of the first three letters of the English alphabet; finite
9. the set consisting of all even natural numbers; infinite
11. the set consisting of all multiples of five; infinite
13. the set consisting of the days of the week; finite
15. {Saturday, Sunday}
17. {3, 6, 9, 12, . . .}
19. { }
21. {0}
23. {4}
25. (a)
27. (c)
29.  [number line: point at 1, between 0 and 2]
31. 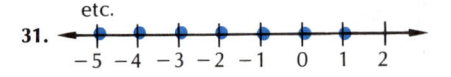 [number line: points at −5, −4, −3, −2, −1, 0, 1, 2, etc.]
33. 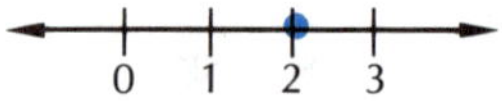 [number line: point at 2, between 0 and 3]
35. integer, rational, real
37. rational, real
39. rational, real
41. (a) true (b) false (c) true
43. (a) true (b) true (c) true
45. (a) true (b) true (c) false

Exercise 1.2

1. $-6 < -3$
3. $x > 0$
5. $0 \leq 3$
7. $<$
9. $>$
11. $<$
13. $>$
15. -3
17. 0
19. -6
21. $-a$
23. m
25. $-5, -4$
27. $-3, -2, -1, 0, 1, 2$
29. $-1, 0, 1, 2, 3$
31. 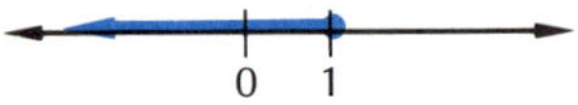 [number line: segment from left to 1, 0 marked]
33. 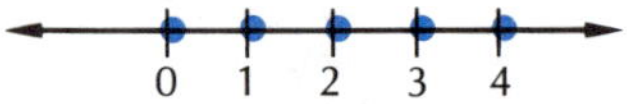 [number line: points at 0, 1, 2, 3, 4]
35.  [number line: segment from −1 to 6]
37. 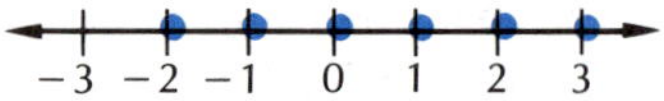 [number line: points at −2, −1, 0, 1, 2, 3]
39. 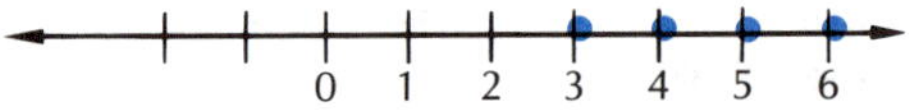 [number line: points at 3, 4, 5, 6]
41.  [number line: segment from −3 to 0, open at −3]
43. 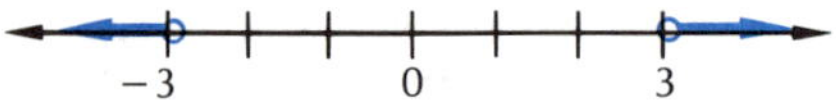 [number line: segment left of −3, open; segment right of 3, open]
45.  [number line: segment from −1 to 4, open at both ends]
47.  [number line: segment from left to 2, open at 2]
49. 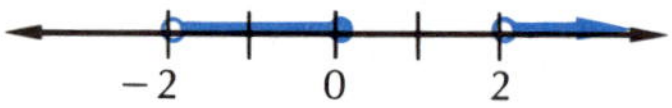 [number line: segment from −2 to 2, open at both ends]
51. 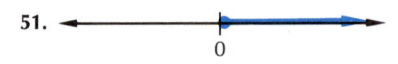 [number line: ray from 0 to right]
53.  [number line: left of −2 open, and right of 0 open]
55. 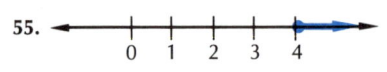 [number line: ray from 4 to right]
57. 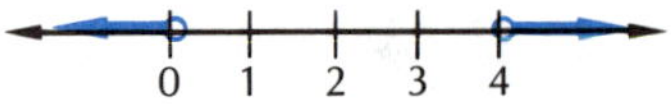 [number line: segment left of 0 open; segment right of 4 open]
59. 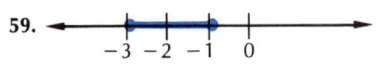 [number line: segment from −3 to 0]

Exercise 1.3

1. 17; basic sum
3. 10; basic sum
5. 14; basic difference
7. 15; basic difference
9. 5; basic sum
11. 4; basic quotient
13. 2; basic difference
15. 68; basic product
17. 14; basic difference
19. 144; basic product
21. 252; basic product
23. $3 \cdot (2 + 2 \cdot 5) = 36$
25. $(3 + 9 \div 3) \cdot 2 = 12$
27. $(8 \div 2 + 4 - 6) \div 2 = 1$
29. $\dfrac{5 - (3 - 1)}{2[(4) - 1]} = \dfrac{1}{2}$
31. $15 \overset{②}{\div} 3 \overset{③}{\cdot} (5 \overset{①}{-} 2) \overset{④}{\cdot} 3 = 45$
33. $4\{16 \overset{④}{-} [(13 \overset{③}{-} 2) \overset{①}{-} 4]\} \overset{②}{=} 36$
35. 44; basic sum
37. 2; basic quotient
39. 18; basic product
41. 106
43. 8.85
45. 3.99
47. .108
49. 93900
51. 19.5

Exercise 1.4

1. commutative property for multiplication
3. commutative property for addition
5. associative property for addition
7. multiplication by zero
9. additive inverse
11. multiplicative inverse
13. additive inverse
15. addition property of equality
17. symmetric property of equality
19. substitution property
21. additive inverse
23. additive inverse
25. $5c + 5d$
27. $8m + 4n$
29. $10abc + 5abd + 15abe$
31. $\sqrt{2}(a + b)$
33. $4a(b + 2c + 2)$
35. $\dfrac{1}{4}$
37. 3
39. $\dfrac{2}{7}$
41. (a) no   (b) yes
43. (a) no   (b) no
45. (a) true   (b) true
47. (a) true   (b) false
49. (a) false   (b) false
51. (1) commutative property for addition
    (2) associative property for addition
    (3) additive inverse
    (4) additive identity

Exercise 1.5

1. -2
3. -7
5. -11
7. -2
9. 18
11. 0
13. -11
15. 64
17. -2
19. 4
21. -96
23. 4
25. -264
27. -24
29. -13
31. 63
33. 5
35. property C
37. property E
39. property B
41. $10x + 15y - 20z$
43. $-2abx + 4aby - 6abz$
45. $-2ax + 2ay + 4az$
47. $4(x - 2y + 3)$
49. $3(9xy - 2x - 4y)$
51. $\dfrac{-a}{b} = \dfrac{-1 \cdot a}{b}$ opposite law
$= (-1 \cdot a) \cdot \dfrac{1}{b}$ division
$= -1\left(a \cdot \dfrac{1}{b}\right)$ associative
$= -1\left(\dfrac{a}{b}\right)$ division
$= -\left(\dfrac{a}{b}\right)$ opposite law

53. -3.18
55. 49.9
57. -189

Exercise 1.6

1. -1
3. -2
5. 5
7. 216
9. 32
11. -10
13. 8
15. -4
17. 1
19. 5
21. 2^7
23. 5^5
25. x^7
27. x^0 or 1
29. x^8
31. $-4x^5y^3$
33. a^7b^4
35. x^{10}
37. x^7y^5
39. $-3x^4y$
41. $-xz^2$
43. x^6y^4
45. 2
47. $-27x^5$
49. -18
51. $\dfrac{1}{9}$
53. -664
55. 2280
57. 9.35
59. $-.0506$

Exercise 1.7

1. 40
3. -3
5. 40
7. 26
9. $7x^2 + 2x + 1$
11. $-2x^2 - 7x - 2$
13. $2x^2 - 6x + 3$
15. $3x^2 - 14xy$
17. $14x^4 + 2x^3 - 12$
19. $-7x^2 + 9x$
21. $-2x + 1$
23. $-y$
25. $-23x + 12$
27. $-4x + y + 4$
29. $x^2 + 2x - y + y^2$
31. $-ab + 12b^2$
33. $2x^2 + 2x - 2$
35. $6x^2 + 5x - 4$
37. $x^4 - 1$
39. $x^3 + 4x^2 + 8x + 15$
41. $3x^3 - 5x^2 + 8x - 4$
43. $x^3 + 1$
45. $x^4 - x^2 + 4x - 4$
47. $-36x^7y^4$
49. $3x^3y^2 - 4x^2y^3 + 6x^2y^2 + 2xy^3$
51. $6x^6 + 2x^3$

Exercise 2.1

1. identity
3. conditional
5. $x = 13$
7. $y = 0$
9. $y = -2$
11. $m = 3$
13. $x = -2$
15. $n = -2$
17. $n = -2$
19. $x = 0$
21. $y = -\dfrac{7}{2}$
23. no solution
25. $a = 10$
27. $a = -\dfrac{10}{3}$
29. $x = -\dfrac{33}{5}$
31. $x = -\dfrac{29}{4}$
33. $x = 24$
35. $x = \dfrac{10}{9}$
37. $x = 3$
39. $x = \dfrac{1}{6}$
41. $x = 5$
43. $x = \dfrac{11}{21}$
45. $x = -1$
47. $y = -\dfrac{1}{5}$

Exercise 2.2

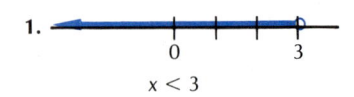

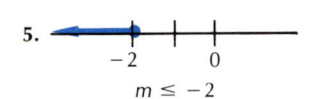

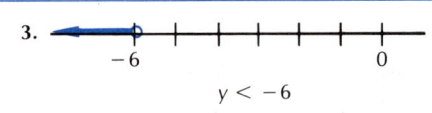

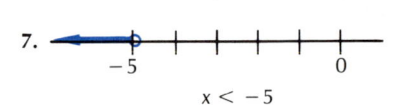

1. $x < 3$
3. $y < -6$
5. $m \leq -2$
7. $x < -5$

ANSWERS TO ODD-NUMBERED EXERCISES   A-13

9. [number line] $a > -4$

11. [number line] $x \geq -3$

13. [number line] $y \leq -2$

15. no solution

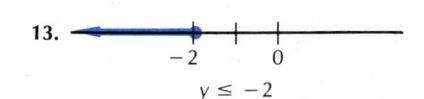

17. [number line] $x < 3$

19. $y > -\frac{3}{7}$
21. $x < \frac{22}{5}$
23. $y > -\frac{1}{3}$
25. $x < 2$
27. $x \geq 3$
29. $x > 6$
31. $x < -14$
33. $x < -\frac{7}{4}$
35. $x \leq \frac{1}{5}$
37. $0 \leq x \leq 9$
39. $-6 < y < 4$
41. $-4 < x < 2$
43. $-1 \leq x \leq 1$
45. $0 < x < 4$

Exercise 2.3

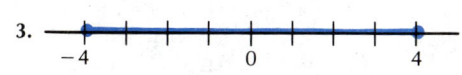

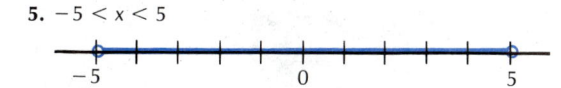

1. $|x| = \begin{cases} x \text{ if } x \geq 0 \\ -x \text{ if } x < 0 \end{cases}$
3. solutions are 1 and -1
5. solution is 0
7. solutions are 11 and -5
9. solutions are $\frac{11}{3}$ and $-\frac{13}{3}$
11. solution is $\frac{3}{4}$
13. solutions are $\frac{3}{4}$ and $\frac{1}{4}$
15. solutions are 4 and -12
17. solutions are 15 and -15
19. no solution
21. solutions are $\frac{3}{2}$ and $\frac{5}{2}$
23. solutions are -4 and 5
25. solutions are 0 and -3
27. no solution
29. solutions are 0 and $-\frac{3}{2}$
31. solutions are 5 and -1
33. solutions are -1 and -3
35. solutions are 6, -8, -4, and 2

Exercise 2.4

1. [number line with points at -3 and 3]
3. [number line from -4 to 4]
5. $-5 < x < 5$ [number line]
7. $-3 \leq z \leq 3$ [number line]
9. $-4 < x < 2$ [number line]

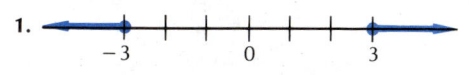

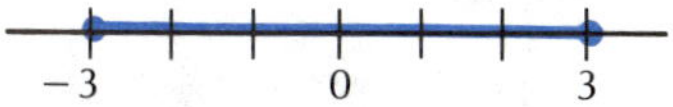

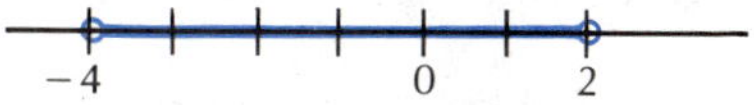

11. $-5 < z < 9$ 13. no solution

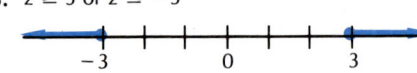

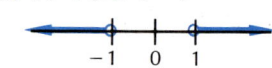

15. $z \geq 3$ or $z \leq -3$ 17. $x > 1$ or $x < -1$

19. $z > 8$ or $z < -4$

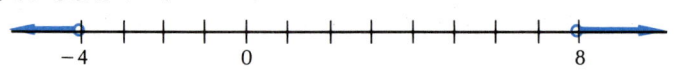

21. $-\dfrac{4}{3} < x < 2$ 23. $0 \leq x \leq 4$ 25. $-2 < x < 3$

27. $x \leq -3$ or $x \geq 9$ 29. $x \leq -3$ or $x \geq 7$ 31. $x < -2$ or $x > 10$

33. $-\dfrac{7}{6} < x < \dfrac{5}{6}$ 35. $-4 < x < 8$

Exercise 2.5

1. 87.5 miles   3. 35° Celsius   5. $132.80

7. $t = \dfrac{D}{r}$ 9. $r = \dfrac{I}{pt}$ 11. $b = \dfrac{2A}{h}$

13. $r = \dfrac{C}{2\pi}$ 15. $w = \dfrac{P - 2l}{2}$ 17. $c = P - a - b$

19. $E = IR$ 21. $h = \dfrac{2A}{b + c}$ 23. $F = \dfrac{9C + 160}{5}$

25. $n = \dfrac{l - a + d}{d}$ 27. $r = \dfrac{S - a}{S}$ 29. $x = \dfrac{y - b}{m}$

31. $f = \dfrac{pq}{q + p}$ 33. $s = \dfrac{550tHP}{F}$ 35. $a = \dfrac{2c}{k^2 - 2}$

37. $a = \dfrac{S(1 - r)}{1 - r^n}$ 39. $a = \dfrac{m + n}{1 - x + y}$

Exercise 2.6

1. $6x$ 3. $x - 5$ 5. $10 - 2x$
7. $x + (x + 2)$ 9. $x + .10x$ 11. 18
13. 25   15. -6 17. length 24 feet
                                   width 16 feet
19. 18, 20, 22   21. 24, 25, 26   23. 25
25. length 31 inches   27. 32 feet wide   29. $1,800 at 6%
    width 14 inches   56 feet long $2,150 at 8%
31. 90 miles per hour   33. 240 miles   35. 180 nautical miles
37. $17\dfrac{1}{7}$ gallons of 25% antifreeze   39. 3 gallons of 30% alcohol
                                                     2 gallons of 80% alcohol
    $12\dfrac{6}{7}$ gallons of 60% antifreeze
41. 7 dimes,   43. 33 nickels,
    8 quarters   11 dimes,
                    7 quarters
45. at least $50,000   47. between 125 and 185

Exercise 3.1

1. trinomial; 2
3. binomial; 1
5. monomial; 0
7. $5x^5 + 3x^3 - 2x - 4$
9. $-x^4 + x^3 + x^2 - 3$
11. $P(2) = 9$
    $P(-2) = 21$
13. $P(-3) = -4$
    $P(2) = 6$
    $P(3) = 14$
15. $P(1) = -3$
    $P(-2) = -24$
    $P(0) = 2$
17. $P(1) = 4$
    $P(-3) = -44$
19. $7x - 5$
21. $2x^2 - 5x + 5$
23. $x^2 + 8x + 6$
25. $-x^2 + 2x$
27. $3y^3 - y^2 + 8y + 11$
29. $3x^2 + 3x + 4$
31. 20.2
33. -429
35. 40.2

Exercise 3.2

1. $16a^5b^4c^3$
3. $-10a^4b^4$
5. $6t^4y^4$
7. $12x^3y^2$
9. $-x^3yz^5$
11. $-4x^4y + 12x^3y - 8x^2y$
13. $6x^3y^4 + 9x^4y^2 - 12x^3y^2$
15. $6x^5 - 8x^6 + 16x^4 - 10x^3$
17. $2x^2 + 5x + 3$
19. $4x^2 + 4x + 1$
21. $12a^3 + 7a^2 - 7a - 2$
23. $y^3 + 8$
25. $x^3 + 3x^2 + 3x + 1$
27. $3x^3 + 5x^2 - 2x - 9x^2y - 15xy + 6y$
29. $m^4 - 2m^3 + 2m^2 - m - 2$
31. $3x - 4$
33. $3x^2 + x + 4 + \dfrac{7}{x - 2}$
35. $x^3 - x^2 + x - 1 + \dfrac{2}{x + 1}$
37. $x^4 + x^3 + x^2 + x + 1$
39. $2x - 1 + \dfrac{-2x}{3x^2 + 2x - 1}$

Exercise 3.3

1. $x^2 + 12x + 27$
3. $5x^2 + 17x + 6$
5. $y^2 + 3y - 10$
7. $t^2 - 8t + 12$
9. $5x^2 - 7xy + 2y^2$
11. $y^2 - 10y + 25$
13. $4n^2 + 12n + 9$
15. $9a^2 + 48ab + 64b^2$
17. $6x^2 + 5x - 4$
19. $8y^2 + 2y - 3$
21. $9x^2 - 12xy + 4y^2$
23. $8x^2 + 26x + 15$
25. $36x^2 - y^2$
27. $3t^2 - 19t + 20$
29. $x^4 + 7x^2 + 12$
31. $x^6 - 2x^3 - 24$
33. $8x^6 + 18x^3 - 5$
35. $x^2y^4 + 8xy^2 - 20$
37. $a^2 + 2ab + b^2 + 7a + 7b + 12$
39. $x^2 + 2xy + y^2 - 25$
41. $a^2 + 4a + 4 + 2ab + 4b + b^2$
43. $x^4 - 6x^2 + 9 - 2x^2y + 6y + y^2$
45. $3x^3 + 9x^2 - 2x - 6$
47. $5x^3 + 2x^2 - 5xy^2 - 2y^2$

Exercise 3.4

1. $10(x + 2)$
3. $3y(2y + 1)$
5. $7(a + 3)$
7. $4xy^2(2 - x)$
9. $ab(4 - 5b)$
11. $5a(x + 3a)$
13. $2x^4y^3(7x^2 - 3y)$
15. $xy(x - y + xy)$
17. $2b^3(b + 2b^3 - 4)$
19. $12xy(2xy + 3x - 4)$
21. $3mn^2(1 - 2m)$
23. $a(y^2 + by + b)$
25. $(a - b)(5x - 3y)$
27. $(x - y)^2(7x - 7y + 11)$
29. $(x + y)(4a - 1)$
31. $(a + b)(4 - a - b)$
33. $(x + y)(1 + 7a)$
35. $(y - 2)(y^2 + 5)$
37. $(x^2 + 3)(x - 7)$
39. $(a + b)(m + n)$
41. $(4a^4 + b^2)(2a^2 + b^2)(2a^2 - b^2)$
43. $(a + b)(x^2 + y^2)$
45. $(a^2 + 1)(b^2 + 4)$

Exercise 3.5

1. $(a + 7)(a - 7)$
3. $(2 + 3y)(2 - 3y)$
5. $9(x + 2)(x - 2)$
7. $\left(\dfrac{1}{2} + 3x\right)\left(\dfrac{1}{2} - 3x\right)$
9. $x(x + 1)(x - 1)$
11. not factorable
13. $x(x + 6)(x - 6)$
15. $(x - 1)(x^2 + x + 1)$
17. $(2x + y)(4x^2 - 2xy + y^2)$
19. $(x - 3y)(x^2 + 3xy + 9y^2)$
21. $(x - 2 + y)(x - 2 - y)$
23. $5(x + 5)(x - 5)$
25. $(2x - 3y)(4x^2 + 6xy + 9y^2)$
27. $(x - 1 + y)(x^2 - 2x + 1 - xy + y + y^2)$
29. $(s + t + 3a)(s + t - 3a)$
31. $2x^3(2x + 1)(2x - 1)$
33. $(x + 1 + 6y)(x + 1 - 6y)$
35. $(4x^4 + 9y^6)(2x^2 + 3y^3)(2x^2 - 3y^3)$
37. $(2a^2 + b^2)(2a^2 - b^2)$
39. $(2a - 5b)(4a^2 + 10ab + 25b^2)$

Exercise 3.6

1. $(x + 2)(x + 3)$
3. $(a + 12)(a + 1)$
5. $(b - 6)(b + 3)$
7. $(y - 5)(y - 4)$
9. $(5x + 4)(x + 1)$
11. $(7y + 2)(y + 3)$
13. $(5b + 6)(2b - 3)$
15. $(3a + 2)(a - 9)$
17. $(1 + 8y)^2$
19. $(2x + 1)(x - 5)$
21. $(x + 2)(x - 2)(x + 1)(x - 1)$
23. $x^2(x - 7)(x + 1)$
25. $(x + 3)(x - 3)(x^2 + 3)$
27. $(5m + 3n)(2m - n)$
29. $x^6(8x + 7)(2x - 1)$
31. $(3x^2 - 5)(2x^2 + 5)$
33. $3(3x + 2)(2x - 3)$
35. $(5xy - 2)^2$
37. $(7x + b)(7x - b)$
39. $x(x - 5)(x + 2)$
41. $(x + 2y)(x^2 - 2xy + 4y^2)$
43. $(x - y)(2a + b)$
45. $(x^2 + 4y^2)(x + 2y)(x - 2y)$
47. $3y^2(y - 6)(y + 5)$
49. $(1 - x)(1 + x + x^2)$
51. $(n + 1)(n - 1)(m + 2)(m^2 - 2m + 4)$

Exercise 3.7

1. solutions are 4 and 2
3. solutions are 0 and $\dfrac{1}{2}$
5. solutions are 0 and 1
7. solutions are 4 and 1
9. solutions are 2 and $-\dfrac{1}{2}$
11. solution is 3
13. solution is $\dfrac{2}{3}$
15. solutions are 4 and 2
17. solutions are 5 and $-\dfrac{3}{2}$
19. solutions are 4 and 1
21. solutions are $\dfrac{3}{2}$ and $-\dfrac{3}{2}$
23. solutions are 1 and $\dfrac{1}{2}$
25. solutions are 4 and -3
27. solutions are 0 and $\dfrac{5}{2}$
29. solutions are 0, 2, and -3
31. solutions are 5 and 1
33. solutions are 2, -2, 3 and -3
35. the number is either 4 or -5
37. the numbers are -5 and -2 or 2 and 5
39. the numbers are 4 and 6 or $-\dfrac{12}{5}$ and $-\dfrac{34}{5}$
41. length 7 inches, width 3 inches

ANSWERS TO ODD-NUMBERED EXERCISES A-17

Exercise 4.1

1. all real numbers except 4
3. all real numbers
5. all real numbers except -2
7. all real numbers except 2 and -2
9. all real numbers except 4 and -7
11. $\dfrac{-x+1}{-4}$ or $\dfrac{1-x}{-4}$
13. $\dfrac{-x+3}{-x+4}$ or $\dfrac{3-x}{4-x}$
15. $\dfrac{-a-b}{-a+b}$ or $\dfrac{-a-b}{b-a}$
17. $\dfrac{-b^3}{2}$
19. $\dfrac{5b^2}{4}$
21. $x+3$
23. $y-2$
25. not reducible
27. $\dfrac{x+2}{2}$
29. -1
31. $-a-3$
33. $b-a$
35. $\dfrac{1}{x-1}$
37. $\dfrac{2-x}{4-x}$
39. $\dfrac{x+2y}{x+y}$
41. not reducible
43. $\dfrac{x^2-xy+y^2}{x-y}$
45. $4x$
47. $\dfrac{3y-x}{y+x}$
49. $\dfrac{-3x}{2y(2x+1)}$
51. $\dfrac{-3-2x}{4-x}$
53. $\dfrac{c+d}{b+c}$
55. $x+3$
57. $\dfrac{x^2-2xy+4y^2}{2y-x}$

Exercise 4.2

1. $\dfrac{10x^2y}{9}$
3. $\dfrac{x^4}{6y^3z}$
5. $\dfrac{b}{3a^2}$
7. $\dfrac{9}{x^3}$
9. $\dfrac{y^2}{xz^2}$
11. $\dfrac{x}{25y}$
13. $\dfrac{x^6}{27}$
15. $\dfrac{x+1}{x+5}$
17. $\dfrac{30x}{7}$
19. $\dfrac{4}{5}$
21. $\dfrac{3}{x+2}$
23. $\dfrac{x-y}{4}$
25. $y(x-y)$
27. 2
29. $\dfrac{xy-1}{x^2}$
31. $\dfrac{2(x+2)}{y(x-2)}$
33. $\dfrac{y+2x}{y-2x}$
35. $-1-x$
37. $\dfrac{3(y-1)}{2x(x-1)}$
39. $(x+5)(2x+1)$
41. $(x+1)^2$
43. $\dfrac{1-x}{x+1}$
45. $\dfrac{x^4+5x^2+9}{(x+3)(x^2-x+3)}$
47. $\dfrac{(x-y)^2}{3}$
49. $\dfrac{x-y}{x^2}$
51. $\dfrac{x-1}{3}$
53. $\dfrac{a+5}{a+4}$
55. $\dfrac{x-1}{x-3}$
57. $\dfrac{-(b+a)(2a-b)}{2ab(4-b)}$

Exercise 4.3

1. 5
3. $\dfrac{13}{2r}$
5. $\dfrac{x-1}{3}$
7. $\dfrac{-x-4}{6}$
9. $\dfrac{10x^2+33}{36x^4}$
11. $\dfrac{y-56}{7y^2}$
13. $\dfrac{5x+9y^2}{3y^2}$
15. $\dfrac{11}{10(a-b)}$
17. $\dfrac{-3}{x-3}$
19. $\dfrac{-x+33}{2(x-3)(x+3)}$
21. $\dfrac{11-3x}{x^3-8}$
23. $\dfrac{a^2(1+b)(1-b)}{b^2(a-b)}$
25. $\dfrac{9x-20}{(x-3)(x+4)}$
27. $\dfrac{-5(y+1)}{(2y+1)(y-2)}$
29. $\dfrac{3}{b^2}$
31. $\dfrac{-2(3x+2)}{(x-1)(x+4)(x-4)}$
33. $\dfrac{3x^2+10x+16}{(x+1)(x+4)(x-4)}$
35. $\dfrac{a^2-12}{a-4}$
37. $\dfrac{7x+13}{(x+3)(x-3)(x+7)}$
39. $\dfrac{b^2-2ab}{a(a-b)}$
41. $\dfrac{96-9b-40b^2}{24b^3}$
43. $\dfrac{2t^2-t-5}{t^2(t+1)(t-1)}$
45. $\dfrac{x^3+x^2-3x-9}{x(x+3)(x-3)}$
47. $\dfrac{y^2+2y-1}{y(y+1)(y-1)}$
49. $\dfrac{1}{2t-1}$
51. $\dfrac{y^2+6y+1}{(y-1)(y+1)^2}$

Exercise 4.4

1. $\dfrac{9}{10}$
3. $-\dfrac{5}{8}$
5. $2x^3$
7. $\dfrac{x(y-1)}{y+1}$
9. $b-1$
11. $\dfrac{x+1}{x-1}$
13. $\dfrac{x+2}{4(x-1)}$
15. $\dfrac{1}{4x}$
17. $\dfrac{x+y}{x-y}$
19. $\dfrac{x+y}{3x(x-y)}$
21. $\dfrac{y-6}{y+5}$
23. $\dfrac{b}{a-2b}$
25. $\dfrac{3x}{2}$
27. $\dfrac{y}{3y+x}$
29. xy
31. $b-a$
33. $\dfrac{x^2-x-3}{x(3x+8)}$
35. $\dfrac{-(2x+a)}{x^2(x+a)^2}$
37. 1
39. $\dfrac{x(2x-1)}{2x+1}$
41. $\dfrac{y+1}{y(1+y-y^2)}$
43. $1-x$

Exercise 4.5

1. $x=-2$
3. all real numbers except 2
5. no solution
7. no solution
9. $y=2$ or $y=-\dfrac{3}{4}$
11. $x=\dfrac{1}{3}$
13. $a=9$ or $a=5$
15. $t=8$
17. $p=-9$
19. $t=1$
21. $x=-1$
23. no solution
25. $y=-\dfrac{6}{5}$
27. $x=\dfrac{3}{2}$

29. $a = \dfrac{1}{2}$
31. $t = -3$
33. $y = -\dfrac{13}{17}$
35. $x = \dfrac{12}{5}$
37. $x = \dfrac{y + 1}{y}$
39. $r = \dfrac{A - P}{Pt}$
41. $r = \dfrac{st}{t + 2s}$
43. $a = \dfrac{Rr + br}{1 - b - R}$
45. $x = a + b$
47. $r = \dfrac{S - a}{S - t}$
49. $t = a + b$
51. $A = \dfrac{aC}{\pi a - \pi C}$
53. $p = \dfrac{s + 1}{s}$

Exercise 4.6

1. $\dfrac{19}{29}$
3. $\dfrac{9}{5}$ and $\dfrac{18}{5}$
5. 18 and 40
7. -4
9. 4 and 5
11. 400 miles per hour
13. 50 miles per hour
15. 28 miles per hour
17. 2,400 miles
19. $\dfrac{12}{7}$ hours
21. 18 hours
23. $\dfrac{28}{3}$ hours
25. $\dfrac{4}{3}$ hours
27. 36 and 66

Exercise 5.1

1. -5
3. -16
5. not a real number
7. -5
9. -4
11. 3
13. -4
15. $-\dfrac{5}{6}$
17. 2
19. $-\dfrac{4}{5}$
21. 3
23. $|y|^3$
25. $|x| \cdot y^4$
27. $-2|x| \cdot |y|^9$
29. $\dfrac{3}{4}|x|$
31. $a^3 b^6 c^8$
33. not a real number
35. $x^2 \cdot |y|$
37. $-9x$
39. $|x| \cdot |y|^3$
41. $6x^2 y$
43. $12|x + y|$
45. $5|x| \cdot y^2$
47. $-13a^4 |b|^5$
49. $6|x| \cdot |y|$
51. $-9x^2 y^3 z^4$

Exercise 5.2

1. x^{10}
3. x^4
5. -4
7. $\dfrac{1}{x^{12}}$
9. $\dfrac{1}{64y^6}$
11. $4x^4 y$
13. $\dfrac{8}{x^4}$
15. $14x^3$
17. 1
19. 2
21. 3^2 or 9
23. 4^3 or 64
25. $\dfrac{1}{16}$
27. 3^6 or 729
29. 2.51×10^8

31. 1.16×10^{10} **33.** 3.68547×10^3 **35.** 3.78×10^{-4}
37. .000312 **39.** 7140 **41.** .005
43. 1×10^4 **45.** 6×10^0 **47.** 2.32×10^{-3}
49. $\dfrac{1}{200}$ or .005 **51.** $\dfrac{1}{5}$ or 0.2 **53.** $\dfrac{3y^6z^2}{4x^4}$
55. $\dfrac{y+x}{xy}$ **57.** $\dfrac{y+x}{y-x}$ **59.** $\dfrac{ab}{b-a}$
61. $\dfrac{a^2b^2}{(a+b)(a^2+b^2)}$ **63.** 8.87×10^{-3} **65.** 1.01×10^0
67. 5.79×10^{-3} **69.** 1.05×10^5 **71.** 8.19×10^{-2}
73. 1.34×10^2

Exercise 5.3

1. 8 **3.** -3 **5.** $\dfrac{9}{5}$

7. 8 **9.** 5 **11.** $-\dfrac{1}{3}$

13. $\dfrac{17}{4}$ **15.** $\dfrac{5}{9}$ **17.** 9

19. $\dfrac{9}{2}$ **21.** $\dfrac{4}{9}$ **23.** 8

25. $\dfrac{1}{625}$ **27.** not a real number **29.** $a^{3/7}$

31. $x^{1/3} + y^{1/3}$ **33.** $x^{2/5} \cdot y^{8/5}$ **35.** $\sqrt[5]{a^2}$

37. $2\sqrt{x} \cdot \sqrt[3]{y}$ **39.** $\sqrt[3]{x^2} \cdot \sqrt{y}$ **41.** $\dfrac{1}{\sqrt{x^2+y^2}}$

43. 10.64 **45.** 0.39 **47.** 0.76
49. 43.24

Exercise 5.4

1. $2\sqrt{10}$ **3.** $2\sqrt[3]{5}$ **5.** $2\sqrt[3]{3}$
7. $18\sqrt{2}$ **9.** $10\sqrt{3}$ **11.** $-8\sqrt{5}$
13. $-5\sqrt[3]{2}$ **15.** $2\sqrt{13}$ **17.** not a real number
19. $10|x|$ **21.** $5m\sqrt{m}$ **23.** $-2x\sqrt{5}$
25. $3x^4\sqrt{2}$ **27.** $10\sqrt{70}$ **29.** $2ab^2\sqrt[3]{5ab}$
31. $3xy^2\sqrt{2xy}$ **33.** $2m^2|n|\sqrt{6m}$ **35.** $2xy\sqrt[3]{4x^2y}$

37. $\dfrac{\sqrt{x}}{|y|}$ **39.** $\dfrac{4x^2}{5}$ **41.** $\dfrac{6\sqrt{2}}{5}$

43. $\dfrac{\sqrt{3}}{4}$ **45.** $2xy\sqrt[3]{xy^2}$ **47.** 5

49. a **51.** $2\sqrt{3}$ **53.** 8

55. $|x|$ **57.** $\dfrac{3}{x}$

ANSWERS TO ODD-NUMBERED EXERCISES A-21

Exercise 5.5

1. $13\sqrt{3}$
3. $29\sqrt{3}$
5. $15\sqrt[3]{2}$
7. $11\sqrt{5}$
9. $\dfrac{-15\sqrt{3}}{2}$
11. $9\sqrt{7}$
13. $-3\sqrt[3]{3}$
15. $3x\sqrt[3]{xy^2}$
17. $-31x\sqrt{2}$
19. $23\sqrt[3]{5x}$
21. 18
23. $8 + 2\sqrt{3}$
25. $5\sqrt{2} - 5$
27. $6x^3y^2\sqrt{x}$
29. -7
31. $\sqrt{2}$
33. $13 + 4\sqrt{3}$
35. 2
37. $16x + 8\sqrt{x} + 1$
39. $9x + 12\sqrt{xy} + 4y$
41. $26 - 8\sqrt{35}$
43. $\sqrt[3]{4} - 36$
45. $\dfrac{4 + \sqrt{2}}{4}$
47. $\dfrac{10 + \sqrt{5}}{2}$
49. $\dfrac{3 + 2\sqrt{3}}{2}$
51. 5

Exercise 5.6

1. $\sqrt{3}$
3. $\dfrac{\sqrt{7}}{7}$
5. $\dfrac{\sqrt{15}}{10}$
7. $\dfrac{\sqrt{2}}{2}$
9. $\dfrac{\sqrt[3]{10}}{2}$
11. $\dfrac{x^2\sqrt[3]{y^2}}{y}$
13. $\dfrac{2b\sqrt[3]{3c^2}}{3c}$
15. $\sqrt[3]{3}$
17. $2\sqrt{2} + 1$
19. $\dfrac{2\sqrt{3} + \sqrt{10}}{2}$
21. $\dfrac{6a^4\sqrt{2b}}{b^2}$
23. $\dfrac{\sqrt{xy}}{y}$
25. $3(2 - \sqrt{3})$
27. $2(2 + \sqrt{3})$
29. $-2 + 2\sqrt{3}$
31. $\sqrt{2} - \sqrt{6} - \sqrt{3} + 3$
33. $\dfrac{\sqrt{30}}{6}$
35. $6 + \sqrt{35}$
37. $\dfrac{x + \sqrt{xy}}{x - y}$
39. $\dfrac{3\sqrt{x} - 2}{9x - 4}$
41. $\dfrac{11\sqrt{3}}{3}$
43. $\dfrac{14\sqrt{10}}{5}$
45. $3\sqrt[3]{9}$
47. $\dfrac{-\sqrt{3}}{3}$
49. $\dfrac{x - \sqrt{xy}}{x - y}$
51. 2.81
53. -3.03
55. 11.1

Exercise 5.7

1. 36
3. 14
5. $\dfrac{11}{2}$
7. 0
9. 2
11. 1 and -4
13. -3 and -2
15. 1
17. 12
19. 16
21. 0 and 4
23. 5
25. 3 and -1
27. $\dfrac{1}{4}$
29. -8
31. $\dfrac{23}{4}$
33. 5
35. 1

Exercise 6.1

1. $x^2 + (-3)x + (-5) = 0$; $a = 1, b = -3, c = -5$
3. $2x^2 + 3x + (-1) = 0$; $a = 2, b = 3, c = -1$
5. 0 and $\frac{5}{3}$
7. -1 and $\frac{1}{3}$
9. -3 and $\frac{2}{5}$
11. $-\frac{4}{3}$ and $\frac{2}{5}$
13. -3 and $\frac{2}{5}$
15. -3 and 2
17. -2 and 3
19. -1 and 4
21. $-\frac{2}{3}$ and 1
23. 1 and 3
25. -3 and 1
27. $-\frac{4}{3}$ and $-\frac{1}{6}$
29. -2 and 2
31. $-3, 3, -1,$ and 1
33. -2 and 2
35. $-\frac{1}{10}$ and 1
37. $-\frac{1}{3}$ and $\frac{1}{3}$
39. $\frac{2}{3}$ and $\frac{3}{2}$
41. -4 and 1
43. 6
45. $\frac{4}{9}$ and 16
47. 8
49. 5

Exercise 6.2

1. 9
3. $\frac{1}{16}$
5. ± 6
7. $\pm 3\sqrt{2}$
9. -7 and 1
11. $4 \pm \sqrt{3}$
13. $\frac{-4 \pm 5\sqrt{2}}{3}$
15. no real solutions
17. $\frac{3 \pm 4\sqrt{3}}{4}$
19. 1 and -3
21. -1 and 7
23. $3 \pm \sqrt{15}$
25. $\frac{1 \pm \sqrt{3}}{2}$
27. $\frac{-1 \pm \sqrt{5}}{2}$
29. $3 \pm 2\sqrt{2}$
31. $\frac{-3 \pm \sqrt{41}}{4}$
33. no real solutions
35. no real solutions
37. $\frac{3 \pm \sqrt{22}}{2}$
39. -5 and -3 or 3 and 5
41. 4
43. $\frac{-5 \pm \sqrt{53}}{2}$; $1.14, -6.14$
45. $\frac{1 \pm \sqrt{32001}}{40}$; $4.50, -4.45$
47. $\frac{2 \pm \sqrt{19}}{3}$; $2.12, -0.79$

Exercise 6.3

1. -1 and 6
3. no real solutions
5. $\frac{5 \pm \sqrt{37}}{6}$
7. $\frac{1 \pm \sqrt{11}}{2}$
9. $\frac{7 \pm \sqrt{29}}{2}$
11. $\frac{1 \pm \sqrt{29}}{2}$
13. $\frac{1 \pm \sqrt{6}}{5}$
15. $\frac{1 \pm \sqrt{3}}{3}$
17. $1 \pm \sqrt{7}$
19. no real solutions
21. $\frac{-1 \pm \sqrt{13}}{5}$
23. $-\frac{3}{2}$
25. $2 \pm 2\sqrt{5}$
27. $\frac{3 \pm \sqrt{21}}{6}$
29. $\frac{7 \pm \sqrt{10}}{3}$

31. $D = 13$; two real solutions **33.** $D = 121$; two real solutions
35. $D = 0$; one real solution **37.** ± 8
39. $\dfrac{25}{4}$ **41.** $\dfrac{-7 \pm \sqrt{29}}{10}$ **43.** 6 and 8
45. 3.18, -1.03 **47.** 0.11, -0.50 **49.** 2.20, -0.38

Exercise 6.4

1. $\{x \mid -2 < x < 3\}$ **3.** $\{x \mid 2 \leq x \leq 3\}$ **5.** $\{x \mid y < -2 \text{ or } y > 3\}$
7. $\{y \mid y \leq -1 \text{ or } y \geq 4\}$ **9.** $\{t \mid -2 < t < 4\}$ **11.** $\{a \mid 0 \leq a \leq 5\}$
13. $\{x \mid x \in R\}$ **15.** $\left\{y \mid y < -4 \text{ or } y > \dfrac{3}{2}\right\}$ **17.** $\{x \mid 1 < x < 2\}$
19. $\left\{x \mid x < -1 \text{ or } x > \dfrac{1}{3}\right\}$ **21.** $\left\{t \mid -\dfrac{2}{3} \leq x \leq 5\right\}$ **23.** $\left\{y \mid y < \dfrac{1}{2} \text{ or } y > 4\right\}$
25. $\left\{y \mid y \leq -\dfrac{2}{5} \text{ or } y \geq 1\right\}$ **27.** $\left\{t \mid t < -1 \text{ or } t > -\dfrac{3}{4}\right\}$ **29.** $\{x \mid x < 4 \text{ but } x \neq 2\}$
31. $\{a \mid -3 < a < 2 \text{ or } a > 5\}$ **33.** $\{x \mid x \leq -2 \text{ or } x \geq 3\}$
35. $\{x \mid x \leq 0 \text{ or } x > 1\}$ **37.** $\{x \mid x < 5\}$
39. $\left\{x \mid -1 < x < -\dfrac{2}{3}\right\}$

Exercise 6.5

1. false **3.** false **5.** false
7. $2 + 5i$; real part 2, imaginary part 5.
9. $-\dfrac{1}{2} + (-1)i$; real part $-\dfrac{1}{2}$, imaginary part -1
11. $-1 + (\sqrt{2})i$; real part -1, imaginary part $\sqrt{2}$
13. -1 **15.** 1 **17.** -1
19. $5i$ **21.** 2 **23.** $28i$
25. $2i\sqrt{10}$ **27.** $4i\sqrt{3}$ **29.** $-6i$
31. $-5\sqrt{6}$ **33.** $i\sqrt{2}$ **35.** 5
37. $\dfrac{3}{2}$ **39.** $1 + i\sqrt{2}$ **41.** $-1 - i\sqrt{5}$
43. $20i$

Exercise 6.6

1. $4 + 9i$ **3.** $8 + 3i$ **5.** $-4 - i$
7. $-3 + i$ **9.** $-5 + 3i$ **11.** $10 - 10i$
13. $7 + 17i$ **15.** $1 - 2i$ **17.** $6 + 2i\sqrt{3}$
19. $8 + 6i$ **21.** $32 - 24i$ **23.** $-\dfrac{3}{2}i$
25. $\dfrac{3 - i}{10}$ **27.** $\dfrac{9 + 3i}{5}$ **29.** i
31. $\dfrac{-2 + 9i}{17}$ **33.** 0 **35.** $-2 - 2i$

37. $0 \pm 5i$ or $\pm 5i$ **39.** $3 \pm 2i$ **41.** $\dfrac{1 \pm i\sqrt{11}}{2}$

43. $\dfrac{-5 \pm i\sqrt{11}}{6}$ **45.** $\dfrac{2 + i\sqrt{6}}{2}$ **47.** $2 \pm 2i$

49. $-4 - 15i$ **51.** $\dfrac{-1 - 2i}{10}$

Exercise 7.1

1. domain: $\{-4, -1, 4, 0\}$; range: $\{3, 2, 0\}$
3. domain: $\{-3, 0, 3, 1\}$; range: $\{0, -2, 3, -3\}$
5. domain: $\{2, 3, 4\}$; range: $\{4\}$
7. domain: $\{x \mid x \in R\}$; range: $\{y \mid y \geq 1\}$
9. domain: $\{x \mid x \in R\}$; range: $\{y \mid y \in R\}$
11. domain: $\{x \mid x \geq 0\}$; range: $\{y \mid y \geq 0\}$
13. domain: $\{x \mid x \geq 4\}$; range: $\{y \mid y \geq 0\}$
15. domain: $\{x \mid x \geq 0\}$; range: $\{y \mid y \in R\}$
17. domain: $\{x \mid x \geq 0\}$; range: $\{y \mid y \in R\}$

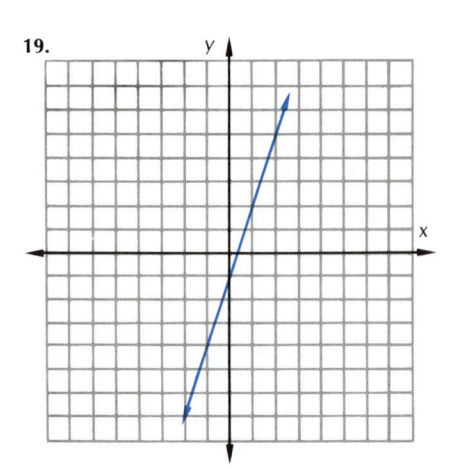

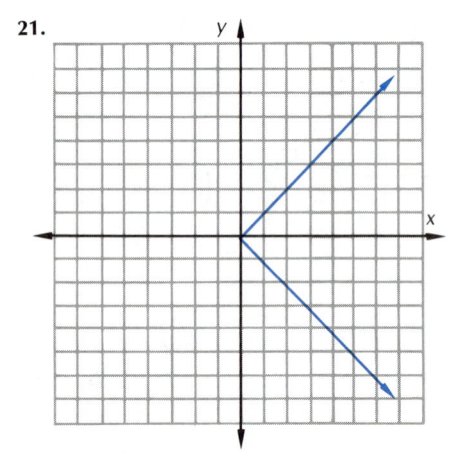

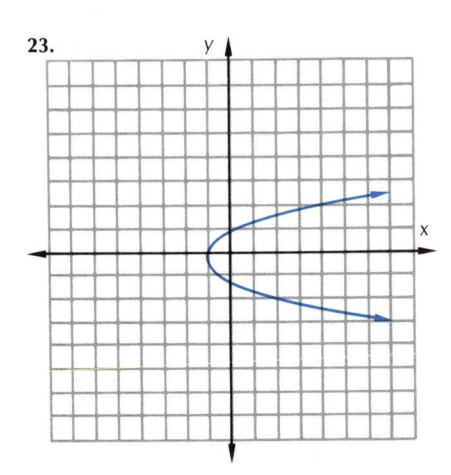

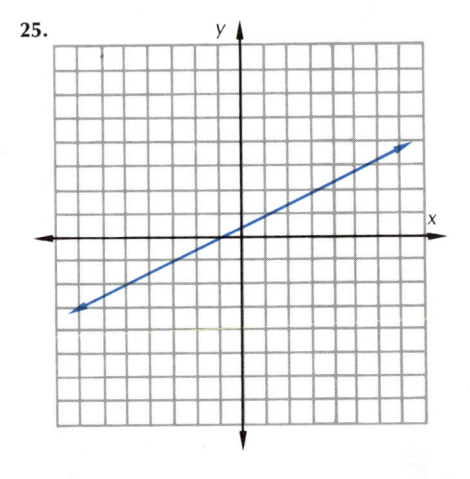

19.

21.

23.

25.

A-25

27. 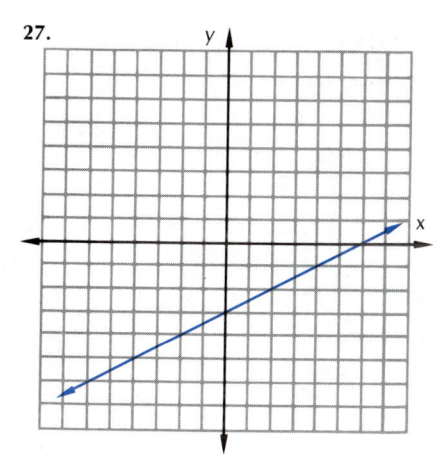
domain: $\{x \mid x \in R\}$
range: $\{y \mid y \in R\}$

29. 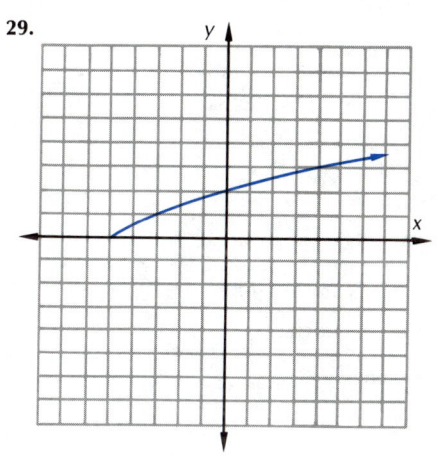
domain: $\{x \mid x \geq -5\}$
range: $\{y \mid y \geq 0\}$

31. 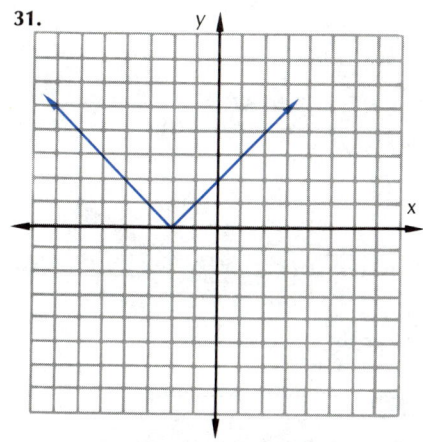
domain: $\{x \mid x \in R\}$
range: $\{y \mid y \geq 0\}$

33. 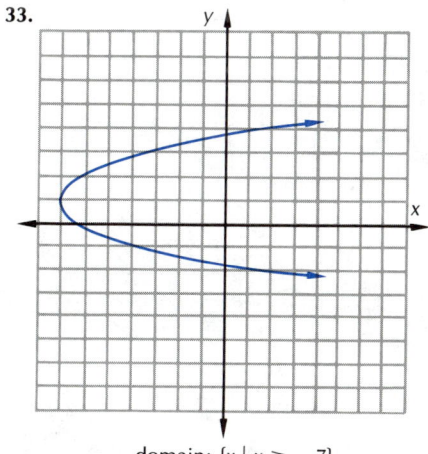
domain: $\{x \mid x \geq -7\}$
range: $\{y \mid y \in R\}$

35. 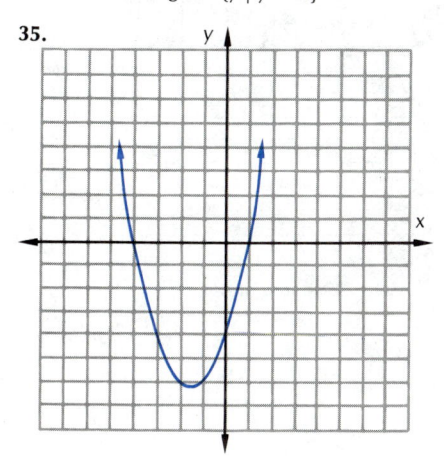
domain: $\{x \mid x \in R\}$
range: $\{y \mid y \geq -\frac{25}{4}\}$

37. 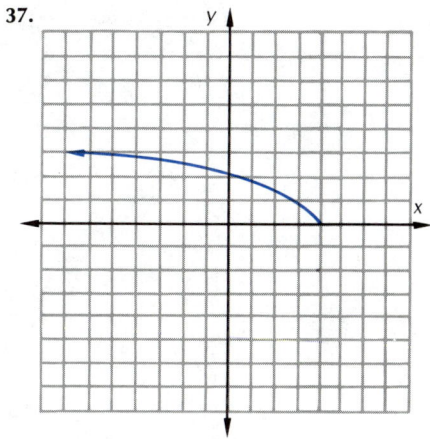
domain: $\{x \mid x \leq 4\}$
range: $\{y \mid y \geq 0\}$

Exercise 7.2

1. function
   domain: $\{x \mid x \in R\}$
   range: $\{y \mid y \geq -4\}$

3. function
   domain: $\{x \mid x \in R\}$
   range: $\{y \mid y \in R\}$

5. function
   domain: $\{x \mid x \in R\}$
   range: $\{y \mid y \geq -3\}$

7. not a function

9. function
   domain: $\{x \mid x \in R\}$
   range: $\{y \mid y \in R\}$

11. function

13. function

15. function

17. not a function

19. $f(x) = -3x + 8$
    $f(3) = -1$
    $f(-2) = 14$

21. $f(x) = x^2 - 2$
    $f(3) = 7$
    $f(-2) = 2$

23. $f(x) = \sqrt[3]{x}$
    $f(3) = \sqrt[3]{3}$
    $f(-2) = \sqrt[3]{-2}$

25. $f(0) = -1$

27. $h(-3)$ does not exist

29. $g(2) + h(1) = 4$

31. $g(1) - h(0) = -6$

33. $g(a + h) = 3a^2 + 6ah + 3h^2 - 4a - 4h - 3$

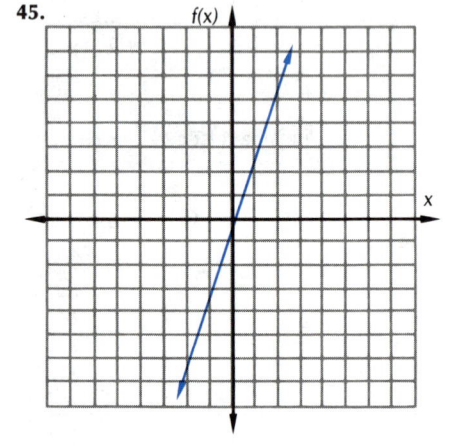

35. $f\left(\dfrac{3}{4}\right) + f\left(\dfrac{1}{4}\right) = 2$

37. $\dfrac{2}{5}$

39. -5 and 5

41. no real zeros

43. -3 and $\dfrac{1}{2}$

45.

domain: $\{x \mid \in R\}$
range: $\{y \mid \in R\}$

47.

domain: $\{x \mid x \geq 3\}$
range: $\{y \mid y \geq 0\}$

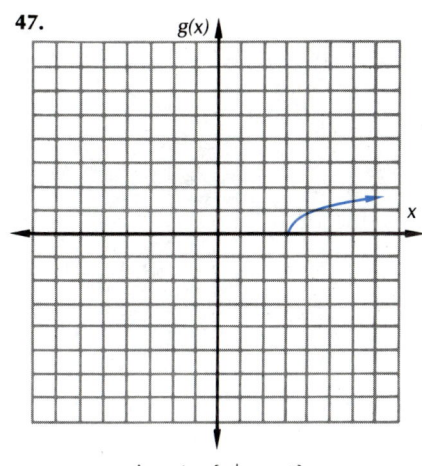

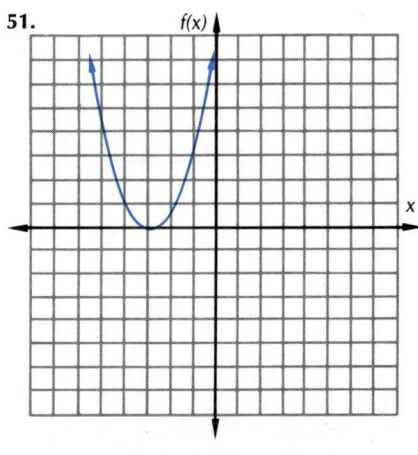

49.

domain: $\{x \mid x \leq 4\}$
range: $\{y \mid y \geq 0\}$

51.

domain: $\{x \mid x \in R\}$
range: $\{y \mid y \geq 0\}$

53. -3

55. $2x^2 + 4ah + 2h^2 - 1$

57. $2ah + h^2 - 3h$

Exercise 7.3

1. $\{(0, 3) (1, 1) (2, 0) (3, 4)\}$; function
3. $\{(6, 4) (9, 3) (3, -2)\}$; function
5. $\{(-1, 1) (-2, 2) (-3, 3) (-4, 4)\}$; function

7.

9.

11.

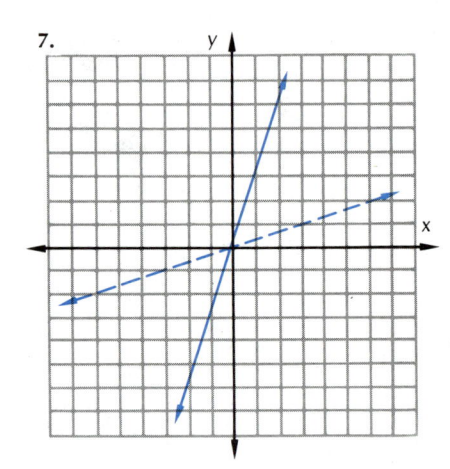

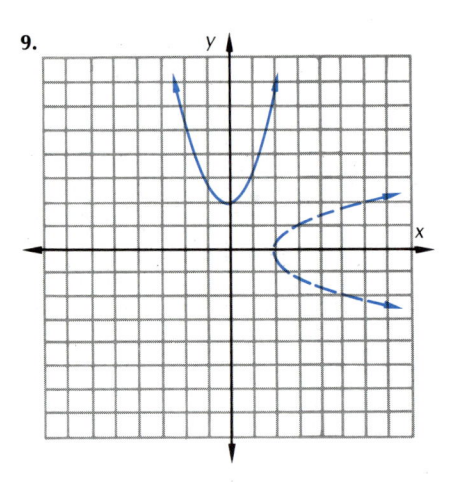

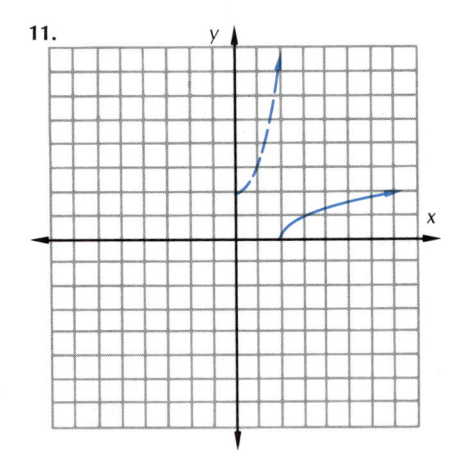

13. not a function

15. function

17. function

19. function

21. $f^{-1}(x) = \dfrac{x+2}{3}$
domain: $\{x \mid x \in R\}$
range: $\{f^{-1}(x) \mid f^{-1}(x) \in R\}$

23. $f^{-1}(x) = \dfrac{x}{2}$
domain: $\{x \mid x \in R\}$
range: $\{f^{-1}(x) \mid f^{-1}(x) \in R\}$

25. $f^{-1}(x) = 3x - 1$
domain: $\{x \mid x \in R\}$
range: $\{f^{-1}(x) \mid f^{-1}(x) \in R\}$

27. $f^{-1}(x) = \dfrac{12 - 2x}{3}$
domain: $\{x \mid x \in R\}$
range: $\{f^{-1}(x) \mid f^{-1}(x) \in R\}$

29. $f^{-1}(x) = 2x + 4$
domain: $\{x \mid x \in R\}$
range: $\{f^{-1}(x) \mid f^{-1}(x) \in R\}$

31. $f^{-1}(x) = x^2 + 1$
domain: $\{x \mid x \geq 0\}$
range: $\{f^{-1}(x) \mid f^{-1}(x) \geq 1\}$

33. $f^{-1}(x) = x^2 - 2$
domain: $\{x \mid x \geq 0\}$
range: $\{f^{-1}(x) \mid f^{-1}(x) \geq -2\}$

35. $f^{-1}(x) = \sqrt{x} + 1$
domain: $\{x \mid x \geq 0\}$
range: $\{f^{-1}(x) \mid f^{-1}(x) \geq 1\}$

37. $f^{-1}(x) = \dfrac{3x + 1}{4}$
domain: $\left\{x \mid 1 \leq x \leq \dfrac{11}{3}\right\}$
range: $\{f^{-1}(x) \mid 1 \leq f^{-1}(x) \leq 3\}$

ANSWERS TO ODD-NUMBERED EXERCISES A-29

39. $f^{-1}(x) = \dfrac{x-1}{3}$; $f^{-1}(3) = \dfrac{2}{3}$, $f^{-1}(0) = -\dfrac{1}{3}$, $f^{-1}(-2) = -1$
41. $f^{-1}(x) = x^2 + 1$; $f^{-1}(3) = 10$, $f^{-1}(0) = 1$, $f^{-1}(-2)$ does not exit
43. -1 **45.** -3

---

1. x-intercept is 4
y-intercept is -8

3. x-intercept is $\dfrac{4}{3}$
y-intercept is -4

5. x-intercept is $\dfrac{8}{5}$
y-intercept is $\dfrac{8}{3}$

7. x-intercept is 6
y-intercept is -4

9. $m = -\dfrac{3}{4}$

11. $m = \dfrac{1}{7}$

13. $m = -\dfrac{4}{7}$

15. $m = 0$

17. $m = -\dfrac{21}{10}$

19.

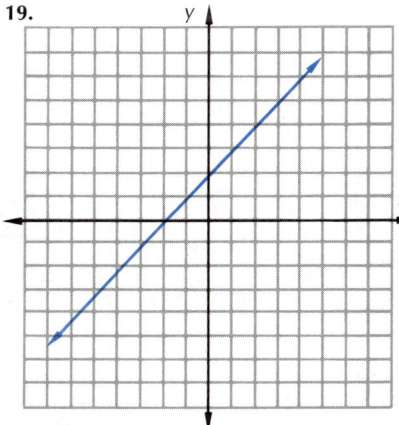

$m = 2$

21.

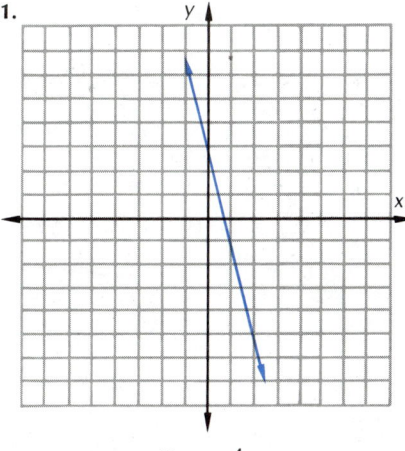

$m = -4$

23.

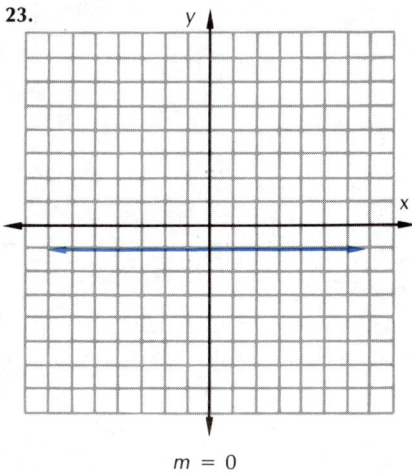

$m = 0$

25.

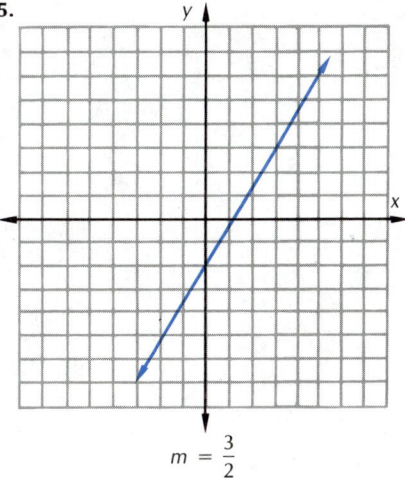

$m = \dfrac{3}{2}$

27.

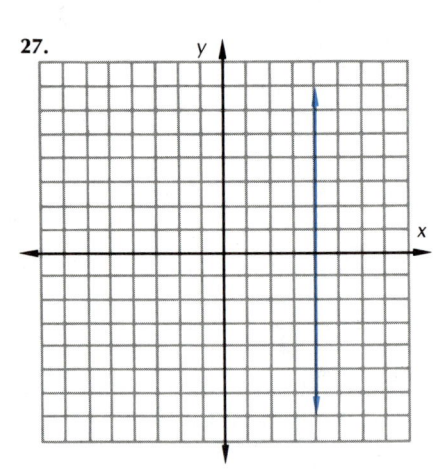

slope does not exist

29. $L_1: m = \dfrac{2}{3}$
$L_2: m = -1$

31. $L_1: m = -3$
$L_2: m = 1$

33. neither

35. parallel

37. perpendicular

39. parallel

41. neither

Exercise 7.5

1. $-3x + y - 9 = 0$
3. $x + y + 3 = 0$
5. $-3x + 4y + 2 = 0$
7. $4x + 5y - 15 = 0$
9. $4x + y - 10 = 0$
11. $y = 4$
13. $12x + y - 10 = 0$
15. $m = -2$
y-intercept is -1
17. $m = -1$
y-intercept is 3
19. $m = 0$
y-intercept is 3
21. $m = -10$
y-intercept is 6
23. $-3x + y - 10 = 0$
25. $-2x + y - 14 = 0$
27. $x + 2y - 16 = 0$
29. $x + 3y - 19 = 0$
31. $3x + 5y - 15 = 0$
33. $2x + 3y - 1 = 0$

35.

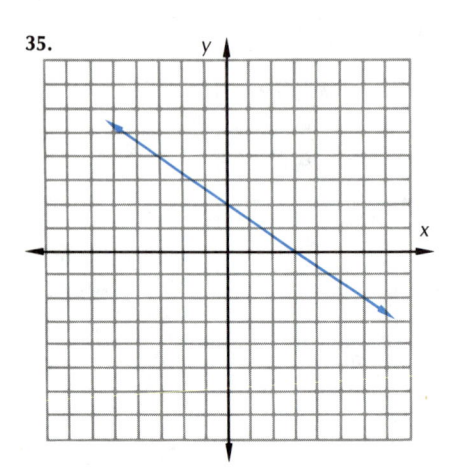

ANSWERS TO ODD-NUMBERED EXERCISES   A-31

37.

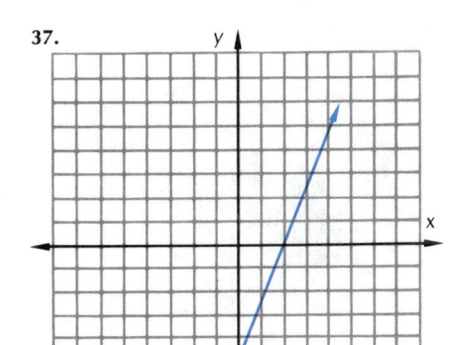

39.

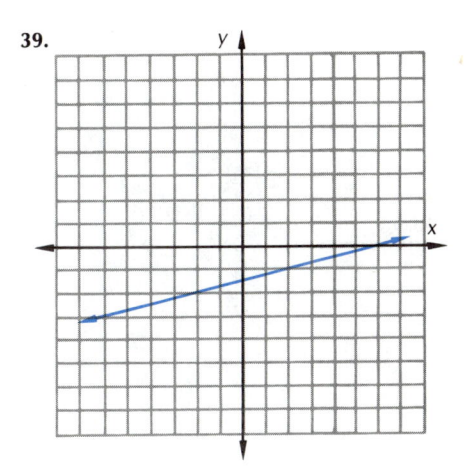

Exercise 7.6

1. does not belong
3. belongs
5. 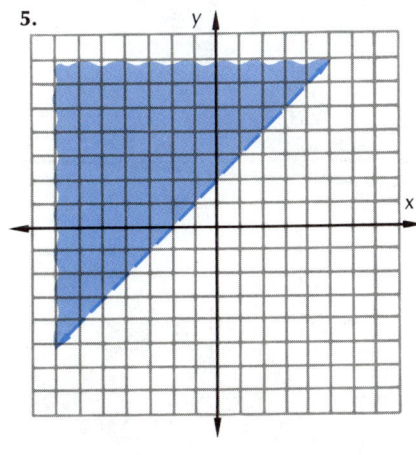
7. 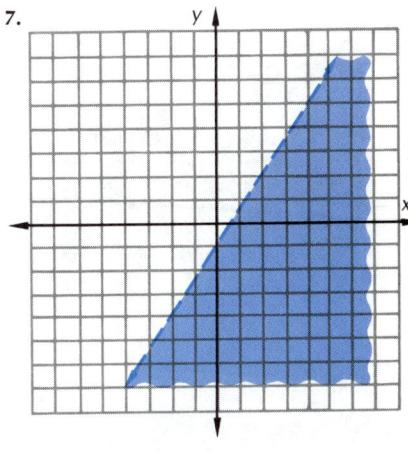
9. 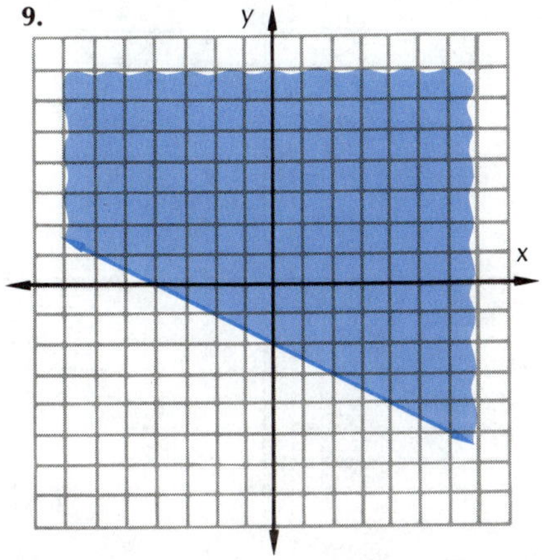
11.

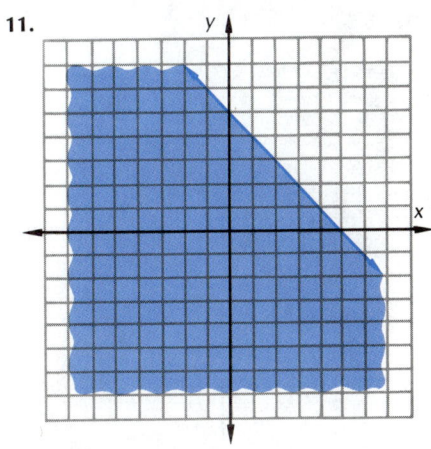

ANSWERS TO ODD-NUMBERED EXERCISES

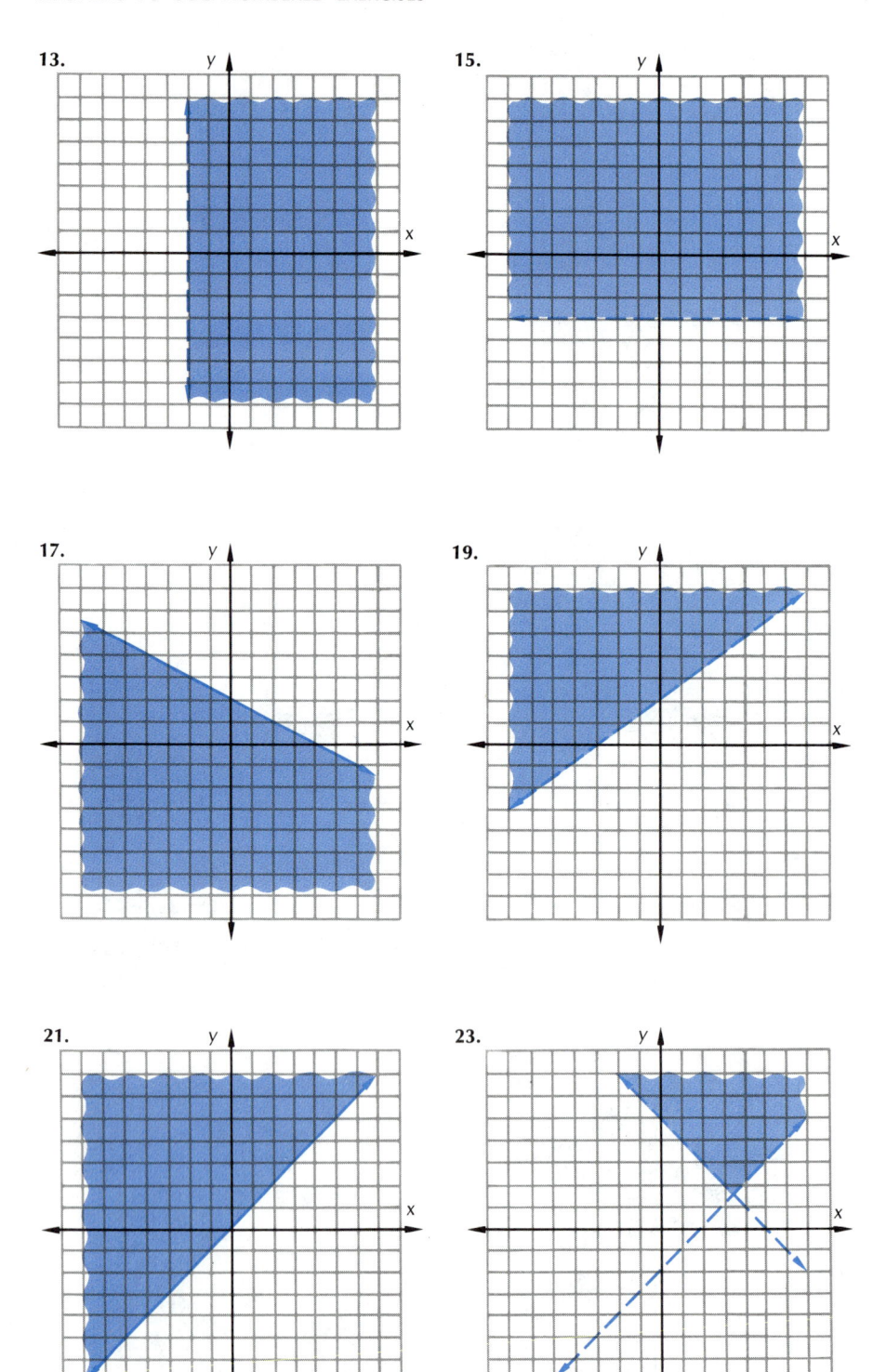

13.

15.

17.

19.

21.

23.

ANSWERS TO ODD-NUMBERED EXERCISES   A-33

25. 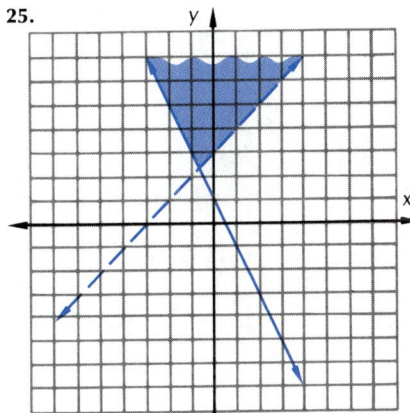
27. 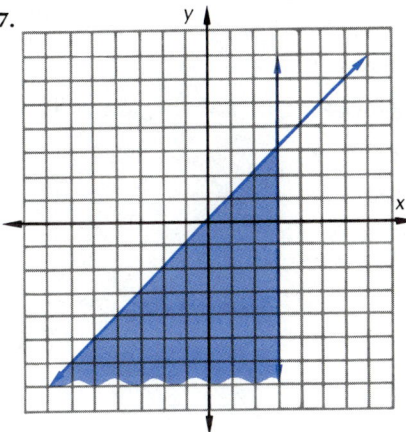
29. 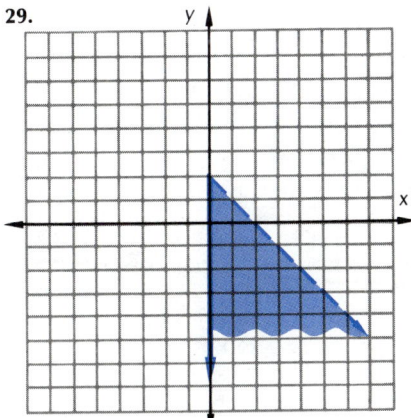
31.

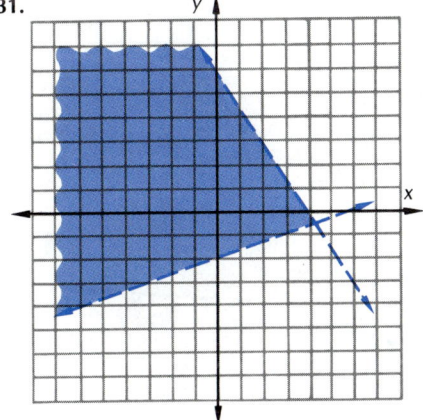

Exercise 7.7

1. $P = k\sqrt{L}$
3. $s = \dfrac{k}{d}$
5. $y = 28$
7. $x = \dfrac{5}{6}$
9. $x = \dfrac{35}{2}$
11. $y = \dfrac{18}{25}$
13. $C = 3\pi$
15. 256 feet
17. $28\sqrt{2}$ miles
19. $1,600
21. 1 ohm
23. $12\dfrac{1}{2}$ pounds per square foot
25. $2\dfrac{4}{5}$ pounds
27. 540 pounds
29. 8.4 grams
31. 10 horsepower
33. 9
35. one-half as large
37. $\dfrac{4}{3}$ as great
39. 4.92

Exercise 8.1

1. yes
3. consistent
5. dependent
7. (6, 2)
9. inconsistent; no solution
11. (2, −3)
13. (0, 0)
15. $\left(-\frac{7}{5}, \frac{13}{5}\right)$
17. (3, −2)
19. (4, 0)
21. (4, 3)
23. (1, 1)
25. (1, 2)
27. $\left(\frac{1}{2}, \frac{2}{3}\right)$
29. (−5, 4)
31. (3, −2)
33. inconsistent; no solution
35. $\left(\frac{6}{13}, -\frac{23}{26}\right)$
37. (−2.498, −1.255)
39. (1.092, 2.567)

Exercise 8.2

1. (1, −1, 0)
3. (−1, 5, −2)
5. (0, 0, 0)
7. (−2, 5, 6)
9. (1, −2, 3)
11. (3, 2, 1)
13. (2, 1, −1)
15. (3, −2, 2)
17. (2, −1, 3)
19. no solution
21. (1, 1, 1)
23. (1, −1, 2)
25. (5, −4, 6)
27. (1, 3, 5)

Exercise 8.3

1. 9
3. −10
5. 0
7. $\frac{7}{16}$
9. 0
11. 4
13. −1
15. 18
17. −14
19. 40
21. 96
23. 5
25. 10
27. 20
29. $2a - 2c$
31. 0
33. −14
35. $\frac{9}{2}$
37. $\frac{27}{4}$
39. 6
41. $-2, -\frac{7}{4}$

Exercise 8.4

1. (−1, 3)
3. $\left(0, \frac{5}{3}\right)$
5. $\left(3, -\frac{5}{2}\right)$
7. (2, −2)
9. $\left(-\frac{15}{43}, -\frac{27}{43}\right)$
11. (−4, 0)
13. (−2, 3, 1)
15. $\left(\frac{1}{4}, \frac{1}{4}, \frac{1}{2}\right)$
17. $\left(\frac{1}{2}, \frac{1}{2}, 1\right)$
19. (−2, 5, 6)
21. (5, 1, 1)
23. (2, −1, 3)

Exercise 8.5

1. 65° and 115°
3. 7 and 12
5. 15
7. $4.50 and $5.50
9. 20 dimes
11. 300
13. 13%
15. 44 meters by 52 meters

ANSWERS TO ODD-NUMBERED EXERCISES A-35

17. 16 inches by 14 inches

19. 7 kilometers per hour for swimmer
1 kilometer per hour for current

21. 600 miles per hour for plane
100 miles per hour for wind

23. 3, 4, and 11

25. 2 quarters

27. Harry has $40
Tom has $200
Dick has $120

Exercise 9.1

1. x-intercept is ±2, y-intercept is −4
3. no x-intercept, y-intercept is 3
5. x-intercepts are 5 and 1, y-intercept is −10
7. x-intercept is 3, y-intercepts are 3 and 1
9. up
11. right
13. left
15. down

17.

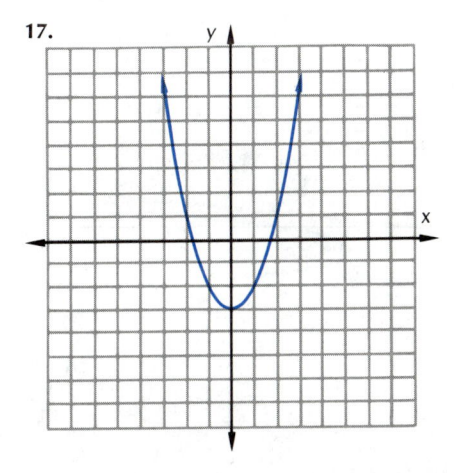

19.

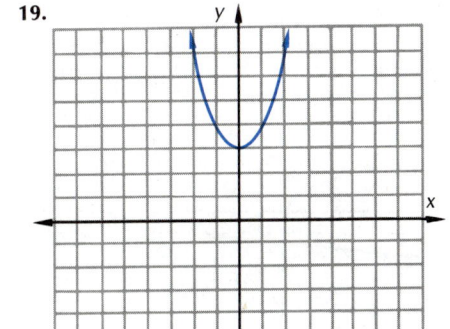

21.

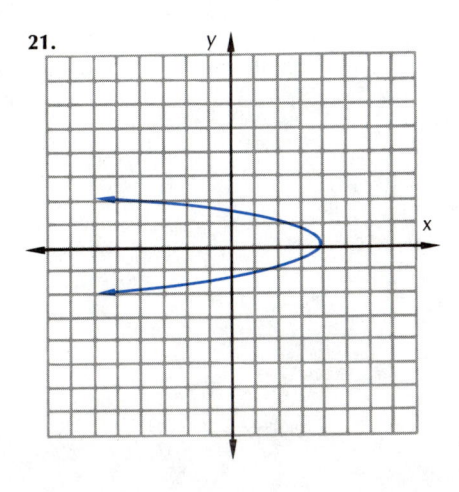

23.

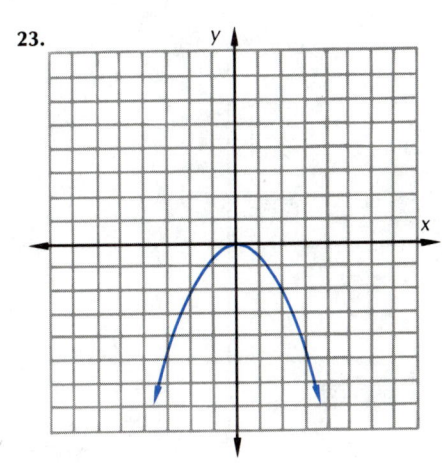

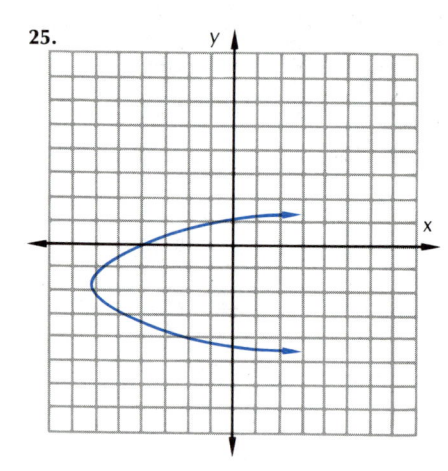

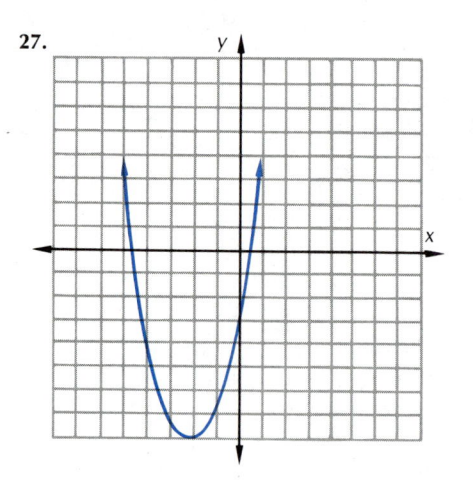

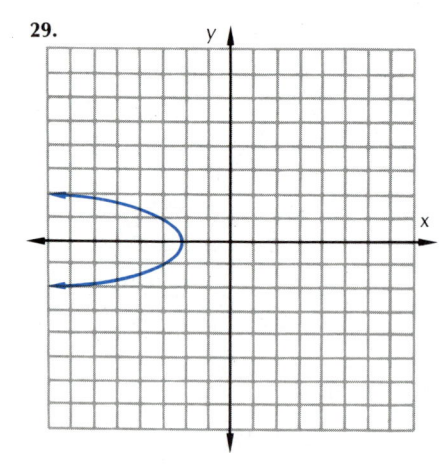

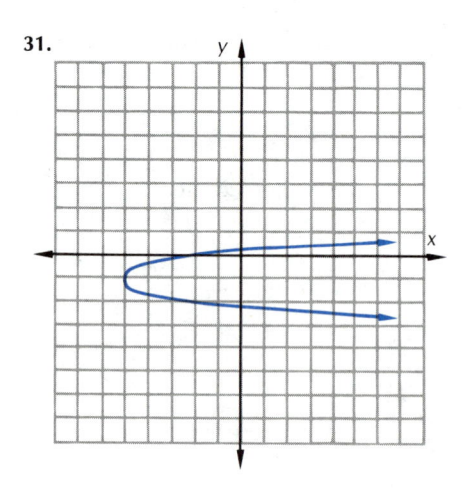

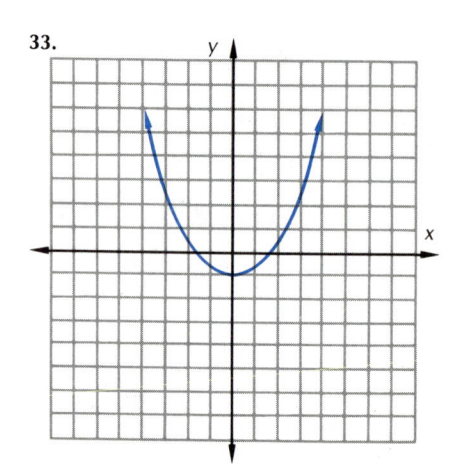

25.

27.

29.

31.

33.

ANSWERS TO ODD-NUMBERED EXERCISES

Exercise 9.2

1. concave up; minimum

3. concave up; minimum

5. x-intercept: 3
   y-intercept: 9

7. x-intercepts: 0, $-\frac{3}{2}$
   y-intercept: 0

9. x-intercepts: 1, 4
   y-intercept: -4

11. $(2, -2)$
    minimum value: -2

13. $\left(-\frac{3}{2}, -\frac{15}{2}\right)$
    minimum value: $-\frac{15}{2}$

15. $\left(\frac{1}{2}, \frac{1}{4}\right)$
    maximum value: $\frac{1}{4}$

17. $\left(-\frac{3}{2}, \frac{15}{2}\right)$
    maximum value: $\frac{15}{2}$

19. $\left(\frac{5}{2}, -\frac{9}{4}\right)$
    minimum value: $-\frac{9}{4}$

21. $\left(-\frac{3}{10}, -\frac{49}{20}\right)$
    minimum value: $-\frac{49}{20}$

23.

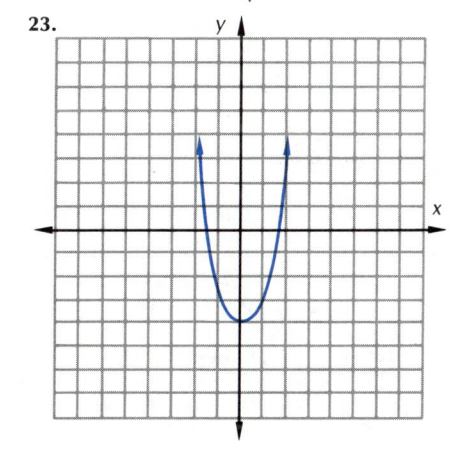

25.

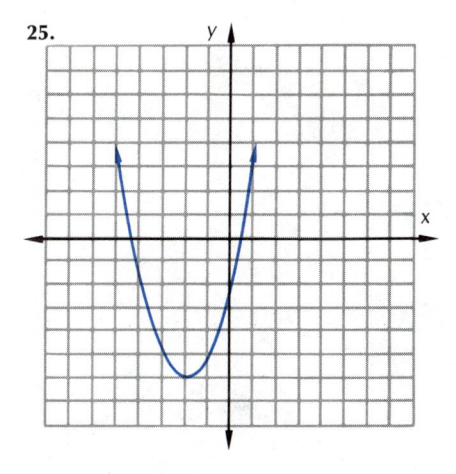

27.

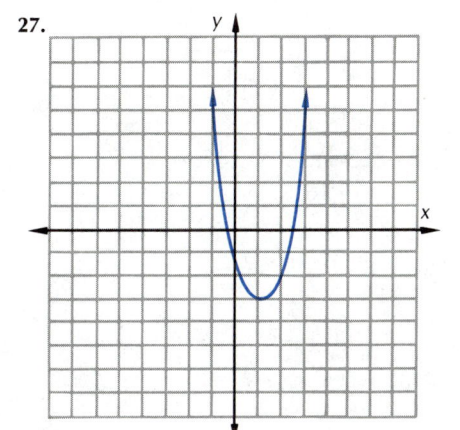

29. domain is $\{x \mid x \in R\}$
    range is $\{y \mid y \leq 0\}$

31. domain is $\{x \mid x \in R\}$
    range is $\{y \mid y \leq 7\}$

33. domain is $\{x \mid x \in R\}$
    range is $\{y \mid y \geq -5\}$

35. 8 and 8

37. 160,000 feet

39. 800 square feet

Exercise 9.3

1. $2\sqrt{10}$
3. $2\sqrt{13}$
5. $5\sqrt{5}$
7. $|x - y|$
9. $12\sqrt{2}$
11. $\sqrt{58} + \sqrt{74} + 2\sqrt{26}$

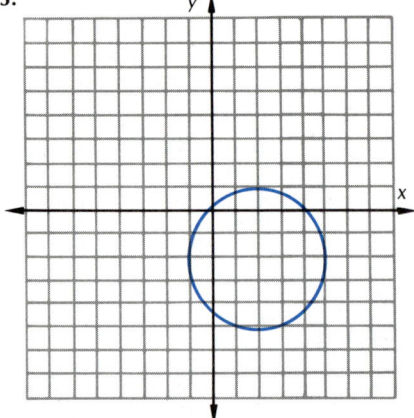

13.

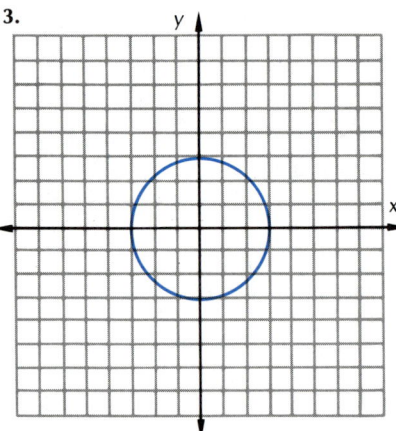

center: $(0, 0)$ radius: $r = 3$

15.

center: $(2, -2)$ radius: $r = 3$

17.

center: $(1, 5)$ radius: $r = \sqrt{10}$

19. $(x - 2)^2 + (y + 4)^2 = 256$

21. $(x + 1)^2 + (y - 4)^2 = 4$
23. center: $(-1, 4)$
    radius: $r = 4$
25. center: $(5, -5)$
    radius: $r = 2\sqrt{7}$
27. center: $\left(\dfrac{5}{2}, -1\right)$
    radius: $r = \dfrac{\sqrt{77}}{2}$
29. a right triangle since $(\sqrt{160})^2 = (\sqrt{32})^2 + (\sqrt{128})^2$
31. 2 or 4
33. $(x + 1)^2 + (y - 1)^2 = 20$

Exercise 9.4

1. x-intercepts are ± 2
   y-intercepts are ± 1
3. x-intercepts are ± 5
   y-intercepts are ± 10
5. x-intercepts are ± 3
   y-intercepts are ± 5
7. x-intercepts are $\pm 2\sqrt{5}$
   y-intercepts are ± 2

ANSWERS TO ODD-NUMBERED EXERCISES   A-39

9.

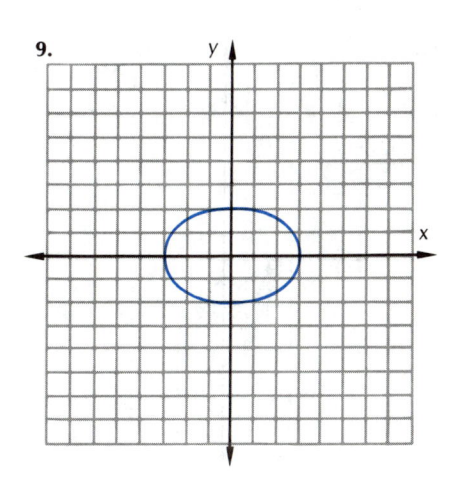

11.

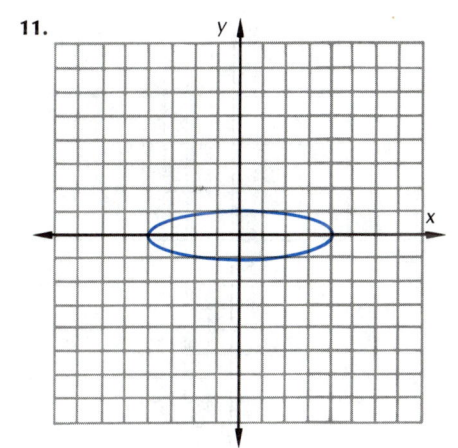

13.

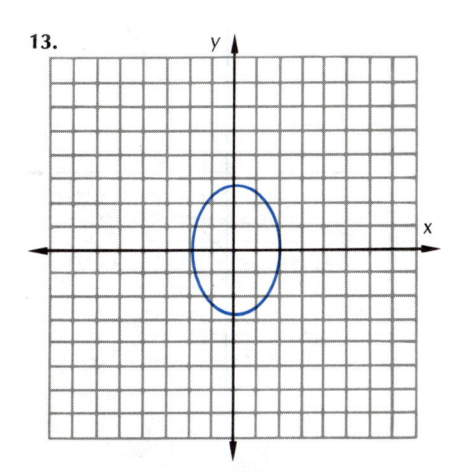

15.

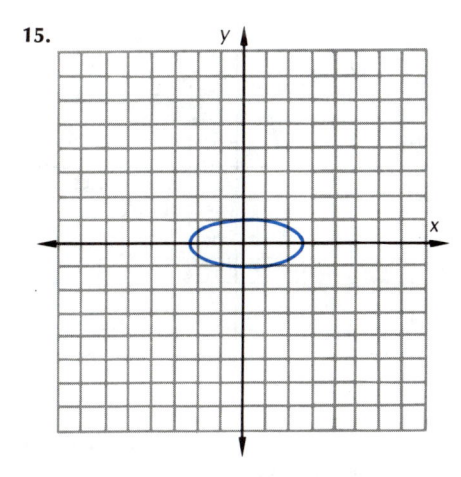

17.

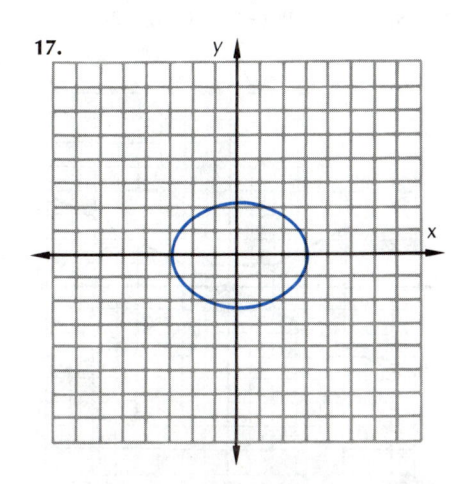

19.

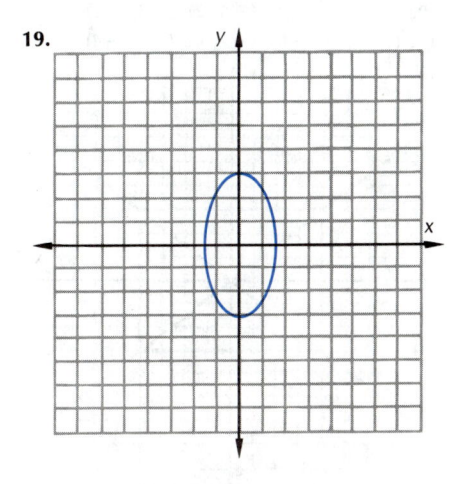

A-40 ANSWERS TO ODD-NUMBERED EXERCISES

21.

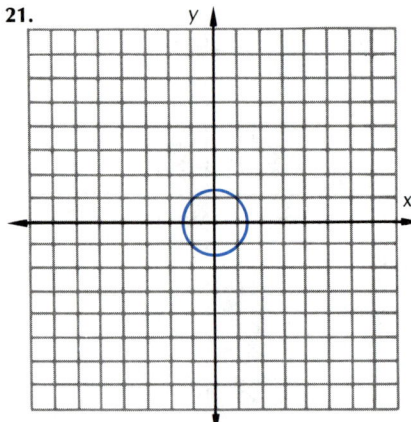

23.

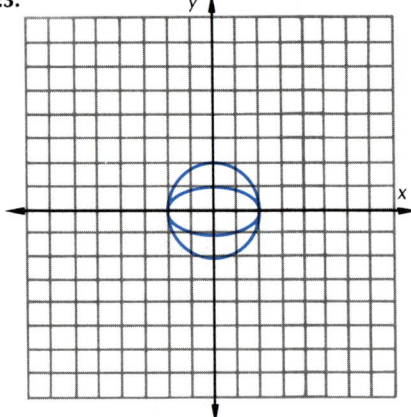

25.

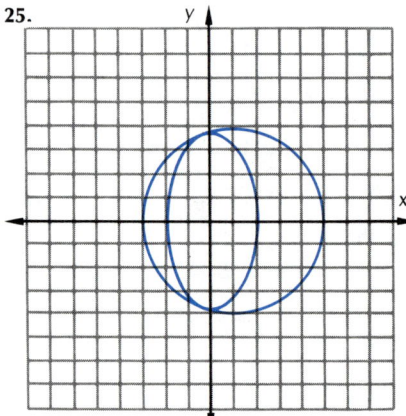

27.

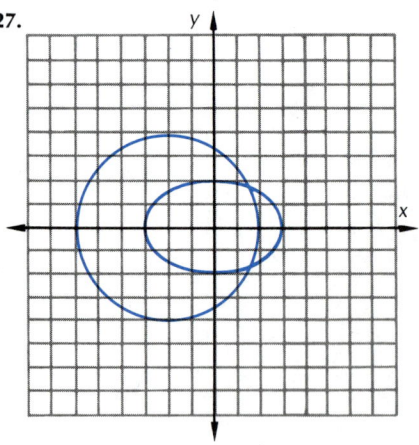

Exercise 9.5

1. x-intercepts are ± 4
3. no x- or y-intercepts
5. y-intercepts are $\pm 2\sqrt{3}$
7. y-intercepts are $\pm\sqrt{2}$
9. $(3, 2), (3, -2), (-3, 2), (-3, -2)$
11. $(5, 2), (5, -2), (-5, 2) (-5, -2)$
13. $(1, 4), (1, -4), (-1, 4), (-1, -4)$

15.

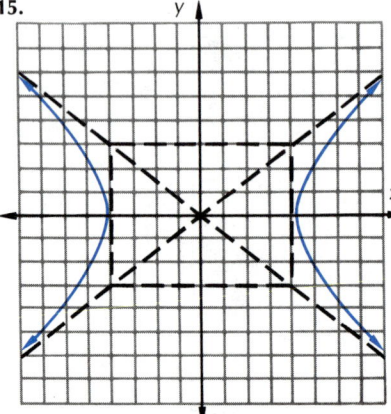

17.

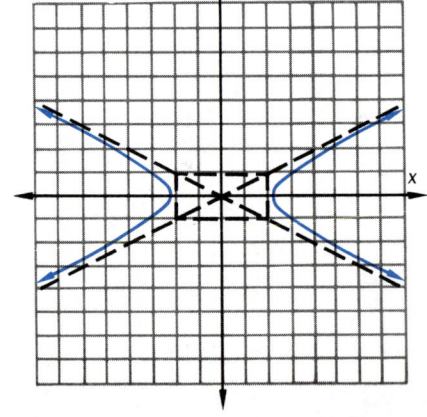

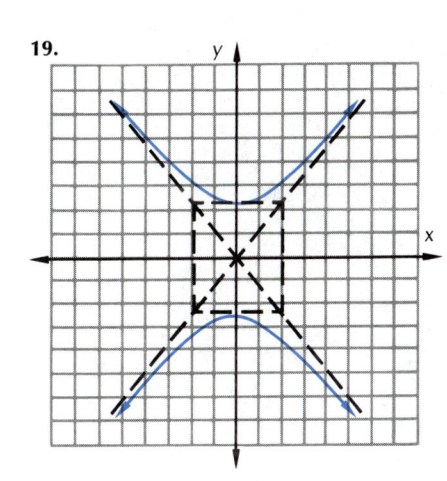

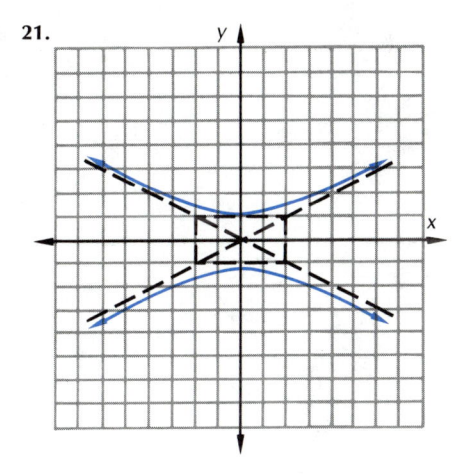

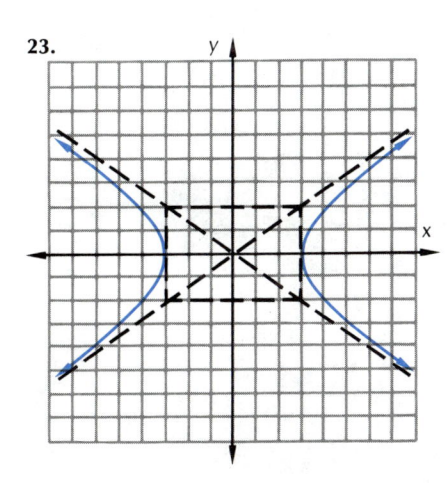

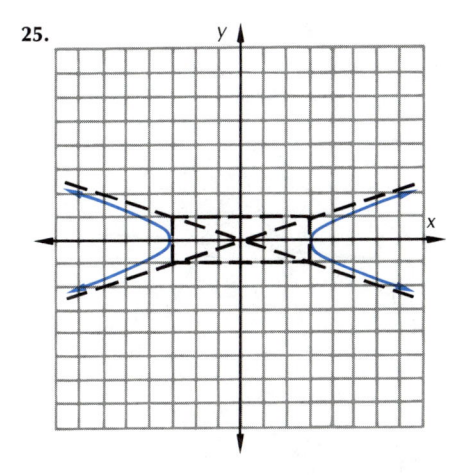

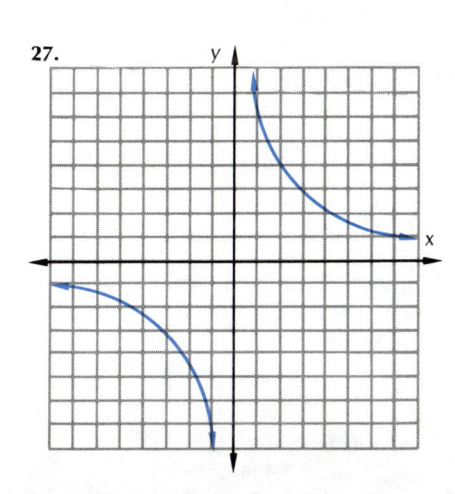

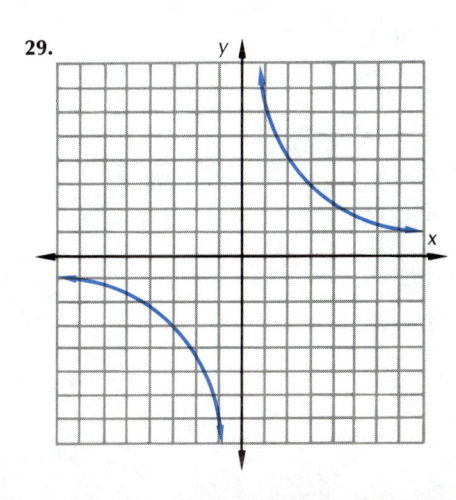

19.

21.

23.

25.

27.

29.

31.

33.

35.

37.

Exercise 9.6

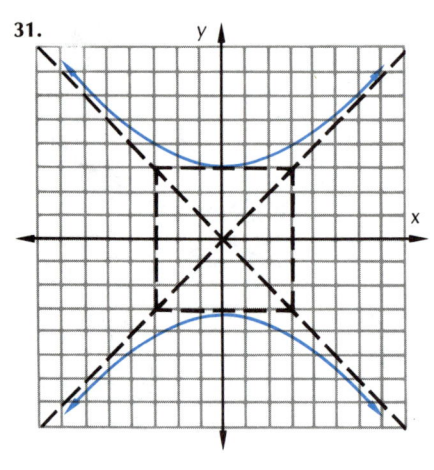

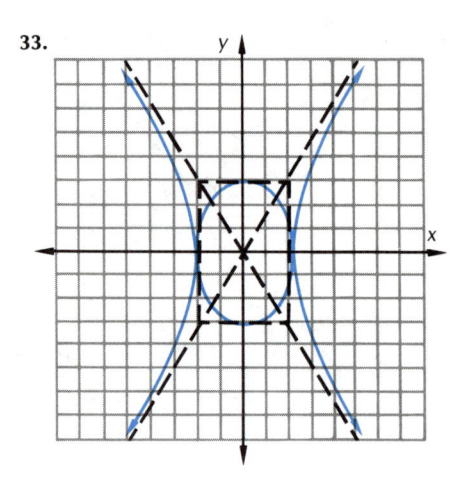

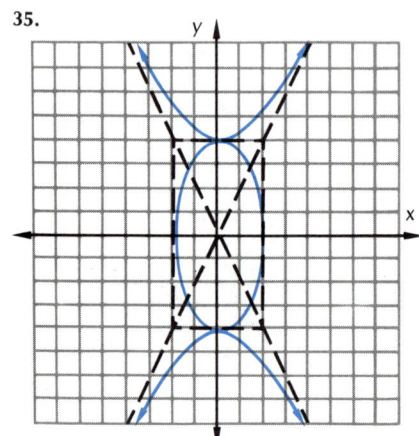

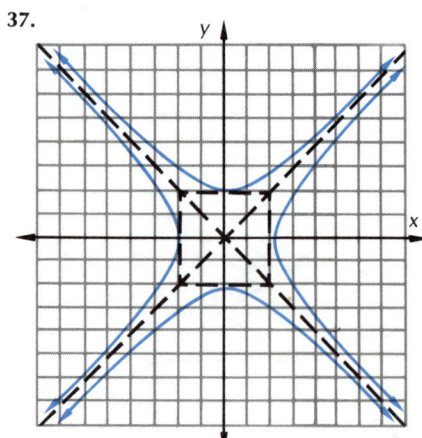

1. hyperbola; $\frac{x^2}{1} - \frac{y^2}{9} = 1$; opens horizontally, x-intercepts are ± 1

3. ellipse; $\frac{x^2}{3} + \frac{y^2}{9} = 1$; x-intercepts are $\pm\sqrt{3}$, y-intercepts are ± 3

5. hyperbola; $\frac{x^2}{9} - \frac{y^2}{81} = 1$; opens horizontally, x-intercepts are ± 3

7. line; $x - y - 1 = 0$; x-intercept is 1; y-intercept is -1

9. hyperbola; $\frac{y^2}{4} - \frac{x^2}{1} = 1$; opens vertically, y-intercepts are ± 2

11. circle; $(x - 0)^2 + (y - 0)^2 = \frac{1}{4}$; center at (0, 0), radius is $\frac{1}{2}$

13. parabola; $y = x^2 - 9$; concave up, vertex at $(0, -9)$

15. circle; $(x - 1)^2 + (y + 2)^2 = 4$; center at $(1, -2)$, radius is 2

17. line; $-x + y + 4 = 0$; x-intercept is 4, y-intercept is -4

19. parabola; $y = -3x^2$; concave down, vertex at (0, 0)

21. domain: $\{x \mid x \in R\}$
    range: $\{y \mid y \geq -4\}$

23. domain: $\{x \mid -3 \leq x \leq 3\}$
    range: $\{y \mid -4 \leq y \leq 4\}$

ANSWERS TO ODD-NUMBERED EXERCISES   A-43

25. domain: $\{x \mid -2\sqrt{3} \leq x \leq 2\sqrt{3}\}$
    range: $\{y \mid -2\sqrt{3} \leq y \leq 2\sqrt{3}\}$
27. domain: $\{x \mid x \in R\}$
    range: $\{y \mid y \geq 2 \text{ or } y \leq -2\}$
29. domain: $\{x \mid -5 \leq x \leq 5\}$
    range: $\{y \mid -1 \leq y \leq 1\}$
31. domain: $\{x \mid x \in R\}$
    range: $\{y \mid y \geq 3 \text{ or } y \leq -3\}$

Exercise 9.7

1. (3, 2) (−2, −3)
3. (−1, −6) (6, 1)
5. (2, 0)
7. (0, −5) (−3, 4)
9. (1, 1) (2, 0)
11. $(-1, -2)\left(6, \dfrac{1}{3}\right)$
13. (3, 4) (2, 6)
15. $\left(\dfrac{5}{2}, -4\right)$ (2, −5)
17. (2, 0)
19. (−5, 15) (3, −1)
21. (1, 2)
23. $(\sqrt{2}, 1)(\sqrt{2}, -1)(-\sqrt{2}, 1)(-\sqrt{2}, -1)$
25. $(2\sqrt{2}, \sqrt{3})(2\sqrt{2}, -\sqrt{3})(-2\sqrt{2}, \sqrt{3})(-2\sqrt{2}, -\sqrt{3})$
27. (5, 0) (−5, 0)
29. $(3, 0)\left(-\dfrac{9}{5}, \dfrac{12}{5}\right)$
31. (0, −3) ($\sqrt{5}$, 2) (−$\sqrt{5}$, 2)
33. $(2\sqrt{3}, 4)(2\sqrt{3}, -4)(-2\sqrt{3}, 4)(-2\sqrt{3}, -4)$

Exercise 9.8

1. 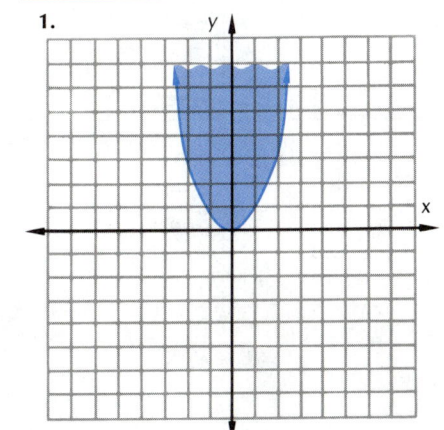
3. 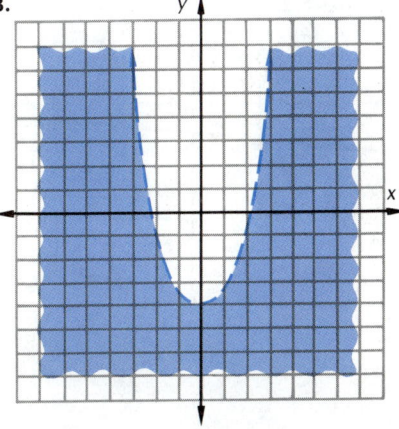
5. 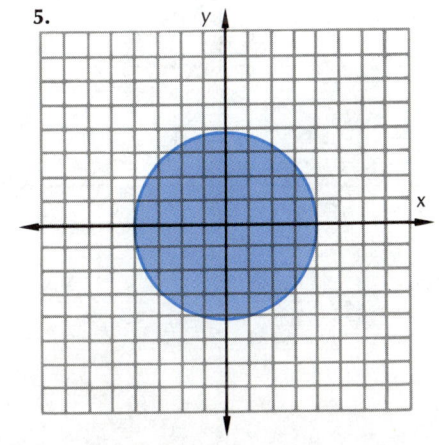
7.

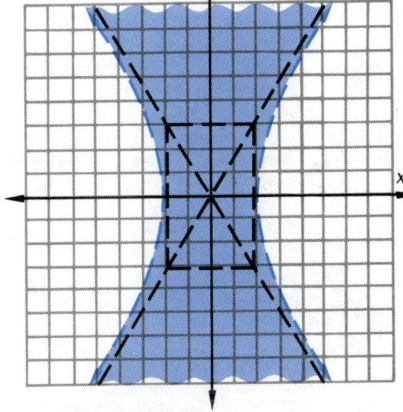

ANSWERS TO ODD-NUMBERED EXERCISES

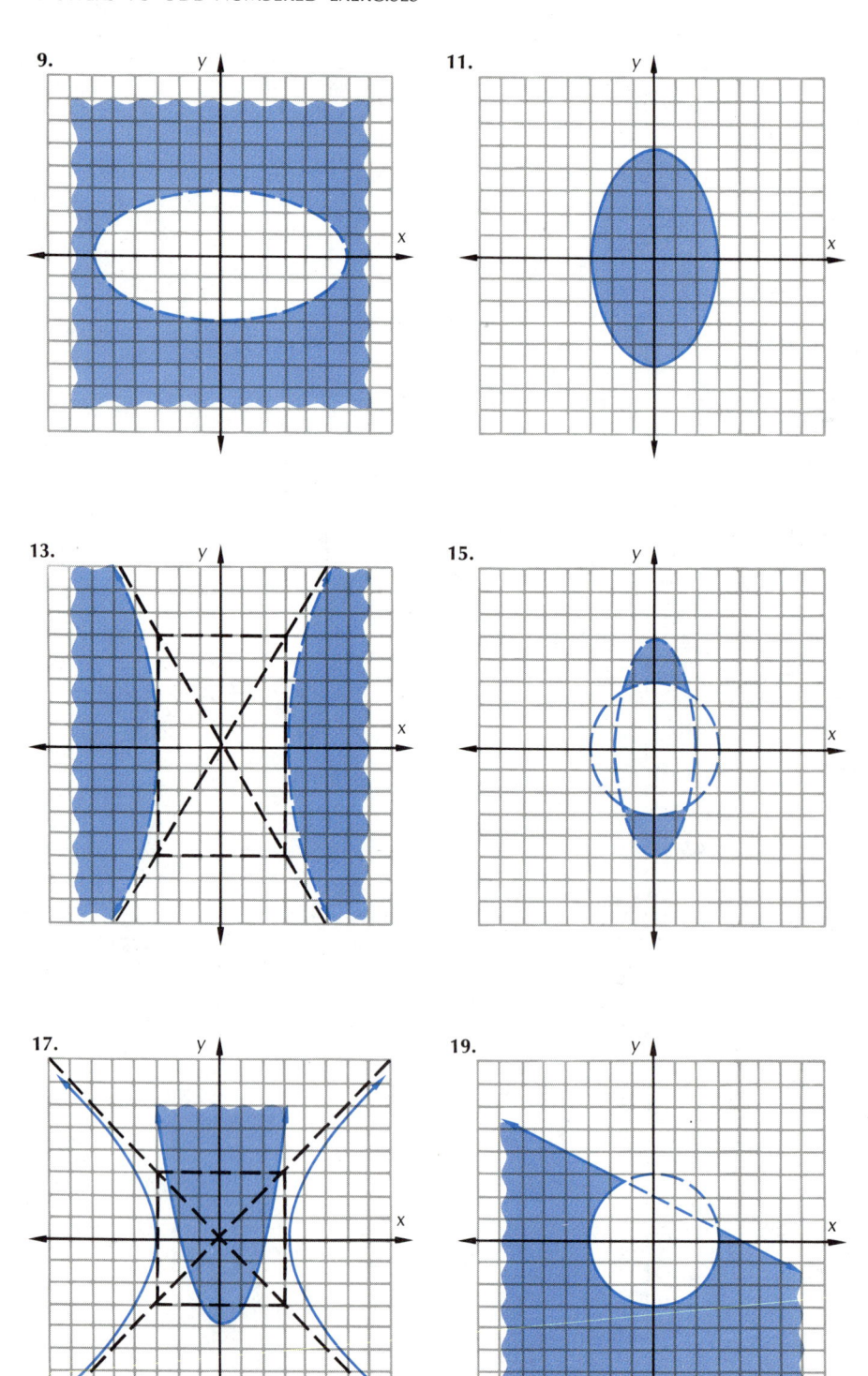

9.

11.

13.

15.

17.

19.

ANSWERS TO ODD-NUMBERED EXERCISES A-45

21.

23.

25.

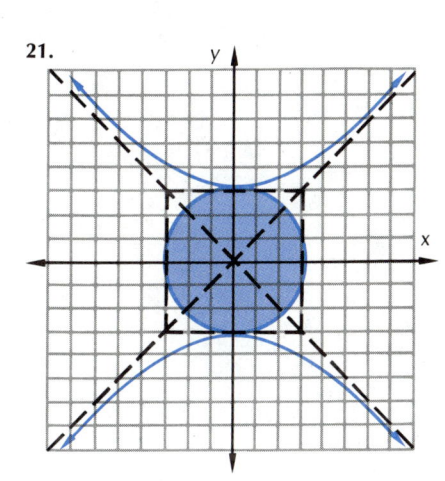

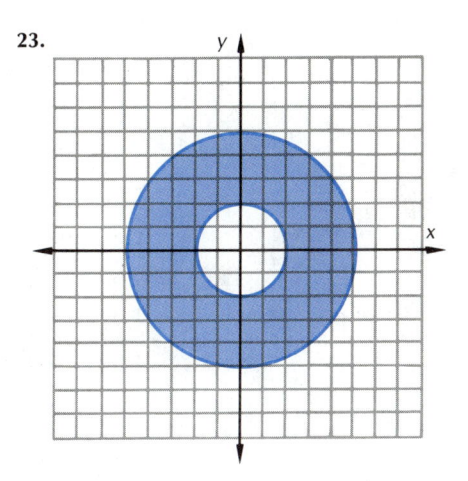

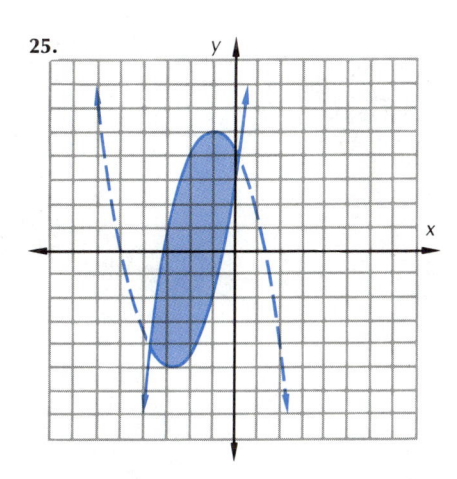

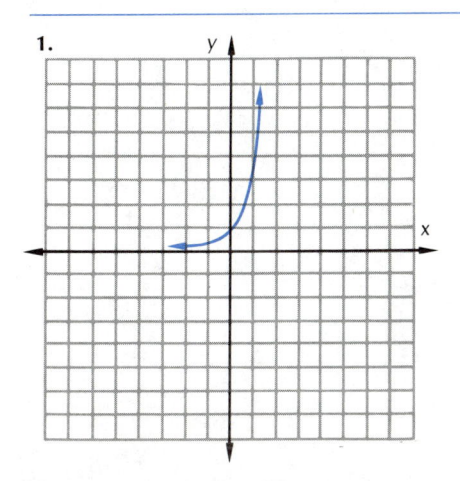

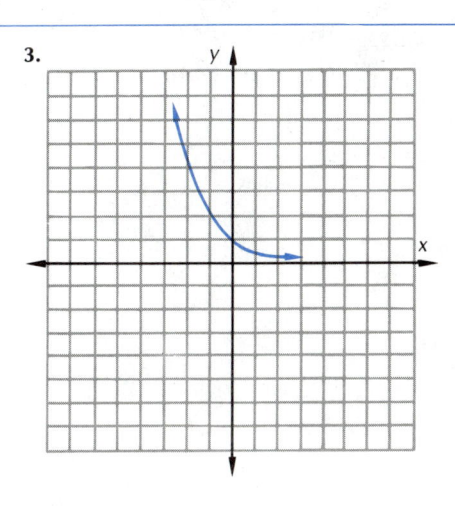

Exercise 10.1

1.

3.

5.

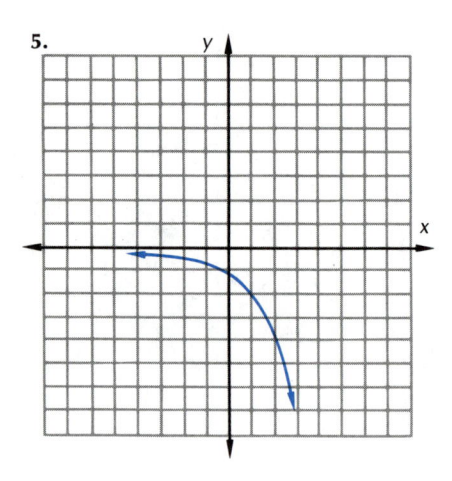

7. domain: $\{x \mid x \in R\}$
range: $\{f(x) \mid f(x) > 0\}$
increasing

9. domain: $\{x \mid x \in R\}$
range: $\{f(x) \mid f(x) > 0\}$
decreasing

11. domain: $\{x \mid x \in R\}$
range: $\{f(x) \mid f(x) > 0\}$
increasing

13. $y = \left(\dfrac{1}{2}\right)^x$

15. $y = 5^x$

17. $x = 2$

19. $x = \dfrac{5}{2}$

21. $x = -2$

23. $x = 1$

25. $x = 2$

27. $x = -2$

29. $x = 4$

31. $x = -3$

33. a) 200   b) 400   c) 800   d) 10

Exercise 10.2

1. a) 3.3201   b) 3.3201169
3. a) 14.585   b) 14.585093
5. a) 0.198   b) 0.1978987
7. a) 3.456   b) 3.4556135
9. $252.50
11. $5,049.91
13. $5,657.24
15. $1,774.82
17. 30.0 inches of mercury
19. 27.0 inches of mercury
21. 14.7 pounds per square inch
23. 13.2 pounds per square inch
25. 65,949
27. 358 million
29. $550.54
31. $1,011.56
33. 3.85 micrograms
35. 1.45 micrograms
37. .99 grams

Exercise 10.3

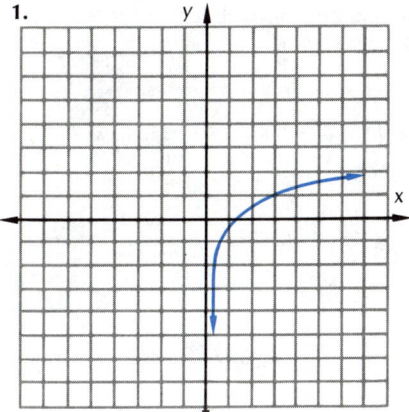

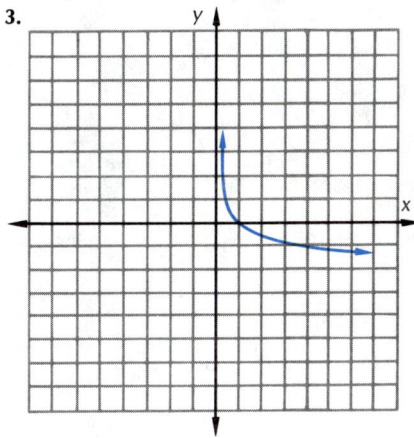

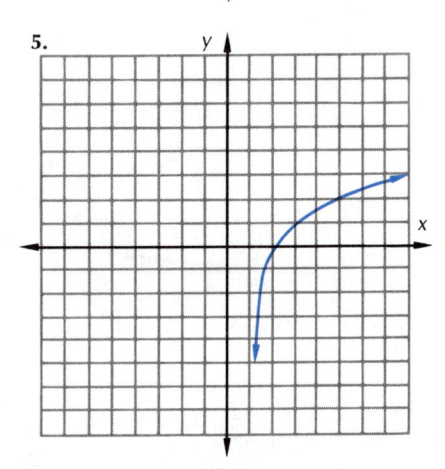

1. [graph]

3. [graph]

5. [graph]

7. Domain: $\{x \mid x > 0\}$
   Range: $\{y \mid y \in R\}$
   Increases

9. $\log_2 64 = 6$

11. $\log_4 64 = 3$

13. $\log_4 \frac{1}{16} = -2$

15. $\log_{10} .01 = -2$

17. $5^2 = 25$

19. $2^{-1} = \frac{1}{2}$

21. $3^3 = 27$

23. $8^{1/3} = 2$

25. 2

27. $\frac{1}{2}$

29. 1

31. $\frac{1}{3}$

33. -1

35. $\frac{5}{2}$

37. $\frac{5}{2}$

39. $x = \frac{1}{2}$

41. $b = \frac{1}{5}$

43. $x = 3$

45. $x = 8$

47. $x = \frac{3}{2}$

ANSWERS TO ODD-NUMBERED EXERCISES

Exercise 10.4

1. $\log_4 7 + \log_4 x$
3. $\log_3 2 + \log_3 \pi + \log_3 r$
5. $\log_{10} R - \log_{10} V$
7. $\log_2 x - \log_2 y - \log_2 z$
9. $5 \log_{10} x$
11. $\frac{1}{2} \log_{10} 7$
13. $1 + \log_4 x - \log_4 y$
15. $\log_{10} 5 - 6 \log_{10} 2 - 3 \log_{10} x$
17. $\frac{1}{2} \log_5 x + \frac{1}{2} \log_5 z - \frac{1}{2} \log_5 y$
19. $\frac{1}{3} \log_3 x + \frac{3}{4} \log_3 y - \frac{1}{5} \log_3 z$
21. $\frac{1}{2} \log_{10} s + \frac{1}{2} \log_{10} (s - a) + \frac{1}{2} \log_{10} (s - b) + \frac{1}{2} \log_{10} (s - c)$
23. $\log_2 \frac{7\sqrt{5}}{15^2}$
25. $\log_6 \frac{x^3}{t^2 y}$
27. $\log_b \frac{(3xy)^2(2y)}{3x^3y^2}$ or $\log_b \frac{6y}{x}$
29. $\log_b \sqrt{\frac{xy}{z^4}}$ or $\log_b \frac{\sqrt{xy}}{z^2}$
31. 1.4771
33. 1.4313
35. .7781
37. 1.5562
39. 2.3010
41. .5881
43. Let $\log_a a = x$
Change to exponential form: $a^x = a^1$
Use equality property for exponentials: $x = 1$
Thus, $\log_a a = 1$

Exercise 10.5

1. characteristic: 1
mantissa: .7419
3. characteristic: -1
mantissa: .5314
5. 0.1271
7. $0.6776 - 2$
9. 4.0969
11. $0.2553 - 1$
13. $0.9595 - 1$
15. 3.32×10^2 or 332
17. 6.25×10^3 or 6250
19. 2.46×10^{-1} or .246
21. 5.81×10^1 or 58.1
23. 2.29×10^{-1} or .229
25. 42.3
27. 42.9
29. .929
31. 16.1
33. 5.51
35. .0406
37. .0991
39. 129

Exercise 10.6

1. -1.7386
3. -2.1176
5. -3.5712
7. characteristic: -3
mantissa: .3817
9. characteristic: -2
mantissa: .4173
11. characteristic: -1
mantissa: .7368
13. 1.7658175
15. -1.1472761
17. 2.5416807
19. -0.2231435
21. -3.192871
23. 2.7945368
25. 0.0249229
27. 6.82653
29. 67.782226
31. 0.00208257

33. 176928.33
37. 16.277961
41. 1.386154
45. 4.2920297
49. 0.2956839

35. 0.0026595
39. 2.4428614
43. 2.1463924
47. −2.261781

Exercise 10.7

1. 2.6309
7. 0.9925
13. 1.1527
19. $x = 2$ or 5
25. $x = 5$
31. 14.2 years
37. 8.5
43. 8.84 miles
49. 10 times

3. 1.4307
9. −0.5370
15. 1.3912
21. $x = 4$
27. $x = 2$
33. 617,355
39. 2.0
45. 3.81 miles

5. 2.3383
11. 0.8928
17. 1.8765
23. $x = \dfrac{100}{3}$
29. 8.8 years
35. 3.4%
41. 3.98×10^{-4}
47. 2.00 times

Exercise 11.1

1. 1, 3, 5, 7
7. $2, \dfrac{3}{2}, \dfrac{4}{3}, \dfrac{5}{4}$
13. 44
19. $\dfrac{40}{81}$
25. $\sum_{i=1}^{3} \dfrac{i}{i+1}$

3. 8, 16, 32, 64
9. −3, 5, −7, 9
15. −2
21. $\sum_{i=1}^{6} i^2$
27. $\sum_{i=1}^{5} \dfrac{\sqrt{i}}{i+1}$

5. $\dfrac{2}{3}, \dfrac{3}{4}, \dfrac{4}{5}, \dfrac{5}{6}$
11. 38
17. $-\dfrac{7}{24}$
23. $\sum_{i=1}^{4} \dfrac{1}{2i}$

Exercise 11.2

1. $d = 3$; 14, 17
7. $a_n = 3n - 4$
13. 7
19. 40
25. $\dfrac{23}{3}$
31. $\dfrac{23}{2}$

3. $d = 5$; $a + 15$, $a + 20$
9. 42
15. $\dfrac{65}{2}$
21. 7260
27. 29
33. 165

5. $a_n = 4n - 2$
11. $\dfrac{39}{4}$
17. 495
23. 95,760
29. 40
35. $11\dfrac{2}{3}, 18\dfrac{1}{3}$

Exercise 11.3

1. -2
3. $-\dfrac{1}{3}$
5. $a_n = -6(2)^{n-1}$
7. $a_n = \dfrac{1}{9}(3)^{n-1}$
9. $\dfrac{1}{8}$
11. -96
13. $\dfrac{1}{162}$
15. $\dfrac{1}{243}$
17. 1092
19. 4372
21. 10
23. $-\dfrac{484}{243}$
25. 12
27. 32
29. $\dfrac{1}{3}$
31. $\dfrac{8}{3}$
33. $\dfrac{7}{9}$
35. $\dfrac{37}{33}$
37. $\dfrac{31}{99}$
39. $\dfrac{11}{9}$
41. 162
43. 40 or -20

Exercise 11.4

1. $n+1$
3. $4;\ 3$
5. $x^4 + 4x^3y + 6x^2y^2 + 4xy^3 + y^4$
7. $x^4 - 4x^3 + 6x^2 - 4x + 1$
9. $1 - 5x + 10x^2 - 10x^3 + 5x^4 - x^5$
11. $x^4 - 4x^3y^3 + 6x^2y^6 - 4xy^9 + y^{12}$
13. $32x^5 - 80x^4 + 80x^3 - 40x^2 + 10x - 1$
15. $8x^6 - 12x^4y + 6x^2y^2 - y^3$
17. $x^5 - \dfrac{5x^4}{2} + \dfrac{5x^3}{2} - \dfrac{5x^2}{4} + \dfrac{5x}{16} - \dfrac{1}{32}$
19. $x^6 + 3x^4y^2 + 3x^2y^4 + y^6$
21. $x^4 - 4x^2 + 6 - \dfrac{4}{x^2} + \dfrac{1}{x^4}$
23. $x^{13} + 13x^{12} + 78x^{11} + 286x^{10} + \ldots$
25. $1 + 24x^3 + 252x^6 + 1512x^9 + \ldots$
27. 6
29. $\dfrac{1}{2700}$
31. $\dfrac{1}{6}$
33. $120x^7y^3$
35. $240x^6$
37. $-4320x^6$
39. $\dfrac{35y^8}{8}$

INDEX

Abscissa, 255, 310
Absolute value, 11, 53, 71
   equations, 71–75, 96
   inequalities, 12–13, 76–79, 96
Addition
   applied problems, 85
   associative properties of, 22–23, 52
   closure properties for, 22, 52
   commutative properties of, 22, 52
   of complex numbers, 242, 248
   of rational expressions, 149–52, 169
   with signed real numbers, 27–29
Addition property, 21, 52
Addition property of exponents, 38–39
Addition property of logarithms, 433, 457
Addition-subtraction property
   of equalities, 59, 96
   of inequalities, 66, 96
Additive inverse, 23
Algebraic expressions, evaluating, 45–49
Antilogarithm(s), 441, 457
   and calculators, 447
A.P.   See *Arithmetic progression.*
Applied problems
   and exponential functions, 423–25
   and linear equations and inequalities, 85–93
   and rational expressions, 164–67
   and systems of linear equations, 346–51

Arithmetic progression(s), 470–74, 491
Associative properties of addition and
   multiplication, 22–23
   and variables, 33
Asymptotes, of hyperbola, 386, 408
Axis, of parabola, 366

Base, logarithm, 448–49
Base, natural, 424
Binomial expansion, 483–90
   numerical coefficients of, 485
   r^{th} term of a, 489
Binomial(s), 134
   special products of, 111–15
Binomial squared shortcut, 114, 135
Binomial theorem, 486
   with factorial notation, 488

Calculators, operations with, 18–19
   and antilogarithms, 447
   and common logarithms, 445–47
   exponents and roots, 43–44
   negative exponents, 184
   polynomials, 104
Cartesian coordinate system, 254
Circle(s), 373–78, 392, 393
   definition of, 375
   equation of a, 376

Closure properties for addition and multiplication, 22, 52
Cofactor of an element, 337, 353
Common ratio, 476
Commutative properties of addition and multiplication, 22, 52
and variables, 33
Complex number(s), 248
operating with, 241–46
set of, 236–40
Compound inequalities, 69–70
Compound interest, 423, 452–53, 458
Compound statements, 9–10
Conditional equations, 58, 96
Conic sections, 393–95, 408
circle, 373–78, 392, 393
ellipse, 380–83, 392, 393
hyperbola, 384–91, 392–93
parabola, 360–64, 366, 392, 393
straight line, 393, 394
Conjugate, 200, 248
of a complex number, 244
Constant function, 281, 311
Continuous growth or decay, 453
Coordinate axes, 254, 310
Coordinate(s), on number line, 3
Cramer's Rule, 341–45, 354
Critical value, 230, 248

Decay, continuous, 453
Degree of a polynomial in one variable, 102, 134
Denominator(s)
rationalizing, 198–201
Descartes, Rene, 254
Determinant(s), 335–39, 353, 354
Difference of two cubes, factoring, 120–23
Difference of two squares, factoring, 119–20
Dimension, of matrix, 335
Direct variation, 303, 311
Discriminant of the quadratic equation, 228, 247
Distance, problems involving, 165–66
Distance formula, 375, 408
Distributive property of exponents, 41
Distributive property of multiplication, 24
extending, 24–25, 52
with respect to addition, 24
with respect to subtraction, 34
Divison
applied problems, 86
of complex numbers, 244–45, 248
definition of, 53
of polynomials, 106–10
of rational expressions, 145–47, 169
with signed real numbers, 32
with variables, 34–35
by and of zero, 25
Division property of radicals, 192–93
Domain
of a function, 267–68
of a quadratic function, 371–72
of a relation, 256, 310
Double opposite law, 25, 52

e, 424, 457
Elements
of matrix, 335
of sets, 2
Elimination, 323
Ellipse(s), 380–83, 392, 393
equation of, in standard position, 381
Empty set, 2
Entries, of matrix, 335
ϵ, 2
Equality
and complex numbers, 242
properties of, 21–22
Equality property for exponentials, 457
Equality property for logarithms, 437, 457
Equality property of absolute value, 72, 97
Equation(s), 96
of the form $x = k$, 282
Equivalent equations, 58–59, 96
Exponential equations, 450–51, 452–54
Exponential functions, 416–21, 423–25, 457
Exponent(s), 51
negative, 178–81
properties of, 37–41, 52
rational, 186–89
scientific notation, 182–84
Expressions
classifying, 16–18
radical, 194–97
Extraneous solutions, 159

Factorial(s), 486, 491
zero, 487
Factoring
binomials, 119–24
common factors, 116–18
solving equations by, 130–33
solving quadratic equations by, 212–15
trinomials, 124–29

Factoring *(continued)*
  types, 128
Factors, 51
Finite series, 468, 491
FOIL shortcut, 111, 112, 134–35
Formulas, 80–83, 96
  for arithmetic progressions, 474
  containing rational expressions, 161–62
  for geometric progressions, 482
Fractions
  fundamental principle of, 141, 169
  least common denominator (LCD), 149, 169
  rationalizing the denominator, 198–201
  reducing to lowest terms, 141
Functional notation, 311
Function(s), 262–68, 311
  constant, 281, 311
  exponential, 416–21, 423–25, 457
  inverse, 271–75, 310
  linear, 278–88
  logarithmic, 428–31, 457
  quadratic, 365–72
  sequence, 466–68
  variation, 303–7
  vertical line test for, 264

Geometric progression(s), 476–82, 492
G.P. See *Geometric progressions.*
Graphs and graphing
  circles, 373–78
  conic sections, 392–95
  constructing, 257
  ellipse, 380–83
  exponential functions, 416–17
  hyperbola, 384–91
  intervals, 13
  inverse functions, 274–75
  ordered pairs, 254–55
  parabola, 360–64
  quadratic equations in two variables, 360–64
  of a relation, 257
  second degree inequalities, 402–5
Growth, continuous, 453

Horizontal sweep, 258
Hyperbola(s), 384–91, 392, 393
  equation of, in standard position, 385

Identity, properties of, 23, 52
  and variables, 33
Identity equations, 58, 96

Imaginary number(s), 236, 248
Imaginary unit(s), 236, 248
Inequality(ies), 7–9, 96
  absolute value, 12, 96
  compound, 69–70, 96
  linear, 65–69, 96, 296–302
  quadratic, 230–35
  second degree, 401–5
Inequality properties
  for absolute value ($>$), 76–77, 97
  for absolute value ($<$), 77–78, 97
Infinite series, 468, 491
Integers, 4
Interest, compound, 423, 452–53, 458
Intervals, 13
Inverse functions, 271–75, 311
Inverse properties, 23, 52
Inverse relations, 270–71, 311
Inverse variation, 304, 311
Irrational numbers, 4–5

Joint variation, 305, 311

Least common denominator (LCD), 149, 169
Linear equations, 59
  applied problems, 346–51
  Cramer's Rule for solving, 341–45
  properties of, 59
  solving, 58–64
  solving by substitution, 327
  systems, in three variables, 330–34, 353
  systems, in two variables, 322–28, 353
Linear functions, 278–88
Linear inequalties, 67, 311
  graphing, 296–302
  solving, 65–69
Logarithmic equations, 431, 451–53
Logarithmic functions, 428–31, 457
Logarithms
  and calculators, 445–49
  change of base, 448–49
  common, 439, 457
  computing with common, 439–44
  natural, 448, 457
  properties of, 433–37

Mantissa, 440, 457
Matrix, 335, 353
Minor of an element, 337, 353
Monomial, 134

Multiplication
  applied problems, 86
  associative properties of, 22–23, 52
  of binomials, 111–15
  closure properties for, 22, 52
  commutative properties of, 22, 52
  of complex numbers, 243, 248
  of polynomials, 106–10
  of rational expressions, 144–45, 169
  of signed real numbers, 31
  of variables, 33
  by zero, 25, 31, 52
Multiplication–division property
  of equalities, 59, 96
  of inequalties, 66, 97
Multiplication property, 21–22, 207

Natural base, 424
Natural number(s), 4
  exponents, 37–41
  roots, 42–43
  square roots of, 237
Negative exponents, 178–81
Non-linear systems of equations, 396–400
Number line, 3
Numbers, 4–5
Numerical relationships, 164

Operations
  order of, 15–16, 43
  with signed real numbers, 27–35
Opposite law, 33, 52
Order, of matrix, 335
Ordered pair(s), 254–55, 310
Ordered triple(s), 330
Ordinate, 255, 310
Origin, 254, 310

Pair(s), ordered, 254–55, 310
Parabola(s), 360–64, 366, 392, 393
  vertex of, 366, 368
Perfect square trinomial(s), 217
pH of a solution, finding, 454
Point-slope form, 289, 290, 311
Polynomial(s), 102–4, 134
  common factors, 116–18
Power property of equations, 203, 207
Power property of logarithms, 435, 457
Powers, variation of, 305
Power to a power property of exponents, 41

Principal square root, 174
Progression(s), arithmetic, 470–74
Pythagorean theorem, 373, 408

Quadrant(s), 255, 310
Quadratic equation(s), 247
  discriminant of, 228, 247
  graphing in two variables, 360–64
  quadratic formula, 222–29, 247
  solving by completing the square, 216–21
  solving by factoring, 212–15
Quadratic formula, 222–29, 247
Quadratic function, 365–70
  domain of, 371–72
Quadratic inequalities, solving, 230–35, 248

Radical expressions, 194–97
Radicals, 174–77
  division property of, 192–93
  simplifying, 190–93
  solving equations containing, 203–5
Range
  of a function, 267–68
  of a relation, 256, 310
Ratio, common, 476
Rational exponents, 186–89
Rational expressions, 169
  addition of, 149–52, 169
  applied problems, 164–67
  division of, 145–47, 169
  equations containing, 158–61
  formulas containing, 161–62
  multiplication of, 144–45, 169
  reduction of, 140–42
  subtraction of, 149–52, 169
Rationalizing the denominator, 198–201
Rational numbers, 4
Real numbers
  operations with signed, 27–35
  properties of, 22–24
  set of, 5
  theorems of, 24–25
Rectangular coordinate system, 254, 310
Reducing fractions, 141
Reflexive property, 21, 52
Relation(s), 256, 310
  domain of, 256, 310
  graph of a, 257, 310
  inverse, 270–71, 311
  range of, 256, 310
Replacement set, 2–3

Root(s), 42–43, 96
  properties of, 176
  and radicals, 174–77
  variation of, 304, 305

Scientific notation, 182–84
Second degree inequalities, 401–5
Second order determinants, 336, 354
Sequence, 466, 491
Sequence function, 466–68
Series, 468
  finite, 468
  infinite, 468
Set(s), 2
Σ, 468
Sigma notation, 468
Slope-intercept form, 289, 292, 311
Slope of a line, 282–86, 311
  of parallel lines, 287, 311
  of perpendicular lines, 287–88, 311
  point-slope form, 289, 290, 311
  slope-intercept form, 289, 292, 311
  of straight lines, 286
Special product(s), of binomials, 111–15
Square matrix, 335, 353
Square root property of equations, 218, 248
Square roots, of negative numbers, 237
Substitution, 323, 327
Substitution property, 21, 52
Subtraction
  applied problems, 86
  and complex numbers, 242, 248
  definition of, 30, 52
  of rational expressions, 149–52, 169
  with signed real numbers, 29–31
Subtraction property of exponents, 39–40
Subtraction property of logarithms, 434, 457
Summation notation, 468
Sum of two cubes, factoring, 120–22
Sum times a difference shortcut, 113, 135
Symbols, 51
  ϕ (epsilon), 2
  inequality, 8
  $n!$ (n factorial), 486
  radical, 175
  removing group, 48
  Σ (sigma), 468
Symmetric property, 21, 52

Term, of a sequence, 466, 491
  finite, 468
  general, 467
  infinite, 468
Third order determinants, 336
Time, problems involving, 165–66
Trinomial(s), 134
  factoring, 124–29
  perfect square, 217–18

Variable(s), 33
  dependent, 256, 310
  independent, 256, 310
  multiplying, 33–34
Variation, 303–7
  direct, 303, 311
  inverse, 304, 311
  joint, 305, 311
  of powers and roots, 304, 305
Vertex, of parabola, 366, 368
Vertical line test for a function, 264, 311
Vertical sweep, 258

Whole numbers, 4
Word problems, 88–93

x-axis, 254
x-intercept, 279, 311

y-axis, 254
y-intercept, 279, 311

Zero
  division by and of, 32
  multiplication by, 25, 31, 52
Zero factorial, 487
Zero factor property, 130, 135, 248
  solving equations using, 131–32
  solving quadratic equations using, 213
Zeros of a function, 266, 311

WE VALUE YOUR OPINION—PLEASE SHARE IT WITH US

Merrill Publishing and our authors are most interested in your reactions to this textbook. Did it serve you well in the course? If it did, what aspects of the text were most helpful? If not, what didn't you like about it? Your comments will help us to write and develop better textbooks. We value your opinions and thank you for your help.

Text Title _____ Edition _____

Author(s) _____

Your Name (optional) _____

Address _____

City _____ State _____ Zip _____

School _____

Course Title _____

Instructor's Name _____

Your Major _____

Your Class Rank _____ Freshman _____ Sophomore _____ Junior _____ Senior

_____ Graduate Student

Were you required to take this course? _____ Required _____ Elective

Length of Course? _____ Quarter _____ Semester

1. Overall, how does this text compare to other texts you've used?

    _____ Superior _____ Better Than Most _____ Average _____ Poor

2. Please rate the text in the following areas:

	Superior	Better Than Most	Average	Poor
Author's Writing Style	_____	_____	_____	_____
Readability	_____	_____	_____	_____
Organization	_____	_____	_____	_____
Accuracy	_____	_____	_____	_____
Layout and Design	_____	_____	_____	_____
Illustrations/Photos/Tables	_____	_____	_____	_____
Examples	_____	_____	_____	_____
Problems/Exercises	_____	_____	_____	_____
Topic Selection	_____	_____	_____	_____
Currentness of Coverage	_____	_____	_____	_____
Explanation of Difficult Concepts	_____	_____	_____	_____
Match-up with Course Coverage	_____	_____	_____	_____
Applications to Real Life	_____	_____	_____	_____

3. Circle those chapters you especially liked:
   1 2 3 4 5 6 7 8 9 10 11 12 13 14 15 16 17 18 19 20
   What was your favorite chapter? _____
   Comments:

4. Circle those chapters you liked least:
   1 2 3 4 5 6 7 8 9 10 11 12 13 14 15 16 17 18 19 20
   What was your least favorite chapter? _____
   Comments:

5. List any chapters your instructor did not assign. _____

6. What topics did your instructor discuss that were not covered in the text?_____
   _____

7. Were you required to buy this book? _____ Yes _____ No

   Did you buy this book new or used? _____ New _____ Used

   If used, how much did you pay? _____

   Do you plan to keep or sell this book? _____ Keep _____ Sell

   If you plan to sell the book, how much do you expect to receive? _____

   Should the instructor continue to assign this book? _____ Yes _____ No

8. Please list any other learning materials you purchased to help you in this course (e.g., study guide, lab manual).
   _____

9. What did you like most about this text? _____
   _____

10. What did you like least about this text? _____
    _____

11. General comments:

    May we quote you in our advertising? _____ Yes _____ No

    Please mail to: Boyd Lane
                      College Division, Research Department
                      Box 508
                      1300 Alum Creek Drive
                      Columbus, Ohio 43216

    Thank you!